New Trends in Robotics and Automation

New Trends in Robotics and Automation

Editors

Jian Wu
Xiangkun He
Guangfei Xu

MDPI

Basel • Beijing • Wuhan • Barcelona • Belgrade • Novi Sad • Cluj • Manchester

Editors
Jian Wu
School of Mechanical and Automotive Engineering
Liaocheng University
Liaocheng
China

Xiangkun He
School of Mechanical and Aerospace Engineering
Nanyang Technological University
Singapore
Singapore

Guangfei Xu
School of Mechanical and Automotive Engineering
Liaocheng University
Liaocheng
China

Editorial Office
MDPI AG
Grosspeteranlage 5
4052 Basel, Switzerland

This is a reprint of articles from the Special Issue published online in the open access journal *Machines* (ISSN 2075-1702) (available at: www.mdpi.com/journal/machines/special_issues/7DR1AL0KQ5).

For citation purposes, cite each article independently as indicated on the article page online and using the guide below:

Lastname, A.A.; Lastname, B.B. Article Title. *Journal Name* **Year**, *Volume Number*, Page Range.

ISBN 978-3-7258-2058-0 (Hbk)
ISBN 978-3-7258-2057-3 (PDF)
https://doi.org/10.3390/books978-3-7258-2057-3

Contents

About the Editors

Jian Wu

Jian Wu received his Ph.D. in vehicles from the Nanjing University of Aeronautics and Astronautics, Nanjing, China, in 2015. He was a Postdoctoral Researcher at Tsinghua University, Beijing, China, from 2016 to 2018. He is a Professor at Liaocheng University, Liaocheng, China. His research interests include driver–vehicle system dynamics, driver–vehicle automation collaboration, and shared control.

Xiangkun He

Dr. Xiangkun is currently a Research Fellow at Nanyang Technological University, Singapore. From 2019 to 2021, he served as a Senior Research Scientist at Huawei Noah's Ark Lab. He received his Ph.D. in 2019 from the School of Vehicle and Mobility, Tsinghua University, China. He has contributed over 50 papers, including TPAMI, TITS, TIV, TRC, and Engineering, and holds eight granted patents. His research interests involve reinforcement learning, trustworthy AI, autonomous vehicles, and robotics. He received many awards or honors, including but not limited to the Tsinghua University Outstanding Doctoral Dissertation Award in 2019, Best Paper Finalist at IEEE ICMA 2020, Huawei Major Technological Breakthrough Award in 2021, Best Paper Runner-Up Award at CVCI 2022, and Runner-Up at Intelligent Algorithm Final of 2022 Alibaba Global Future Vehicle Challenge. He serves as a reviewer for over 50 renowned journals and conferences and was also a member of the award committee for the Autonomous Driving Control Benchmark Challenge at IEEE CDC 2023.

Guangfei Xu

Guangfei Xu received his Ph.D. in vehicles from the Shandong University of Technology, Zibo, China. He is a lecturer at Liaocheng University, Liaocheng, China. His research interests include intelligent agricultural machinery systems, vehicle dynamics control and driver–vehicle system dynamics.

Preface

Robotics and automation technologies have been extensively applied across various domains, including agriculture, healthcare, and transportation, yielding significant societal benefits and impacts. As these technologies drive the Fourth Industrial Revolution, it is essential to thoroughly explore emerging trends to ensure their sustainable development. With the rapid advancements in artificial intelligence (AI), digital twins (DT), the Internet of Things (IoT), and human–computer interaction (HCI), the field of robotics and automation presents a vast landscape for further discussion and innovation.

This Special Issue is a platform for researchers and practitioners to share their latest theoretical and engineering breakthroughs. Topics included but not limited to the following:

1. Intelligent control in robotics;
2. Intelligent connected vehicles;
3. Trustworthy artificial intelligence;
4. AI applications in intelligent systems;
5. Modeling and control of human–machine systems;
6. System state information acquisition and parameter identification;
7. Applications of linear and nonlinear analysis and control;
8. Decision-making and control using multi-source information;
9. Decision-making and control for multi-agent systems.

We extend our heartfelt appreciation to the authors for their exemplary research contributions and to the editors for their invaluable support and guidance.

Jian Wu, Xiangkun He, and Guangfei Xu

Editors

machines

MDPI

Review

A Survey on Path Planning for Autonomous Ground Vehicles in Unstructured Environments

Nan Wang [1], Xiang Li [1], Kanghua Zhang [1], Jixin Wang [1,*] and Dongxuan Xie [2,*]

1 Key Laboratory of CNC Equipment Reliability, Ministry of Education, School of Mechanical and Aerospace Engineering, Jilin University, Changchun 130022, China; nwang22@mails.jlu.edu.cn (N.W.); lixiang_sub@163.com (X.L.); zhangkh20@mails.jlu.edu.cn (K.Z.)

2 School of Computer Science and Engineering, Changchun University of Technology, Changchun 130012, China

* Correspondence: jxwang@jlu.edu.cn (J.W.); xiedongxuan@ccut.edu.cn (D.X.)

Abstract: Autonomous driving in unstructured environments is crucial for various applications, including agriculture, military, and mining. However, research in unstructured environments significantly lags behind that in structured environments, mainly due to the challenges posed by harsh environmental conditions and the intricate interactions between vehicles and terrains. This article first categorizes unstructured path planning into hierarchical and end-to-end approaches and then the special parts compared to structured path planning are emphatically reviewed, such as terrain traversability analysis, cost estimation, and terrain-dependent constraints. This article offers a comprehensive review of the relevant factors, vehicle–terrain interactions, and methods of terrain traversability analysis. The estimation methods of safety cost, energy cost, and comfort cost are also emphatically summarized. Moreover, the constraints caused by the limits of terrains and vehicles are discussed. The applications of algorithms in recent articles for path planners are reviewed. Finally, crucial areas requiring further research are analyzed in unstructured path planning

Keywords: unstructured environments; terrain traversability analysis; cost estimation; path planning; autonomous ground vehicles; autonomous driving technology

Citation: Wang, N.; Li, X.; Zhang, K.; Wang, J.; Xie, D. A Survey on Path Planning for Autonomous Ground Vehicles in Unstructured Environments. *Machines* **2024**, *12*, 31. https://doi.org/10.3390/machines12010031

Academic Editors: Jian Wu, Xiangkun He and Guangfei Xu

Received: 28 November 2023
Revised: 29 December 2023
Accepted: 30 December 2023
Published: 2 January 2024

1. Introduction

In recent years, significant advancements have been made in autonomous driving technology, with notable breakthroughs and accomplishments in perception, planning, and control in structured environments [1–3]. However, despite these achievements, the progress in dealing with unstructured environments lags behind that of structured environments [4,5].

At present, the definition of structured environments and unstructured environments need further research and discussion. Structured environments typically encompass urban roads and highways characterized by regular road geometries, paved road surfaces, and lane markings [6], as depicted in Figure 1. In light of the existing literature [5,7] and practical applications, unstructured environments, which are also referred to using similar keywords such as off-road and terrain, are comprehensively defined in this article as encompassing various types of unpaved roads, as illustrated in Figure 2a, and paved roads where the lane boundaries become indiscernible due to the presence of water, snow, or other coverings, as depicted in Figure 2b. Real-world applications of autonomous driving in unstructured environments have been diverse, with notable implementations in agriculture [8,9], military [10], mining [11], rescue [12], and planetary exploration [13,14].

Figure 1. Example of structured environments generated by PanoSim [15].

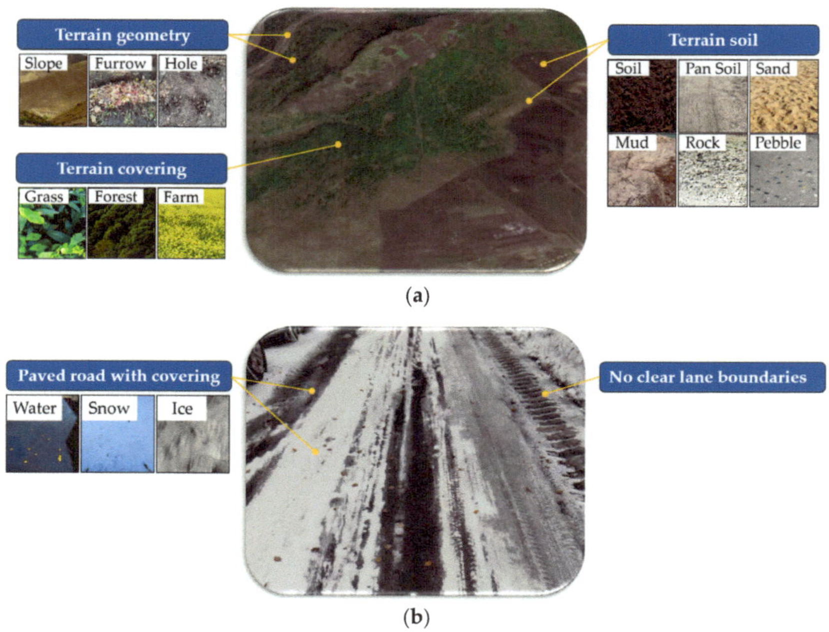

Figure 2. Some examples of terrains in unstructured environments: (**a**) unpaved road environments where the original geomorphological capture comes from Baidu Maps; (**b**) paved roads where the lane boundaries are indistinguishable due to water, snow, or other coverings.

At present, due to the development of machine learning and sensor technology, the research on perception in unstructured environments has made great progress [16,17]. However, there are still many challenges in the path planning of unstructured environments [7,18].

In path planning, it is essential to take into account road conditions and traffic environments, as highlighted in Table 1. The road conditions in unstructured environments present formidable challenges due to their intricate and limitless variety, inconsistent friction distribution, and perpetually changing dynamic nature [19–21]. Additionally, various states and ratios of water, sand, and air within different soil conditions display distinct properties, which are significantly influenced by volatile weather conditions, such as rain, snow, and sandstorms.

The path planning process encompasses both global and local path planning. In structured environments, global path planning relies on sensor information and high-precision maps to determine the most efficient route from the start to the destination. As a general rule, the primary objective of global path planning is to determine the route that results in the shortest distance or the shortest time of travel. Subsequently, local path planning is carried out in response to real-time traffic situations [22], such as lane changes, intersections, and obstacle avoidance, while also considering traffic regulations.

Table 1. Comparison of environmental characteristics and path planning between unstructured and structured environments.

Contrast Item		Structured Environment	Unstructured Environment
Environment characteristics	Road environment	Simple variety, less changes, normed geometries, even friction distribution	Intricate and boundless variety, frequent changes, normed and terrain geometries, uneven friction distribution
	Traffic environment	Lane markings, lane boundaries	No lane markings, no lane boundaries
Global path planning	Road information	Sensing information, high-precision map	Sensing information, geographic map, topographic map, TTA information
	Planning goal	Shortest distance or shortest time	Lower cost of safety, energy, or comfort
Local path planning	Consideration	Traffic regulation, vehicle dynamics, controlling target	Terrain, traffic regulation, vehicle dynamics, controlling target
	Obstacles avoidance	Normal vehicles, pedestrians	Normal and special vehicles, pedestrians and animals

In unstructured path planning, the foremost considerations should focus on the conditions of the road and soil, as well as the adaptability of various vehicles within stochastic environments. Moreover, the driving capacity of diverse vehicles, along with the same vehicles under different loads, is variable. Therefore, global path planning in unstructured environments typically relies on integrated information from sensors, geographic maps, and topographic maps. Obtaining high-precision maps can be challenging in unstructured environments due to factors such as the quality of the GPS signal, military protection, and dynamic changes. Following this, terrain traversability analysis (TTA) is carried out based on the collected data. The purpose of path planning varies among vehicles in different applications. For instance, in agricultural applications, the primary concerns are energy consumption and task completion; meanwhile, in military applications, traversability is paramount. In mining operations, both safety and operator comfort are of utmost importance. Therefore, path planning in unstructured environments is complicated and challenging.

2. Related Works and Survey Boundaries

In structured environments, global path planning and local path planning algorithms are well established in both scientific research and market applications. However, these algorithms may not be well suited for unstructured environments and often require adjustments. In recent years, the integration of machine learning has led to significant progress in TTA research. Various surveys and reviews were conducted to evaluate path planning algorithms and TTA methods in unstructured environments, as summarized in Table 2. Various surveys and reviews have been conducted to evaluate path planning algorithms and TTA methods.

Table 2. The existing review articles that have covered path planning of unstructured environments, including this article.

Study	Area Covered	Content Contained	Technology Focused
Wang et al., 2021 [23]	Wheeled vehicles	Perception, navigation	Sensors, collision avoidance methods
Guastella et al., 2021 [24]	Autonomous ground vehicles	Perception, navigation	Learning-based methods of TTA and end-to-end navigation
Chakraborty et al., 2022 [8]	Agricultural ground vehicles	Path planning	Path planning algorithms and application
Bai et al., 2023 [9]	Agricultural ground vehicles	Perception, navigation	Sensors, obstacle avoidance, TTA, location, path planning algorithms
Wijayathunga et al., 2023 [25]	Autonomous ground vehicles	Perception, scene understanding, path planning	Sensor fusion, location, map creation, path planning algorithms
This article	Autonomous ground vehicles	Path planning	TTA, cost estimation, constraints of terrain and vehicles, path planning algorithms

Wang et al. [23] mainly reviewed the perception and navigation methods of collision avoidance. However, obstacle avoidance is only a part of path planning in unstructured environments. Guastella et al. [24] mainly focused on learning-based methods of perception and navigation for ground vehicles, including methods of terrain traversability analysis and end-to-end navigation algorithms. But for both TTA and end-to-end navigation, there are other methods besides the learning-based ones. Chakraborty et al. [8] and Bai et al. [9] focused on path planning in agriculture. Chakraborty et al. [8] mainly reviewed the path planning algorithms and the application characteristics of these methods, such as the application scenarios and limitations. Bai et al. [9] considered obstacle avoidance and TTA of path planning from the perspective of vision methods. Wijayathunga et al. [25] summarized the path planning algorithms encompassing location and map creation methods.

The focus of the above articles was primarily directed towards the design and implementation of path planning algorithms, along with the application of TTA methods. Nevertheless, these articles did not provide a comprehensive discussion of the various factors that influence TTA and did not particularly consider the cost estimation in path planning, along with the constraints imposed by terrain and vehicles. The cost estimation process in unstructured environments is of paramount importance, as it directly impacts the safety risks and economic costs associated with vehicle traversability. Furthermore, the constraints imposed by the terrain and the vehicles themselves must be taken into consideration in order to ensure the practical feasibility of the generated path.

The main purpose of this article is to provide a comprehensive overview and assessment of several unique aspects compared to structured path planning, such as TTA, cost estimation, and constraints. Concerning TTA, a detailed analysis and summary of the influential factors of terrain, vehicle, and vehicle–terrain interaction will be conducted in this article. When discussing cost evaluation, safety costs, energy consumption, and comfort costs will be the main focus. Moreover, the analysis and handling methods of terrain-related constraints will be highlighted as a critical aspect. Additionally, the application of path planning algorithms over the past five years will be reviewed.

Path planning in unstructured environments is categorized into two main broad categories in this article, presented in Figure 3. The first category involves the hierarchical path planning approach, which encompasses terrain traversability analysis, cost estimation, global path planning, constraint analysis, and local path planning [13,26]. Generally, the outcomes of terrain traversability analysis and cost estimation are utilized as input data for global path planners [27,28]. Subsequently, the local path is generated with the selected global paths and the limits of the constraints. The second category encompasses the end-to-end path planning approach, which directly outputs the path based on the learning methods according to the input of sensor data. The article is structured as follows: Section 3 introduces the terrain traversability analysis methods. Section 4 analyzes the cost estimation of safety, energy, and comfort. Section 5 summarizes the algorithms for global path planners. Section 6 concludes with the constraints and algorithms for local path planners. Section 7 reviews the learning methods of end-to-end path planning. Section 8 contains the conclusions. Section 9 proposes some suggestions for future work.

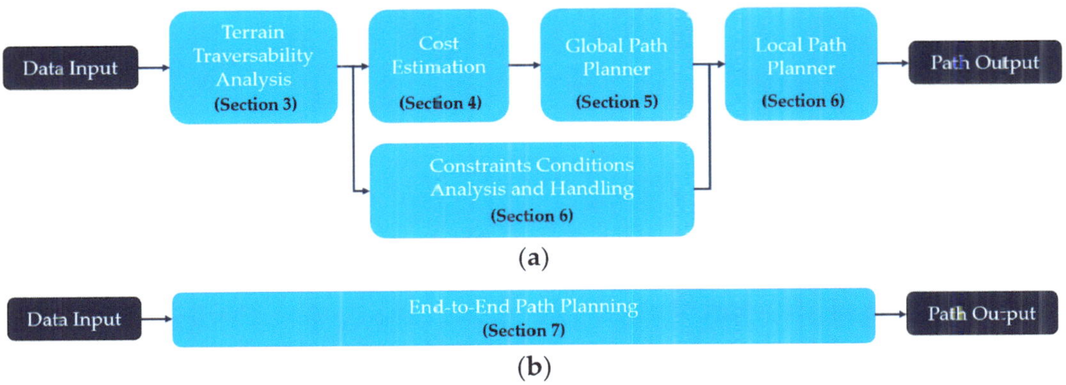

Figure 3. Path planning approaches in unstructured environments: (**a**) hierarchical path planning approach; (**b**) end-to-end path planning approach.

3. Terrain Traversability Analysis

Traversability refers to the capacity of a ground vehicle to traverse a particular terrain area while satisfying predefined objectives and criteria [29]. This capacity of vehicles has also been referred to with keywords like drivability, navigability, trafficability, and mobility [10,30].

Predicting traversability in advance is fundamental for autonomous path planning in unstructured environments. Inaccurate or inefficient information regarding traversability can result in the generation of substandard paths which not only consume excessive energy and time but also pose unnecessary risks of damaging both equipment and the environment [29]. The global path planner is responsible for selecting the path and optimizing it based on the outputs obtained from the TTA results, such as the traversability map [31,32], traversability cost model [33,34], or terrain classifier [35].

3.1. TTA Relevant Factors

Terrain traversability of autonomous driving vehicles depends on the geometrical and physical properties of the terrain as well as the dynamics and structure parameters of the vehicle [24,30]. Simultaneously, there are complex interactions between the vehicle and the terrain. Figure 4 shows partial relevant factors of TTA.

Figure 4. Partial relevant factors of TTA.

3.1.1. Terrain Geometrical and Physical Factors

The geometric features of terrain related to TTA encompass physiognomy and static obstacles. Physiognomy is characterized by elevation, slope, roughness, and curvature [29]. Elevation changes and slope characteristics, including vertical and lateral slopes, are fundamental factors in TTA models [36], impacting climbing ability and potential rollover [10]. The research presented by [37] shows simulation and comparison of traversability under different slope conditions. The TTA model also incorporates roughness [24,28], which describes the coarseness and surface irregularity of the ground plane traversed by the vehicle [38]. Heterogeneous roughness surfaces, such as vegetation, pebbles, sand, mud, and rock, pose particular challenges for vehicle driving. In [39], different levels of roughness are classified based on traversability cost. Furthermore, the curvature conditions of the road, constrained by the minimum turning radius of the vehicle, significantly affect traversability. Surface obstacles include crossable features like grass and low tree roots, as well as uncrossable obstacles like thick trees and cliffs. These surface obstacles are further categorized into two morphological groups: continuous ones, such as rivers, ditches, and banks, and scattered ones, such as pits, mounds, and craters.

Terrain physical factors generally comprise terrain types and soil properties. Different weather conditions, such as rain, snow, sandstorms, and frost, can result in the same terrain exhibiting varying characteristics. Temperature also has a notable impact. For instance, hot weather can cause soil drying, while cold weather can lead to freezing. Terrain types are commonly categorized as rigid and deformable [40]. Soil properties encompass permanent characteristics and transient characteristics [41]. Permanent characteristics, such as particle size, particle shape, composition, specific gravity, and consistency limits, remain relatively stable despite external stress fields or short-term environmental effects. Transient characteristics, including moisture, density, and permeability, are subject to change based on environmental and loading conditions. Moreover, soil properties engender intricate interactions between vehicles and the ground, consequently influencing traversability.

3.1.2. Vehicle Dynamics and Structure Parameters

The trafficability of a vehicle is greatly influenced by a range of vehicle dynamics parameters, including those related to the dynamic system, powertrain system, and drivetrain system. One example of this influence can be seen in the vehicle's ability to overcome the maximum tilt slope, which is dependent on the dynamic system and powertrain system. This ability is also closely linked to the vehicle's control strategy, which involves factors such as throttle opening, brake operation, and gear selection. Furthermore, different driving mechanisms, such as wheels and tracks, contribute to varying levels of traversability. Tracked vehicles, in contrast to wheeled vehicles, exhibit superior climbing ability and are less prone to getting stuck or slipping on wet mud or soft soil. Conversely, wheeled vehicles are better suited for navigating narrow spaces due to their flexible steering and good traversability. The key parameters affecting the performance of wheels include tire diameter, pattern, material, and pressure, while the main parameters of tracks are their length and width.

The traversability of a vehicle is influenced by various structural size parameters, such as length, width, and height, which determine its ability to navigate narrow roads. Furthermore, the wheelbase and center of gravity position play a crucial role in the vehicle's ability to corner effectively. In addition, the vehicle's ability to overcome trenches is closely related to geometric parameters, including tire diameter and wheelbase. Furthermore, parameters such as minimum ground clearance, brake-over angle, approach angle, and departure angle under full load conditions significantly impact the vehicle's ability to navigate obstacles without collision. Additionally, fording depth determines the vehicle's traversability in wading areas. The mass and load distribution of the vehicle also affects its acceleration performance, braking distance, and steering stability.

3.1.3. Vehicle–Terrain Interaction

The traditional path planners, as noted in prior research, are designed for operation within structured and homogeneous environments, meaning they do not require explicit consideration of the vehicle–terrain interaction [16]. However, recent studies on TTA have underscored the critical significance of vehicle–terrain interaction [42–44]. This interaction can give rise to variations such as wheel slip, wheel entrapment, and vehicle rollover, which consequently influence the overall traversability of the terrain.

The analysis of vehicle–terrain interaction is commonly conducted through the terramechanics model, which can be categorized into three main types [45]: empirical models, physics-based models, and semi-empirical methods. The authors of [46] provided a detailed review of the theory, characteristics, and applications of these models. Empirical models, while simple, lack generalizability beyond the experimental conditions in which they were developed [47]. On the other hand, physics-based models, such as finite and discrete element models, offer the highest fidelity, but their extensive computational requirements make them unsuitable for real-time tire force prediction and limit their applicability for real-time terrain estimation [42]. In contrast, semi-empirical models strike a balance between computational efficiency and fidelity, making them better suited for real-time estimation and control [46,48]. Among the semi-empirical methods, Bekker-based models have emerged as the most widely used [28,49]. These models calculate stresses over the contact patch between a rigid tire and deformable terrain, which are then integrated to determine the forces acting on the tire [50]. To accurately capture the complex stress distribution at the contact patch, Bekker-based models rely on numerous parameters that describe terrain characteristics, such as cohesion and internal friction angle [44].

3.2. TTA Methods

Sensing data are required in TTA to produce various results based on application requirements and method characteristics, as depicted in Figure 5.

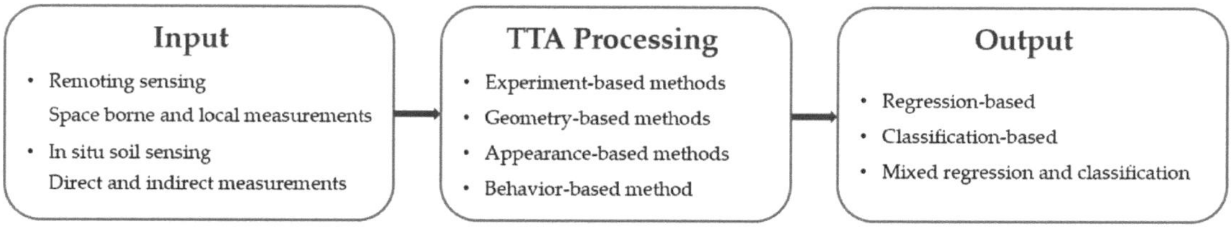

Figure 5. The types of methods for TTA input, processing, and output.

According to the sensing types, TTA can be classified into two main groups: remote sensing and in situ soil sensing: remote sensing and in situ soil sensing [48]. Remote sensing encompasses space-borne measurements, such as satellite and thermal spectroscopic, along with local measurements, such as cameras and radars. On the other hand, in situ soil sensing includes both direct and indirect measurements. Commonly used in direct measurements are sub-surface drilling, penetrometers, and bevameters [43]. In the case of indirect measurements, semi-empirical models, simulation models, and principle experiments are employed. The simulation aspect can be carried out using software like Chrono [42,51], Gazebo [18,52], and Open Dynamics Engine [13,35]. Additionally, experiments can be conducted using instrumented wheels [49,53], small smart demo cars [19,54], or real vehicles [31,39].

The early TTA was primarily conducted through experiments. One example is the cone index method, initially proposed by the Waterways Experiment Station (WES) of the U.S. military, to assess the traversability of vehicles in various soil conditions. This method has successfully correlated the soil rating cone index with vehicle performance in loose sand and wet soil [41]. Despite its ability to comprehensively reflect the mechanical and physical properties of soil, the method is not suitable for unknown terrains and necessitates substantial experimental costs. Nevertheless, it remains widely utilized in military and agricultural applications. In contrast, recent mainstream TTA methods have harnessed computer technology and can be categorized into geometry-based, appearance-based, and behavior-based approaches [32]. Geometry-based methods first develop a geometric representation of the terrain, evaluating traversability by comparing features such as height, roughness, slope, curvature, and width with vehicle parameters [30]. Although independent of lighting conditions, smoke, shadows, or adverse weather, this approach based on geomorphology overlooks the physical characteristics of the terrain. Appearance-based methods treat traversability analysis as an image-processing problem, which distinguishes different types of soil and vegetation with distinct costs [29]. While cost-effective in terms of camera usage, this method is susceptible to lighting conditions, smoke, shadows, or poor weather. Its primary application lies in discrete terrain class assessment. A new trend in TTA processing involves behavior-based methods, which evaluate traversability using experts' driving trajectories and learning-based techniques [55]. This approach incorporates human driving experience and vehicle dynamics but may yield inconsistent assessment results due to variations in drivers and vehicles. The summary and comparison of these TTA methods are shown in Table 3.

The outputs of the TTA delivered to the path planner may be regression-based outputs [56], classification-based outputs [44,57], or mixed regression–classification-based outputs [14,58]. Classification-based results are generally discrete classes of terrain type, including binary classification and multi-classification. Regression-based results are traversability maps or functions with continuous scores. The types of outputs can be selected according to the following path planning algorithms.

Table 3. Summary and comparison of TTA methods.

Method	Main Mean	Basis Input	Advantage	Disadvantage	Application
Cone index method	Experiment	Soil state, vehicle parameters	Comprehensively reflects the mechanical and physical properties of soil	Not suitable for unknown terrains, substantial experimental costs	Traversability assessment on soft soil, widely used in military and agriculture
Geometry-based method	Computer technology	Terrain geometric features, vehicle parameters	Not affected by lighting conditions, smoke, shadows, or poor weather	No consideration of terrain physical characteristics	Traversability assessment based on geomorphology
Appearance-based method	Computer technology	Images of terrain, vehicle parameters	Low cost of assessment equipment	Affected by lighting conditions, smoke, shadows, or poor weather	Discrete terrain classes assessment
Behavior-based method	Computer technology	Driving trajectories, vehicle parameters	Driving experience is taken into account	Inconsistent assessment and difficult to repeat	Traversability assessment based on driving operations

To obtain accurate and reliable analysis results, some TTA methods took into ac-count the factors discussed above, including terrain characteristics, vehicle parameters, and vehicle–terrain interactions. Table 4 shows different methods proposed in recent years with the details of sensing type, processing type, factors considered, and output type.

Table 4. Overview of different TTA methods published in recent years.

Study	Sensing Type	Processing Type	Factor Considered	Output Type
Mei et al., 2018 [59]	Remoting	Appearance-based	Terrain, vehicle	Classification-based
Chavez-Garcia et al., 2018 [35]	In situ	Geometry-based	Terrain	Classification-based
Dallas et al., 2020 [47]	In situ	Behavior-based	Terrain, vehicle, vehicle–terrain interaction	Mixed regression and classification
Zhu et al., 2020 [32]	Remoting	Behavior-based	Terrain, vehicle	Regression-based
Palazzo et al., 2020 [54]	Remoting	Appearance-based	Terrain	Mixed regression and classification
Oliveira et al., 2021 [39]	Remoting, in situ	Mixed behavior and geometry	Terrain, vehicle	Mixed regression and classification
Wallin et al., 2022 [29]	In situ	Geometry-based	Terrain, vehicle	Regression-based
Lee et al., 2023 [60]	Remoting	Mixed behavior and geometry	Terrain, vehicle	Regression-based
Xue et al., 2023 [61]	Remoting, in situ	Geometry-based	Terrain	Regression-based
Bae et al., 2023 [62]	Remoting	Mixed appearance and geometry	Terrain	Regression-based
Guan et al., 2023 [63]	Remoting	Mixed appearance and geometry	Terrain	Mixed regression and classification

4. Cost Estimation

The TTA mainly determines whether the vehicle can pass through the area to eliminate the untraversable area. However, for traversable areas, vehicles may pass with some costs, such as safety costs [64,65], energy costs [51,66], and comfort costs [39,67]. Generally, the costs are estimated based on the previous TTA output and then the estimation results such as cost function [68] or cost map [69,70] can be obtained and passed to the path planner [34].

4.1. Safety Cost

Traditional path planning mainly considers obstacle avoidance in terms of safety [71], but for unstructured environments, rollover and wheel slip caused by terrain characteristics should also be considered [67]. For example, for traversable areas with negative obstacles and steep slopes, the vehicle may pass after some good operations in a certain direction or at a certain speed [52], but there may be speed limitations or experience requirements for control operation, so there is still a safety risk of rollover in this condition. Negative obstacles are those with a negative height below the ground [5], such as pits and holes. For traversable areas, such as grainy roads, slippery roads, and icy roads, there is a risk of wheel slip, so vehicles have to pass under certain speed or power limitations. Table 5 summarizes the safety cost considerations and methods presented in the recent literature.

Table 5. Overview of safety cost considerations and methods published in recent years.

Study	Cost Considered	Method Estimated
Gao et al., 2010 [72]	Obstacle avoidance, Wheel slip	Cost function
Ono et al., 2015 [27]	Obstacle avoidance	Rollover model and cost function
Brandao et al., 2016 [73]	Wheel slip	Slippage model and cost function
Li et al., 2019 [67]	Obstacle avoidance, rollover	Rollover model and cost function
Sakayori et al., 2021 [13]	Rollover of pitch and roll	Cost function
Xiong et al., 2021 [64]	Obstacle avoidance, rollover	Cost function
Oliveira et al., 2021 [39]	Rollover of pitch and roll	Cost function and cost map
Lee et al., 2023 [60]	Obstacle avoidance, rollover	Cost function
Dang et al., 2023 [74]	Obstacle avoidance	Cost function

The static environment obstacles have been considered in TTA, but there is still a collision risk for the transition zone between the traversable and untraversable areas. This risk can be assessed by setting continuous or discrete safety costs. The safety cost of transition and uncertain zones can be probabilistically set higher compared to normal. The authors of [72] established a safety cost model for obstacle avoidance, which takes into account vehicle size and speed.

Modeling can be used to estimate the safety cost of rollover by determining the probability of driving risk, which is mainly assessed through roll and pitch evaluation. In the case of off-road vehicles, particularly engineering and agricultural machinery, the estimation modeling of rollover heavily relies on the positioning of the center of gravity [43,75]. The authors of [43] provide a detailed calculation method for determining the center of gravity position on slopes. Dynamic rollover models encompass two distinct types: the full-car model [76] and the curvilinear trajectory rollover model [77]. The full-car model offers the benefit of characterizing the roll produced by the impact of the terrain on the vehicle, whereas the curvilinear trajectory rolling model only characterizes the roll produced by steering [78]. The pitch and roll angle derived from the modeling process are essential for estimating the rollover cost [43,67].

The estimation of the safety cost of wheel slip is typically carried out through slip modeling, taking into account factors such as vehicle load, tire type, road type, and adhesion condition. The accuracy of the tire model plays a critical role in this estimation. The Pacejka model is widely used in wheel slip estimation [72,75] because it has been shown applicable to off-road conditions [42]. Various methods for estimating tire–road friction coefficient have been examined in [79], encompassing off-board sensors-based, vehicle dynamics-based, and data-driven approaches. Moreover, a review in [80] outlines slippage estimation methods for planetary exploration rovers, including proprioceptive-based, model-based, and exteroceptive-based strategies. Depending on the safety cost requirements, the results of wheel slip estimation can be discrete or continuous. For instance, in [44], wheel slip is discretized into categories of low slip, moderate slip, and high slip. Furthermore, regression evaluation of wheel slip has been performed using machine learning methods such as multi-layer perceptron (MLP), random forest (RF), and extreme gradient boosting (XGB) [81].

Notably, the estimation model proposed in [81] demonstrates strong performance in both discrete and regression prediction of wheel slip.

4.2. Energy Cost

In unstructured environments, the variations in energy consumption among different paths are substantial due to the diverse terrain conditions. For resource-constrained contexts, such as military, agriculture, planetary rover, and disaster relief missions, prioritizing the reduction in energy costs is crucial for the successful accomplishment of the tasks [51,66,82–84]. In pursuit of minimizing energy expenses and enhancing operational efficiency, the selection of the shortest distance or time-optimal path is often favored [85,86]. However, it is noteworthy that in unstructured environments, the path with the shortest distance does not necessarily translate to the one with the least energy consumption [87]. Energy costs are influenced by various factors, including terrain characteristics (such as terrain types, slope, friction, curvature, stiffness, and softness) [51,88] and vehicle parameters (such as engine speed, torque, and gear ratio) [11,89].

The energy cost estimation methods of unstructured environments typically fall into two categories: physics model-based methods and data-driven methods [83], as shown in Table 6.

Table 6. Overview of energy cost approaches and considerations in recent years.

Study	Approach Type	Factor Considered
Ganganath et al., 2015 [87]	Physics model-based	Terrain slope, friction
Higa et al., 2019 [90]	Mixed of data-driven and physics model	Terrain slope, robot's location, heading angle
Wallace et al., 2019 [91]	Physics model-based	Terrain slope, slippage, vehicle speed
Quann et al., 2020 [88]	Data-driven	Terrain type, slope, viscosity, vehicle–terrain interaction
Sakayori et al., 2021 [13]	Physics model-based	Terrain slope, heading angle, vehicle–terrain interaction
Huang et al., 2022 [92]	Data-driven	Terrain slope, driving mode
Goulet et al., 2022 [66]	Physics model-based	Terrain elevation, vehicle dynamics, vehicle–terrain interaction
Visca et al., 2022 [51]	Data-driven-	Terrain type, elevation, driving operation

4.2.1. Physics Model-Based Approaches

Several physics-based models have been developed to accurately estimate driving energy consumption in natural terrains [83,93]. A physics-based motion energy cost model was proposed in [94], which considered terrain attributes such as slope and friction, as well as the effects of configuration changes on operational energy costs. One challenging issue of physics-based approaches is to estimate the interaction between the vehicle and terrain [95]. To address this problem, a self-supervised approach is presented in [95], which considers terrain geometry and soil types to predict energy consumption. The authors of [13] examined the vehicle–terrain interaction and vehicle dynamic model and subsequently establishes the power consumption model through a neural network.

However, physics-based approaches in terrain modeling often exhibit only an approximate representation of the mechanical behavior of various types of terrain [96]. Furthermore, these approaches typically rely on terrain-dependent parameters that are frequently unknown a priori. Moreover, the utilization of these methods may result in substantial computational costs and may be unsuitable for time-sensitive planning frameworks [97].

4.2.2. Data-Driven Approaches

Data-driven approaches have been proposed as a solution to the challenges associated with physics-based approaches [83,98]. One type of these approaches is to classify and identify the environment into a set of different categories. Then, different data-driven energy models, such as semi-empirical function [92,99], table lookup [100], or neural network [101,102], are used to link the terrain geometry to the drive energy consumption of each terrain type. However, this type of approach is not suitable for terrain with uncertainty.

Probabilistic methods can capture and clarify the uncertainty in energy prediction [51]. The authors of [82] used Gaussian process regression (GPR) and known vehicle modeling information to predict future path energy costs. A framework based on multi-task Gaussian process regression (MTGP) to share terrain information across multiple robots was developed to predict power consumption [88]. Meta learning was used in [51,83] to model the existing energy consumption uncertainties and effectively adapt probabilistic predictions based on a small number of local measurements.

However, data-driven approaches are specific to one individual and generally cannot be directly applied to different individuals or the same individual under different operating conditions (e.g., tire pressure, weight distribution) [88].

4.3. *Comfort Cost*

When vehicles drive on uneven ground in unstructured environments, vibration occurs, impacting the performance and lifespan of vehicle parts. Additionally, this vibration affects the integrity of the transported goods. Consequently, it is essential to evaluate the comfort cost when planning the vehicle's path in unstructured terrains to ensure stable driving [11,103].

Comfort is closely related to different parameters such as the natural frequency of the vehicle system, road profile, and system characteristics. The elevation variance of the ground nodes was used in [52,104] to describe the terrain unevenness. The authors of [43] studied vibration responses of different models to different ground conditions, including quarter-car model, bicycle–car model, half-car model, and full-car model. The relationship among vehicle shock, ground roughness, and vehicle speed has been studied in [103] using machine learning. Based on the above influencing factors and their correlation, cost functions of comfort were established. The cost function of comfort in [67] mainly considered the terrain smoothness and path consistency. In addition, path vertical variation was also taken into account in [105].

5. Global Path Planner

The global path planner is mainly used to generate the cursory path which is based on the traversability and cost estimation. The traditional algorithms for global path planners have been improved and applied in unstructured environments, such as graph-search-based, sampling-based, and artificial potential field (APF) methods, and optimal methods [7,106,107], as depicted in Table 7.

Table 7. Overview of global path planning algorithms for unstructured environments.

Type	Algorithm	Study	Application	Limitation
Graph-search-based	A*, Dijkstra	[39,70,108–110]	Widely used in global path, global optimal solution, vehicle constraints considered	Not suitable for real-time and complex transformation environments
Sampling-based	RRT, PRM	[91,109,111,112]	Probabilistic completeness	Difficult to find the optimal solution
APF	APF	[113]	High real-time obstacles avoidance	Falling into the local minimum, not necessarily global optimal solution
Optimal method	GDA, PSO, GA	[11,85]	Multi-objective optimization, search efficiency improved	Taking a lot of computing resources and time, falling into local minimum solutions

Graph search-based methods discretize the configuration space as a graph and then search for a minimum-cost path within it [7]. The A* algorithm [108] and Dijkstra algorithm [39,114] and their improvements are the most common approaches. The obtained cost map of TTA can easily be integrated into a classical graph search algorithm [24]. For this method, the global optimal solution can be obtained, complex terrain and obstacles can be dealt with, and the dynamic constraints of the vehicle can be taken into account. However, this method takes a long time to search and is not suitable for real-time path planning. Moreover, for dynamic environments, map information needs to be updated constantly, which increases the computational load and complexity.

For sampling-based methods, a set of nodes or other forms are used to randomly sample the environment to obtain the final path [52]. Rapidly exploring Random Tree (RRT) [109] and the Probabilistic Road Map (PRM) [91] are the two most common methods. These methods have probabilistic completeness, but it is difficult to find the optimal solution in unstructured environments [7,64].

For artificial potential field methods, the global path is planned by setting the gravitational field of the starting and ending points and the repulsive force field of obstacles [67], which is simple and easy to understand, and can be implemented in the case of high real-time requirements for path planning. However, this method is prone to falling into local minima, and the planned path is not necessarily the global optimal solution [64,67].

In optimal methods, the path optimization problem can be solved by gradient descent algorithm (GDA) or intelligent optimization algorithm such as particle swarm optimization (PSO) [85] and genetic algorithm (GA) [11]. This method can deal with multi-objective optimization problems and can improve search efficiency by introducing heuristic methods and strategies based on prior knowledge of the problem. However, it needs high computational complexity and is easy to fall into local minimum solutions.

Due to the complexity of the unstructured environment, several methods are often combined to plan the optimal path. For example, some articles discretize vehicle states by using such as sampling-based methods and search for a least-cost path by using the Dijkstra or A* algorithm [13,69]. The artificial potential field method and optimal path methods were combined in [113] to search for the globally optimal path.

6. Local Path Planner

Local path planning optimizes the path from global path planning to ensure smoothness and safety, considering vehicle and terrain limits, path feasibility, and dynamic environment collision avoidance.

6.1. Constraints

Local path planners must account for various constraints related to vehicle and terrain limitations. Table 8 provides an overview of the constraints considered in recent articles.

For position constraints, the optimized trajectories must pass through some specific points whose positions are known during the planning process, such as the start point and the goal point [52,115]. Continuity constraints are mainly required for the connection points of each segment trajectory. To ensure the smoothness of the trajectory, the continuity of the movement, and the requirement of comfort, it is necessary to ensure that the derivative of the connection point is continuous [13]. Kinematic constraints contain inequality constraints and equality constraints. For inequality constraints, an autonomous vehicle has to satisfy the limits of physical restrictions on the values of velocity, acceleration, and steering angle within their corresponding bounds [11,116]. According to the chosen vehicle model, an autonomous vehicle should also satisfy the kinematic equality constraints. Dynamics constraints are established according to the vehicle dynamic model and driving conditions. For example, the constraints of engine power are set according to the application scenario and vehicle dynamic parameters [11]. Safety constraints contain the collision avoidance constraints [85,104,117] and rollover-free constraints [66,75]. Obstacle avoidance can be met by setting a safe distance from obstacles or limits on speed and acceleration [118]. Rollover can be avoided by limiting the steering angle [75], pitch angle, and roll angle [66]. The speed and acceleration constraints during traveling on a side slope were studied in [43].

Table 8. Overview of constraints considered in recent articles.

Literature	Position	Continuity	Kinematic	Dynamics	Safety
Usami et al., 2020 [119]	Initial, terminal	Curvature, b-spline	Speed, acceleration	Vehicle dynamics, Motor/Engine–Torque	Wheel slip, Rollover
Tian et al., 2021 [11]	Terrain-dependent	/	Speed, acceleration, terrain-dependent	Engine Power	/
Zhou et al., 2022 [52]	Initial, terminal	Bezier curve, derivative continuity	Speed, acceleration	/	Collision avoidance
Guo et al., 2022 [85]	Initial, terminal	/	Speed, acceleration, steering angle, terrain-dependent	Vehicle dynamics	Collision avoidance
Goulet et al., 2022 [66]	Initial, terminal	/	Speed, acceleration	Vehicle dynamics	Wheel slip, rollover, collision avoidance
Sun et al., 2022 [104]	Initial, terminal	/	Speed, acceleration, steering angle, steering angular speed	Vehicle dynamics	Collision avoidance
Qi et al., 2023 [120]	Initial, terminal	Three-layers continuous optimizer	Speed, acceleration	Vehicle dynamics	Collision avoidance
Chen et al., 2023 [31]	Initial, terminal	Curvature	Speed, acceleration, angular speed, angular acceleration	Vehicle dynamics	Collision avoidance

There are also constraints related to terrain [11,52,85]. For example, when going up and down slopes or around corners, speed and acceleration have different limits depending on the terrain [11]. In addition, when the vehicle approaches the shoulder edge, the vehicle steering is restricted to protect the vehicle tire, and ignoring the constraints will cause blowout and tire cord breakdown [85,121]. For field operations, such as mining areas, the vehicle has to enter a narrow area, but due to environmental restrictions, the vehicle sometimes cannot turn and can only go backward. To effectively handle the terrain-dependent constraints, an approximation formulation for these constraints was designed to eliminate the non-differentiability of these constraints [85]. Some terrain constraints can be set as a series of subgoals based on different terrain features, and then decompose the whole planning problem into a series of subproblems with different constraints, which are widely applied to mountain slopes [119,122].

6.2. Dynamic Obstacles Avoidance Algorithms

Local path planning strategies generally utilize sensor data about their surroundings to respond to the changing dynamics of the environment [25]. Dynamic obstacle avoidance is crucial to ensure safety, which makes obstacle avoidance decisions based on the perception of the constantly changing local environment [123]. Table 9 presents various algorithms employed for dynamic obstacle avoidance in unstructured local path planning.

Table 9. Overview of global path planning algorithms for dynamic obstacle avoidance in unstructured environments.

Algorithm	Study	Application	Limitation
Dynamic Window Approach	[31,124,125]	High real-time performance	High computational complexity
APF	[18,126,127]	Unknown obstacles avoidance	Falling into the local minimum
Time Elastic Banding Algorithm	[74,128]	Continuity, smoothness	Poor adaptability for complex situation

The dynamic window approach (DWA) is planned in the velocity space to generate a reference point motion speed that satisfies the vehicle constraints and ensures obstacle avoidance. It is suitable for real-time path planning and it can also optimize the algorithm's performance by adjusting the window parameters. However, it is sensitive to changes in the environment and requires frequent updates of obstacle information around the mobile robot [18,129].

The artificial potential field method is utilized in local path planning to establish a repulsive force field around the robot's obstacles, thereby guiding the robot to avoid obstacles and reach its target via a safe path. This method is particularly advantageous in the avoidance of unknown obstacles. Nonetheless, in complex environments, it may readily become trapped in local minima and is susceptible to noise interference [64,129].

The Time Elastic Banding Algorithm is the improvement based on the elastic band algorithm, which uses the concept of elastic body in mechanics to apply force to points on the path, and realize obstacle avoidance through the action of force. It considers the continuity of time, making the path smoother, but it cannot deal with complex scenes [129].

7. End-to-End Path Planning

For end-to-end path planning, the output path is generated directly according to the input of sensor data based on the learning methods which include the consideration of terrain and vehicle without separate procedures of TTA and cost estimation. With the development of machine learning, different kinds of learning-based algorithms were applied in end-to-end path planning, as shown in Table 10. To improve the adaptation of different vehicles, a CNN (Convolutional Neural Network) was used in [4] with a stereo camera for off-road path prediction. Apprenticeship learning was used in [19] by the sensor data from cameras and proprioceptive sensors for the adaptation navigation of different terrain types, such as concrete, grass, mud, pebble, and rock. The authors of [130] proposed an online adaptive self-supervised learning for off-road driving in rough terrain to realize a smooth, collision-free navigation policy that was suitable for plentiful vegetation, rock, and sandy environments. Inverse reinforcement learning using soft value iteration networks was proposed in [131] to solve the difficult path planning problems of planetary rovers with satellite maps. To improve planning time and planning success rate, deep reinforcement learning based on Rainbow architecture was used in [18] to navigate unknown rough terrain. Gaussian process regression (GPR) and decision tree theory were combined in [132] to build a proactive method to autonomously adapt the behavior to a priori unknown environments. [133] was based on Gaussian process regression to study the navigation in dynamic uncertain and unseen terrain. The path planning of terrain with different slop, friction coefficient, and roughness conditions was studied in [60] by Gaussian process regression according to the inputs of RGB camera and LiDAR.

The end-to-end path planning approach has stronger adaptability with different terrains than the hierarchical approach, especially for unknown environments. However, it depends on the dataset which is insufficient in unstructured environments at present. In addition, the end-to-end approach requires a lot of computing power, so it does not perform well in real-time planning.

Table 10. Overview of learning-based algorithms applied in end-to-end path planning.

Literature	Algorithm	Sensor Data	Application
Holder et al., 2018 [4]	CNN	Stereo camera	Off-road
Siva et al., 2019 [19]	Apprenticeship learning	Camera, wheel odometry, motor speed, IMU	Concrete, grass, mud, pebble, rock
Pflueger et al., 2019 [131]	Inverse reinforcement learning	Satellite map	Planetary rover
Manderson et al., 2020 [130]	Self-supervised learning	Camera, aerial image	Plentiful vegetation, rock, and sandy
Josef et al., 2020 [18]	Deep reinforcement learning	IMU, elevation map	Unknown rough terrain
Gall et al., 2021 [132]	Gaussian process regression, decision tree	Simulation data from Gazebo	Priori unknown, rough and slippery terrain
Siva et al., 2022 [133]	Gaussian process regression	Color camera, LiDAR, GPS, IMU, wheel odometry	Dynamic uncertain grass terrain, unseen forest terrain
Lee et al., 2023 [60]	Gaussian process regression	RGB camera, LiDAR	Terrain with different slop, friction coefficient, roughness conditions

8. Conclusions

Methods of unstructured path planning were reviewed in detail in this article, including the path planning framework, the relevant factors, the analysis methods, the necessary models, and algorithms. This article focused on the unique aspects compared to structured path planning, such as TTA, cost estimation, and terrain-dependent constraints.

- This article categorized path planning in unstructured environments into hierarchical and end-to-end approaches. The hierarchical path planning approach involves TTA, cost estimation, global path planner, constraints analysis and handling, and local path planner. The end-to-end path planning approach directly outputs the path based on the learning methods according to the input of sensor data. The recent articles on end-to-end path planning were summarized in terms of algorithm, sensor data, and application. The end-to-end path planning approach has stronger adaptability, but it needs a large quantity of dataset and computing power.
- The relevant factors influencing TTA were meticulously summarized and thoroughly discussed. These factors encompass terrain geometrical and physical attributes, vehicle dynamics, structural parameters, and the interaction between the vehicle and the terrain. This thorough analysis of relevant factors forms the foundation of TTA. Incorporating a more comprehensive consideration of these factors is instrumental in delivering a more accurate TTA. The inputs for TTA were categorized into remote sensing and in situ soil sensing, and the corresponding sensors and sensing techniques were introduced. For TTA processing, there are four methods: the cone index method was proposed early and is mainly based on experimental; the other methods are based on computer technology, including geometry-based, appearance-based, and behavior-based ones. These four methods were compared and summarized in detail. The output of the TTA delivered to the path planner may be regression-based, classification-based, or mixed regression and classification.
- For cost estimation, the models and methods were summarized, including safety cost, energy cost, and comfort cost. Among them, the energy cost was analyzed and discussed emphatically. The physics model-based and data-driven energy cost approaches were introduced and compared. In addition, the considerations and estimation methods of safety cost and comfort cost were reviewed.
- The constraints considered in recent articles were summarized, including position constraints, continuity constraints, kinematic constraints, dynamics constraints, and safety constraints. Among them, the terrain-dependent constraints and their handling methods were introduced emphatically.
- The global and local path planning methods suitable for unstructured environments in recent articles were reviewed and summarized.

9. Perspectives

In the future, to achieve autonomous driving in unstructured environments, several important research areas remain to be explored:

- Multiple uncertainties in path planning. In practical applications, it is often challenging to acquire an accurate prior map. Even in familiar environments, there can be randomness in different weather conditions. Especially in agriculture and rescue operations, the actual environment is often dynamic. Additionally, for operational vehicles, the load is often variable. These interactive uncertainties increase the complexity that should be further considered in TTA and path planning.
- The reliability and coverage of TTA. High-fidelity terramechanics models are essential for reliable TTA. However, most terramechanics models currently used in unstructured path planning are simplified, making it challenging to accurately represent real-world conditions. Additionally, off-road vehicles often operate on submerged ground, such as agricultural machinery and amphibious vehicles. It is undoubtedly more challenging for vehicles to traverse submerged soil compared to normal soil. Nevertheless, there is currently limited consideration of this in TTA.
- Vehicle instability evaluation. Vehicle instability evaluation is fundamental for unstructured path planning. Accurate failure analysis is necessary to make a reasonable estimation of vehicle traversability and cost. Currently, numerous studies have been conducted on the models and evaluation criteria of failure analysis, but their coverage and accuracy fall short of meeting the requirements of unstructured environment path planning.
- The effect of vehicle posture on path planning. Under the same terrain and conditions, the traversability and cost of vehicles with different postures are often different. For example, the risk severity of a vehicle moving forward and backward at the same speed is different. Besides, on the same slope, the energy consumption of vehicles going uphill and downhill is also different. In addition, for some scenarios, the vehicle cannot traverse in the forward direction but can successfully pass through by adjusting the heading angle or with a series of maneuvers. However, there is limited consideration of vehicle posture in current research.
- Safety redundancy in path planning. Safety is a crucial aspect of path planning, and indeed, it is necessary to achieve higher safety goals. To ensure safety, redundancy often exists in actual path planning. However, for vehicles, especially military vehicles operating in more demanding conditions, excessive safety redundancy can affect the maneuverability range. Therefore, the precise boundaries of safety need to be further explored.
- Adaptation and real-time requirements in path planning. As discussed above, the present hierarchical path planning does not perform well in adaptation and the end-to-end path planning cannot meet the real-time requirements. However, adaptability and real-time requirements are usually simultaneously needed. Therefore, whether the two frameworks can be combined, or a new framework can be proposed to meet the requirements should be further studied.
- Completeness verification for path planning methods. Real vehicle validation in unstructured environments is sometimes unsafe and does not cover all scenarios. Simulation and small smart demo vehicle testing can partially replace real vehicle testing. However, two issues need to be addressed in the alternative tests. One is the high fidelity of the alternative tests, including the reproduction of environments, vehicles, and their incentives and responses. The other is the effective and targeted scene design. The test scenario set can be generated according to the characteristics of application scenarios and vehicle instability assessment.
- Applications for passenger cars. Current market applications for autonomous driving in unstructured environments tend to center on agricultural and engineering vehicles. While most driver assistance or autonomous driving features on passenger cars cannot be utilized in unstructured environments, there is a certain demand among passenger

vehicle customers for autonomous driving capabilities in snow, ice, desert, or other unstructured environments. However, there is limited research on the path planning for this purpose.

Author Contributions: Conceptualization, N.W., J.W. and D.X.; methodology, N.W. and J.W.; formal analysis, N.W. and X.L.; investigation, N.W., X.L. and J.W.; writing—original draft preparation, N.W., X.L., K.Z., J.W. and D.X.; writing—review and editing, N.W., X.L., J.W. and D.X.; supervision, J.W. and D.X.; project administration, J.W.; funding acquisition, J.W. All authors have read and agreed to the published version of the manuscript.

Funding: This research was funded by National Natural Science Foundation of China, grant number 52272434.

Data Availability Statement: Not applicable.

Conflicts of Interest: The authors declare no conflicts of interest.

References

1. Cheng, C.; Sha, Q.; He, B.; Li, G. Path Planning and Obstacle Avoidance for AUV: A Review. *Ocean Eng.* **2021**, *235*, 109355. [CrossRef]
2. Li, L.; Li, J.; Zhang, S. Review Article: State-of-The-Art Trajectory Tracking of Autonomous Vehicles. *Mech. Sci.* **2021**, *12*, 419–432. [CrossRef]
3. He, X.; Wu, J.; Huang, Z.; Hu, Z.; Wang, J.; Sangiovanni-Vincentelli, A.; Lv, C. Fear-Neuro-Inspired Reinforcement Learning for Safe Autonomous Driving. *IEEE Trans. Pattern Anal. Mach. Intell.* **2023**, *46*, 267–279. [CrossRef] [PubMed]
4. Holder, C.J.; Breckon, T.P. Learning to Drive: Using Visual Odometry to Bootstrap Deep Learning for Off-Road Path Prediction. In Proceedings of the 2018 IEEE Intelligent Vehicles Symposium (IV), Changshu, China, 26–30 June 2018; pp. 2104–2110.
5. Islam, F.; Nabi, M.M.; Ball, J.E. Off-Road Detection Analysis for Autonomous Ground Vehicles: A Review. *Sensors* **2022**, *22*, 8463. [CrossRef] [PubMed]
6. Albrecht, A.; Heide, N.F.; Frese, C.; Zube, A. Generic Convoying Functionality for Autonomous Vehicles in Unstructured Outdoor Environments. In Proceedings of the 2020 IEEE Intelligent Vehicles Symposium (IV), Las Vegas, NV, USA, 19 October–13 November 2020; pp. 1949–1955.
7. Li, Z.; Zhao, P.; Jiang, C.; Huang, W.; Liang, H. A Learning-Based Model Predictive Trajectory Planning Controller for Automated Driving in Unstructured Dynamic Environments. *IEEE Trans. Veh. Technol.* **2022**, *71*, 5944–5959. [CrossRef]
8. Chakraborty, S.; Elangovan, D.; Govindarajan, P.L.; ELnaggar, M.F.; Alrashed, M.M.; Kamel, S. A Comprehensive Review of Path Planning for Agricultural Ground Robots. *Sustainability* **2022**, *14*, 9156. [CrossRef]
9. Bai, Y.; Zhang, B.; Xu, N.; Zhou, J.; Shi, J.; Diao, Z. Vision-Based Navigation and Guidance for Agricultural Autonomous Vehicles and Robots: A Review. *Comput. Electron. Agric.* **2023**, *205*, 107584. [CrossRef]
10. Heštera, H.; Pahernik, M. Physical-Geographic Factors of Terrain Trafficability of Military Vehicles According to Western World Methodologies. *Hrvat. Geogr. Glas.* **2018**, *80*, 5–31. [CrossRef]
11. Tian, F.; Zhou, R.; Li, Z.; Li, L.; Gao, Y.; Cao, D.; Chen, L. Trajectory Planning for Autonomous Mining Trucks Considering Terrain Constraints. *IEEE Trans. Intell. Veh.* **2021**, *6*, 772–786. [CrossRef]
12. Morales, J.; Vázquez-Martín, R.; Mandow, A.; Morilla-Cabello, D.; García-Cerezo, A. The UMA-SAR Dataset: Multimodal Data Collection from a Ground Vehicle during Outdoor Disaster Response Training Exercises. *Int. J. Robot. Res.* **2021**, *40*, 835–847. [CrossRef]
13. Sakayori, G.; Ishigami, G. Energy-Aware Trajectory Planning for Planetary Rovers. *Adv. Robot.* **2021**, *35*, 1302–1316. [CrossRef]
14. Zhou, R.; Feng, W.; Yang, H.; Gao, H.; Li, N.; Deng, Z.; Ding, L. Predicting Terrain Mechanical Properties in Sight for Planetary Rovers with Semantic Clues. *arXiv* **2020**. [CrossRef]
15. Zhang, L.; Du, Z.; Zhao, S.; Zhai, Y.; Shen, Y. Development and Verification of Traffic Confrontation Simulation Test Platform Based on PanoSim. In Proceedings of the 2020 IEEE 4th Information Technology, Networking, Electronic and Automation Control Conference (ITNEC), Chongqing, China, 12–14 June 2020; pp. 1814–1818.
16. Xiao, X.; Biswas, J.; Stone, P. Learning Inverse Kinodynamics for Accurate High-Speed Off-Road Navigation on Unstructured Terrain. *IEEE Robot. Autom. Lett.* **2021**, *6*, 6054–6060. [CrossRef]
17. Zhong, C.; Li, B.; Wu, T. Off-Road Drivable Area Detection: A Learning-Based Approach Exploiting LiDAR Reflection Texture Information. *Remote Sens.* **2023**, *15*, 27. [CrossRef]
18. Josef, S.; Degani, A. Deep Reinforcement Learning for Safe Local Planning of a Ground Vehicle in Unknown Rough Terrain. *IEEE Robot. Autom. Lett.* **2020**, *5*, 6748–6755. [CrossRef]
19. Siva, S.; Wigness, M.; Rogers, J.; Zhang, H. Robot Adaptation to Unstructured Terrains by Joint Representation and Apprenticeship Learning. In Proceedings of the Robotics: Science and Systems, Freiburg im Breisgau, Germany, 12–17 July 2019.
20. Yu, K.; Xu, C.; Ma, J.; Fang, B.; Ding, J.; Xu, X.; Bao, X.; Qiu, S. Automatic Matching of Multimodal Remote Sensing Images via Learned Unstructured Road Feature. *Remote Sens.* **2022**, *14*, 4595. [CrossRef]

21. Guo, C. Analysis of Unstructured Road Information Recognition Methods for Unmanned Vehicles. In Proceedings of the 2019 International Conference on Intelligent Transportation, Big Data & Smart City (ICITBS), Changsha, China, 12–13 January 2019; pp. 84–89.
22. He, X.; Yang, H.; Hu, Z.; Lv, C. Robust Lane Change Decision Making for Autonomous Vehicles: An Observation Adversarial Reinforcement Learning Approach. *IEEE Trans. Intell. Veh.* **2023**, *8*, 184–193. [CrossRef]
23. Wang, Y.; Li, X.; Zhang, J.; Li, S.; Xu, Z.; Zhou, X. Review of Wheeled Mobile Robot Collision Avoidance under Unknown Environment. *Sci. Prog.* **2021**, *104*, 00368504211037771. [CrossRef]
24. Guastella, D.C.; Muscato, G. Learning-Based Methods of Perception and Navigation for Ground Vehicles in Unstructured Environments: A Review. *Sensors* **2021**, *21*, 73. [CrossRef] [PubMed]
25. Wijayathunga, L.; Rassau, A.; Chai, D. Challenges and Solutions for Autonomous Ground Robot Scene Understanding and Navigation in Unstructured Outdoor Environments: A Review. *Appl. Sci.* **2023**, *13*, 9877. [CrossRef]
26. He, X.; Chen, H.; Lv, C. Robust Multiagent Reinforcement Learning toward Coordinated Decision-Making of Automated Vehicles. *SAE Int. J. Veh. Dyn. Stab. NVH* **2023**, *7*, 475–488. [CrossRef]
27. Ono, M.; Fuchs, T.J.; Steffy, A.; Maimone, M.; Yen, J. Risk-Aware Planetary Rover Operation: Autonomous Terrain Classification and Path Planning. In Proceedings of the 2015 IEEE Aerospace Conference, Big Sky, MT, USA, 7–14 March 2015; pp. 1–10.
28. Mayuku, O.; Surgenor, B.W.; Marshall, J.A. A Self-Supervised Near-to-Far Approach for Terrain-Adaptive Off-Road Autonomous Driving. In Proceedings of the 2021 IEEE International Conference on Robotics and Automation (ICRA), Xi'an, China, 30 May–5 June 2021; pp. 14054–14060.
29. Wallin, E.; Wiberg, V.; Vesterlund, F.; Holmgren, J.; Persson, H.J.; Servin, M. Learning Multiobjective Rough Terrain Traversability. *J. Terramech* **2022**, *102*, 17–26. [CrossRef]
30. Papadakis, P. Terrain Traversability Analysis Methods for Unmanned Ground Vehicles: A Survey. *Eng. Appl. Artif. Intell.* **2013**, *26*, 1373–1385. [CrossRef]
31. Chen, D.; Zhuang, M.; Zhong, X.; Wu, W.; Liu, Q. RSPMP: Real-Time Semantic Perception and Motion Planning for Autonomous Navigation of Unmanned Ground Vehicle in off-Road Environments. *Appl. Intell.* **2023**, *53*, 4979–4995. [CrossRef]
32. Zhu, Z.; Li, N.; Sun, R.; Zhao, H.; Xu, D. Off-Road Autonomous Vehicles Traversability Analysis and Trajectory Planning Based on Deep Inverse Reinforcement Learning. In Proceedings of the 2020 IEEE Intelligent Vehicles Symposium (IV), Las Vegas, NV, USA, 19 October–13 November 2020; pp. 971–977.
33. Graf, U.; Borges, P.; Hernández, E.; Siegwart, R.; Dubé, R. Optimization-Based Terrain Analysis and Path Planning in Unstructured Environments. In Proceedings of the 2019 International Conference on Robotics and Automation (ICRA), Montreal, QC, Canada, 20–24 May 2019; pp. 5614–5620.
34. Zhang, K.; Niroui, F.; Ficocelli, M.; Nejat, G. Robot Navigation of Environments with Unknown Rough Terrain Using Deep Reinforcement Learning. In Proceedings of the 2018 IEEE International Symposium on Safety, Security, and Rescue Robotics (SSRR), Philadelphia, PA, USA, 6–8 August 2018; pp. 1–7.
35. Chavez-Garcia, R.O.; Guzzi, J.; Gambardella, L.M.; Giusti, A. Learning Ground Traversability from Simulations. *IEEE Robot. Autom. Lett.* **2018**, *3*, 1695–1702. [CrossRef]
36. Saarilahti, M. *Development of a Protocol for Ecoefficient Wood on Sensitive Sites (Ecowood)*; University of Helsinki: Turku, Finland, 2002.
37. Green, A.R.; Rye, D. Sensible Planning for Vehicles Operating Over Difficult Unstructured Terrains. In Proceedings of the 2007 IEEE Aerospace Conference, Big Sky, MT, USA, 3–10 March 2007; pp. 1–8.
38. Howard, A.; Seraji, H. Vision-Based Terrain Characterization and Traversability Assessment. *J. Robot. Syst.* **2001**, *18*, 577–587. [CrossRef]
39. Oliveira, F.G.; Neto, A.A.; Howard, D.; Borges, P.; Campos, M.F.M.; Macharet, D.G. Three-Dimensional Mapping with Augmented Navigation Cost through Deep Learning. *J. Intell. Robot. Syst.* **2021**, *101*, 50. [CrossRef]
40. Serban, R.; Taylor, M.; Negrut, D.; Tasora, A. Chrono::Vehicle: Template-Based Ground Vehicle Modelling and Simulation. *Int. J. Veh. Perform.* **2019**, *5*, 18–39. [CrossRef]
41. Kejian, Z. *Vehicle-Terrainmechanics*; National Defense Industry Press: Beijing, China, 2002.
42. Dallas, J.; Weng, Y.; Ersal, T. Combined Trajectory Planning and Tracking for Autonomous Vehicles on Deformable Terrains. In Proceedings of the ASME 2020 Dynamic Systems and Control Conference, Pittsburgh, PA, USA, 5–7 October 2020; pp. 1–8.
43. Taghavifar, H.; Mardani, A. *Off-Road Vehicle Dynamics: Analysis, Modelling and Optimization*, 1st ed.; Studies in Systems, Decision and Control; Springer: Berlin/Heidelberg, Germany, 2019; Volume 70.
44. Gonzalez, R.; Iagnemma, K. DeepTerramechanics: Terrain Classification and Slip Estimation for Ground Robots via Deep Learning. *arXiv* **2018**. [CrossRef]
45. Dallas, J.; Cole, M.P.; Jayakumar, P.; Ersal, T. Neural Network Based Terramechanics Modeling and Estimation for Deformable Terrains. *arXiv* **2020**. [CrossRef]
46. Taheri, S.; Sandu, C.; Taheri, S.; Pinto, E.; Gorsich, D. A Technical Survey on Terramechanics Models for Tire-Terrain Interaction Used in Modeling and Simulation of Wheeled Vehicles. *J. Terramech* **2015**, *57*, 1–22. [CrossRef]
47. Dallas, J.; Jain, K.; Dong, Z.; Cole, M.P.; Jayakumar, P.; Ersal, T. Online Terrain Estimation for Autonomous Vehicles on Deformable Terrains. *J. Terramech* **2020**, *91*, 11–22. [CrossRef]

48. Chhaniyara, S.; Brunskill, C.; Yeomans, B.; Matthews, M.C.; Saaj, C.; Ransom, S.; Richter, L. Terrain Trafficability Analysis and Soil Mechanical Property Identification for Planetary Rovers: A Survey. *J. Terramech* **2012**, *49*, 115–128. [CrossRef]
49. Tsubaki, H.; Ishigami, G. Experimental Study on Wheel-Soil Interaction Mechanics Using in-Wheel Sensor and Particle Image Velocimetry Part I: Analysis and Modeling of Normal Stress of Lightweight Wheeled Vehicles. *J. Terramech* **2021**, *93*, 23–39. [CrossRef]
50. Smith, W.C. *Modeling of Wheel-Soil Interaction for Small Ground Vehicles Operating on Granular Soil*; The University of Michigan: Ann Arbor, MI, USA, 2014.
51. Visca, M.; Powell, R.; Gao, Y.; Fallah, S. Meta-Conv1D Energy-Aware Path Planner for Mobile Robots in Unstructured Terrains. In Proceedings of the 2022 7th International Conference on Robotics and Automation Engineering, Singapore, 18–20 November 2022; pp. 150–157.
52. Zhou, B.; Yi, J.; Zhang, X.; Chen, L.; Yang, D.; Han, F.; Zhang, H. An Autonomous Navigation Approach for Unmanned Vehicle in Outdoor Unstructured Terrain with Dynamic and Negative Obstacles. *Robotica* **2022**, *40*, 2831–2854. [CrossRef]
53. Brooks, C.A.; Iagnemma, K. Vibration-Based Terrain Classification for Planetary Exploration Rovers. *IEEE Trans. Robot.* **2005**, *21*, 1185–1191. [CrossRef]
54. Palazzo, S.; Guastella, D.C.; Cantelli, L.; Spadaro, P.; Rundo, F.; Muscato, G.; Giordano, D.; Spampinato, C. Domain Adaptation for Outdoor Robot Traversability Estimation from RGB Data with Safety-Preserving Loss. In Proceedings of the 2020 IEEE/RSJ International Conference on Intelligent Robots and Systems (IROS), Las Vegas, NV, USA, 24 October–24 January 2020; pp. 10014–10021.
55. Zhang, Y.; Wang, W.; Bonatti, R.; Maturana, D.; Scherer, S. Integrating Kinematics and Environment Context into Deep Inverse Reinforcement Learning for Predicting Off-Road Vehicle Trajectories. *arXiv* **2018**. [CrossRef]
56. Wulfmeier, M.; Ondruska, P.; Posner, I. Maximum Entropy Deep Inverse Reinforcement Learning. *arXiv* **2015**. [CrossRef]
57. Chiodini, S.; Torresin, L.; Pertile, M.; Debei, S. Evaluation of 3D CNN Semantic Mapping for Rover Navigation. In Proceedings of the 2020 IEEE 7th International Workshop on Metrology for AeroSpace (MetroAeroSpace), Pisa, Italy, 22–24 June 2020; pp. 32–36.
58. Suryamurthy, V.; Raghavan, S.; Laurenzi, A.; Tsagarakis, N.G.; Kanoulas, D. Terrain Segmentation and Roughness Estimation Using RGB Data: Path Planning Application on the CENTAURO Robot. In Proceedings of the 2019 IEEE-RAS 19th International Conference on Humanoid Robots (Humanoids), Toronto, ON, Canada, 15–17 October 2019; pp. 1–8.
59. Mei, J.; Yu, Y.; Zhao, H.; Zha, H. Scene-Adaptive Off-Road Detection Using a Monocular Camera. *IEEE Trans. Intell. Transp. Syst.* **2018**, *19*, 242–253. [CrossRef]
60. Lee, H.; Kwon, J.; Kwon, C. Learning-Based Uncertainty-Aware Navigation in 3D Off-Road Terrains. In Proceedings of the 2023 IEEE International Conference on Robotics and Automation (ICRA), London, UK, 29 May–2 June 2023; pp. 10061–10068.
61. Xue, H.; Fu, H.; Xiao, L.; Fan, Y.; Zhao, D.; Dai, B. Traversability Analysis for Autonomous Driving in Complex Environment: A LiDAR-Based Terrain Modeling Approach. *J. Field Robot.* **2023**, *40*, 1779–1803. [CrossRef]
62. Bae, J.; Seo, J.; Kim, T.; Jeon, H.G.; Kwak, K.; Shim, I. Self-Supervised 3D Traversability Estimation With Proxy Bank Guidance. *IEEE Access* **2023**, *11*, 51490–51501. [CrossRef]
63. Guan, T.; He, Z.; Song, R.; Zhang, L. TNES: Terrain Traversability Mapping, Navigation and Excavation System for Autonomous Excavators on Worksite. *Auton. Robots* **2023**, *47*, 695–714. [CrossRef]
64. Xiong, L.; Fu, Z.; Zeng, D.; Leng, B. An Optimized Trajectory Planner and Motion Controller Framework for Autonomous Driving in Unstructured Environments. *Sensors* **2021**, *21*, 4409. [CrossRef] [PubMed]
65. He, X.; Huang, W.; Lv, C. Toward Trustworthy Decision Making for Autonomous Vehicles: A Robust Reinforcement Learning Approach with Safety Guarantees. *Engineering* **2023**. [CrossRef]
66. Goulet, N.; Ayalew, B. Energy-Optimal Ground Vehicle Trajectory Planning on Deformable Terrains. *IFAC-PapersOnLine* **2022**, *55*, 196–201. [CrossRef]
67. Li, X.; Tang, B.; Ball, J.; Doude, M.; Carruth, D.W. Rollover-Free Path Planning for off-Road Autonomous Driving. *Electronics* **2019**, *8*, 614. [CrossRef]
68. Valencia-Murillo, R.; Arana-Daniel, N.; López-Franco, C.; Alanís, A.Y. Rough Terrain Perception through Geometric Entities for Robot Navigation. In Proceedings of the 2nd International Conference on Advances in Computer Science and Engineering, Los Angeles, CA, USA, 1–2 July 2013; pp. 309–314.
69. Roy, J.; Wan, N.; Goswami, A.; Vahidi, A.; Jayakumar, P.; Zhang, C. A Hierarchical Route Guidance Framework for Off-Road Connected Vehicles. *J. Dyn. Syst. Meas. Control* **2018**, *140*, 071011. [CrossRef]
70. Pfrunder, A.; Borges, P.V.K.; Romero, A.R.; Catt, G.; Elfes, A. Real-Time Autonomous Ground Vehicle Navigation in Heterogeneous Environments Using a 3D LiDAR. In Proceedings of the 2017 IEEE/RSJ International Conference on Intelligent Robots and Systems (IROS), Vancouver, BC, Canada, 24–28 September 2017; pp. 2601–2608.
71. He, X.; Liu, Y.; Lv, C.; Ji, X.; Liu, Y. Emergency Steering Control of Autonomous Vehicle for Collision Avoidance and Stabilisation. *Veh. Syst. Dyn.* **2019**, *57*, 1163–1187. [CrossRef]
72. Gao, Y.; Lin, T.; Borrelli, F.; Tseng, E.; Hrovat, D. Predictive Control of Autonomous Ground Vehicles With Obstacle Avoidance on Slippery Roads. In Proceedings of the ASME 2010 Dynamic Systems and Control Conference, Cambridge, MA, USA, 12–15 September 2010; pp. 265–272.

73. Brandao, M.; Shiguematsu, Y.M.; Hashimoto, K.; Takanishi, A. Material Recognition CNNs and Hierarchical Planning for Biped Robot Locomotion on Slippery Terrain. In Proceedings of the 2016 IEEE-RAS 16th International Conference on Humanoid Robots (Humanoids), Cancun, Mexico, 15–17 November 2016; pp. 81–88.
74. Dang, T.V. Autonomous Mobile Robot Path Planning Based on Enhanced A* Algorithm Integrating With Time Elastic Band. *MM Sci. J.* **2023**, 6717–6722. [CrossRef]
75. Liu, J.; Jayakumar, P.; Stein, J.L.; Ersal, T. A Nonlinear Model Predictive Control Formulation for Obstacle Avoidance in High-Speed Autonomous Ground Vehicles in Unstructured Environments. *Int. J. Veh. Mech. Mobil.* **2018**, *56*, 853–882. [CrossRef]
76. Sandu, C. Vehicle Dynamics: Theory and Applications. *J. Guid. Control. Dyn.* **2010**, *33*, 287–288. [CrossRef]
77. Andrzejewski, R.; Awrejcewicz, J. *Nonlinear Dynamics of a Wheeled Vehicle*; Springer: Berlin/Heidelberg, Germany, 2005.
78. Eick, A. *A Nonlinear Model Predictive Control Algorithm for an Unmanned Ground on Variable Terrain*; Auburn University: Auburn, AL, USA, 2016.
79. Wang, Y.; Hu, J.; Wang, F.; Dong, H.; Yan, Y.; Ren, Y.; Zhou, C.; Yin, G. Tire Road Friction Coefficient Estimation: Review and Research Perspectives. *Chin. J. Mech. Eng.* **2022**, *35*, 6. [CrossRef]
80. Gonzalez, R.; Iagnemma, K. Slippage Estimation and Compensation for Planetary Exploration Rovers. State of the Art and Future Challenges. *J. Field Robot.* **2018**, *35*, 564–577. [CrossRef]
81. Basri, M.; Karapetyan, A.; Hassan, B.; Khonji, M.; Dias, J. A Hybrid Deep Learning Approach for Vehicle Wheel Slip Prediction in Off-Road Environments. In Proceedings of the 2022 IEEE International Symposium on Robotic and Sensors Environments (ROSE), Abu Dhabi, United Arab Emirates, 14–15 November 2022; pp. 1–7.
82. Quann, M.; Ojeda, L.; Smith, W.; Rizzo, D.; Castanier, M.; Barton, K. Off-Road Ground Robot Path Energy Cost Prediction through Probabilistic Spatial Mapping. *J. Field Robot.* **2020**, *37*, 421–439. [CrossRef]
83. Visca, M.; Powell, R.; Gao, Y.; Fallah, S. Probabilistic Meta-Conv1D Driving Energy Prediction for Mobile Robots in Unstructured Terrains. *IEEE Access* **2022**, *10*, 107913–107928. [CrossRef]
84. Lacaze, A.; Moscovitz, O.Y.; Declaris, N.; Murphy, K. Path Planning for Autonomous Vehicles Driving over Rough Terrain. In Proceedings of the 1998 IEEE International Symposium on Intelligent Control (ISIC) held jointly with IEEE International Symposium on Computational Intelligence in Robotics and Automation (CIRA) Intell, Gaithersburg, MD, USA, 17 September 1998; pp. 50–55.
85. Guo, Y.; Yao, D.; Li, B.; He, Z.; Gao, H.; Li, L. Trajectory Planning for an Autonomous Vehicle in Spatially Constrained Environments. *IEEE Trans. Intell. Transp. Syst.* **2022**, *23*, 18326–18336. [CrossRef]
86. Li, B.; Wang, K.; Shao, Z. Time-Optimal Maneuver Planning in Automatic Parallel Parking Using a Simultaneous Dynamic Optimization Approach. *IEEE Trans. Intell. Transp. Syst.* **2016**, *17*, 3263–3274. [CrossRef]
87. Ganganath, N.; Cheng, C.T.; Tse, C.K. A Constraint-Aware Heuristic Path Planner for Finding Energy-Efficient Paths on Uneven Terrains. *IEEE Trans. Industr Inform.* **2015**, *11*, 601–611. [CrossRef]
88. Quann, M.; Ojeda, L.; Smith, W.; Rizzo, D.; Castanier, M.; Barton, K. Power Prediction for Heterogeneous Ground Robots through Spatial Mapping and Sharing of Terrain Data. *IEEE Robot. Autom. Lett.* **2020**, *5*, 1579–1586. [CrossRef]
89. Kamal, M.A.S.; Mukai, M.; Murata, J.; Kawabe, T. Ecological Vehicle Control on Roads with Up-down Slopes. *IEEE Trans. Intell. Transp. Syst.* **2011**, *12*, 783–794. [CrossRef]
90. Higa, S.; Iwashita, Y.; Otsu, K.; Ono, M.; Lamarre, O.; Didier, A.; Hoffmann, M. Vision-Based Estimation of Driving Energy for Planetary Rovers Using Deep Learning and Terramechanics. *IEEE Robot. Autom. Lett.* **2019**, *4*, 3876–3883. [CrossRef]
91. Wallace, N.D.; Kong, H.; Hill, A.J.; Sukkarieh, S. Energy Aware Mission Planning for WMRs on Uneven Terrains. *IFAC-PapersOnLine* **2019**, *52*, 149–154. [CrossRef]
92. Huang, G.; Yuan, X.; Shi, K.; Liu, Z.; Wu, X. A 3-D Multi-Object Path Planning Method for Electric Vehicle Considering the Energy Consumption and Distance. *IEEE Trans. Intell. Transp. Syst.* **2022**, *23*, 7508–7520. [CrossRef]
93. Buse, F.; Lichtenheldt, R.; Krenn, R. SCM-A Novel Approach for Soil Deformation in a Modular Soil Contact Model for Multibody Simulation. In Proceedings of the 4th Joint International Conference on Multibody System Dynamics, Montréal, QC, Canada, 29 May–2 June 2016; pp. 1–14.
94. Wallace, N.D.; Kong, H.; Hill, A.J.; Sukkarieh, S. Motion Cost Characterisation of an Omnidirectional WMR on Uneven Terrains. *IFAC-PapersOnLine* **2019**, *52*, 31–36. [CrossRef]
95. Otsu, K.; Kubota, T. Energy-Aware Terrain Analysis for Mobile Robot Exploration. In *Field and Service Robotics*; Springer: Cham, Switzerland, 2016; Volume 113, pp. 373–388. [CrossRef]
96. Johnson, J.B.; Duvoy, P.X.; Kulchitsky, A.V.; Creager, C.; Moore, J. Analysis of Mars Exploration Rover Wheel Mobility Processes and the Limitations of Classical Terramechanics Models Using Discrete Element Method Simulations. *J. Terramech.* **2017**, *73*, 61–71. [CrossRef]
97. He, R.; Sandu, C.; Khan, A.K.; Guthrie, A.G.; Schalk Els, P.; Hamersma, H.A. Review of Terramechanics Models and Their Applicability to Real-Time Applications. *J. Terramech.* **2019**, *81*, 3–22. [CrossRef]
98. He, X.; Lv, C. Toward Intelligent Connected E-Mobility: Energy-Aware Cooperative Driving With Deep Multiagent Reinforcement Learning. *IEEE Veh. Technol. Mag.* **2023**, *18*, 101–109. [CrossRef]
99. Gruning, V.; Pentzer, J.; Brennan, S.; Reichard, K. Energy-Aware Path Planning for Skid-Steer Robots Operating on Hilly Terrain. In Proceedings of the 2020 American Control Conference (ACC), Denver, CO, USA, 1–3 July 2020; pp. 2094–2099.
100. Fallah, S.; Yue, B.; Vahid-Araghi, O.; Khajepour, A. Energy Management of Planetary Rovers Using a Fast Feature-Based Path Planning and Hardware-in-the-Loop Experiments. *IEEE Trans. Veh. Technol.* **2013**, *62*, 2389–2401. [CrossRef]

101. Sakayori, G.; Ishigami, G. Energy Efficient Slope Traversability Planning for Mobile Robot in Loose Soil. In Proceedings of the 2017 IEEE International Conference on Mechatronics (ICM), Churchill, VIC, Australia, 13–15 February 2017; pp. 99–104.
102. Visca, M.; Bouton, A.; Powell, R.; Gao, Y.; Fallah, S. Conv1D Energy-Aware Path Planner for Mobile Robots in Unstructured Environments. In Proceedings of the 2021 IEEE International Conference on Robotics and Automation (ICRA), Xi'an, China, 30 May–5 June 2021; pp. 2279–2285.
103. Stavens, D.; Thrun, S. A Self-Supervised Terrain Roughness Estimator for Off-Road Autonomous Driving. *arXiv* **2012**. [CrossRef]
104. Sun, C.; Li, Q.; Li, B.; Li, L. A Successive Linearization in Feasible Set Algorithm for Vehicle Motion Planning in Unstructured and Low-Speed Scenarios. *IEEE Trans. Intell. Transp. Syst.* **2022**, *23*, 3724–3736. [CrossRef]
105. Hu, X.; Chen, L.; Tang, B.; Cao, D.; He, H. Dynamic Path Planning for Autonomous Driving on Various Roads with Avoidance of Static and Moving Obstacles. *Mech. Syst. Signal Process* **2018**, *100*, 482–500. [CrossRef]
106. Patle, B.K.; Babu, L.G.; Pandey, A.; Parhi, D.R.K.; Jagadeesh, A. A Review: On Path Planning Strategies for Navigation of Mobile Robot. *Def. Technol.* **2019**, *15*, 582–606. [CrossRef]
107. Mohanan, M.G.; Salgoankar, A. A Survey of Robotic Motion Planning in Dynamic Environments. *Rob. Auton. Syst.* **2018**, *100*, 171–185. [CrossRef]
108. Min, H.; Xiong, X.; Wang, P.; Yu, Y. Autonomous Driving Path Planning Algorithm Based on Improved A* Algorithm in Unstructured Environment. *Proc. Inst. Mech. Eng. Part D J. Automob. Eng.* **2021**, *235*, 513–526. [CrossRef]
109. Krüsi, P.; Furgale, P.; Bosse, M.; Siegwart, R. Driving on Point Clouds: Motion Planning, Trajectory Optimization, and Terrain Assessment in Generic Nonplanar Environments. *J. Field Robot.* **2017**, *34*, 940–984. [CrossRef]
110. Jiang, J.; Han, Z.; Li, J.; Wang, Y.; Wang, J.; Xu, S. Global Path Planning of UGVs in Large-Scale Off-Road Environment Based on Improved A-Star Algorithm and Quadratic Programming. In Proceedings of the 2023 IEEE Intelligent Vehicles Symposium (IV), Anchorage, AK, USA, 4–7 June 2023; pp. 1–7.
111. Yin, J.; Hu, Z.; Mourelatos, Z.P.; Gorsich, D.; Singh, A.; Tau, S. Efficient Reliability-Based Path Planning of Off-Road Autonomous Ground Vehicles Through the Coupling of Surrogate Modeling and RRT*. *IEEE Trans. Intell. Transp. Syst.* **2023**, *24*, 15035–15050. [CrossRef]
112. Yin, J.; Li, L.; Mourelatos, Z.P.; Liu, Y.; Gorisich, D.; Singh, A.; Tau, S.; Hu, Z. Reliable Global Path Planning of Off-Road Autonomous Ground Vehicles Under Uncertain Terrain Conditions. *IEEE Trans. Intell. Veh.* **2023**, *8*, 1–14. [CrossRef]
113. Zhu, S.; Zhu, W.; Zhang, X.; Cao, T. Path Planning of Lunar Robot Based on Dynamic Adaptive Ant Colony Algorithm and Obstacle Avoidance. *Int. J. Adv. Robot. Syst.* **2020**, *17*, 1–14. [CrossRef]
114. Quann, M.; Ojeda, L.; Smith, W.; Rizzo, D.; Castanier, M.; Barton, K. Chance Constrained Reachability in Environments with Spatially Varying Energy Costs. *Rob. Auton. Syst.* **2019**, *119*, 1–12. [CrossRef]
115. He, X.; Lv, C. Toward Personalized Decision Making for Autonomous Vehicles: A Constrained Multi-Objective Reinforcement Learning Technique. *Transp. Res. Part C Emerg. Technol.* **2023**, *156*, 104352. [CrossRef]
116. Ji, X.; He, X.; Lv, C.; Liu, Y.; Wu, J. Adaptive-Neural-Network-Based Robust Lateral Motion Control for Autonomous Vehicle at Driving Limits. *Control Eng. Pract.* **2018**, *76*, 41–53. [CrossRef]
117. Pinkovich, B.; Rivlin, E.; Rotstein, H. Monocular Sensing for Predictive Driving in an Unstructured Scenario. In Proceedings of the 2019 27th Mediterranean Conference on Control and Automation (MED), Akko, Israel, 1-4 July 2019; pp. 606–611.
118. Nestlinger, G.; Rumetshofer, J.; Solmaz, S. Leader-Based Trajectory Following in Unstructured Environments—From Concept to Real-World Implementation. *Electronics* **2022**, *11*, 1866. [CrossRef]
119. Usami, R.; Kobashi, Y.; Onuma, T.; Maekawa, T. Two-Lane Path Planning of Autonomous Vehicles in 2.5D Environments. *IEEE Trans. Intell. Veh.* **2020**, *5*, 281–293. [CrossRef]
120. Qi, Y.; He, B.; Wang, R.; Wang, L.; Xu, Y. Hierarchical Motion Planning for Autonomous Vehicles in Unstructured Dynamic Environments. *IEEE Robot. Autom. Lett.* **2023**, *8*, 496–503. [CrossRef]
121. Rudny, D.F.; Sallmann, D.W. Analysis of Accidents Involving Alleged Road Surface Defects (i.e., Shoulder Drop-Offs, Loose Gravel, Bumps and Potholes). SAE Technical Papers Series. In Proceedings of the SAE International Congress and Exposition, Detroit, MI, USA, 26–29 February 1996; 1996; p. 960654. [CrossRef]
122. Cherif, M. Motion Planning for All-Terrain Vehicles: A Physical Modeling Approach for Coping with Dynamic and Contact Interaction Constraints. *IEEE Trans. Robot. Autom.* **1999**, *15*, 202–218. [CrossRef]
123. Gao, X.; Li, J.; Fan, L.; Zhou, Q.; Yin, K.; Wang, J.; Song, C.; Huang, L.; Wang, Z. Review of Wheeled Mobile Robots' Navigation Problems and Application Prospects in Agriculture. *IEEE Access* **2018**, *6*, 49248–49268. [CrossRef]
124. Missura, M.; Bennewitz, M. Predictive Collision Avoidance for the Dynamic Window Approach. In Proceedings of the 2019 International Conference on Robotics and Automation (ICRA), Montreal, QC, Canada, 20–24 May 2019; pp. 8620–8626.
125. Chang, L.; Shan, L.; Jiang, C.; Dai, Y. Reinforcement Based Mobile Robot Path Planning with Improved Dynamic Window Approach in Unknown Environment. *Auton. Robots* **2021**, *45*, 51–76. [CrossRef]
126. Hao, K.; Zhao, J.; Li, Z.; Liu, Y.; Zhao, L. Dynamic Path Planning of a Three-Dimensional Underwater AUV Based on an Adaptive Genetic Algorithm. *Ocean. Eng.* **2022**, *263*, 112421. [CrossRef]
127. Rousseas, P.; Bechlioulis, C.P.; Kyriakopoulos, K.J. Optimal Robot Motion Planning in Constrained Workspaces Using Reinforcement Learning. In Proceedings of the 2020 IEEE/RSJ International Conference on Intelligent Robots and Systems (IROS), Las Vegas, NV, USA, 24 October–24 January 2020; pp. 6917–6922.

128. Nguyen, L.A.; Pham, D.; Ngo, D.; Truong, X.T. A Proactive Trajectory Planning Algorithm for Autonomous Mobile Robots in Dynamic Social Environments. In Proceedings of the 2020 17th International Conference on Ubiquitous Robots (UR), Kyoto, Japan, 22–26 June 2020; pp. 309–314.
129. Qin, H.; Shao, S.; Wang, T.; Yu, X.; Jiang, Y.; Cao, Z. Review of Autonomous Path Planning Algorithms for Mobile Robots. *Drones* **2023**, *7*, 211. [CrossRef]
130. Manderson, T.; Wapnick, S.; Meger, D.; Dudek, G. Learning to Drive Off Road on Smooth Terrain in Unstructured Environments Using an On-Board Camera and Sparse Aerial Images. In Proceedings of the 2020 IEEE International Conference on Robotics and Automation (ICRA), Paris, France, 31 May–31 August 2020; pp. 1263–1269.
131. Pflueger, M.; Agha, A.; Sukhatme, G.S. Rover-IRL: Inverse Reinforcement Learning with Soft Value Iteration Networks for Planetary Rover Path Planning. *IEEE Robot. Autom. Lett.* **2019**, *4*, 1387–1394. [CrossRef]
132. Gall, C.; Bezzo, N. Gaussian Process-Based Interpretable Runtime Adaptation for Safe Autonomous Systems Operations in Unstructured Environments. In Proceedings of the 2021 IEEE/RSJ International Conference on Intelligent Robots and Systems (IROS), Prague, Czech Republic, 27 September–1 October 2021; pp. 123–129.
133. Siva, S.; Wigness, M.; Rogers, J.G.; Quang, L.; Zhang, H. NAUTS: Negotiation for Adaptation to Unstructured Terrain Surfaces. In Proceedings of the 2022 IEEE/RSJ International Conference on Intelligent Robots and Systems (IROS), Kyoto, Japan, 23–27 October 2022; pp. 1733–1740.

machines

MDPI

Article

Path Tracking Control Based on T-S Fuzzy Model for Autonomous Vehicles with Yaw Angle and Heading Angle

Yelin He, Jian Wu *, Fuxing Xu, Xin Liu, Shuai Wang and Guanjie Cui

School of Automotive Engineering, Liaocheng University, Liaocheng 252000, China; 2320230204@stu.lcu.edu.cn (Y.H.); 2220230103@stu.lcu.edu.cn (F.X.); 2120230102@stu.lcu.edu.cn (X.L.); 2120230112@stu.lcu.edu.cn (S.W.); 2120230115@stu.lcu.edu.cn (G.C.)
* Correspondence: wujian@lcu.edu.cn

Abstract: Existing vehicle-road models used for road tracking do not take into account the side slip angle, which leads to a reduction in road tracking accuracy in scenarios where the vehicle is at a large side slip angle, such as an emergency lane change. Consequently, this study presents a path-tracking control technique based on the T-S fuzzy model of heading angle vehicle autonomy. In this paper, based on the yaw angle-based vehicle tracking model, a heading angle-based tracking model considering the side slip angle is constructed. Second, since the vehicle speed varies with time, this paper selects the membership function of the vehicle speed to establish the T-S fuzzy model of autonomous vehicle based on the yaw angle and heading angle, respectively, and ensures the robustness and stability over the whole parameter space by the linear parameter variation robust H_∞ controller. Then, cost functions based on the yaw angle and heading angle augmented error systems are created separately to optimize the system's overall performance. Ultimately, simulation and experimentation confirm that the algorithm for control, which is based on the fuzzy model of the heading angle vehicle, has superior autonomous trajectory performance.

Keywords: side slip angle; T-S fuzzy model; heading angle model; path tracking control strategy

Citation: He, Y.; Wu, J.; Xu, F.; Liu, X.; Wang, S.; Cui, G. Path Tracking Control Based on T-S Fuzzy Model for Autonomous Vehicles with Yaw Angle and Heading Angle. *Machines* **2024**, *12*, 375. https://doi.org/10.3390/machines12060375

Academic Editor: Nasser L. Azad

Received: 29 April 2024
Revised: 19 May 2024
Accepted: 27 May 2024
Published: 29 May 2024

1. Introduction

With automobile ownership increasing year by year and problems such as traffic congestion and accidents becoming increasingly severe, the development of traditional automobiles is facing significant challenges [1]. With the integration of the new generation of information and communication, new energy, and other technologies with the automobile industry, the development situation of the world's automobile industry is facing significant changes. Automobile automatic driving has become the focus and hotspot of research in automobile technology at this stage [2,3]. Automated automobile driving is essential in improving road capacity and reducing traffic congestion [4].

One of the fundamental components of autonomous vehicle driving is the automatic steering system, and the main focus of current research on these systems is trajectory tracking control [5,6]. Since the accuracy of trajectory tracking control correlates significantly with vehicle and road models [7], thoroughly discussing the vehicle-road model is essential. In most vehicle-road models established so far, the heading angle, also known as the yaw angle, is the angle formed by the longitudinal axis of the vehicle and the reference horizontal line. On the other hand, the heading angle, which is the sum of the yaw and side slip angles, is the angle that exists between the vehicle's traveling direction and the reference horizontal line [8]. When a vehicle is at a large side slip angle, like during an emergency lane change, the accuracy of road tracking decreases since the modeling does not account for the side slip angle's effect. To help increase path-tracking accuracy, it is crucial to investigate the side slip angle and enhance the vehicle-road model's accuracy.

Owing to the wheel subsystem's rapid convergence, a path-tracking model, a friction model in both the longitudinal and transverse directions, and a reduced vehicle motion model based on singular perturbation are constructed [9]. These models are combined to create the final control model. In [10], a path-tracking model is obtained by modeling the dynamics of a four-wheel-independent-steered (4WIS) vehicle and equating the path-tracking problem to minimizing the lateral deviation of the control vehicle concerning the angular deviation of the heading. This paper formulates the vehicle path-tracking task as a constraint-following problem in a novel way [11]. It uses it to represent a model of path-following constraints based on the vehicle model. In [12], an active front steering system's dynamic model is developed. Unmeasured model perturbations and parameter perturbations in the system are estimated in real-time by introducing a radial-based forward neural network. While the far-precession point model adjusts the steering wheel for anticipated changes in road curvature concerning the lane, the near-precession point model adjusts the steering wheel to estimate the lane center and vehicle state errors. These two steering models are developed independently in [13]. Different models have been developed in the above literature to improve the vehicle path tracking performance, where the heading angle used is the angle between the reference horizontal line and the direction of the vehicle's longitudinal axis. Still, the actual heading angle is the angle between the vehicle's direction of travel and the reference horizontal line, and the heading angle is the sum of the yaw angle and the side slip angle [8]. The effect is more pronounced when the vehicle is in a situation with a giant slip angle since this lowers the precision of the vehicle-road model, which, in turn, reduces the accuracy of the route tracking. To improve the accuracy of the model and to study the effect of lateral inclination angle, we propose a vehicle model based on traverse angle and heading angle and draw conclusions by comparing these two models.

Vehicles in the process of driving, vehicle speed, and other vehicle parameters will continue to change, and other elements of the external environment, such as the road surface adhesion coefficient, will also alter. A vehicle two-degree-of-freedom model and nonlinear tires are modeled in [14] with fixed parameters but do not accurately describe the vehicle and the road. Due to the uncertainty in the vehicle's longitudinal speed and the road friction coefficient, a robust H_∞ state feedback controller with linear parameters and temporal variation is devised. In the paper, a rectangular polyhedron is built and the linear matrix inequality (LMI) at the four vertices is solved to realize the feedback gain with different parameter variations [15] Two non-independent parameters, the longitudinal velocity and the reciprocal of the longitudinal velocity, were chosen as variable parameters to design an improved trapezoidal polytope to improve the correlation and to reduce the area of the region in the trapezoidal polytope that is inaccessible to the real state of the vehicle [16]. In order to boost vehicle stability and enhance trajectory tracking precision, adaptive preview times are calculated based on the lateral deviation and road boundaries during vehicle travel [17]. In fact, the modeling of parameter time-varying has been widely studied. In this paper, a T-S fuzzy model is developed for the variation of longitudinal speeds to realize the controller and give the optimal control law at different speeds of the vehicle.

In [18], a fuzzy fast terminal sliding mode control (FFTSM) approach is proposed, which is mainly for multiple-input multiple-output (MIMO) systems. The system state is realized to attain equilibrium more quickly in a finite amount of time by using robust control techniques, and linear matrix inequality (LMI) determines the control parameters. A state feedback controller is suggested in [19] to maintain system performance and stability in the face of time-varying disturbances and system uncertainty. A method of learning-based control for autonomous cars is presented in this research. Using input and output data from an existing controller and the adaptive neuro-fuzzy inference system (ANFIS) method, an explicit Takagi–Sugeno (TS) controller is trained [20]. The hierarchical controller is used, where the upper layer handles the control objectives, the intermediate layer generates the longitudinal slip reference, and the low layer handles the braking of each wheel [21]. In order to provide route tracking control of a vehicle, this study suggests a linear parameter time-varying robust controller. This controller solves the state feedback gain to determine the front wheel steering angle of the vehicle autopilot.

In summary, in order to study the effect of the side slip angle, this paper proposes a vehicle fuzzy model based on the yaw angle and heading angle. It establishes the dynamics and kinematics mathematical models of the two models. On this basis, two robust H_∞ coordinated control strategies based on the T-S fuzzy model of yaw angle and heading angle vehicles are proposed according to the change in vehicle speed. More effective control schemes are obtained through comparison to improve the accuracy and driving stability of road tracking for autopilot vehicles. The main work of this paper includes the following aspects.

(1) To improve the accuracy of the model and study the influence of the side slip angle on vehicle path tracking, the vehicle path model based on the yaw angle or heading angle is established by taking the line connecting the intersection point of the vehicle traveling direction or body direction and the target path as the origin, respectively.
(2) Aiming at solving the problem of time-varying longitudinal speed of vehicles, the membership function of vehicle speed is selected, the T-S fuzzy model of vehicles is established, and the gain scheduling robust H_∞ control algorithm considering the variation of the vehicle speed parameter is proposed so as to obtain the optimal control law.
(3) A cost function based on the yaw angle and heading angle augmented error system is defined in order to optimize the system's overall performance and increase the accuracy of the vehicle path tracking. Additionally, the system's eigenvalues are limited to a fixed circular region in order to enhance the dynamic performance.

The remainder of this paper is structured as follows. Section 2 presents the T-S fuzzy modeling of vehicle-road systems. Section 3 presents the design of the T-S fuzzy controller. Sections 4 and 5 validate and compare the two control algorithms through simulation and hardware-in-the-loop (HIL) experiments.

2. Establishment of Vehicle-Road Model Based on T-S Fuzzy Rules

In this section, we present the vehicle dynamics and kinematics model based on the yaw and heading angles. The longitudinal velocity is used as a fuzzy variable to establish the T-S fuzzy modeling of vehicle-road systems.

2.1. Vehicle Dynamics Modeling Based on Yaw Angle and Heading Angle

The steering system's influence is disregarded, the front wheel angle is taken directly as input, the suspension is not taken into consideration, it is determined that the vehicle's compartment only moves in a planar manner parallel to the ground, and it is assumed that the tire side deflection characteristics are within a linear range, etc., when choosing the vehicle model. On this basis, the vehicle's two-degree-of-freedom (2DOF) model is selected in this paper. The vehicle dynamics model based on the yaw angle is shown in Figure 1, where the state quantity is the yaw rate, which is the angular velocity in the direction of the vehicle body; it does not take into account the vehicle's side slip angle. Here, the yaw rate is replaced as the heading rate, and then the heading angle-based vehicle dynamics model is developed, as shown in Figure 2. Figures 1 and 2 show that the vehicle dynamics equations based on the yaw and heading angles can be described as in [22–25].

$$\begin{bmatrix} \dot{v}_y \\ \dot{\gamma} \end{bmatrix} = \begin{bmatrix} a_{11} & a_{12} \\ a_{21} & a_{22} \end{bmatrix} \begin{bmatrix} v_y \\ \gamma \end{bmatrix} + \begin{bmatrix} b_1 \\ b_2 \end{bmatrix} [\delta], \tag{1}$$

$$\begin{bmatrix} \dot{v}_y \\ \ddot{\psi} \end{bmatrix} = \begin{bmatrix} a_{11} & a_{12} \\ a_{21} & a_{22} \end{bmatrix} \begin{bmatrix} v_y \\ \dot{\psi} \end{bmatrix} + \begin{bmatrix} b_1 \\ b_2 \end{bmatrix} [\delta], \tag{2}$$

where m is mass of the vehicle, β is the vehicle side slip angle, δ is the front wheel angle, I_z is yaw inertia of the vehicle, v_x is the vehicle longitudinal velocity, v_y is the vehicle lateral velocity, γ is vehicle yaw rate, $\dot{\psi}$ is vehicle heading rate, a and b are the distances from the front and rear axles of the vehicle center of gravity, respectively, C_f and C_r are values of the front and rear tire cornering stiffness, respectively, and the other parameters are defined as follows:

$$a_{11} = \frac{C_f + C_r}{mv_x}, a_{12} = \frac{aC_f - bC_r}{mv_x} - v_x, a_{21} = \frac{aC_f - bC_r}{I_z v_x},$$
$$a_{22} = \frac{a^2 C_f + b^2 C_r}{I_z v_x}, b_1 = -\frac{C_f}{m}, b_2 = -\frac{aC_f}{I_z}.$$

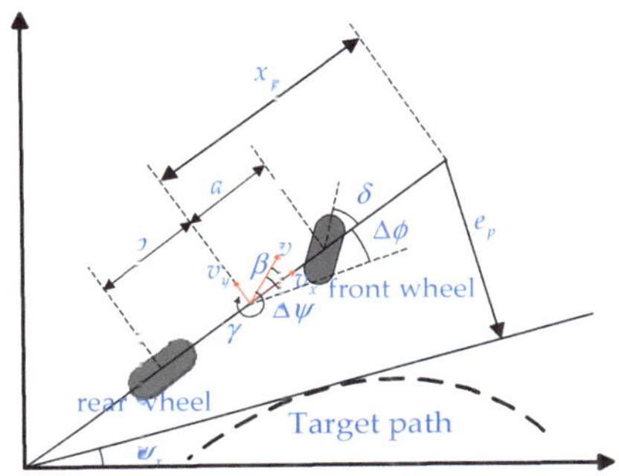

Figure 1. Schematic of vehicle-road system model based on yaw angle.

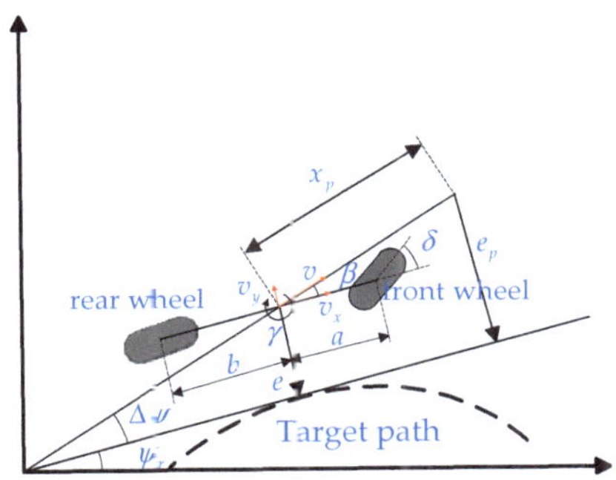

Figure 2. Schematic of vehicle-road system model based on heading angle.

2.2. Vehicle Kinematics Modeling Based on Yaw and Heading Angles

The lateral deviation of the preview point and the deviation of the yaw or heading angle are selected as the state quantities to guarantee the vehicle's path tracking, and the vehicle kinematics equations based on the yaw and heading angles are expressed through the dynamics parameters of the vehicle, respectively [26–29].

The direction of the preview point for the vehicle kinematics equations based on the yaw angle is the vehicle body direction, which can be described as

$$\begin{cases} \dot{s} = v_x cos(\Delta\phi) - v_y sin(\Delta\phi) \\ \dot{e} = v_y cos(\Delta\phi) + v_x sin(\Delta\phi) \\ e_p = e + x_p sin(\Delta\phi) \\ \Delta\phi = \phi - \psi_r \end{cases}. \tag{3}$$

The simplification by the small angle theorem can be described as

$$\begin{cases} \dot{e}_p = v_y + x_p\gamma + v_x\Delta\phi - x_p\rho v_x \\ \Delta\dot{\phi} = \gamma - \rho v_x \end{cases}. \tag{4}$$

The direction of the preview point for the vehicle kinematics equation based on the heading angle is the direction of vehicle travel, which can be described as

$$\begin{cases} \dot{s} = v_x cos(\Delta\psi - \beta) - v_y sin(\Delta\psi - \beta) \\ \dot{e} = v_y cos(\Delta\psi - \beta) + v_x sin(\Delta\psi - \beta) \\ e_p = e + x_p sin(\Delta\psi) \\ \Delta\psi = \psi - \psi_r \end{cases}. \tag{5}$$

The simplification by the small angle theorem can be described as

$$\begin{cases} \dot{e}_p = v_x \Delta\psi + x_p \dot{\psi} - x_p \rho v_x \\ \Delta\dot{\psi} = \dot{\psi} - \rho v_x \end{cases}. \tag{6}$$

where ϕ is the vehicle yaw angle, ψ_r is the heading angle of the reference path, x_p is the pre-scanning distance, and ρ is the road curvature.

Remark 1. *The intersection point of the vehicle moving direction and the target route direction are taken as the origin of the coordinate system in the heading angle-based vehicle-road system model. It takes the heading angle error $\Delta\psi$ as the state quantity where the heading angle error $\Delta\psi$ is equal to the sum of the yaw angle error $\Delta\phi$ and the side slip angle β. This accounts for the side slip angle's influence on increasing the vehicle-road model's accuracy.*

2.3. Vehicle-Road Systems Based on T-S Fuzzy Modeling

Combining (1)–(3), and (5), the vehicle model based on yaw angle and heading angle can be described as

$$\begin{cases} \dot{x}_1 = Ax_1 + B_1 u_1 + B_2 w \\ y_1 = Cx_1 \end{cases}, \tag{7}$$

$$\begin{cases} \dot{x}_2 = Ex_2 + B_1 u_2 + B_2 w \\ y_2 = Cx_2 \end{cases}, \tag{8}$$

where $x_1 = \begin{bmatrix} v_y & \gamma & e_p & \Delta\phi \end{bmatrix}^T$ represents the state vector base on yaw angle, $x_2 = \begin{bmatrix} v_y & \dot{\psi} & e_p & \Delta\psi \end{bmatrix}^T$ represents the state vector base on heading angle, $u_1 = u_2 = [\delta]$ is the control input, $w = \begin{bmatrix} 0 & 0 & -x_p\rho v_x & -x_p v_x \end{bmatrix}^T$ is the disturbance vector, $y_1 = \begin{bmatrix} \gamma & e_p & \Delta\phi \end{bmatrix}^T$ is the measured output base on yaw angle, $y_2 = \begin{bmatrix} \dot{\psi} & e_p & \Delta\psi \end{bmatrix}^T$ is the measured output base on heading angle, and the other parameters are defined as follows.

$$A = \begin{bmatrix} a_{11} & a_{12} & 0 & 0 \\ a_{21} & a_{22} & 0 & 0 \\ 1 & x_p & 0 & v_x \\ 0 & 1 & 0 & 0 \end{bmatrix}, B_1 = \begin{bmatrix} b_1 \\ b_2 \\ 0 \\ 0 \end{bmatrix}, B_2 = I_{4\times4}, C = \begin{bmatrix} 0 & 1 & 0 & 0 \\ 0 & 0 & 1 & 0 \\ 0 & 0 & 0 & 1 \end{bmatrix} E = \begin{bmatrix} a_{11} & a_{12} & 0 & 0 \\ a_{21} & a_{22} & 0 & 0 \\ 0 & x_p & 0 & v_x \\ 0 & 1 & 0 & 0 \end{bmatrix}.$$

In the process of automobile driving, the longitudinal speed of the vehicle constantly changes to effectively depict the real working conditions of the vehicle in the actual driving, to realize the control system in order to give the optimal control law by judging the current state of the vehicle, and to achieve improvement in the vehicle's path tracking effect, to take the longitudinal speed of the automobile in the process of driving as a fuzzy variable, and to define i $(i = 1, 2 \ldots \varpi)$. Then, from system (7), the i $(i = 1, 2 \ldots \varpi)$ rule based on the fuzzy model of yaw angle can be described as [30–32].

Control Rule i:
IF $v(t)$ is M_1^i
Then

$$\begin{cases} \dot{x}_1(t) = A_i x_1 + B_1^i u_1 + B_2^i w \\ y_1 = C_i x_1 \end{cases}, \tag{9}$$

where $v_x(t) = [v_1(t) \quad v_2(t) \quad v_3(t) \quad v_4(t)]$ are fuzzy variables. The system fuzzy variable is $v_x(t) = 30, 40, 50, 60$ km/h, where $M_1^i(v(t))$ denotes the membership function of $v(t)$ with respect to the fuzzy set M_1^i [33,34] as shown in Figure 3:

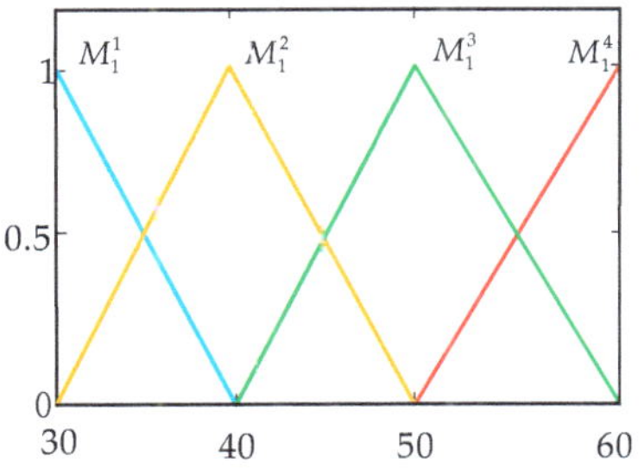

Figure 3. Membership functions of vehicle speed.

Following fuzzy processing, the yaw angle-based T-S fuzzy system state equation can be obtained, and can be described as [35–37]

$$\begin{cases} \dot{x}_1(t) = \sum_{i=1}^{\varpi} h_i(v(t))\left(A_i x_1(t) + B_1^i u_1(t) + B_2^i w(t)\right) \\ y_1(t) = \sum_{i=1}^{\varpi} h_i(v(t))(C_i x_1(t)) \end{cases}. \tag{10}$$

For any t, the following equation holds.

$$h_i(v(t)) \geq 0, i = 1, 2 \ldots, \varpi, \sum_{i=1}^{\varpi} \mu_i(v(t)) = 1,$$

$$h_i(v(t)) = \frac{\mu_i(v(t))}{\sum_{i=1}^{\varpi} \mu_i(v(t))}, \mu_i(v(t)) = \prod_{i=1}^{\varpi} M_1^i(v(t)),$$

Similarly, the state equation of the T-S fuzzy system based on the heading angle obtained from the system (9) can be described as:

$$\begin{cases} \dot{x}_2(t) = \sum_{i=1}^{\varpi} h_i(v(t))\left(E_i x_2(t) + B_1^i u_2(t) + B_2^i w(t)\right) \\ y_2(t) = \sum_{i=1}^{\varpi} h_i(v(t))(C_i x_2(t)) \end{cases}. \tag{11}$$

3. Design of Robust H_∞ Controller Based on T-S Fuzzy Modeling

The path tracking controller for system (10), which is dependent on the yaw angle, can be described as [35]

$$u_1(t) = \sum_{i=1}^{\varpi} h_i(v(t)) K_1^i x_1(t). \tag{12}$$

As a result, the system can be described as

$$\begin{cases} \dot{x}_1(t) = \sum_{i=1}^{\varpi} h_i(v(t))\left(A_m^i x_1(t) + B_2^i w(t)\right) \\ y_1(t) = \sum_{i=1}^{\varpi} h_i(v(t))(C_i x_1(t)) \end{cases}, \tag{13}$$

where

$$A_m^i = A_i + B_1^i K_1^i.$$

Similarly, the system is described by the heading angle-based path tracking controller Equation (14) in the form of Equation (15)

$$u_2(t) = \sum_{i=1}^{\varpi} h_i(v(t)) K_2^i x_2(t), \tag{14}$$

$$\begin{cases} \dot{x}_2(t) = \sum\limits_{i=1}^{\varpi} h_i(v(t)) \left(E_n^i x_2(t) + B_2^i w(t) \right) \\ y_2(t) = \sum\limits_{i=1}^{\varpi} h_i(v(t)) (C_i x_2(t)) \end{cases}, \tag{15}$$

where

$$E_n^i = E_i + B_1^i K_2^i.$$

Remark 2. *The vehicle-road system based on the T-S fuzzy model takes the longitudinal speed v_x of the vehicle as a variable parameter, and different longitudinal speeds correspond to corresponding feedback gains. The feedback gains in the speed range of neighboring endpoints are solved by the feedback gains of the speeds at the neighboring endpoints with the designed vehicle speed membership function.*

In this paper, we consider multiple objectives, including path tracking error, system stability, etc., motivated by the control of the linear quadratic regulator (LQR), and we define the cost function of the system based on the yaw angle augmented error, which can be described as [38,39]

$$J = \int_0^\infty \left(g_1 e_p^2 + g_2 \Delta\phi^2 \right) dt, \tag{16}$$

where q_1, q_2 are the weighting factors and, further, the cost function (16) can be described as

$$J_1 = \int_0^\infty \sum_{i=1}^{\varpi} h_i(v(t)) \left((C_i x_1(t))^T G_1 (C_i x_1(t)) \right) dt, \tag{17}$$

where

$$G_1 = \begin{bmatrix} g_1 & 0 \\ 0 & g_2 \end{bmatrix}.$$

The cost function (17) can be equivalently described as

$$J_1 = \int_0^\infty \sum_{i=1}^{\varpi} h_i(v(t)) \left(\left(G_1^{\frac{1}{2}} C_i x_1(t) \right)^T \left(G_1^{\frac{1}{2}} C_i x_1(t) \right) \right) dt. \tag{18}$$

The auxiliary evaluation signal that defines the system based on the cost function can be described as

$$z_1(t) = \sum_{i=1}^{\varpi} h_i(v(t)) \left(C_m^i x_1(t) \right), \tag{19}$$

where

$$C_m^i = G_1^{\frac{1}{2}} C_i.$$

Rewriting the auxiliary evaluation signals yields the final yaw angle-based state system, which can be described as

$$\begin{cases} \dot{x}_1(t) = \sum_{i=1}^{\varpi} h_i(v(t))\left(A_m^i x_1(t) + B_2^i w(t)\right) \\ z_1(t) = \sum_{i=1}^{\varpi} h_i(v(t))\left(C_m^i x_1(t)\right) \end{cases}. \tag{20}$$

Similarly, the final state system based on heading angle is obtained, which can be described as

$$\begin{cases} \dot{x}_2(t) = \sum_{i=1}^{\varpi} h_i(v(t))\left(E_n^i x_2(t) + B_2^i w(t)\right) \\ z_2(t) = \sum_{i=1}^{\varpi} h_i(v(t))\left(C_n^i x_2(t)\right) \end{cases}. \tag{21}$$

where

$$C_n^i = G_2^{\frac{1}{2}} C_i, G_2 = \begin{bmatrix} g_3 & 0 \\ 0 & g_4 \end{bmatrix}.$$

Remark 3. *The control objectives are weighted according to the requirements of the control task, and the desired control output is obtained. The matrices G_1, G_2 are the weighting matrices which, as shown in Table 1, increases the weights of the lateral deviation e_p and the yaw angle error $\Delta\varphi$ or heading angle error $\Delta\psi$, where the weight of the lateral deviation e_p is valued more highly to realize the improvement of the path tracking accuracy. To ensure that systems (20) and (21) are stabilized, achieve the control objectives, and satisfy the robust H_∞ performance requirements, we introduce the following lemmas*

Table 1. Controller Parameters.

Parameter	Value	Parameter	Value
(q_1, r_1)	(26, 26)	g_3^2	2.5
(q_2, r_2)	(26, 26)	g_4^2	1.5
g_1^2	2.5	λ_1	4.4
g_2^2	1.5	λ_2	5.3

Lemma 1 *For the system that is closed-loop (20), for a given scalar, $\lambda_1 > 0$, $\sum_{i=1}^{4} \kappa_1^i < 0$ holds if there is a positive definite matrix P_1 with the right dimension. Following this, the system maintains stability and fulfills the rigorous performance standards for robustness [40,41]. κ_1^i can be described as*

$$\kappa_1^i = \begin{bmatrix} A_i P_1 + B_1^i V_1^i + \left(A_i P_1 + B_1^i V_1^i\right)^T & B_2^i & \left(C_m^i P_1\right)^T \\ \left(B_2^i\right)^T & -I & \mathbf{0} \\ C_m^i P_1 & \mathbf{0} & -I \end{bmatrix},$$

where

$$K_1^i = V_1^i P_1^{-1}.$$

Lemma 2 *For the system that is closed-loop (21), for a given scalar, $\lambda_2 > 0$, $\sum_{i=1}^{4} \kappa_2^i < 0$ holds if there is a positive definite matrix P_1 with the right dimension. Following this, the system maintains stability and fulfills the rigorous performance standards for robustness. κ_2^i can be described as*

$$\kappa_2^i = \begin{bmatrix} E_i P_2 + B_1^i V_2^i + \left(E_i P_2 + B_1^i V_2\right)^T & B_2^i & \left(C_n^i P_2\right)^T \\ \left(B_2^i\right)^T & -I & \mathbf{0} \\ C_n^i P_2 & \mathbf{0} & -I \end{bmatrix},$$

where

$$K_2^i = V_2^i P_2^{-1}.$$

Lemma 3. *For D-stability, it is necessary to find a positive definite matrix, denoted as X_1, of suitable dimensions. When Equation (22) is fulfilled, all eigenvalues of this matrix are contained within the region defined by the disk (q, r) [42].*

$$\begin{bmatrix} -rX_1 & qX_1 + X_1 A \\ (qX_1 + X_1 A)^T & -rX_1 \end{bmatrix} < 0. \quad (22)$$

Due to the influence of the closed-loop system's state matrix eigenvalues on the controller's transient response, these eigenvalues undergo scrutiny against the D-stability criterion outlined below, aiming to mitigate transient effects.

Corollary 1. *For D-stability, for the system that is closed-loop (20), when $\sum_{i=1}^{4} \kappa_3^i < 0$ is satisfied, there is a positive definite matrix P_1 of suitable dimensions such that each of the matrix's eigenvalues A_m^i reside in the disk (q_1, r_1) area. κ_3^i can be described as*

$$\kappa_3^i = \begin{bmatrix} -r_1 P_1 & q_1 P_1 + A_1^i P_1 + B_1^i V_1^i \\ (q_1 P_1 + A_1^i P_1 + B_1^i V_1^i)^T & -r_1 P_1 \end{bmatrix}.$$

The D-stability criterion for the system (20) obtained from Lemma 3 can be described as

$$\begin{bmatrix} -r_1 X_1 & q_1 X_1 + X_1 A_m^i \\ (q_1 X_1 + X_1 A_m^i)^T & -r_1 X_1 \end{bmatrix} < 0. \quad (23)$$

The matrix A_m^i expanded can be described as

$$\begin{bmatrix} -r_1 X_1 & q_1 X_1 + X_1 (A_i + B_1^i K_1^i) \\ (q_1 X_1 + X_1 (A_i + B_1^i K_1^i))^T & -r_1 X_1 \end{bmatrix} < 0. \quad (24)$$

The left and right sides of the matrix of Equation (24) are multiplied by the matrix X_2, respectively, to obtain Equation (25), which can be described as

$$\begin{bmatrix} -r_1 P_1 & q_1 P_1 + A_1^i P_1 + B_1^i V_1^i \\ (q_1 P_1 + A_1^i P_1 + B_1^i V_1^i)^T & -r_1 P_1 \end{bmatrix} < 0, \quad (25)$$

where

$$X_2 = \begin{bmatrix} X_1^{-1} & 0 \\ 0 & X_1^{-1} \end{bmatrix}, P_1 = X_1^{-1}, V_1^i = K_1^i P_1.$$

The proof of Corollary 1 is completed here.

Corollary 2. *Similarly, for D-stability, for the system that is closed-loop (21), when $\sum_{i=1}^{4} \kappa_4^i < 0$ is satisfied, there is a positive definite matrix P_2 of suitable dimensions such that each of the matrix's eigenvalues A_n^i reside in the disk (q_2, r_2) area. κ_4^i can be described as*

$$\kappa_4^i = \begin{bmatrix} -r_2 P_2 & q_2 P_2 + E_1^i P_2 + B_1^i V_2^i \\ (q_2 P_2 + E_1^i P_2 + B_1^i V_2^i)^T & -r_2 P_2 \end{bmatrix}.$$

Combining Lemma 1 with the linear matrix inequality (LMI) in Corollary 1, the robust H_∞ state feedback gain matrix coordinated controller based on the yaw angle fuzzy model can be obtained. Combining Lemma 2 with the linear matrix inequality (LMI) in Corollary

2, the robust H_∞ state feedback gain matrix coordinated controller based on the heading angle fuzzy model can be obtained.

4. Simulation Setup and Results

This part uses MATLAB/Simulink 2021CarSim 8.02 co-simulation to verify and compare the robust path-tracking control algorithms' efficacy for the two models presented in this paper

In MATLAB/Simulink, the fuzzy model of the vehicle is constructed based on the yaw and heading angles. The optimization issue is then addressed for the aforementioned linear matrix inequality (LMI) using YALMIP, a MATLAB toolbox feature. To finish building the joint simulation model, the vehicle's computed front wheel angle is sent into CarSim, and the state quantity parameters of the vehicle that CarSim produces are fed into MATLAB/Simulink. Table 2 displays the parameter list for the entire vehicle model.

Table 2. Parameters for an autonomous vehicle.

Symbol	Parameter	Value	Units
m	Vehicle Total mass	1110	kg
I_z	Vehicle yaw moment of inertia	1343.1	kg·m^2
C_f	Equivalent Cornering Stiffness of front tire	42,000	Nm/rad
C_r	Equivalent Cornering Stiffness of rear tire	95,000	Nm/rad
a	Distance of CG from front axle	1.04	m
b	Distance of CG from rear axle	1.56	m

The maximum permitted deviation for trajectory tracking is one example of the data provided by the literature [35]. Using these parameters, we calculate the weights G_1 and G_2 of the cost function, as indicated in Table 1. which also includes the pole configuration parameters and the recommended attenuation level.

Using these parameters, by resolving the optimization issue, we obtain the control gains of the two models at the endpoint speeds, and the feedback control gains at different speeds can be obtained by the affiliation function of the vehicle speeds in Figure 3. To confirm whether the controllers, which use fuzzy models for heading and yaw angles, are effective, additional robust H_∞ controllers based on the yaw angle and heading angle models were added, and the controllers of the four models were evaluated for their performance in the following scenarios: vehicles are subjected to path-following experiments under serpentine and double lane change with high pavement adhesion coefficients μ= 0.85 at different speeds.

4.1. Serpentine Road Path Tracking

The speed variation of the autonomous vehicle for path tracking on a serpentine road is shown in Figure 4; the speed is increased from the initial 30 km/h, after 60 s increments to 60 km/h, and after that, the speed is uniform. In this paper, the curvature of the target path is selected to change in a certain range, and there is a continuous curve of the serpentine line road. The curvature of the serpentine line road changes, as shown in Figure 5, and the curvature changes in $(-0.005, 0.01)$.

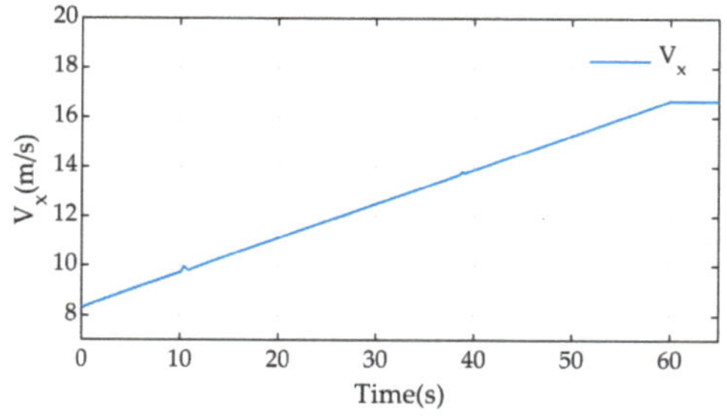

Figure 4. Vehicle speed changes on a serpentine road.

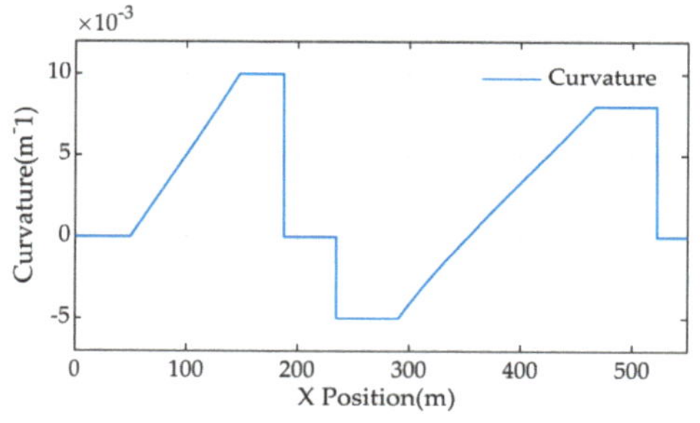

Figure 5. Serpentine road curvature.

The path tracking and lateral deviation e_p comparison results for the four model control algorithms are given in Figures 6 and 7c. In the figure below the initial end of the arrow shows the position of the corresponding curve, and the end points to a magnified view at that position. Based on the figures, it is evident that the maximum lateral deviation of the algorithm for control of the yaw angle model does not exceed 0.3 m, the maximum lateral deviation of the algorithm for control of the heading angle model does not exceed 0.27 m, the maximum lateral deviation of the algorithm for control based on the fuzzy model of the yaw angle does not exceed 0.25 m, the maximum lateral deviation of the algorithm for control based on the fuzzy model of the heading angle does not exceed 0.23 m, and the tracking accuracies of the paths of the four control algorithms are increased in order.

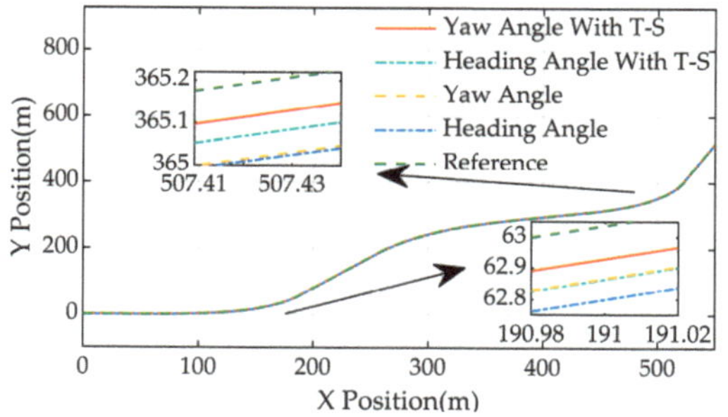

Figure 6. The simulation results of path tracking on a serpentine road are compared.

The comparative results of lateral velocity v_y, yaw velocity γ and heading velocity $\dot{\psi}$, and front-wheel steering angle δ for the four model control algorithms are given in Figure 7a,b and Figure 8. As illustrated in the figure, the four model control algorithms have the same variation in the above three parameters when tracking a vehicle's path.

Figure 7d gives the results of the comparison of the yaw angle error $\Delta\phi$ and heading angle error $\Delta\psi$. From the figure, it is evident that the magnitude of the yaw angle error $\Delta\phi$ is close to and does not exceed 2.8° for the two models, the yaw angle fuzzy model based on the yaw angle and the yaw angle model based on the yaw angle. The magnitude of the heading angle error $\Delta\psi$ of both models, the heading angle fuzzy-based model and heading angle-based model, is close to and does not exceed 2.1°. It is evident that the heading angle error $\Delta\psi$ is less than the yaw angle error $\Delta\phi$.

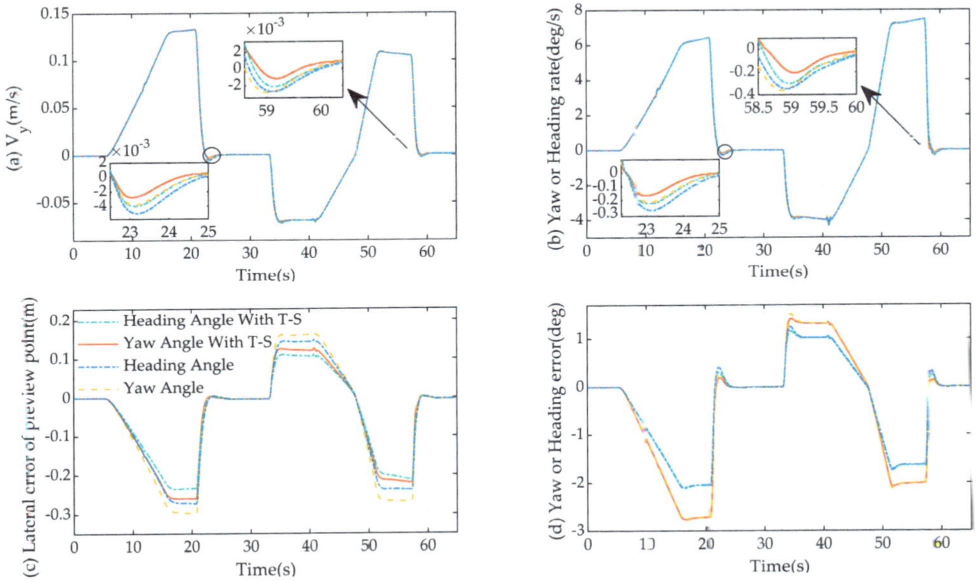

Figure 7. The simulation results of the state vector on a serpentine road are compared. (**a**) Results of lateral velocity simulations are compared; (**b**) results of yaw velocity and heading velocity simulations are compared; (**c**) results of lateral deviation simulations are compared; (**d**) results of the yaw angle error and heading angle error simulations are compared.

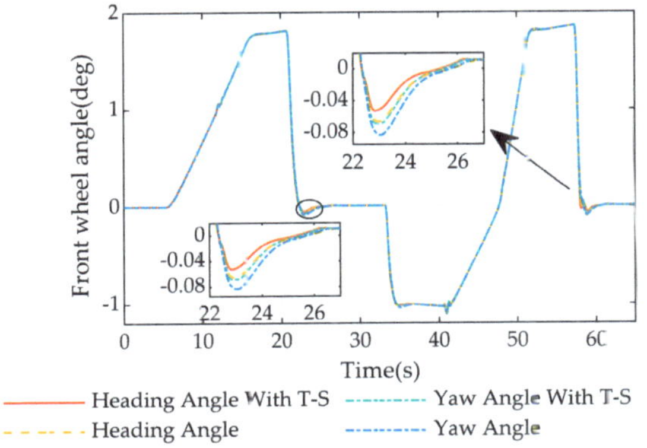

Figure 8. The simulation results of the front-wheel steering angle on a serpentine road are compared.

From the analysis of the above results, it is obtained that the algorithm for control based on the vehicle fuzzy model has a more excellent automatic trajectory performance. Comparing the algorithm for control based on the yaw angle fuzzy model with the algorithm for control based on the heading angle fuzzy model, the algorithm for control based on the heading angle fuzzy model has better automatic trajectory performance.

4.2. Double Lane Change Path Tracking

The speed variation of the autonomous vehicle for path tracking on a double lane change is shown in Figure 9; the speed is increased from the initial 30 km/h, after 10 s increments to 50 km/h, and after that the speed is uniform. The curvature of the double lane change changes, as shown in Figure 10, and the curvature changes in (−0.04, 0.04).

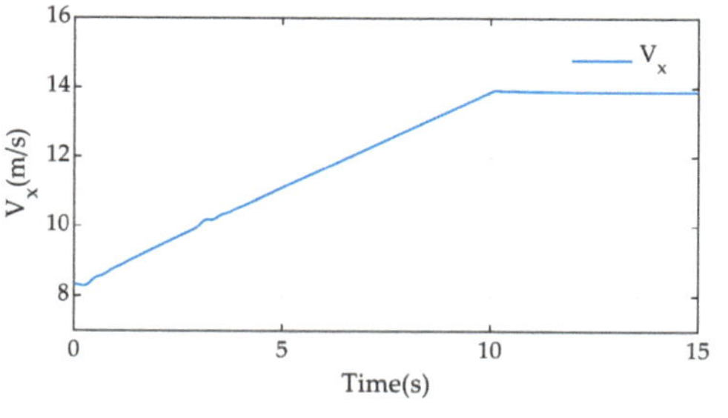

Figure 9. Vehicle speed changes on a double lane change.

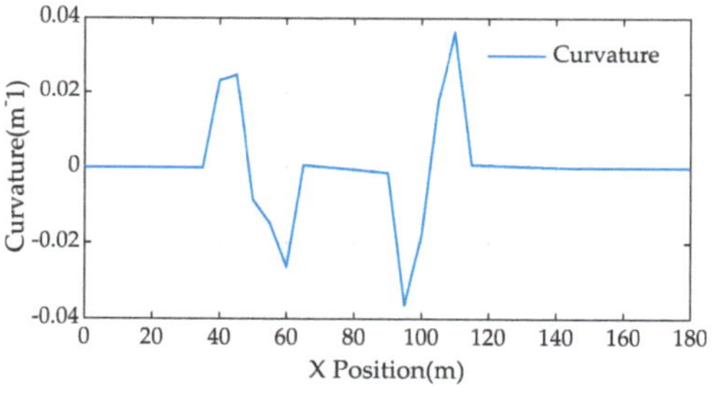

Figure 10. Double lane change curvature.

The path tracking and lateral deviation e_p comparison results for the four model control algorithms are given in Figures 11 and 12c. Based on the figure, it is evident that the maximum lateral deviation of the algorithm for control of the yaw angle model does not exceed 0.61 m, the maximum lateral deviation of the algorithm for control of the heading angle model does not exceed 0.51 m, the maximum lateral deviation of the algorithm for control based on the fuzzy model of the yaw angle does not exceed 0.47 m, the maximum lateral deviation of the algorithm for control based on the fuzzy model of the heading angle does not exceed 0.41 m, and the tracking accuracies of the paths of the four control algorithms are increased in order.

The comparative results of lateral velocity v_y, yaw velocity γ and heading velocity $\dot{\psi}$, and front-wheel steering angle δ for the four model control algorithms are given in Figure 12a,b and Figure 13. As illustrated in the figure, the four model control algorithms have the same variation in the above three parameters when tracking a vehicle's path. Figure 12d gives the results of the comparison of the yaw angle error $\Delta\phi$ and heading angle error $\Delta\psi$. From the figure, it is evident that the magnitude of the yaw angle error $\Delta\phi$ is close to and does not exceed 7.3° for the two models, the yaw angle fuzzy model based on the yaw angle and the yaw angle model based on the yaw angle. The magnitude of the heading angle error $\Delta\psi$ of both models, the heading angle fuzzy-based model and heading angle-based model, is close to and does not exceed 6.2°. It is evident that the heading angle error $\Delta\psi$ is less than the yaw angle error $\Delta\phi$.

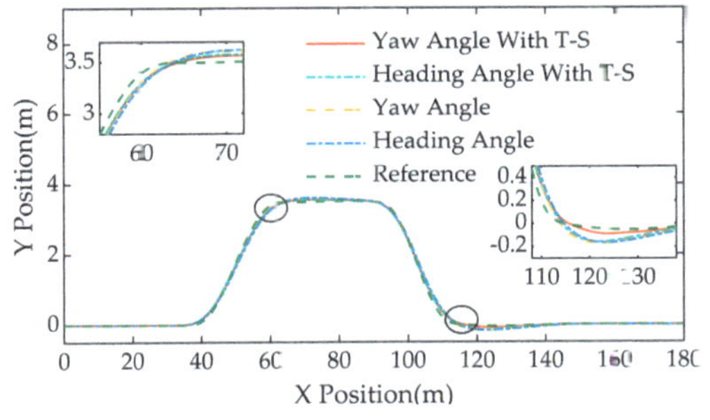

Figure 11. The simulation results of path tracking on a double lane change are compared.

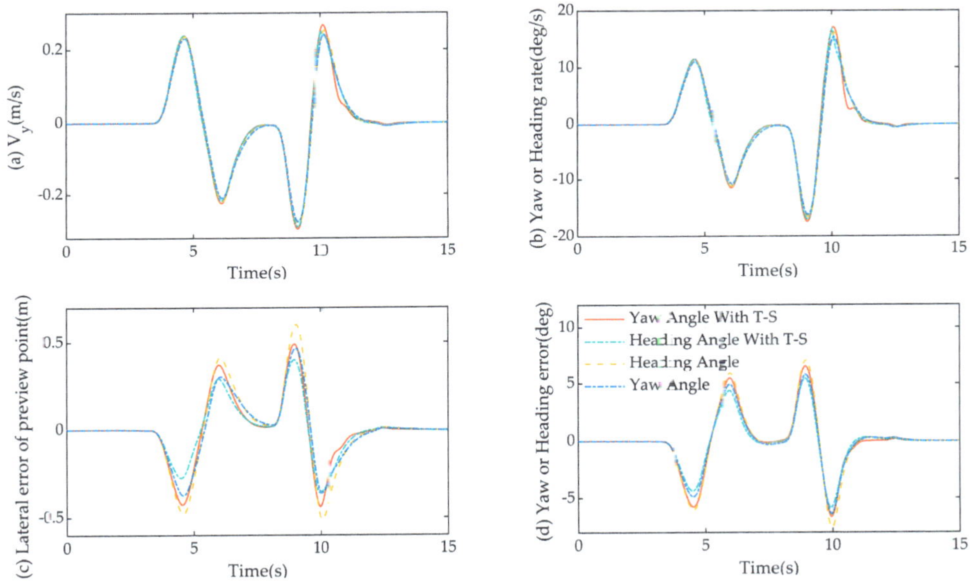

Figure 12. The simulation results of the state vector on a double lane change are compared. (**a**) Results of lateral velocity simulations are compared; (**b**) results of yaw velocity and heading velocity simulations are compared; (**c**) results of lateral deviation simulations are compared; (**d**) results of the yaw angle error and heading angle error simulations are compared.

By analyzing the above results in conjunction with the results of path tracking on serpentine roads by autonomous vehicles, it is demonstrated that the control algorithm based on the vehicle fuzzy model has more excellent automatic trajectory performance. Comparing the algorithm for control based on the yaw angle fuzzy model with the algorithm for control based on the heading angle fuzzy model, the algorithm for control based on the heading angle fuzzy model has better automatic trajectory performance.

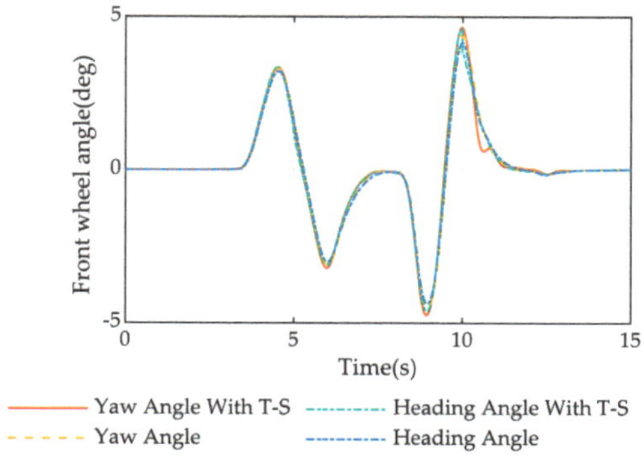

Figure 13. The simulation results of the front-wheel steering angle on a double lane change are compared.

5. Experimental Setup and Results

In this section, to validate the real-time performance of the two model control algorithms, hardware-in-the-loop (HIL) experiments are conducted using a serpentine road as an example in this paper. Figure 14 shows the structure of the control scheme based on the yaw or heading angle model. Figure 15 shows the steering control experimental bed used for the hardware-in-the-loop (HIL) experiments, and the related hardware includes the upper computer, the lower computer-NI/PXI system, and the steering system hardware.

The upper computer controls the program via the LabView RT 2017 software. The lower computer-NI/PXI system is embedded in the real-time simulation scenario provided by CarSim (Ann Arbor, MI, USA), receives the actual steering angle transmitted by the steering system hardware, obtains the vehicle's state quantities and other parameters in CarSim, obtains the target steering angle calculated by the algorithm of this paper through the above parameters, and finally sends the target steering angle to the steering system hardware. The steering system hardware includes items such as the steering wheel, sensors, and steering motor. One of the sensors used is the Hella sensor, which obtains information such as steering angle and steering. The steering motor receives the control signals sent by the bottom controller and controls them. The design of the bottom steering controller mainly relies on the 5744 chip, which is developed based on C language using the S32 environment. The bottom controller receives the target turning angle and sends the required control signals according to the target steering angle to drive the steering motor to rotate. All of the above hardware uses CAN for real-time information exchange.

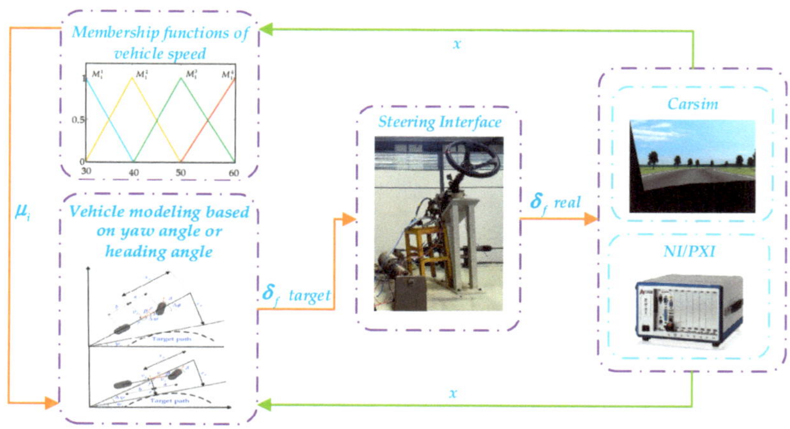

Figure 14. Structure of the control scheme based on the model of the yaw angle or heading angle.

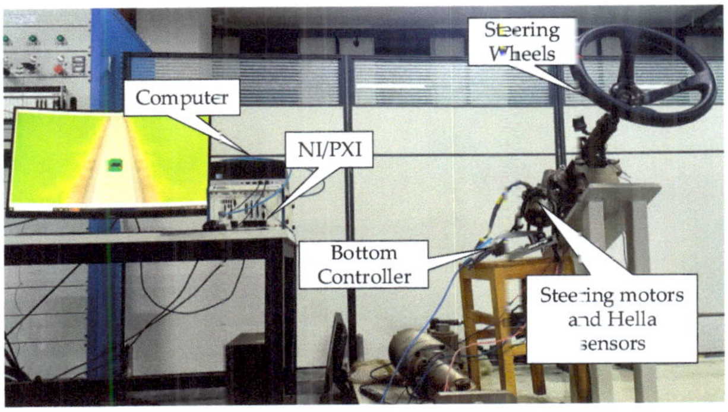

Figure 15. Hardware-in-the-loop (HIL) test rig.

To verify the conclusions obtained above, the four control algorithms mentioned above are experimented with path tracking at varying speeds, respectively.

Experiment-Path Tracking with Varying Vehicle Speed.

The path tracking and lateral deviation e_p comparison results for the four model control algorithms are given in Figures 16 and 17c. In the figure below the initial end of the arrow shows the position of the corresponding curve, and the end points to a magnified view at that position. Based on the figure, it is evident that the maximum lateral deviation of the algorithm for control of the yaw angle model does not exceed 0.57 m, the maximum lateral deviation of the algorithm for control of the heading angle model does not exceed 0.54 m, the maximum lateral deviation of the algorithm for control based on the fuzzy model of the yaw angle does not exceed 0.5 m, the maximum lateral deviation of the algorithm for control based on the fuzzy model of the heading angle does not exceed 0.47 m, and the tracking accuracies of the paths of the four control algorithms are increased in order.

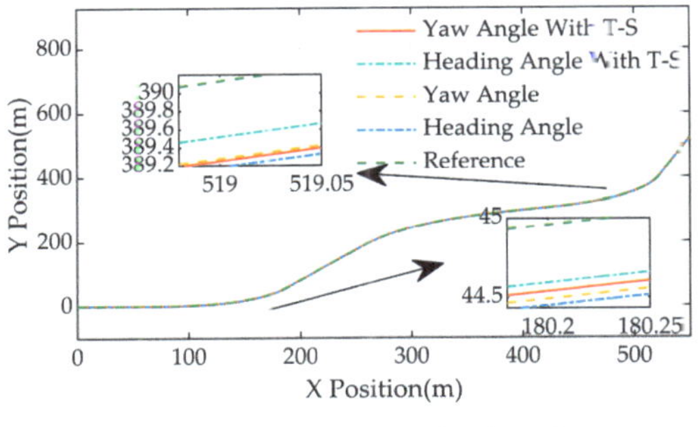

Figure 16. The experimental results of path tracking on a serpentine road are compared.

Figure 17d gives the results of the comparison of the yaw angle error $\Delta\varphi$ and heading angle error $\Delta\psi$. From the figure, it is evident that the magnitude of the yaw angle error $\Delta\phi$ is close to and does not exceed 2.88° for the two models, the yaw angle fuzzy model based on the yaw angle and the yaw angle model based on the yaw angle. The magnitude of the heading angle error $\Delta\psi$ of both models, the heading angle fuzzy-based model and heading angle-based model, is close to and does not exceed 2.45°. It is evident that the heading angle error $\Delta\psi$ is less than the yaw angle error $\Delta\phi$. The comparative results of lateral velocity v_y, yaw velocity γ and heading velocity $\dot{\psi}$, and front-wheel steering angle δ for the four model control algorithms are given in Figure 17a,b, and Figure 18. As illustrated in the figure, the four model control algorithms have the same variation in the above three parameters when tracking a vehicle's path.

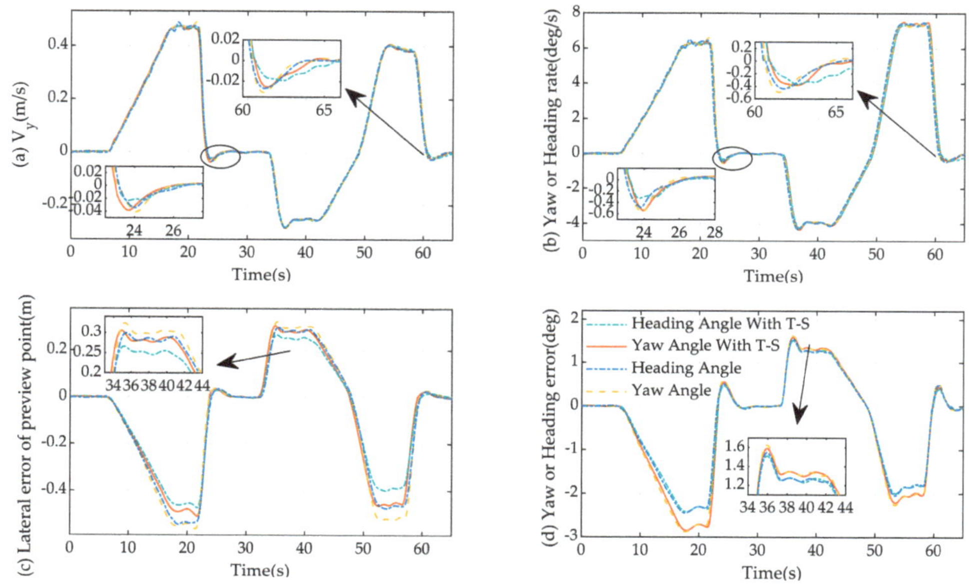

Figure 17. The experimental results of the state vector on a serpentine road are compared. (**a**) Results of lateral velocity experimental are compared; (**b**) results of yaw velocity and heading velocity experimental are compared; (**c**), results of lateral deviation experimental are compared; (**d**) results of the yaw angle error and heading angle error experimental are compared.

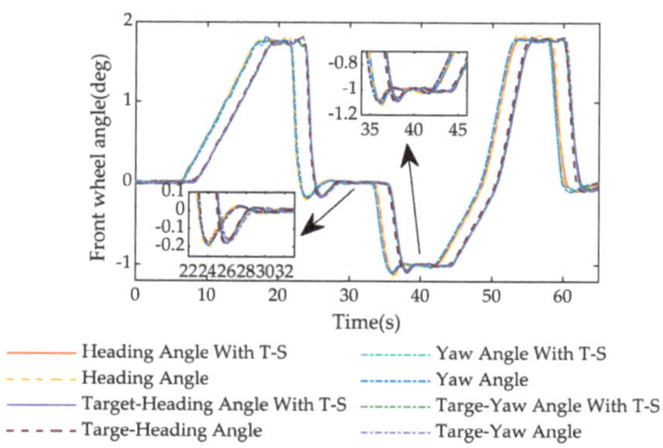

Figure 18. The experimental results of front-wheel steering angle on a serpentine road are compared.

Where the target and actual front wheel steering angles for each of the four control methods are displayed in Figure 18, the graphic shows that the target steering angle's size and the actual steering angle follow the same trend, and there exists only a lag of 1.5 s. The actual steering angle is the same as that of the target steering angle.

From the above hardware-in-the-loop (HIL) experimental results, the algorithm for control based on the vehicle fuzzy model has more excellent automatic trajectory performance. Comparing the algorithm for control based on the fuzzy model of yaw angle and the algorithm for control based on the fuzzy model of heading angle, the algorithm for control based on the fuzzy model of heading angle has better lateral stability and automatic trajectory performance.

Following analysis of the outcomes of hardware-in-the-loop (HIL) experiments and co-simulation using MATLAB 2021/Simulink–CarSim 8.02, the two results are similar, and both of them verified that the heading angle fuzzy model based on the fuzzy model has a more excellent automatic trajectory performance.

Since many parameters, such as vehicle or environment, vary over time, more time-varying parameters need to be selected to build the T-S fuzzy model in order to accurately portray the vehicle-road model, which increases the complexity of the control algorithm, thereby increasing the response time, etc., which further leads to an increase in the cost of the product. The accuracy of the control algorithm is inversely proportional to the product cost.

6. Conclusions

This work proposes a path-tracking control approach for autonomous cars based on the T-S fuzzy model of heading angle. First, the autonomous driving vehicle model based on heading angle is established by taking the intersection point between the vehicle traveling direction and the target path direction as the origin of the coordinate system and, in the significant side slip angle instances, the side slip angle is taken into consideration to enhance the precision of the vehicle's road tracking. To guarantee that the control system can provide the best control law, the longitudinal speed of the vehicle is used as a variable parameter, and the domain and affiliation function is identified in order to build further a T-S fuzzy model based on heading angle. Inspired by the linear quadratic regulator (LQR) control, a performance metric function for heading angle state broadening error-based systems is defined to account for multiple control objectives. The system's poles are limited to a fixed circular region to ensure a given dynamic performance. Ultimately, co-simulation with Matlab/Simulink–CarSim and hardware-in-the-loop (HIL) experiments are conducted. The outcomes demonstrate the superior automatic trajectory performance of the algorithm for control based on the fuzzy model of the vehicle with a heading angle. The advancement of automated driving control technology will be impacted by this study.

In future work, the introduction of observers into the model is considered to allow for the observation of parameters such as transverse velocity, which are difficult to measure. If the experimental conditions allow, real vehicle tests can be conducted to test the conditions under which the control algorithm is applicable.

Author Contributions: Conceptualization, J.W., Y.H. and X.L.; writing—original draft preparation, Y.H.; writing—review and editing, J.W., Y.H. and S.W.; methodology, J.W. and Y.H.; data curation, Y.H.; hardware, F.X., Y.H. and G.C.; software, Y.H., X.L. and G.C.; visualization, J.W.; formal analysis, F.X. and S.W. All authors have read and agreed to the published version of the manuscript.

Funding: This research was funded by the Shandong Province Key R and D Program (Major Science and Technology Innovation Project) Tasks, grant number 2023CXGC010214. This research also was funded by Shandong Province Key R and D Program (Competitive Innovation Platform), grant number 2022CXPT025.

Data Availability Statement: Data are contained within the article.

Conflicts of Interest: The authors declare no conflict of interest.

References

1. Mahmud, K.; Gope, K.; Chowdhury, S.M.R. Possible causes & solutions of traffic jam and their impact on the economy of Dhaka City. *J. Mgmt. Sustain.* **2012**, *2*, 112.
2. Gan, L. Globalization of the automobile industry in China: Dynamics and barriers in greening of the road transportation. *Energy Policy* **2003**, *31*, 537–551. [CrossRef]
3. Yuan, X.; Liu, H.; Qi, R. Research on key technologies of autonomous driving platform. *J. Phys. Conf. Ser.* **2021**, *1754*, 012127. [CrossRef]
4. Narla, S.R. The evolution of connected vehicle technology: From smart drivers to smart cars to... self-driving cars. *Ite J.* **2013**, *83*, 22–26.
5. Sorniotti, A.; Barber, P.; De Pinto, S. Path tracking for automated driving: A tutorial on control system formulations and ongoing research. In *Automated Driving: Safer and More Efficient Future Driving*; Springer: Cham, Switzerland, 2017; pp. 71–140.
6. Wang, R.; Ye, Q.; Cai, Y.; Wang, Y.; Xu, X.; Meng, X.; Long, C. Analyzing the influence of automatic steering system on the trajectory tracking accuracy of intelligent vehicle. *Adv. Eng. Softw.* **2018**, *121*, 188–196. [CrossRef]

7. Wu, H.; Li, Z.; Si, Z. Trajectory tracking control for four-wheel independent drive intelligent vehicle based on model predictive control and sliding mode control. *Adv. Mech. Eng.* **2021**, *13*, 16878140211045142. [CrossRef]
8. Li, X.; Chan, C.-Y.; Wang, Y. A reliable fusion methodology for simultaneous estimation of vehicle sideslip and yaw angles. *IEEE Trans. Veh. Technol.* **2015**, *65*, 4440–4458. [CrossRef]
9. Yang, M.; Chen, C.-F.; Shu, M.-L.; Yang, Y.-Y. Path Tracking Control for Uncertainty Vehicles via LMI Optimization. In Proceedings of the 3rd International Conference on Wireless Communication and Sensor Networks (WCSN 2016), Wuhan, China, 10–11 December 2016; pp. 338–344.
10. Hang, P.; Chen, X.; Luo, F.; Fang, S. Robust control of a four-wheel-independent-steering electric vehicle for path tracking. *SAE Int. J. Veh. Dyn. Stab. NVH* **2017**, *1*, 307–316. [CrossRef]
11. Yang, Z.; Huang, J.; Yang, D.; Zhong, Z. Design and optimization of robust path tracking control for autonomous vehicles with fuzzy uncertainty. *IEEE Trans. Fuzzy Syst.* **2021**, *30*, 1788–1800. [CrossRef]
12. Chen, M.; Ren, Y.; Ou, M. Adaptive robust path tracking control for autonomous vehicles considering multi-dimensional system uncertainty. *World Electr. Veh. J.* **2023**, *14*, 11. [CrossRef]
13. Shen, H.; Ling, R.; Mao, J.; Li, S. Steering control strategy guide by two preview vision cues. *Sci. China Technol. Sci.* **2012**, *55*, 2662–2670. [CrossRef]
14. Hiraoka, T.; Nishihara, O.; Kumamoto, H. Automatic path-tracking controller of a four-wheel steering vehicle. *Veh. Syst. Dyn.* **2009**, *47*, 1205–1227. [CrossRef]
15. Hang, P.; Chen, X. Path tracking control of 4-wheel-steering autonomous ground vehicles based on linear parameter-varying system with experimental verification. *Proc. Inst. Mech. Eng. Part I J. Syst. Control. Eng.* **2021**, *235*, 411–423. [CrossRef]
16. Jin, X.J.; Yin, G.; Chen, N. Gain-scheduled robust control for lateral stability of four-wheel-independent-drive electric vehicles via linear parameter-varying technique. *Mechatronics* **2015**, *30*, 286–296. [CrossRef]
17. Hu, H.; Bei, S.; Zhao, Q.; Han, X.; Zhou, D.; Zhou, X.; Li, B. Research on trajectory tracking of sliding mode control based on adaptive preview time. *Actuators* **2022**, *11*, 34. [CrossRef]
18. Mokhtare, Z.; Vu, M.; Mobayen, S.; Fekih, A. Design of an LMI-Based Fuzzy Fast Terminal Sliding Mode Control Approach for Uncertain MIMO Systems. *Mathematics* **2022**, *10*, 1236. [CrossRef]
19. Chatavi, M.; Vu, M.T.; Mobayen, S.; Fekih, A. H∞ robust LMI-based nonlinear state feedback controller of uncertain nonlinear systems with external disturbances. *Mathematics* **2022**, *10*, 3518. [CrossRef]
20. Sheikhsamad, M.; Puig, V. Learning-Based Control of Autonomous Vehicles Using an Adaptive Neuro-Fuzzy Inference System and the Linear Matrix Inequality Approach. *Sensors* **2024**, *24*, 2551. [CrossRef] [PubMed]
21. Chen, C.; Jia, Y.; Shu, M.; Wang, Y. Hierarchical adaptive path-tracking control for autonomous vehicles. *IEEE Trans. Intell. Transp. Syst.* **2015**, *16*, 2900–2912. [CrossRef]
22. Du, H.; Zhang, N.; Naghdy, F. Velocity-dependent robust control for improving vehicle lateral dynamics. *Transp. Res. Part C Emerg. Technol.* **2011**, *19*, 454–468. [CrossRef]
23. Feng, J.; Yin, G.; Liang, J.; Lu, Y.; Xu, L.; Zhou, C.; Peng, P.; Cai, G. A Robust Cooperative Game Theory based Human-machine Shared Steering Control Framework. *IEEE Trans. Transp. Electrif.* 2023, *Early Access*. [CrossRef]
24. Liang, J.; Wang, F.; Feng, J.; Zhao, M.; Fang, R.; Pi, D.; Yin, G. A Hierarchical Control of Independently Driven Electric Vehicles Considering Handling Stability and Energy Conservation. *IEEE Trans. Intell. Veh.* **2023**, *9*, 38–751. [CrossRef]
25. He, X.; Huang, W.; Lv, C. Trustworthy autonomous driving via defense-aware robust reinforcement learning against worst-case observational perturbations. *Transp. Res. Part C Emerg. Technol.* **2024**, *163*, 104632. [CrossRef]
26. He, X.; Liu, Y.; Yang, K.; Wu, J.; Ji, X. Robust coordination control of AFS and ARS for autonomous vehicle path tracking and stability. In Proceedings of the 2018 IEEE International Conference on Mechatronics and Automation (ICMA), Changchun, China, 5–8 August 2018; pp. 924–929.
27. Wu, J.; Kong, Q.; Yang, K.; Liu, Y.; Cao, D.; Li, Z. Research on the steering torque control for intelligent vehicles co-driving with the penalty factor of human–machine intervention. *IEEE Trans. Syst. Man Cybern. Syst.* **2022**, *53*, 59–70. [CrossRef]
28. He, X.; Lou, B.; Yang, H.; Lv, C. Robust decision making for autonomous vehicles at highway on-ramps: A constrained adversarial reinforcement learning approach. *IEEE Trans. Intell. Transp. Syst.* **2022**, *24*, 4103–4113. [CrossRef]
29. Wu, J.; Zhang, J.; Nie, B.; Liu, Y.; He, X. Adaptive control of PMSM servo system for steering-by-wire system with disturbances observation. *IEEE Trans. Transp. Electrif.* **2021**, *8*, 2015–2028. [CrossRef]
30. Tseng, C.-S.; Chen, B.-S.; Uang, H.-J. Fuzzy tracking control design for nonlinear dynamic systems via TS fuzzy model. *IEEE Trans. Fuzzy Syst.* **2001**, *9*, 381–392. [CrossRef]
31. He, X.; Wu, J.; Huang, Z.; Hu, Z.; Wang, J.; Sangiovanni-Vincentelli, A.; Lv, C. Fear-neuro-inspired reinforcement learning for safe autonomous driving. *IEEE Trans. Pattern Anal. Mach. Intell.* **2023**, *46*, 267–279. [CrossRef] [PubMed]
32. He, X.; Huang, W.; Lv, C. Toward Trustworthy Decision-Making for Autonomous Vehicles: A Robust Reinforcement Learning Approach with Safety Guarantees. *Engineering* **2024**, *33*, 77–89. [CrossRef]
33. Wang, C.; Qin, J.; Guo, M.; Xu, Y. Prediction of vehicle trajectory based on Fuzzy colored petri net. *TELKOMNIKA Indones. J. Electr. Eng.* **2013**, *11*, 5824–5832. [CrossRef]
34. He, X.; Yang, H.; Hu, Z.; Lv, C. Robust lane change decision making for autonomous vehicles: An observation adversarial reinforcement learning approach. *IEEE Trans. Intell. Veh.* **2022**, *8*, 184–193. [CrossRef]

35. Wang, J.; Zhang, G.; Wang, R.; Schnelle, S.C.; Wang, J. A gain-scheduling driver assistance trajectory-following algorithm considering different driver steering characteristics. *IEEE Trans. Intell. Transp. Syst.* **2016**, *18*, 1097–1108. [CrossRef]
36. Xie, W. An equivalent LMI representation of bounded real lemma for continuous-time systems. *J. Inequalities Appl.* **2008**, *2008*, 1–8. [CrossRef]
37. Liang, J.; Feng, J.; Fang, Z.; Lu, Y.; Yin, G.; Mao, X.; Wu, J.; Wang, F. An energy-oriented torque-vector control framework for distributed drive electric vehicles. *IEEE Trans. Transp. Electrif.* **2023**, *9*, 4014–4031. [CrossRef]
38. Lee, H.J.; Park, J.B.; Chen, G. Robust fuzzy control of nonlinear systems with parametric uncertainties. *IEEE Trans. Fuzzy Syst.* **2001**, *9*, 369–379.
39. Liang, J.; Tian, Q.; Feng, J.; Pi, D.; Yin, G. A polytopic model-based robust predictive control scheme for path tracking of autonomous vehicles. *IEEE Trans. Intell. Veh.* **2023**, *9*, 3928–3939. [CrossRef]
40. Benloucif, A.; Nguyen, A.-T.; Sentouh, C.; Popieul, J.-C. Cooperative trajectory planning for haptic shared control between driver and automation in highway driving. *IEEE Trans. Ind. Electron.* **2019**, *66*, 9846–9857. [CrossRef]
41. Zhang, H.; Zhang, X.; Wang, J. Robust gain-scheduling energy-to-peak control of vehicle lateral dynamics stabilisation. *Veh. Syst. Dyn.* **2014**, *52*, 309–340. [CrossRef]
42. Taheri, S.; Rakheja, S.; Hong, H. Influence of human driving characteristics on path tracking performance of vehicle. In Proceedings of the Intelligent Robotics and Applications: 5th International Conference, ICIRA 2012, Montreal, QC, Canada, 3–5 October 2012; Proceedings, Part II 5, 2012; pp. 207–216.

machines

MDPI

Article

Collaborative Behavior for Non-Conventional Custom-Made Robotics: A Cable-Driven Parallel Robot Application

Julio Garrido *, Diego Silva-Muñiz, Enrique Riveiro, Josué Rivera-Andrade and Juan Sáez

Research Group on Efficient and Digital Engineering, Automation and System Engineering Department, School of Industrial Engineering, Campus As Lagoas-Marcosende, Universidade de Vigo, 36310 Vigo, Spain; diego.silva.muniz@uvigo.gal (D.S.-M.); enrriveiro@uvigo.gal (E.R.); josueroberto.rivera.andrade@uvigo.gal (J.R.-A.); juansaez@uvigo.gal (J.S.)

* Correspondence: jgarri@uvigo.gal

Abstract: The human-centric approach is a leading trend for future production processes, and collaborative robotics are key to its realization. This article addresses the challenge of designing a new custom-made non-conventional machine or robot involving toolpath control (interpolated axes) with collaborative functionalities but by using "general-purpose standard" safety and motion control technologies. This is conducted on a non-conventional cable-driven parallel robot (CDPR). Safety is assured by safe commands to individual axes, known as safe motion monitoring functionalities, which limit the axis's speed in the event of human intrusion. At the same time, the robot's motion controller applies an override to the toolpath speed to accommodate the robot's path speed to the limitations of the axes. The implementation of a new Pre-Warning Zone prevents unnecessary stops due to the approach of the human operator. The article also details a real experiment that validates the effectiveness of the proposed strategy.

Keywords: collaborative CDPR; industrial safety; motion control; motion control safety; industrial automation safety trends; collaborative robotics

Citation: Garrido, J.; Silva-Muñiz, D.; Riveiro, E.; Rivera-Andrade, J.; Sáez, J. Collaborative Behavior for Non-Conventional Custom-Made Robotics: A Cable-Driven Parallel Robot Application. *Machines* **2024**, *12*, 91. https://doi.org/10.3390/machines12020091

Academic Editors: Jian Wu, Xiangkun He and Guangfei Xu

Received: 27 December 2023
Revised: 23 January 2024
Accepted: 23 January 2024
Published: 25 January 2024

1. Introduction

1.1. New Trends in Non-Conventional Industrial Robots

Industrial applications with several axes working in a coordinated way to perform trajectories are expanding their application range beyond the traditional machining systems (CNC) and conventional industrial robotics [1,2]. From packaging processes to new flexible production lines, they often incorporate toolpath operations into their workflows [3]. Significant advances in machine automation and industrial motion control technologies have driven this expansion. Consequently, industrial machinery designers are increasingly adopting servo-drive technology [4], along with new mechatronic resources, to meet the demanding flexibility and performance needs of modern factories [5]. Current general-purpose motion-control technologies are enabling not just simple movements, but also the execution of complex, flexible kinematic relationships among different mechanical components of machines. The emergence of fast, deterministic, and synchronous digital networks has facilitated the connection of multiple motors to a single controller. Often, servo technology is being chosen over the traditional large single-motor setup, centralizing functionalities within the machine and replacing mechanical systems with software-based control. This approach is commonly known as a "mechatronic solution" [6].

This motion-control technology trend is contributing to the development of a wide range of custom industrial robotic configurations, including several types of parallel robotics such as delta robots, Stewart platforms, and cable-driven parallel robots (CDPRs) [7–9].

1.2. Conventional Collaborative Automation Architectures

Parallel to the development of industrial servo technology, which is the basis for motion control machinery and industrial robots, the latest industry safety regulations facilitate the conception of new advanced machinery with complex interaction with human operators. Specifically for servo-actuated machines, there has been an expansion of safety regulations designed to facilitate their safe and even collaborative operation [10].

In conventional collaborative robotics, specific standards clearly define safe operation practices, allowing for a reduction in the physical separation traditionally required between humans and hazardous processes, as stated in ISO 12100 [11]. However, except in the exceptional case where a collaborative robot, for constructional reasons, can be considered as an inherently safe machine, it is generally regarded as just another very specialized output device (Safety Related System/Control System: SRC/CSc) in the safety chain [12]. Transferring equivalent collaborative functionalities to general-purpose machinery must deal with the complexities of universal safety regulations and the standard safety features of general automation technologies. For example, ISO 12100 also specifies the need to carry out a risk assessment when designing a new machine as well as the risk reduction process required for risks that cannot be significantly reduced by redesigning the machine. Other standards, such as those detailing performance levels (PL) [13] and safety integrity levels (SIL) [14], guide the design of safety-related control system components. These standards have opened the possibility of using electronically programmed technology to make safety-related decisions. However, it remains a fundamental principle in the implementation of safety that logic must be executed in control devices certified for that purpose. PLCs (programmable logic controllers), CNCs (computer numerical controllers), and general-purpose motion controllers do not fall into this category of certified devices.

Therefore, a two-layer decision architecture is required. Figure 1 shows the two levels of hardware and software that are mandatory in industrial machinery (according to ISO 13849-1 [13]): the safety subsystem, also called the Safety Related System/Control System, which includes inputs, logic, safe actuation, and communications; and the operational subsystem, which comprises standard inputs, process logic, standard actuation, and standard communications.

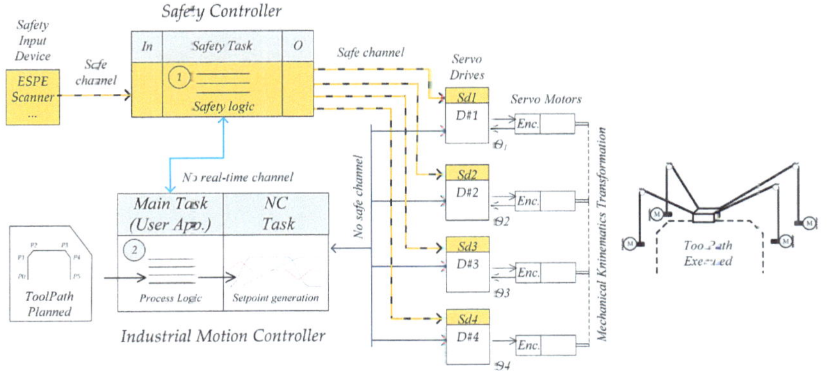

Figure 1. Reference architecture for safe servo-driven industrial systems.

These two decision layers when a machine executes a toolpath are shown in Figure 1: safety logic (1) and process logic (2). In a warning or hazardous situation, the safety controller (1) decides to perform safe stop or limit commands, which are sent to the servo drives ("Sd1" to "Sd4" in Figure 1) via a safe communication channel (represented by yellow and black in Figure 1). In parallel, the process logic (2) receives a planned toolpath specification and generates the motion profiles of the motors (position, speed, and acceleration). At this point, the specific motion commands are sent to the servo drives

("D#1" to "D#4" in Figure 1) via a non-safety fieldbus (highlighted in blue in Figure 1). Although each layer makes its own decisions, they can exchange information to qualify or parameterize their logic via a non-real-time communication channel.

1.3. Collaborative for Non-Standard Robotics: Collaborative CDPRs

The Fifth Industrial Revolution (Industry 5.0) spotlights six key technologies to advance European social and environmental priorities within technological innovation, moving from isolated technologies to a systemic approach [15]. Among these, human-centric solutions and human-machine-interaction technologies highlight the importance of collaborative robots [16]. The trend in robotics is increasingly toward collaborative and user-friendly systems.

Collaborative robotics development must go beyond conventional industrial robots to include other kinds of robotic configurations that may offer functional advantages compared to the more common industry configurations, and that may cover application scenarios where conventional industrial robots are overly specialized. Efforts to enable robots and general-purpose machinery, including CDPRs, to work collaboratively align with the ongoing trends of Industry 5.0.

The study and conceptualization of collaborative behavior in CDPRs is a novel and little-addressed research issue. Most approaches focus on achieving low stiffness during motion. This low stiffness enables the CDPR to absorb energy induced during a collision, thus preventing damage to both the CDPR and the collided object or body part [17]. This has been addressed in different ways, for example, through auxiliary mechanisms attached to the cables [18] or by using architectures with degrees of freedom (DOFs) operated redundantly to achieve a controllable stiffness [19]. The impact of CDPR stiffness on its transparency, which is the robot's ability to move in the direction intended by the user, during co-manipulation tasks involving human-robot interaction is also explored [20]. Other researchers contributed to collaborative CDPR focusing on managing cable interferences and collision detection methods when a human-CDPR physical interaction is needed [21,22]. Additional researchers also contributed to the field by proposing and evaluating advanced control strategies aimed at enhancing physical human-robot collaboration [23].

However, most initiatives overlook the industrial-implementation "standard and normative" issues. As depicted in Figure 1, a general-purpose industrial motion-control-system architecture has a two-layer structure with two decision elements: the safety controller and the motion controller.

On one hand, the safety logic generates commands for the servo axes. Although industrial safety standards have addressed the definition of safety operating modes of servo axes in detail, they do so considering the axes individually. These standards have not considered the implications for the toolpath when these axes are coordinated to operate as a group. This limitation originates in the safety standards themselves, which force that the decision-making architecture must separate safety decisions from toolpath control. As a result, safety decisions require dedicated hardware and software which must have the appropriate certifications.

On the other hand, only the motion controller has all the necessary information to make decisions concerning the toolpath. The individual motion control of each axis depends on the controller, namely its "numerical control task", which generates the setpoints for each axis via its servo drive. However, the servo drives do not know whether they are working alone, for example performing a point-to-point movement, or in coordination with other axes to execute an interpolated movement.

As can be seen from Figure 1, the servo drives can receive commands from two different decision layers. This can lead in some cases to side effects. The article discusses how implementing collaborative functionalities can lead to this situation and provides a standard-compliant solution.

The article addressed the objective of implementing a smooth toolpath movement despite this two-level general-purpose decision architecture when implementing collaborative operation, speed and separation monitoring (SSM), in general-purpose machinery. It further explores the issues related to toolpath speed limitations that arise when human operators and robots share a workspace, alongside the independent role of the safety controller in issuing speed-limiting commands to each axis ("Sd1" to "Sd4" in Figure 1). This is conducted on a CDPR, which is presented in the next section (Section 2.1). CDPRs fall into the category of non-conventional and not widely used mechatronic systems in the industry. Most of their applications are custom-made, and there is no standard commercial development for this type of robot, especially those that involve collaboration with humans. This makes them an excellent subject for researching collaborative functionalities in custom-made trajectory systems. Section 2 also reviews the safety standards for collaborative robots as well as the integration of safety architectures within collaborative operations. Section 3 outlines the toolpath control implemented for the CDPR. Section 4 analyzes an SSM implementation strategy for a CDPR prototype through three use cases: a regular toolpath without human interference, the simultaneous application of SLS and toolpath override, and a new Pre-Warning Zone for smoother toolpath override. This subsystem operates independently to restrict the kinematic parameters of the axes for safety. The developed strategy can be applicable to any robot configuration. The article ends with a discussion and some conclusions concerning the research in Section 5.

2. General-Purpose Motion Control and Safety Resources to Implement Custom-Made Industrial Robots: The CDPR Case

2.1. Cable-Driven Parallel Robots

A CDPR is a kind of parallel robot that uses multiple cables to actuate and control the movement of the robot end effector (i.e., the tool or gripper attached to the robot's endpoint). Parallel robot configurations have been studied for decades. A well-known example is delta robots, introduced by Clavel in 1985 to carry out fast operations for food handling or assembly of electronic circuits, among others [24]. Other relevant parallel configurations are the Stewart platform [25] and the Cappel platform [26]. All these parallel robots involve links that are under tensile and compressive loads. In the mid-1980s, Landsberger and Sheridan proposed an evolution of the Stewart platform but fully actuated by cables (CDPR) [27].

CDPRs typically have a fixed base or frame to support multiple actuators, usually servomotors, to control the cables. The end effector's movement results from the cables' tension and length. Due to the use of cables, CDPRs do not work with compressible loads, only tensile ones [28]. The number of cables and the placement of the motors determine the degrees of freedom, stability, and complexity of the robot. CDPRs whose position cannot change without changing the length of the cables are called fully constrained (see Figure 2a). In general, this architecture requires at least one additional cable than degrees of freedom. An alternative common configuration is the under-constrained or suspended type, in which the robot maintains balance through gravity [29]. Figure 2b shows the suspended 4-cable CDPR developed for this research. Although fully-constrained CDPRs can perform motion trajectories with more precision at high speeds compared to under-constrained CDPRs, the latter has the advantage of reducing the probability of interference between cables and obstacles because the end effector is suspended, so the workspace under the robot is free of cables [30]. CDPRs offer multiple advantages compared to traditional configurations, such as high dynamic performance, a larger workspace, reduced manufacturing and maintenance costs, ease of assembly and disassembly, high modularity and reconfigurability, and a smaller inertia and a higher payload-to-weight ratio [31].

These advantages make them suitable for a variety of application scenarios, including, but not limited to, civil applications such as moving cameras in social event coverage [32], cable-driven configurations on boats [33], the additive manufacturing sector [34], logistics and warehousing [35], painting [36], motion simulation platforms [37], and medical applications such as patient rehabilitation [38]. Within the industrial sector, the application of CDPRs for pick-and-place operations has been among the most reported. For example,

Holland and Cannon [39] patented a suspended cable robot for handling large loads in the maritime and manufacturing industries. Barbazza et al. [30] developed a novel suspended CDPR designed for industrial pick-and-place tasks where a reconfigurable end effector enhanced the adaptability of the robot in avoiding obstacles. Other less conventional designs are, for instance, the high-speed pick-and-place CDPR presented by Zhang et al. in [40] based on the architecture of a delta robot incorporating an extensible limb.

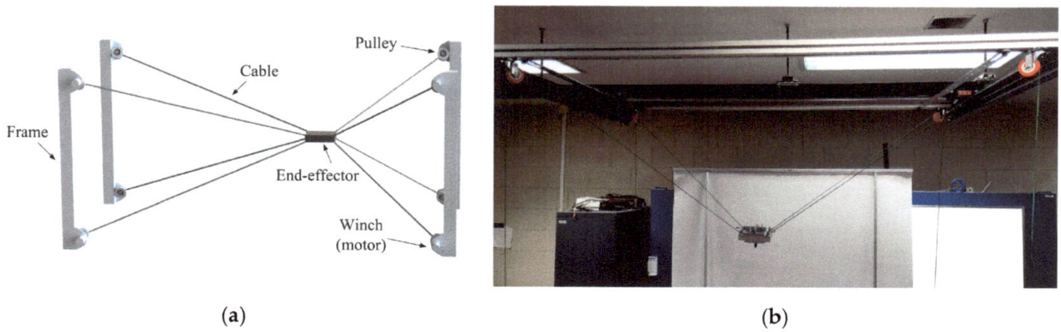

Figure 2. Example of CDPR configurations: (**a**) 8-cables fully constrained (position cannot change without changing the length of the cables); (**b**) Photo of the developed 4-cables suspended (position can be different even subject to the same cable lengths).

2.2. *Safety Standards for Collaborative Robotics*

The development of collaborative robots has emerged as a significant topic within both academic and industrial sectors. Historically, researchers and innovators have propelled this advancement by focusing on various aspects, such as creating new mechanical structures [41], integrating sensors for impact detection [42], and developing control strategies to prevent and reduce injury from collisions [43], among others. Collaborative robots are designed for direct interaction with human workers, sharing workloads, and operating safely without the need for traditional safety enclosures or similar protective barriers [44].

From a regulatory point of view, in addition to the more generic standards concerning robotics (ISO 10218 [45,46]), complementary technical specifications address different collaborative robotics issues, such as ISO/TS 15066 [47]. Together, they define the four types of collaboration between humans and robots: safe monitored Stop (SMS); speed and separation monitoring (SSM); power and force limitation (PFL); and hand guiding (HG). The regulations provide general guidelines about which one is the active mode depending on the operational parameters (human-robot distance, speeds, etc.). Moreover, combinations and alternative implementations of modes are permitted if they achieve the requirements [48]. SMS is the most straightforward collaborative mode and is typically used when humans and robots need to operate within the same workspace but not simultaneously. In this mode, the robot halts if it detects someone nearby.

In HG mode, the robot only moves as a result of direct guiding input from the operator. The operator is in direct contact, and the robot system is under manual control. Both the robot and the operator move at the same time, but the motion is controlled by the person. This allows the operator to move the manipulator freely in the space without exerting significant force and without the need to deactivate the motors [49]. In another operation that involves teaching tasks, the operator instructs the robot by manually guiding it through a sequence of points [50]. The robot then replicates this sequence. In the literature, there are some applications of direct hand guiding such as industrial applications (automotive assembly and production line, automated lifting and moving of heavy items, robotic welding, etc. [51]) and medical applications (rehabilitation, surgery, etc. [52,53]).

PFL represents the most challenging mode, allowing for the possibility of unexpected contact during motion. It characterizes the types of human-robot contact, whether quasi-static or transient. Furthermore, maximum values for biomechanical load limits are given to prevent injury or damage when the robot collides with a different body as well as additional protective measures [54]. A maximum speed limit is set based on the moving mass of the robot and the body part impacted. The robot must then work with human presence detection to decide which of the collaborative modes is active at any moment, and the maximum speed for that. Examples of its use can be found in the recent literature [55,56].

Finally, in the SSM mode, safety is guaranteed by always keeping the minimum protective separation distance between the operator and the robot. The safety mode for SSM in collaborative robots identifies three operating levels (consequently, three zones) based on the proximity of the human to the robot and the consequent speed of the robot [51,57]: full-speed operation when no person is nearby (free zone), speed reduction, using a predefined speed limit, when a person approaches to improve ergonomics and reduce stopping distance (warning zone), and full robot stop when a person enters an area close to the robot (danger zone), respecting the minimum safety distances set in the ISO/TS 15066 standard. This approach ensures safe and efficient interaction between humans and robots in collaborative environments. Few studies have explored the strategy of varying speed dynamically rather than maintaining constant values. In [56], the authors modified the robot's TCP speed based on the separation distance from people, incorporating user-defined parameters that shape the speed change function's profile and intensity [58]. Another significant contribution introduced a method for continuous adaptation of the robot's TCP speed, aiming to maximize productivity [59]. This approach utilized a simplified criterion based on a two-zone distance model. In [60], the authors presented two approaches for setting a robot's TCP speed limit to prevent collisions, focusing on the separation distance from nearby users, as measured by a laser scanner, and considering the robot's direction of motion. These methods allowed for the continuous dynamic adjustment of the robot's speed. Novel research demonstrated a method for dynamically adjusting the protective separation distance, enabling the robot to operate at higher speeds by distinguishing between human and non-human entities using thermal imaging technologies [61]. Finally, another distinct research effort employed fuzzy logic to control a robot's speed by detecting the proximity and interaction velocity between robots and humans [62]. This approach incorporated a human tracking algorithm that dynamically computed the minimum safe distance, facilitating real-time adjustments to the robot's speed with the aid of a quintic polynomial function.

This article presents an SSM collaborative mode implementation using both constant and variable value approaches.

2.3. Safety Architectures and Collaborative Operations: Collaborative Robotics

Collaborative robots are not inherently safe by themselves and must be treated like any industrial device in an automated process. They should be integrated into a safe architecture as components, specifically as output elements. In most cases, a collaborative robot functions as an output device within the safety chain, possessing very particular characteristics.

The safety chain consists of input devices that detect risks, logic devices (safety controllers) that decide on safety actions, and output devices that execute safety responses. These range from simple relays that cut off power to actuators to more complex devices like servo drives that engage with the robots directly.

As collaborative robotics become more prevalent in industries committed to safe operations, a new category of safety-certified equipment has emerged for collaborative workspaces. Such equipment comprises safety controllers or PLCs, laser scanners, light barriers for safety, and emergency stop buttons, among others [63].

These control systems increasingly employ complex electrical, electronic, and programmable devices and systems. Among these, adjustable speed electrical power drive systems (PDS) are notable for their suitability in safety-related applications (PDS/SR) [64].

The latest generation of PDS-SR consists of servo drives with integrated safety functions. These functions control the safe motion of each motor by their corresponding drive, which makes the machine as a whole behave safely. Typically, this is achieved through safe motion monitoring functionalities (SMMF) as specified in IEC 61800-5-2 [64]. Examples of SMMF include, but are not limited to:

- STO: Safe Torque Off
- SLS: Safely Limited Speed
- SLT: Safely Limited Torque
- SLI: Safely Limited Increments

Figure 3 presents two safe motion monitoring functionalities: STO and SLS. Figure 3a illustrates the temporal profile of the initiating signal for the STO safety function, as well as the corresponding motor speed response. When the STO function is triggered, the safety device mandates an uncontrolled stop [65]. Figure 3b depicts the timing of the signal that triggers the SLS safety mechanism alongside a graph of axial speed. This function ensures that operational speeds remain within a preset threshold. If limits are overrun, the system is designed to bring the drive to a safe stop, a process graphically detailed in Figure 3c.

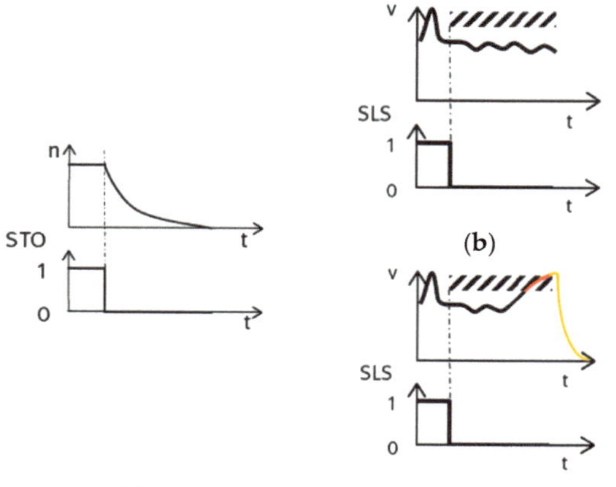

Figure 3. Safe motion monitoring functionalities examples: (**a**) STO; (**b**) SLS with activation of the speed limit; (**c**) SLS with activation and speed limit exceedance: the red line indicates when the speed limit is exceeded and the yellow line indicates when the STO is activated.

These safe monitoring functions (SLx) should be viewed as integral parts of a comprehensive safety system, but if operational thresholds are exceeded, a stopping action (STO) is triggered. Finally, it is important to note that within the safe automation system architecture, no SMMF codes are related specifically to the kinematic parameters of the toolpath as a whole, but only to those of the individual axes.

3. CDPR Toolpath Control with Standard Industrial Motion Control Resources

Several industrial controller manufacturers offer specialized components designed for a variety of robotic systems, including 2D and 3D Cartesian robots, gantry systems, SCARA robots, and delta robots [66,67]. Among others, Beckhoff delivers integrated resources for spatially under-constrained CDPR controlled by three or four cables [68]. These innovations significantly streamline the process of implementing robots with such capabilities and facilitate the coordination of axis groups.

Considering these control resources, Figure 4 illustrates the sequential process for performing a toolpath with the developed suspended CDPR. This diagram is a visual representation of the control software, showing the logic and steps necessary to initialize and operate the CDPR for normal execution of the toolpath. The CDPR was programmed in TwinCAT 3 [69] as is briefly explained below.

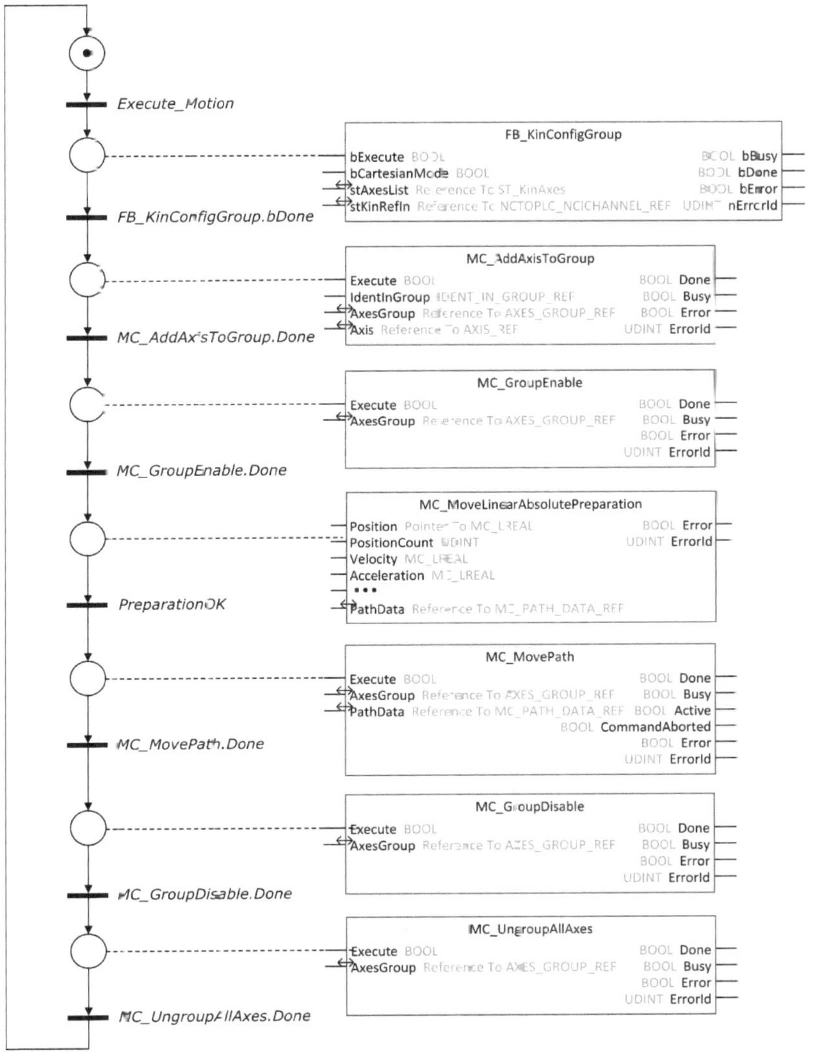

Figure 4. Sequence process control for executing toolpaths in a suspended 4-cable CDPR configuration in TwinCAT 3 with legacy Beckhoff resources.

Once all axes have been powered up, meaning that the brakes are released and the system is in a "Standstill" state, the activation of the "Execute_motion" variable triggers the motion sequence. The flow continues with the initial configuration of the axis groups via the "FB_KinConfigGroup" function block, which configures axes according to the kinematic transformation. In the configuration of the NC task, the type of kinematics ("four-cable robot") and the spatial variables for the position of the end-effector and the pulleys are preconfigured. Additional axes are then added to the group with "MC_AddAxisToGroup", preparing the system for the activation of the axis groups with "MC_GroupEnable". Once enabled, each point-to-point linear motion is prepared in advance with the function block "MC_MoveLinearAbsolutePreparation", defining parameters such as position, velocity, blending transition strategies, etc. This block prepares all movements in a "PathData" structure (of type "MC_PathData_Ref") to be executed by the "MC_MovePath" block. When the movement is finished, the system disables the group of axes with "MC_GroupDisable" and then ungroups them with "MC_UngroupAllAxes", completing the control cycle. The controller drives the sequence in its main processing task by monitoring signals provided by each function block, including "Done", "Busy", "Error", and "ErrorID" signals. Also, Figure 4 illustrates only the normal execution case and omits error management for simplicity.

It is also important to remark here that, after configuring a group with the "MC_GroupEnable" function block, it becomes impossible to issue commands to individual axes in isolation within the motion control application, whether to halt or adjust the speed of a singular axis. From this point onwards, only group-affecting motion commands are valid.

4. Collaborative Behavior Implementation Strategy with General-Purpose Devices

While the standard motion controller takes care of the toolpath control as described in Section 3, the safety controller is responsible for generating SMMF safety commands to the PDS(SR) (inverters or servo drives with safety functions), so that they react accordingly. However, PDS(SR) devices lack the context of their collaboration with other PDS(SR) units in executing the movement of a terminal element. This is the responsibility of the motion controller in the process subsystem.

Despite this dual process architecture of toolpath-related decision-making (safety and standard control process), the safety controller can inform the motion controller that an intrusion has been detected.

The location of this intrusion, represented by variable "PS" in Figure 5, can be provided by the safety detection device itself or by a standard non-safety detector. Finally, the standard controller can decide to limit the speed of the coordinated motion of the axes to, for instance, activate a collaborative speed and separation monitoring (SSM) mode.

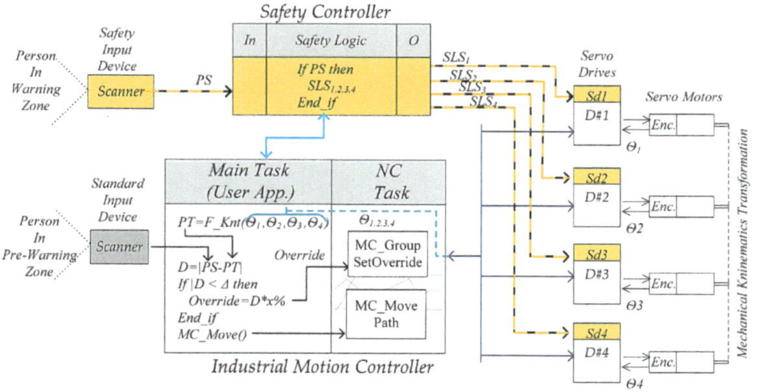

Figure 5. Proposed reference architecture for the experiment. Yellow color indicate safety-related parts.

Industrial motion controllers enable modification of the interpolated motion speed by software resources to adjust the axes group's speed. Industrial motion controller manufacturers offer this capability via the "MC_GroupSetOverride" function block, as specified in the PLCopen Motion Control standard, Part 4 [6]. Although this override value can be modified continuously, the standard indicates that there might be a delay before the motion controller effects this change. To smooth the transition, motion controller manufacturers define transition ramps to go from the actual speed override setting to the new one.

Apart from and independent of this speed correction in the coordinated movement, safety is ensured through the safety subsystem by immediately applying the safety limitation functions to each servo axis (safely limited speed: "SLSx" in Figure 5). With these functions, safe monitoring prevents the drive from exceeding a certain speed, but if the limit value is exceeded, the drive safely stops the axis (Figure 3c). To avoid reaching these thresholds or safe limits so that movement is not suddenly interrupted is the responsibility of the motion controller.

To illustrate this problem and the corresponding proposed strategy to overcome it, three experiments were conducted with the 4-cable suspended CDPR presented in Section 3. For these experiments, a pick-and-place trajectory was planned, and its waypoints are provided in Table 1. The workspace of the developed CDPR laboratory prototype (Figure 2b) covers a volume of 390 × 1520 × 1450 mm (length, width, height). The end effector measures 58 × 103 × 18 mm (length, width, height), is made of aluminum, and weighs 0.602 kg. The aluminum profiles of the structure have a cross-section of 40 × 40 mm. The pulleys used are made of polyurethane and have a radius of 25 mm. The cables used were Dyneema SK75 with a diameter of 1.5 mm. The winches used are made of PLA material with a radius of 45 mm. The electronic components used were from the manufacturer Beckhoff: EL6900 safety controller, CX5130 industrial controller, and four servomotors with holding brakes (series AM8000) and the corresponding servo drives with safety functions (AX5000 series). Finally, the experiment was carried out in a laboratory with a range of temperature and humidity of 20 °C, 75–80%, respectively.

Table 1. Waypoints of the trajectory.

N	X (mm)	Y (mm)	Z (mm)	Velocity (mm/s)
1	−150	−150	400	-
2	−150	−150	700	100
3	−100	−150	800	100
4	0	200	800	200
5	50	200	700	100
6	50	200	400	100

4.1. Case 1: Regular Toolpath without Limitations

The workspace in this experiment comprised the same three safety zones described in Section 2 (Figure 6): free zone (full-speed operation when no person is nearby); warning zone (speed reduction when a person approaches to improve ergonomics and reduce stopping distance); and danger zone (full robot stop when a person enters an area close to the robot). These zones, and the associated functionality, cover the speed and separation monitoring (SSM) collaborative mode.

In this first case, a regular pick-and-place trajectory was conducted (Table 1) without any intrusion in the warning zone, and, therefore, without the need for any toolpath override and without any requestion safe speed limitation to axes. Figure 7a details the five sections of the toolpath through the velocities of the axes. Thus, from the second 0 to 3.23 corresponds to the first vertical toolpath section (row 2 in Table 1); from the second 3.23 to 4.47 to the first oblique section (row 3 in Table 1); then, until approximately the second 6.75, the horizontal displacement section is executed (row 4 in Table 1) to finish with the second descending oblique section between seconds 6.75 and 8.17 (row 5 in Table 1) and the last vertical section until second 11.60 (row 6 in Table 1). A 3D view of the performed toolpath can be seen in Figure 7b.

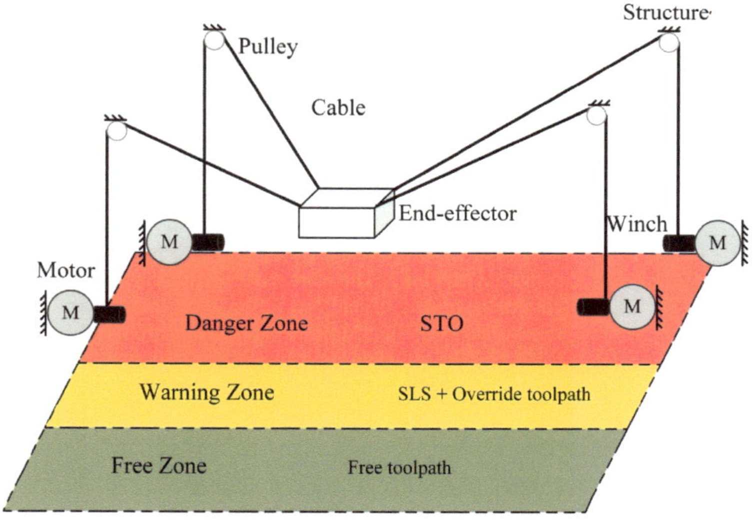

Figure 6. Zones definitions for a collaborative CDPR prototype implementation.

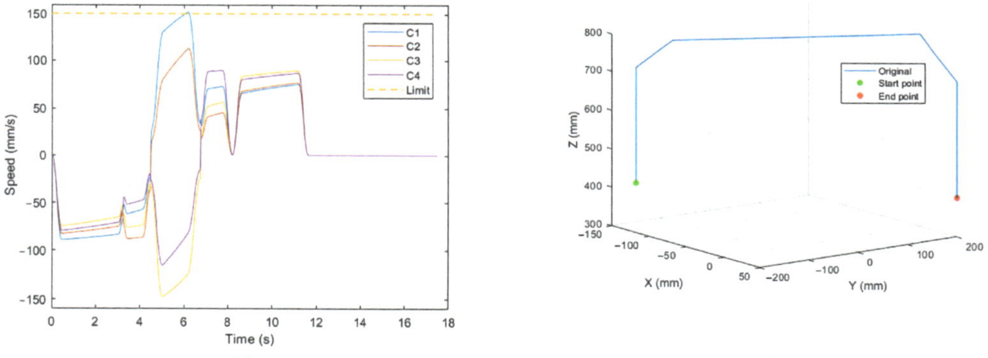

Figure 7. Result of performing the trajectory under normal conditions (with the operator standing in the free zone): (**a**) Time series of the velocity on each axis (cable); (**b**) 3D view of the toolpath. Labels "C1" to "C4" stand for the cables that execute the movement.

4.2. Case 2: Simultaneous Application of SLS and Toolpath Override

This second case shows the result of performing the same toolpath but considering an intrusion in the warning zone (represented by a vertical red line in Figure 8). In response to this intrusion, the safety subsystem triggered an SLS request to the individual servo drives while informing the conventional motion controller to apply a speed and separation monitoring strategy. Consequently, the motion controller modified the speed override of the group of axes. By applying this discrete, pre-calculated override, the individual axes did not exceed the speed limitation and, therefore, they did not perform a safe stop (Figure 8b). The transition to the new value was made by applying a transition ramp (provided by the device manufacturers), thus reducing the possibility of abrupt behavior.

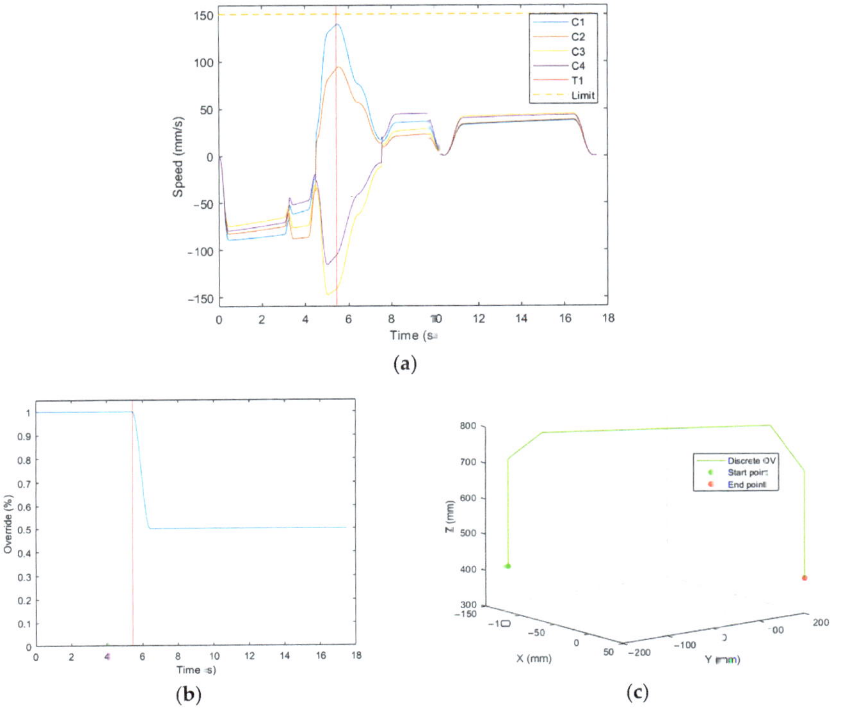

Figure 8. Result of performing the trajectory when a person entered the warning zone, triggering the activation of the SLS safety function and a speed override of the axes group. The vertical red line ("T1") indicates when an operator entered the warning zone. (a) Time evolution of the speed on each axis (cable); (b) Time series of the discrete speed override (level 50%); (c) 3D view of the toolpath Labels "C1" to "C4" stand for the cables that execute the movement.

The maximum safe speed threshold requested from the servo drives and the minimum protective separation distance that establishes their area of application must be pre-calculated according to ISO/TS 15066, and the risk assessment for the specific application [51,61]. Finally, considering the maximum velocity threshold value, the toolpath velocity override was selected so that none of the axes exceeded its individual velocity limit.

However, this strategy does not prevent axes from temporarily surpassing this threshold. It takes some time for the controller to reach the final reduction value, so in the transition, the intermediate levels of the override may not be reached to prevent any axis from exceeding its safe speed limit. This will lead to a safety stop of the axis or axes that have exceeded the limit (commanded by the corresponding safe servo drives), and the consequent stop of the movement of the rest of the axes due to a "path following error" (commanded by the motion controller).

Figure 9 depicts this problem. An intrusion in the warning zone triggered the axes' speed limit (indicated by the red line in Figure 9a) and activated an override on the group of axes (shown in Figure 9b). However, the override failed to reach the target value (50%) in time to prevent the axes from exceeding their maximum safe speed (marked by the blue line in Figure 9a). This caused a safety stop to be produced in the axes that exceeded their limit (in this case, only the axis controlling cable 1), with the consequent stop of the coordinated movement (second 6.07 in Figure 9) and not completing the toolpath (Figure 9c).

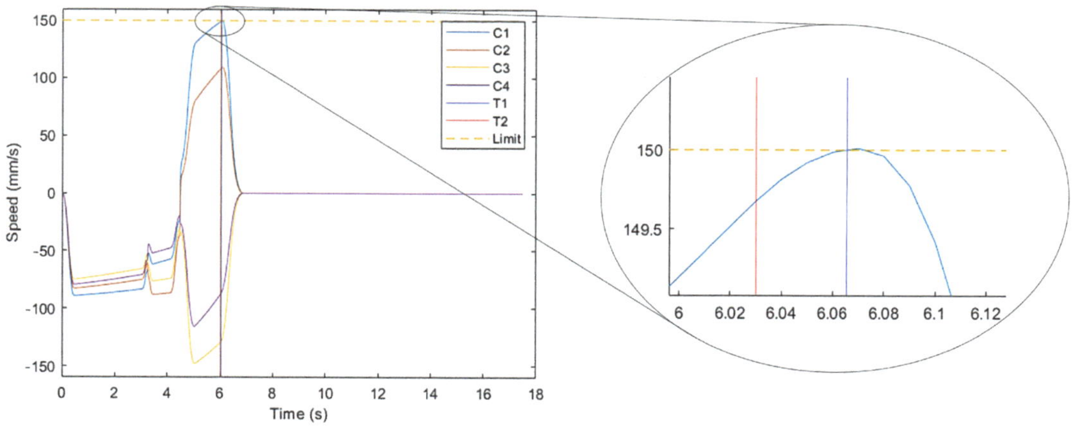

(**a**)

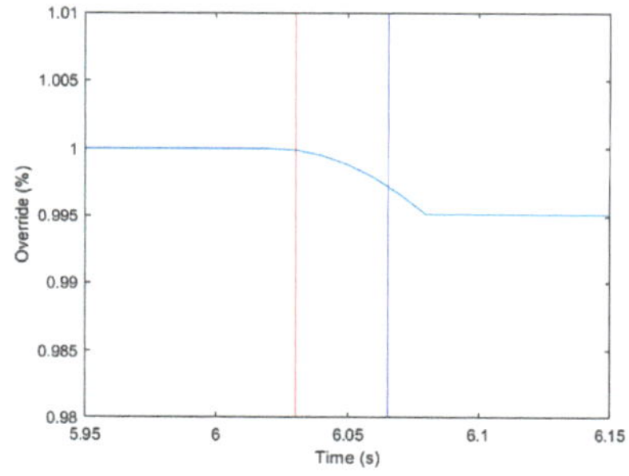

(**b**)

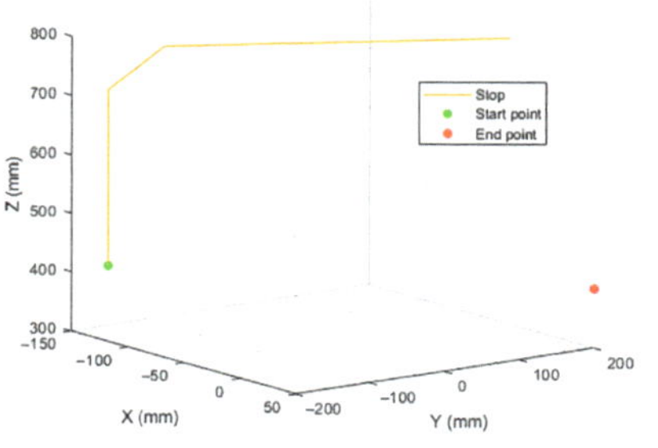

(**c**)

Figure 9. Result of performing the trajectory when a person entered the warning zone, triggering the activation of the SLS safety function and a speed override of the axes group, but one axis exceeded the safe speed limit value. The vertical blue line ("T1") indicates when an operator entered the warning zone and the red one ("T2") represents the time when cable 1 exceeded the safe speed limit: (**a**) Time evolution of the speed on each axis (cable); (**b**) Time series of the discrete speed override; (**c**) 3D view of the toolpath. Labels "C1" to "C4" stand for the cables that execute the movement.

4.3. Case 3: New Pre-Warning Zone for Smooth Application of Toolpath Speed Override

A strategy for progressive override modulation, as the intrusion approaches the warning zone without activating safety limits on the axes, is introduced to avoid the problem presented in the second case. Thus, when the intrusion finally reaches the warning zone, the speed override is already set to its calculated target value. In the transition, the override value depends on the separation between the intrusion and the current robot end-effector position. This separation is the result of subtracting the position of the intrusion, detected by the safety sensor "PS" from the robot's end effector position, determined by forward kinematic calculations, as depicted in Figure 5. Therefore, a new pre-warning zone is proposed between the free zone and the warning zone, as shown in Figure 10.

Finally, the experiment considering the new intermediate Pre-Warning Zone is reproduced below. Figure 11a captures when a human entered this new zone (red line) and transitioned into the Warning Zone (blue line). Figure 11b illustrates how the override transitioned from the Free Zone, where no limitations are required, to the predefined value for the Warning Zone.

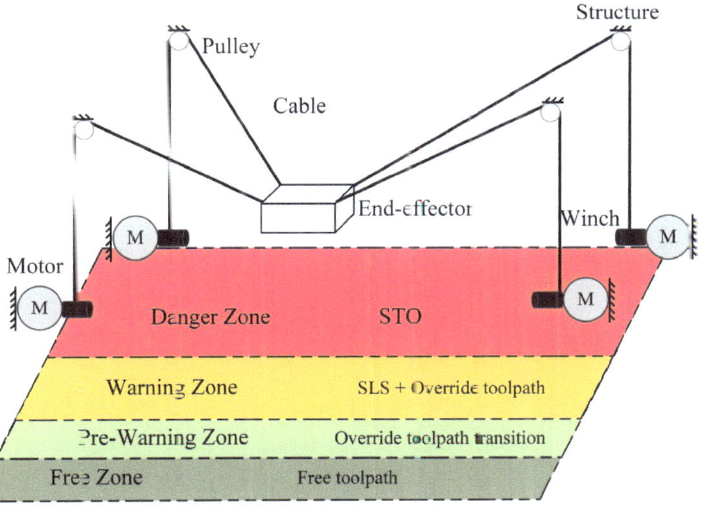

Figure 10. Proposed zones for a collaborative CDPR prototype implementation.

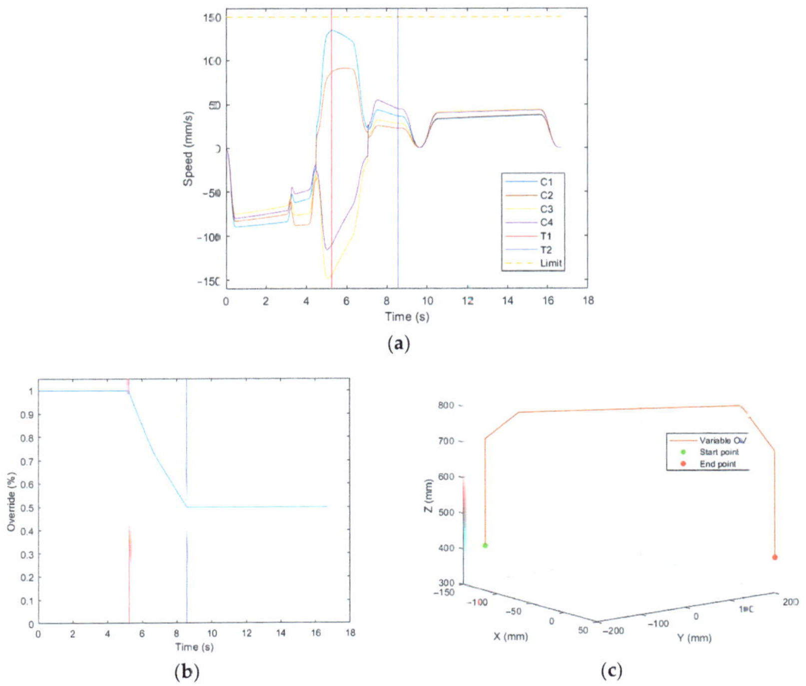

Figure 11. Result of performing the trajectory when a moving person entered the pre-warning zone and subsequently the warning zone, causing the activation of the SLS safety function. The red vertical line ("T1") indicates when this person entered the pre-warning zone and the blue vertical line ("T2") indicates entering the warning zone: (**a**) Time evolution of the speed on each axis (cable); (**b**) Time series of the progressive speed override; (**c**) 3D view of the trajectory. Labels "C1" to "C4" stand for the cables that execute the movement.

The override transition was calculated according with Equation 1 by scaling between the maximum and minimum override value, and the maximum and minimum distance of the new zone, depending on the distance between the robot and the person. In this research, the distance boundary between the Free Zone and Pre Warning Zone (d_{max}) was 3000 mm, the distance boundary between the Pre Warning Zone and the Warning Zone (d_{min}) was 2000 mm, and the maximum (ov_{max}) and minimum (ov_{min}) overrides were 100% and 50%, respectively. The distance (Equation 2) was obtained by performing the 2D-Euclidean distance between the robot's trajectory ("PT") and the diagonal trajectory (worst case) of a person at a speed of 1.6 m/s ("PS"), as stated in ISO/TS 15066 [47]). The location of the intrusion was facilitated by a non-safety laser, as stated in Figure 5. As a result, it has been possible to implement toolpath dynamic speed reductions, thus avoiding stops in the toolpath execution, as shown in Figure 11c.

$$OV = \begin{cases} ov_{min}, & D \leq d_{min} \\ ov_{max}, & D \geq d_{max} \\ \left(\frac{ov_{max} - ov_{min}}{d_{max} - d_{min}} \right) (D - d_{min}) + ov_{min}, & d_{min} < D < d_{max} \end{cases} \quad (1)$$

$$D = \sqrt{(PS_x - PT_x)^2 + (PS_y - PT_y)^2} \quad (2)$$

5. Discussion and Conclusions

The challenge of designing a new custom-made non-conventional machine or robot involving toolpath control (interpolated axes) with collaborative functionalities (SFS) has been faced through a strategy based on the use of general-purpose standard technology for safety and motion control while addressing industrial safety and collaborative robotics regulations.

Safety is guaranteed by the safety subsystem (safe intrusion detection, safe logic in the safety controller, and safe action by certified PDS(SR) devices), which limits the axes' speed in the event of human intrusion. The toolpath speed override mechanism, commanded by the motion controller, and the definition of a new pre-warning zone, prevent unnecessary stops due to the approach of the human operator. The final practical architectures and strategies may vary as they accommodate the technological differences between manufacturers regarding the software functionalities they provide (both safety and motion control functionalities).

The integration of electronic and programmed safety systems has come to increase the flexibility, productivity, and maintainability of machines and processes. Furthermore, even maintaining the strict separation of responsibilities between the safety-related subsystem and standard-process subsystem resulting from the application of international machinery safety regulations when designing a new machine, the increasing integration of both subsystems is a need and a design trend in machines and robot automation.

Developing and applying strategies to provide collaborative capabilities to robots aims not only to increase safety but also to address an industry need. Collaboration adds versatility, allowing work in shared spaces and the creation of more seamless operations. However, this poses significant safety challenges. Effective collaboration requires robots to be designed with a new and more ambitious safety approach that goes beyond traditional physical human-robot separation structures. Innovative safety strategies contribute to the adoption of collaborative functionalities in new workspace configurations.

The article has proposed a collaboration strategy when safety is applied individually to each servo axis despite being a machine that executes interpolated trajectories, which means working with the axes not individually, but in a coordinated manner. This strategy is based on a standard implementation of safety rules through servo drives certified to receive and execute SMMF safety commands. These drives maintain the robot's operational status and motion while implementing both the SMMF functions and a speed reduction via an override to the axes group. This override is progressively scaled considering the distance between the detected intrusion and the end-effector of the robot.

The decision about when triggering individual axis safety limitations depends on the intrusion distance provided by a safe sensor. Simultaneously, the motion controller decides when to modulate the toolpath override based on real-time data. Therefore, it is important to determine the position of the robot's end-effector to modify the toolpath speed override according to the exact distance between the robot and the intrusion. The motion controller performs this adjustment by calculating the forward kinematic transformation in real time.

The article is not about defining new strategies for implementing standard collaborative robots, but rather, given the specific need for developing new robot architectures (as is the case of a CDPR), how to implement collaborative functionalities comparable to those in conventional collaborative robots. The strategy covers one of the collaborative functionalities defined in the standard for collaborative robots, speed and separation monitoring, which is not exempt from complying with the rest of the design safety-related requirements following collaborative robots in general and safe machines in particular.

The proposed strategy can be applied to any general-purpose machine involving axes where enhanced human interaction is desired while maintaining safety. This expands its use to industrial settings and other scenarios involving human interaction, such as ludic or medical applications. Regardless of the application, avoiding physical barriers allows for a better adaptation to various use scenarios. For example, a CDPR used for rehabilitation tasks could interact not only with the patient but also with people nearby, such as therapists.

Despite the advantages of the structurally simple CDPRs, no commercial solutions are available from the main robot manufacturers (ABB, Fanuc...). The responsibility for their configuration (design, assembly, and commissioning) falls to the developers of each specific application. These tasks are complex and far from the end-user orientation in the set-up and programming environments available for other industrial robots. Open contributions from academia and research projects will make these complex tasks progressively easier to address, supporting greater adoption of CDPRs in diverse scenarios.

Author Contributions: Conceptualization, J.G. and J.S.; methodology, J.G. and D.S.-M.; software, J.G., D.S.-M., E.R. and J.R.-A.; validation, J.G. and J.S.; formal analysis, J.G. and D.S.-M.; investigation, J.G. and D.S.-M.; resources, J.G.; data curation, D.S.-M., E.R. and J.R.-A.; writing—original draft preparation, J.G., D.S.-M. and J.S.; writing—review and editing, J.G., D.S.-M., E.R., J.R.-A. and J.S.; visualization, E.R. and J.R.-A.; supervision, J.G. and J.S.; project administration, J.G. All authors have read and agreed to the published version of the manuscript.

Funding: This research received no external funding.

Data Availability Statement: Data are contained within the article.

Acknowledgments: Diego Silva-Muñiz thanks Universidade de Vigo for his predoctoral grant (00VI 131H 6410211-2023).

Conflicts of Interest: The authors declare no conflicts of interest.

References

1. Unnikrishnan, N.; Hull, K.; Nicolson, E. A Review of Challenges in Integrating Robot and Motion Control Into a Single System. In *Proceedings of the Volume 14: Emerging Technologies; Materials: Genetics to Structures, Tampa, FL, USA, 3–9 November 2017*; Safety Engineering and Risk Analysis; American Society of Mechanical Engineers: Tampa, FL, USA, 2017; p. V014T07A017.
2. Wang, H.; Tang, X.; Song, B.; Wang, X. A Novel Architecture of the Embedded Computer Numerical Control System Based on PLCopen Standard. *Proc. Inst. Mech. Eng. Part B J. Eng. Manuf.* **2014**, *228*, 595–605. [CrossRef]
3. Van Der Wal, E.; Simon, R. Recent Developments in Industrial Control Programming. *IFAC Proc. Vol.* **2006**, *39*, 90–94. [CrossRef]
4. Quest TechnoMarketing. *The Future of the Servo Use until 2020 in the German Machinery Industry*; Quest TechnoMarketing: London, UK, 2017; p. 198.
5. Santaliana, D.; Calloni, D.; Laterza, V.; Zanelli, R. *Final Report about Identification of Skills and Needs in the Mechatronics and Metallurgical Sectors' Industries in the 5 Countries*; MEMEVET project reports; Erasmus+ Programme of the European Union: Pordenone, Italy, 2019.

6. PLCopen. *Function Blocks for Motion Control: Part 4—Coordinated Motion*; PLCopen Technical Committee 2—Task Force: Zaltbommel, The Netherlands, 2008.
7. Contreras, J.; Rubio, J.; Martínez, A. PLC Based Control of Robots Using PLCopen Motion Control Specifications. In *Advances in Automation and Robotics Research*; Moreno, H.A., Carrera, I.G., Ramírez-Mendoza, R.A., Baca, J., Banfield, I.A., Eds.; Lecture Notes in Networks and Systems; Springer International Publishing: Cham, Switzerland, 2022; Volume 347, pp. 109–120, ISBN 978-3-030-90032-8.
8. Pott, A.; Mütherich, H.; Kraus, W.; Schmidt, V.; Miermeister, P.; Verl, A. IPAnema: A family of Cable-Driven Parallel Robots for Industrial Applications. In *Cable-Driven Parallel Robots*; Bruckmann, T., Pott, A., Eds.; Mechanisms and Machine Science; Springer Berlin Heidelberg: Berlin/Heidelberg, Germany, 2013; Volume 12, pp. 119–134, ISBN 978-3-642-31987-7.
9. Sancak, K.V.; Bayraktaroglu, Z.Y. Nonlinear Computed Torque Control of 6-Dof Parallel Manipulators. *Int. J. Control Autom. Syst.* **2022**, *20*, 2297–2311. [CrossRef]
10. PLCopen. *PLCopen Safety Software. Part 1: Concepts and Function Blocks*; Technical Specification; PLCopen: Gorinchem, The Netherlands, 2023.
11. *ISO 12100:2010*; Safety of Machinery. General Principles for Design. Risk Assessment and Risk Reduction. International Organization for Standardization: Geneva, Switzerland, 2010.
12. Platbrood, F.; Gornemann, O. *Safe Robotics—Safety in Collaborative Robot Systems*; SICK AG: Waldkirch, Germany, 2018.
13. *ISO 13849-1:2023*; Safety of Machinery. Safety-Related Parts of Control Systems. Part 1: General Principles for Design. International Organization for Standardization: Geneva, Switzerland, 2023.
14. *IEC 62061:2021*; Safety of Machinery—Functional Safety of Safety-Related Control Systems. International Electrotechnical Commission: Geneva, Switzerland, 2021.
15. Breque, M.; De Nul, L.; Petridis, A. Others Industry 5.0: Towards a Sustainable, Human-Centric and Resilient European Industry. *Luxemb. LU Eur. Comm. Dir.-Gen. Res. Innov.* **2021**. [CrossRef]
16. Aheleroff, S.; Huang, H.; Xu, X.; Zhong, R.Y. Toward Sustainability and Resilience with Industry 4.0 and Industry 5.0. *Front. Manuf. Technol.* **2022**, *2*, 951643. [CrossRef]
17. Yeo, S.H.; Yang, G.; Lim, W.B. esign and Analysis of Cable-Driven Manipulators with Variable Stiffness. *Mech. Mach. Theory* **2013**, *69*, 230–244. [CrossRef]
18. Caro, S.; Merlet, J.-P. Failure Analysis of a Collaborative 4-1 Cable-Driven Parallel Robot. In *New Trends in Mechanism and Machine Science*; Pisla, D., Corves, B., Vaida, C., Eds.; Mechanisms and Machine Science; Springer International Publishing: Cham, Switzerland, 2020; Volume 89, pp. 440–447, ISBN 978-3-030-55060-8.
19. Cui, Z.; Tang, X.; Hou, S.; Sun, H. Research on Controllable Stiffness of Redundant Cable-Driven Parallel Robots. *IEEE/ASME Trans. Mechatron.* **2018**, *23*, 2390–2401. [CrossRef]
20. Métillon, M.; Charron, C.; Subrin, K.; Caro, S. Stiffness and Transparency of a Collaborative Cable-Driven Parallel Robot. In *Advances in Robot Kinematics 2022*; Altuzarra, O., Kecskeméthy, A., Eds.; Springer Proceedings in Advanced Robotics; Springer International Publishing: Cham, Switzerland, 2022; Volume 24, pp. 101–109, ISBN 978-3-031-08139-2.
21. Meziane, R.; Cardou, P.; Otis, M.J.-D. Cable Interference Control in Physical Interaction for Cable-Driven Parallel Mechanisms. *Mech. Mach. Theory* **2019**, *132*, 30–47. [CrossRef]
22. Rousseau, T.; Chevallereau, C.; Caro, S. Human-Cable Collision Detection with a Cable-Driven Parallel Robot. *Mechatronics* **2022**, *86*, 102850. [CrossRef]
23. Métillon, M.; Charron, C.; Subrin, K.; Caro, S. Performance and Interaction Quality Variations of a Collaborative Cable-Driven Parallel Robot. *Mechatronics* **2022**, *86*, 102839. [CrossRef]
24. Clavel, R. Device for Displacing and Positioning an Element in Space. *Int. Pat. No WO* **1985**, *87*, 03528.
25. Stewart, D. A Platform with Six Degrees of Freedom. *Proc. Inst. Mech. Eng.* **1965**, *180*, 371–386. [CrossRef]
26. Cappel, K.L. Motion Simulator. United States Patent US 3,295,224, 7 December 1964.
27. Landsberger, S.E. Design and Construction of a Cable-Controlled, Parallel Link Manipulator. Ph.D. Thesis, Massachusetts Institute of Technology, Cambridge, MA, USA, 1984.
28. González-Rodríguez, A.; Martín-Parra, A.; Juárez-Pérez, S.; Rodríguez-Rosa, D.; Moya-Fernández, F.; Castillo-García, F.J.; Rosado-Linares, J. Dynamic Model of a Novel Planar Cable Driven Parallel Robot with a Single Cable Loop. *Actuators* **2023**, *12*, 200. [CrossRef]
29. Zarebidoki, M.; Dhupia, J.S.; Xu, W. A Review of Cable-Driven Parallel Robots: Typical Configurations, Analysis Techniques, and Control Methods. *IEEE Robot. Autom. Mag.* **2022**, *29*, 89–106. [CrossRef]
30. Barbazza, L.; Oscari, F.; Minto, S.; Rosati, G. Trajectory Planning of a Suspended Cable Driven Parallel Robot with Reconfigurable End Effector. *Robot. Comput.-Integr. Manuf.* **2017**, *48*, 1–11. [CrossRef]
31. Qian, S.; Zi, B.; Shang, W.-W.; Xu, Q.-S. A Review on Cable-driven Parallel Robots. *Chin. J. Mech. Eng.* **2018**, *31*, 66. [CrossRef]
32. Zarebidoki, M. The Effects of Air Drag Force on the Efficiency and Control of lightweight higher Speed Robotics. In Proceedings of the 2021 27th International Conference on Mechatronics and Machine Vision in Practice (M2VIP), Shanghai, China, 26–28 November 2021; IEEE: Shanghai, China, 2021; pp. 400–404.
33. Han, G.; Li, J.; Chen, Y.; Wang, S.; Chen, H. Dynamic Modeling and Motion Control Strategy of Cable-Driven Cleaning Robot for Ship Cargo Hold. *J. Mar. Sci. Eng.* **2023**, *11*, 87. [CrossRef]

34. Tho, T.P.; Thinh, N.T. Using a Cable-Driven Parallel Robot with Applications in 3D Concrete Printing. *Appl. Sci.* **2021**, *11*, 563. [CrossRef]
35. Khajepour, A.; Mendez, S.T.; Rushton, M.; Jamshidianfar, H.; Qi, R.; Pazooki, A.; Durali, L.; Soltani, A. A Warehousing Robot: From Concept to Reality. In *Cable-Driven Parallel Robots*; Caro, S., Pott, A., Bruckmann, T., Eds.; Mechanisms and Machine Science; Springer Nature Switzerland: Cham, Switzerland, 2023; Volume 132, pp. 397–406, ISBN 978-3-031-32321-8.
36. Duan, J.; Shao, Z.; Liu, H.; Zhang, Z.; Wang, Y.; Zhao, H. Design Analysis of a Cable-Driven Parallel Robot with Parallel Cables for Ship Side Painting. In Proceedings of the 2023 8th International Conference on Automation, Control and Robotics Engineering (CACRE), Hong Kong, China, 13–15 July 2023; IEEE: Hong Kong, China, 2023; pp. 209–215.
37. Phuoc Tho, T.; Truong Thinh, N. Evaluating Cable Tension Distributions of CDPR for Virtual Reality Motion Simulator. *Mech. Based Des. Struct. Mach.* **2023**, 1–19. [CrossRef]
38. Ben Hamida, I.; Laribi, M.A.; Mlika, A.; Romdhane, L.; Zeghloul, S.; Carbone, G. Multi-Objective Optimal Design of a Cable Driven Parallel Robot for Rehabilitation Tasks. *Mech. Mach. Theory* **2021**, *156*, 104141. [CrossRef]
39. Holland, C.S.; Cannon, D.J. Cable Array Robot for Material Handling. United States Patent US 6,826,452, 30 November 2004.
40. Zhang, Z.; Shao, Z.; Wang, L. ptimization and Implementation of a High-Speed 3-DOFs Translational Cable-Driven Parallel Robot. *Mech. Mach. Theory* **2020**, *145*, 103693. [CrossRef]
41. Matthias, B.; Kock, S.; Jerregard, H.; Kallman, M.; Lundberg, I. Safety of collaborative industrial robots: Certification possibilities for a collaborative assembly robot concept. In Proceedings of the 2011 IEEE International Symposium on Assembly and Manufacturing (ISAM), Tampere, Finland, 25–27 May 2011; IEEE: Tampere, Finland, 2011; pp. 1–6.
42. Povse, B.; Koritnik, D.; Kamnik, R.; Bajd, T.; Munih, M. Industrial robot and human operator collision. In Proceedings of the 2010 IEEE International Conference on Systems, Man and Cybernetics, Istanbul, Turkey, 10–13 October 2010; IEEE: Istanbul, Turkey, 2010; pp. 2663–2668.
43. El Zaatari, S.; Marei, M.; Li, W.; Usman, Z. Cobot Programming for Collaborative Industrial Tasks: An Overview. *Robot. Auton. Syst.* **2019**, *116*, 162–180. [CrossRef]
44. Knudsen, M.; KaiVo-Oja, J. Collaborative Robots: Frontiers of Current Literature. *J. Intell. Syst. Theory Appl.* **2020**, *3*, 13–20. [CrossRef]
45. *ISO 10218-1:2011*; Robots and Robotic Devices—Safety Requirements for Industrial Robots—Part 1: Robots. International Organization for Standardization: Geneva, Switzerland, 2011.
46. *ISO 10218-2:2011*; Robots and Robotic Devices—Safety Requirements for Industrial Robots—Part 2: Robot Systems and Integration. International Organization for Standardization: Geneva, Switzerland, 2011.
47. *ISO/TS 15066:2016*; Robots and Robotic Devices–Collaborative Robots. International Organization for Standardization: Geneva, Switzerland, 2016.
48. Lucci, N.; Lacevic, B.; Zanchettin, A.M.; Rocco, P. Combining Speed and Separation Monitoring with Power and Force Limiting for Safe Collaborative Robotics Applications. *IEEE Robot. Autom. Lett.* **2020**, *5*, 6121–6128. [CrossRef]
49. Vysocky, A.; Novak, P. Human—Robot Collaboration in Industry. *MM Sci. J.* **2016**, *2016*, 903–906. [CrossRef]
50. Mariscal Saldaña, M.Á.; González Pérez, J.; Khalid, A.; Gutiérrez Llorente, J.M.; García Herrero, S. Risks management and cobots. Identifying critical variables. In Proceedings of the 29th European Safety and Reliability Conference (ESREL), Hannover, Germany, 22–26 September 2019. [CrossRef]
51. Villani, V.; Pini, F.; Leali, F.; Secchi, C. Survey on Human–Robot Collaboration in Industrial Settings: Safety, Intuitive Interfaces and Applications. *Mechatronics* **2018**, *55*, 248–266. [CrossRef]
52. Bertelsen, Á.; Scorza, D.; Cortés, C.; Oñativia, J.; Escudero, Á.; Sánchez, E.; Presa, J. Collaborative Robots for Surgical Applications. In *ROBOT 2017: Third Iberian Robotics Conference*; Ollero, A., Sanfeliu, A., Montano, L., Lau, N., Cardeira, C., Eds.; Advances in Intelligent Systems and Computing. Springer International Publishing: Cham, Switzerland, 2018; Volume 694, pp. 524–535, ISBN 978-3-319-70835-5.
53. Gherman, B.; Burz, A.; Jucan, D.; Bara, F.; Carbone, G.; Pisla, D. Upper Limb Rehabilitation with A Collaborative Robot. *ACTA Tech. Napoc.—Ser. Appl. Math. Mech. Eng.* **2019**, *62*, 323–330.
54. Matthias, B.; Reisinger, T. Example Application of ISO/TS 15066 to a Collaborative Assembly Scenario. In Proceedings of the ISR 2016: 47st International Symposium on Robotics, Munich, Germany, 21–22 June 2016; pp. 1–5.
55. Magrini, E.; Ferraguti, F.; Ronga, A.J.; Pini, F.; De Luca, A.; Leali, F. Human-Robot Coexistence and Interaction in Open Industrial Cells. *Robot. Comput.-Integr. Manuf.* **2020**, *61*, 101846. [CrossRef]
56. Svarny, P.; Rozlivek, J.; Rustler, L.; Hoffmann, M. 3D Collision-Force-Map for Safe Human-Robot Collaboration. In Proceedings of the 2021 IEEE International Conference on Robotics and Automation (ICRA), Xi'an, China, 30 May–5 June 2021; IEEE: Xi'an, China, 2021; pp. 3829–3835.
57. Marvel, J.A. Performance Metrics of Speed and Separation Monitoring in Shared Workspaces. *IEEE Trans. Autom. Sci. Eng.* **2013**, *10*, 405–414. [CrossRef]
58. Lasota, P.A.; Rossano, G.F.; Shah, J.A. Toward safe close-proximity human-robot interaction with standard industrial robots. In Proceedings of the 2014 IEEE International Conference on Automation Science and Engineering (CASE), New Taipei, Taiwan, 18–22 August 2014; IEEE: Taipei, Taiwan, 2014; pp. 339–344.
59. Zanchettin, A.M.; Ceriani, N.M.; Rocco, P.; Ding, H.; Matthias, B. Safety in Human-Robot Collaborative Manufacturing Environments: Metrics and Control. *IEEE Trans. Autom. Sci. Eng.* **2016**, *13*, 882–893. [CrossRef]

60. Byner, C.; Matthias, B.; Ding, H. Dynamic Speed and Separation Monitoring for Collaborative Robot Applications – Concepts and Performance. *Robot. Comput.-Integr. Manuf.* **2019**, *58*, 239–252. [CrossRef]
61. Himmelsbach, U.B.; Wendt, T.M.; Hangst, N.; Gawron, P.; Stiglmeier, L. Human–Machine Differentiation in Speed and Separation Monitoring for Improved Efficiency in Human–Robot Collaboration. *Sensors* **2021**, *21*, 7144. [CrossRef]
62. Costanzo, M.; De Maria, G.; Lettera, G.; Natale, C. A Multimodal Approach to Human Safety in Collaborative Robotic Workcells. *IEEE Trans. Autom. Sci. Eng.* **2022**, *19*, 1202–1216. [CrossRef]
63. Karagiannis, P.; Kousi, N.; Michalos, G.; Dimoulas, K.; Mparis, K.; Dimosthenopoulos, D.; Tokçalar, Ö.; Guasch, T.; Gerio, G.P.; Makris, S. Adaptive Speed and Separation Monitoring Based on Switching of Safety Zones for Effective Human Robot Collaboration. *Robot. Comput.-Integr. Manuf.* **2022**, *77*, 102361. [CrossRef]
64. *IEC 61800-5-2:2016*; Adjustable Speed Electrical Power Drive Systems—Part 5-2: Safety Requirements—Functional. International Electrotechnical Commission: Geneva, Switzerland, 2016.
65. *IEC 60204-1:2016*; Safety of Machinery—Electrical Equipment of Machines—Part 1: General Requirements. International Electrotechnical Commission: Geneva, Switzerland, 2021.
66. OMRON Corporation. *NJ Robotics CPU Unit*; NJ_series; OMRON Corporation: Kyoto, Japan, 2022.
67. Siemens AG. *S7-1500T Kinematics functions V6.0 as of STEP 7 V17*; S7-1500T Motion Control; Siemens: Nürnberg, Germany, 2021.
68. Beckhoff Automation GmbH & Co. KG. *TwinCAT 3 | Kinematic Transformation. TF5110–TF5113*; Beckhoff Automation GmbH & Co. KG: Verl, Germany, 2022; pp. 36–42.
69. Beckhoff Automation GmbH & Co. KG. TwinCAT | Automation Software. Available online: https://www.beckhoff.com/es-es/products/automation/twincat/ (accessed on 22 September 2023).

machines

MDPI

Article

Deep Reinforcement Learning-Based Torque Vectoring Control Considering Economy and Safety

Huifan Deng, Youqun Zhao *, Fen Lin and Qiuwei Wang

College of Energy and Power Engineering, Nanjing University of Aeronautics and Astronautics, Nanjing 210016, China; dhf@nuaa.edu.cn (H.D.); flin@nuaa.edu.cn (F.L.); richardwang@nuaa.edu.cn (Q.W.)
* Correspondence: yqzhao@nuaa.edu.cn

Abstract: This paper presents a novel torque vectoring control (TVC) method for four in-wheel-motor independent-drive electric vehicles that considers both energy-saving and safety performance using deep reinforcement learning (RL). Firstly, the tire model is identified using the Fibonacci tree optimization algorithm, and a hierarchical torque vectoring control scheme is designed based on a nonlinear seven-degree-of-freedom vehicle model. This control structure comprises an active safety control layer and a torque allocation layer based on RL. The active safety control layer provides a torque reference for the torque allocation layer to allocate torque while considering both energy-saving and safety performance. Specifically, a new heuristic random ensembled double Q-learning RL algorithm is proposed to calculate the optimal torque allocation for all driving conditions. Finally, numerical experiments are conducted under different driving conditions to validate the effectiveness of the proposed TVC method. Through comparative studies, we emphasize that the novel TVC method outperforms many existing related control results in improving vehicle safety and energy savings, as well as reducing driver workload.

Keywords: torque vectoring control; coordinated control; electric ground vehicles; reinforcement learning (RL); vehicle motion dynamics

Citation: Deng, H.; Zhao, Y.; Lin, F.; Wang, Q. Deep Reinforcement Learning-Based Torque Vectoring Control Considering Economy and Safety. *Machines* **2023**, *11*, 459. https://doi.org/10.3390/machines11040459

Academic Editor: Zheng Chen

Received: 28 February 2023
Revised: 28 March 2023
Accepted: 3 April 2023
Published: 6 April 2023

1. Introduction

The automotive industry has witnessed significant advancements in electric vehicle (EV) technology over the past decade [1,2]. EVs are gaining popularity due to their environmental friendliness and low operating costs. Conventional EVs use a centralized drive system, which transfers the power from the battery to the wheels through a transmission system. However, this system has several limitations such as limited efficiency and reduced stability and control [3]. The four-in-wheel-motor independent-drive electric vehicle (4MIDEV), which places an electric motor in each wheel of the vehicle, has emerged as a promising solution to overcome these limitations [4].

Torque vectoring control (TVC) refers to the ability to distribute the torque to each wheel of the vehicle independently. The torque allocation can be controlled by adjusting the electric motor output torque to each wheel, which is achieved by using an electronic control unit. Several factors affect TVC, including vehicle speed, acceleration, road conditions, and motor efficiency [5]. Hence, it is important to research the TVC of 4MIDEV considering both economy and safety according to the current vehicle state [6].

The TVC is usually classified into holistic and hierarchical structures according to the control framework. The holistic TVC structure typically employs model predictive control (MPC) [7]. However, this approach suffers from certain drawbacks, including a large system size, substantial computational effort, and implementation challenges [8]. Therefore, in practical applications, TVC generally adopts a hierarchical control structure. The active safety control layer generates reference control quantities based on the reference model, while the torque allocation layer generates control commands for each actuator, such as tire longitudinal force and torque.

Many scholars use sliding mode controllers (SMC) to solve the torque allocation problem. Reference [9] presents an SMC with its stability being proven using a Lyapunov function. The Lyapunov control method proposed in the article overcomes the limitations of SMC by improving control accuracy and dynamic tracking performance while effectively reducing chattering and abrupt changes in the control system. In reference [10], a new SMC law is designed to reduce chattering and make state variables converge faster. Reference [11] presents an adaptive and robust controller that combines the driving characteristics of professional drivers to improve vehicle stability and maneuverability and relieve the driver's workload. Experimental results demonstrate the effectiveness of the proposed controller in reducing the peak steering angle, reducing the driver's operating load, and addressing the chattering issue in conventional SMC. Reference [12] presents a coordination control strategy for the vehicle stability control system and differential drive-assisted steering, which considers tire sideslip and aims to improve vehicle stability in different driving conditions and presents CarSim simulations and vehicle experiments verifying its effectiveness. Moreover, some researchers have focused on developing fault-tolerant control strategies that can ensure the safety and performance of the 4MIDEV even in the presence of actuator faults. Reference [13] proposes a fault-tolerant control method for 4MIDEV that considers both vehicle safety and motor power consumption.

However, traditional TVC methods often require complex and computationally expensive control algorithms, which limit their effectiveness. To overcome these challenges, reinforcement learning (RL) offers a machine learning approach that enables an agent to learn how to make optimal decisions by interacting with an environment and receiving rewards as feedback [14]. RL has shown significant potential for improving control performance while reducing complexity in many areas, such as robot control [15] and autonomous driving [16]. Reference [17] proposed a direct torque allocation algorithm that employs a deep RL technique to enhance the safety and fuel economy of the vehicle. This paper uses the integrated control framework to train the RL directly. However, the vehicle's dynamics model is complex, making it difficult to apply RL directly. Additionally, the RL is usually applied to the problem in which objects are difficult to model, or the model is too complex. For stability control, the active safety control layer already has a detailed dynamic model; the required additional yaw moment and longitudinal force can be well solved using model-based control methods such as optimal control or sliding mode control. Therefore, it is not appropriate to use direct RL control in the active safety control layer. Motivated by the above issues, a novel TVC method based on RL for 4MIDEV is proposed in this paper. The contributions of this study are summarized as follows:

- Unlike the reference [18], this paper proposes a TVC method that takes into account both economy and safety. Specifically, the torque allocation layer based on deep RL adaptively adjusts the torque of each wheel according to the current vehicle state.
- An improved heuristic randomized ensembled double Q-learning (REDQ) algorithm is introduced for EV control, which reduces the training complexity of RL compared to existing RL algorithms for direct motor torque control.
- The Fibonacci tree-based tire model identification method is employed, which achieves higher identification accuracy than the genetic algorithm (GA) [19] and the particle swarm optimization (PSO) algorithm [20].

2. The TVC Framework and System Model

2.1. The TVC Framework

Figure 1 illustrates the block diagrams of the TVC system, which is composed of four modules: tire model identification, the vehicle reference model, the active safety control layer, and the RL-based torque allocation layer. The tire mathematical model is obtained through experimental data based on the Fibonacci tree optimization (FTO) algorithm. The active safety control layer is designed to generate longitudinal force F_x and additional yaw moment M_z. The torque allocation layer employs the heuristic REDQ algorithm with integrated consideration of both economy and safety to distribute the four-wheel torque.

Moreover, the average allocation method gives the RL algorithm heuristic training and reduces its training time.

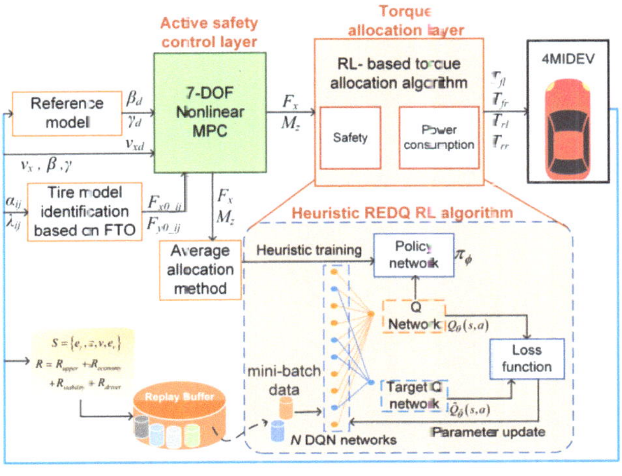

Figure 1. The block diagrams of the TVC system.

2.2. Tire Model Identification

The tire is a complex system that interacts with the road surface and the vehicle's suspension system. The tire's behavior can significantly affect the vehicle's dynamics, including acceleration, braking, cornering, and ride comfort. Therefore, accurate modeling of the tire is essential for vehicle control, design, and optimization. A good tire model can provide reliable predictions of the tire forces and moments, which are critical inputs to vehicle dynamic control algorithms. Additionally, tire models can help understand the effects of different tire designs and operating conditions on vehicle performance and handling.

The magic formula tire model is a semi-empirical tire model that has become one of the most widely used tire models in vehicle dynamics and control research. Additionally, the model includes several parameters that can be tuned to match experimental data or represent different tire designs. According to the magic formula, the longitudinal force F^0_{xij}, and lateral force F^0_{yij} of the tire model are calculated by:

$$
\begin{aligned}
F^0_{xij} &= D_x \sin[C_x \arctan(E_x \phi_{xij})] + S_{vx} \\
\phi_{xij} &= (1 - E_x)(\lambda_{ij}) + (E_x / B_x)\arctan(B_x(\lambda_{ij})) \\
F^0_{yij} &= D_y \sin[C_y \arctan(B_y \phi_{yij})] + S_{vy} \\
\phi_{yij} &= (1 - E_y)(\alpha_{ij}) + (E_y / B_y)\arctan(B_y(\alpha_{ij}))
\end{aligned}
\tag{1}
$$

where F_z is the vertical load of the tire and λ_{ij} and α_{ij} tire slip angle or tire longitudinal slip ratio, respectively. $ij \in \{fl, fr, rl, rr\}$. $B_x = BCD_x/(C_x \cdot D_x)$, $C_x = b_0$, $D_x = b_1 F_z^2 + b_2 F_z$, $BCD_x = (b_3 F_z^2 + b_4 F)e^{-b_5 F_z}$, $E_x = b_6 F_z^2 + b_7 F_z + b_8$, $B_y = BCD_y/(C_y \cdot D_y)$, $C_y = a_0$, $D_y = a_1 F_z^2 + a_2 F_z$, $BCD_y = a_3 \sin(2\arctan(F_z/a_4))$, $E_y = a_6 F_z + a_7$.

The raw data for the FTO tire identification algorithm is obtained through a comprehensive tire mechanics test bench. The powertrain drives the simulated road surface at a translational speed of 0.3 m/s from one side of the test bench to the other. The three-dimensional force sensors of the test bench are used to measure the lateral and longitudinal forces. The friction coefficient is 0.8, the slip ratio is 0–0.7, and the lateral slip angle is 0–20 degrees. The test was conducted at three vertical loads: 10 kN, 15 kN, and 20 kN, and the experimental data are shown in Figure 2. Each point in Figure 2 represents a test result. The FTO algorithm is a computational optimization algorithm based on the golden section method and the Fibonacci optimization principle. It optimizes the global search and local search alternately and iteratively constructs the Fibonacci tree structure to find the best solution. The FTO algorithm consists of two stages for each search: a global search and a local search.

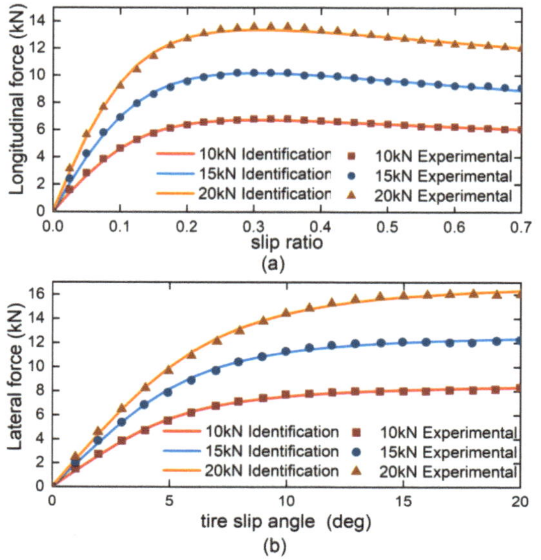

Figure 2. Tire model identification results. (**a**) Longitudinal force recognition results. (**b**) Lateral force recognition results.

In the global search stage, global nodes N_i are randomly generated within the global scope. According to the fitness of each node, global trial nodes $W_a(a = 1, 2, \cdots, F_i)$ are generated as follows:

$$W_a = \begin{cases} N_i + \frac{F_{i-1}}{F_i} \cdot (B_{ij} - N_i), \text{ if } Fit_{best}(N_i, B_{ij}) = N_i \\ B_{ij} + \frac{F_{i-1}}{F_i} \cdot (N_i - B_{ij}), \text{ otherwise} \end{cases} \tag{2}$$

F_i is the Fibonacci series, and the general formula is as follows:

$$F_i = \frac{1}{\sqrt{5}}\left[\left(\frac{1+\sqrt{5}}{2}\right)^i - \left(\frac{1-\sqrt{5}}{2}\right)^i\right] \tag{3}$$

where $Fit_{best}(N_i, B_{ij})$ represents the better adaptability of N_i and B_{ij}, which is calculated according to the objective functions (6) and (7).

In the local search stage, the local trial nodes V_b are generated according to the best node of the current node B_{i1} and the current node B_{ij}:

$$\begin{aligned} B_{i1} &= Fit_{best}(B_{ij}) \\ V_b &= B_{i1} + \frac{F_{i-1}}{F_i}(B_{ij} - B_{i1}) \end{aligned} \tag{4}$$

Finally, W_a, V_b, and the current node B_{ij} are sorted according to fitness, and the best $F_{(i+1)}$ nodes are retained as the next generation nodes $B_{(i+1)j}$:

$$B_{(i+1),j} = Fit_{best}(W_a, V_b, B_{i,j}) \tag{5}$$

The algorithm is expressed as Algorithm 1.

Algorithm 1. FTO algorithm

---

1. Set the depth of Fibonacci tree N and the number of identification parameters n;
2. Randomly generate an initial node B_{11}: randomly generate a global random node N_1;
3. **Repeat:**
4. Generate F_i global trial nodes $W_1 \sim W_{Fi}$ according to the global random node N_i and the node B_{ij};
5. Generate F_{i-1} local trial nodes $V_1 \sim V_{Fi-1}$ according to the best adaptable element B_{i1} in the current node and the remaining nodes;
6. Get the next generation node B_{i+1j};
7. Update the node set. Incorporate the newly generated trial nodes into the current node set S, calculate the fitness function and sort, and retain the first F_{i+1} nodes;
8. **Until** $F_{i+1} \geq F_N$
9. Output the optimal node;

The objective function for first-level parameter identification is as follows:

$$
\begin{aligned}
f1_x &= \sum_{i=1}^{N_x} \{D_x \sin(C_x \text{atan}(B_x\lambda_i - E_x(B_x - \arctan(B_x\lambda_i)))) - F_{xi}^*\}^2 \\
f1_y &= \sum_{i=1}^{N_y} \{D_y \sin(C_y \text{atan}(B_y\alpha_i - E(B_y - \arctan(B_y\alpha_i)))) - F_{yi}^*\}^2
\end{aligned} \tag{6}
$$

where, $F_{xi}{}^*$, $F_{yi}{}^*$, λ_i, and α_i are the experimental data of longitudinal force, lateral force, slip rate, and side slip angle under different vertical loads, respectively. N_x = 29 and N_y = 20 represent the number of tests.

The objective function for second-level parameter identification is as follows:

$$
\begin{array}{ll}
f2_{x1} = \sum\limits_{j=1}^{M} (b_0 - C_{xj}^*)^2 & f2_{y1} = \sum\limits_{j=1}^{M} (a_0 - C_{yj}^*)^2 \\
f2_{x2} = \sum\limits_{j=1}^{M} [(b_1 F_{zj}^2 + b_2 F_{zj}) - D_{xj}^*]^2 & f2_{y2} = \sum\limits_{j=1}^{M} [(a_1 F_{zj}^2 + a_2 F_{zj}) - D_{yj}^*]^2 \\
f2_{x3} = \sum\limits_{j=1}^{M} [(b_3 F_{zj}^2 + b_4 F_{zj}) e^{-b_5 F_{zj}} - BCD_{xj}^*]^2 & f2_{y3} = \sum\limits_{j=1}^{M} [a_3 \sin(2\arctan(F_{zj}/a_4)) - BCD_{yj}^*]^2 \\
f2_{x4} = \sum\limits_{j=1}^{M} [(b_6 F_{zj}^2 + b_7 F_{zj} + b_8) - E_{xj}^*]^2 & f2_{y4} = \sum\limits_{j=1}^{M} (a_6 F_{zj} + a_7) - E_{yj}^*]^2 \\
BCD_{xj}^* = B_{xj}^* \cdot C_{xj}^* \cdot D_{xi}^* & BCD_{yj}^* = B_{yj}^* \cdot C_{yj}^* \cdot D_{yj}^*
\end{array} \tag{7}
$$

where B_{xj}^*, C_{xj}^*, D_{xj}^*, E_{xj}^*, B_{yj}^*, C_{yj}^*, D_{yj}^*, E_{yj}^* are the first-level parameter values identified by the vertical load applied by group j. After classifying the parameters of the tire model into first-level identification parameters and second-level identification parameters, the hierarchical identification approach is found to not only ensure the accuracy of the tire mathematical model but also improve its identification efficiency compared to the non-hierarchical approach.

The recognition result is shown in Figure 2. To further investigate the performance of the FTO algorithm, a comparative analysis with the commonly used GA and PSO optimization algorithms was conducted, and the results are presented in Table 1. The algorithm of GA is from [19], and the algorithm of PSO is from [20]. The depth of the Fibonacci tree is set to 7, while the parameters of the GA are adopted from [19]. The generation gap is set to 0.6, the population size is set to 50, and the crossover probability and mutation probability are set to 0.9 and 0.0097, respectively. Similarly, the parameters of the PSO algorithm are selected from [20], with the population size set to 40 and the inertia weight set to 0.7298, constant $c_1 = 2$, and constant $c_2 = 2$.

Table 1. First-level parameters identification results.

Item	10 kN	15 kN	20 kN
FTO relative residual of longitudinal force	1.40%	1.37%	1.40%
GA relative residual of longitudinal force	4.61%	2.21%	1.73%
PSO relative residual of longitudinal force	3.44%	1.74%	1.66%
FTO relative residual of lateral force	1.48%	1.37%	1.39%
GA relative residual of lateral force	1.78%	1.63%	1.48%
PSO relative residual of lateral force	1.65%	1.69%	1.56%

The relative residual e is expressed as:

$$e = \sqrt{\sum (F^{MF} - F^*)^2 / \sum (F^*)^2} \times 100\% \tag{8}$$

where F^{MF} is the tire force value identified by the FTO algorithm, and F^* is the experimental value of the tire force.

Based on the results presented in Table 1, it is observed that the FTO algorithm outperforms the conventional GA and PSO algorithms in terms of parameter recognition accuracy and relative residual error. The first-level parameter identification has a global relative residual of 1.48%. Moreover, for complex parameter identification problems, it is beneficial to improve the efficiency of parameter identification by classifying the parameters and reducing the number of independent variables of the single identification objective function. After obtaining the identified parameters, we can obtain the mathematical model of the tire, and the longitudinal and lateral forces of the tire can be calculated by substituting different vertical loads.

2.3. Vehicle Reference Model

In vehicle control, achieving both good tracking performance and disturbance rejection is critical for safe and efficient operation. A 2-degree-of-freedom (DOF) reference model is a popular choice for designing controllers that provide these performance objectives. The 2-DOF vehicle model is expressed as:

$$\begin{aligned} mv_x\left(\dot{\beta} + \gamma\right) &= 2C_f\left(\delta_f - \beta - a\gamma/v_x\right) + 2C_r(-\beta + b\gamma/v_x) \\ I_z\dot{\gamma} &= 2aC_f\left(\delta_f - \beta - a/v_x\right) - 2bC_r(-\beta + b/v_x) \end{aligned} \tag{9}$$

where m is the total mass of the vehicle, v_x is the longitudinal velocity, β is the side slip angle of the vehicle center of gravity (CG), γ is the yaw rate, and C_f and C_r are the front and rear tire cornering stiffness, respectively. a and b are the distance from the front and rear wheel axles to the CG, respectively. δ_f is the steering angle of the front wheels, and I_z is the yaw mass moment of inertia.

Applying the Laplace transform to Equation (9) we can obtain the following transfer function:

$$\frac{\gamma(s)}{\delta_f(s)} = G_\gamma \frac{\omega_n^2(\xi_1 s + 1)}{s^2 + 2\omega_n\xi_2 \cdot s + \omega_n^2} \tag{10}$$

where $\omega_n = \frac{2(a+b)}{v_x}\sqrt{\frac{C_f C_r\left(1+Kv_x^2\right)}{mI_z}}$, $G_\gamma = \frac{v_x}{(a+b)\left(1+Kv_x^2\right)}$, $K = -\frac{m\left(aC_f - bC_r\right)}{2C_f C_r (a+b)^2}$, $\xi_1 = \frac{mv_x a}{2C_r(a+b)}$, $\xi_2 = \frac{m\left(C_f a^2 + C_r b^2\right) + I_z\left(C_f + C_r\right)}{2(a+b)\sqrt{mI_z C_f C_r\left(1+Kv_x^2\right)}}$.

The steady-state value of the yaw rate γ^* can be obtained from Equation (10) as:

$$\gamma^* = G_\gamma \delta_f \tag{11}$$

Considering the road adhesion coefficient, γ_d is limited by the following equation:

$$a_y = \gamma_d v_x \le \mu g \tag{12}$$

where μ is the road adhesion coefficient, and g is the acceleration of gravity.

Then, the reference yaw rate is expressed as

$$\gamma_d = \min\left\{ |\gamma^*|, \left|\frac{\mu g}{v_x}\right| \right\} \ \mathrm{sgn}\left(\delta_f\right) \tag{13}$$

We note that vehicle safety is guaranteed when the sideslip angle β varies within a small range for the normal operating condition of the vehicle. Following related works on vehicle dynamics control [21,22], we conservatively set the desired sideslip angle $\beta_d = 0$ to maintain vehicle safety under extreme operating conditions.

2.4. Vehicle 7-DOF Dynamic Model

Vehicle models are used to predict vehicles' behavior under different driving conditions and to design control systems that achieve desired performance objectives. Briefly, 2-DOF vehicle models are widely used due to their simplicity and ease of use, but they have limitations in representing vehicles' motion under complex driving conditions. Additionally, nonlinear 7-DOF models of vehicles provide a more accurate representation of vehicles' motion and are essential for advanced control system design, simulation, and testing [23]. Therefore, a nonlinear 7-DOF vehicle dynamics model is introduced. The vehicle dynamics model, as illustrated in Figure 3, serves as a foundation for the TVC in this research.

$$\begin{aligned} m\dot{v}_x &= F_{xrl} + F_{xrr} + \left(F_{xfl} + F_{xfr}\right)\cos\delta_f - \left(F_{yfl} + F_{yfr}\right)\sin\delta_f + m v_y \gamma \\ m\dot{v}_y &= F_{yrl} + F_{yrr} + \left(F_{xfl} + F_{xfr}\right)\sin\delta_f + \left(F_{yfl} + F_{yfr}\right)\cos\delta_f + m v_x \gamma \\ I_z\dot{\gamma} &= a\left(F_{yfl} + F_{yfr}\right)\cos\delta_f - b\left(F_{yrl} - F_{yrr}\right) + \tfrac{d_w}{2}\left(F_{yfl} - F_{yfr}\right)\sin\delta_f - M_z \end{aligned} \tag{14}$$

$$I_{\omega ij}\dot{\omega}_{ij} = -F_{xij}R_\omega + T_{tij} \tag{15}$$

$$\begin{aligned} M_z = {} & F_{xfl}\left(a\sin\delta_f - d_w\cos\delta_f/2\right) \\ & -F_{xfr}\left(a\sin\delta_f + d_w\cos\delta_f/2\right) + d_w/2(F_{xrr} - F_{xrl}) \end{aligned} \tag{16}$$

where v_y is the vehicle longitudinal velocity. d_w is the track width. ω_{ij} is the wheel rotation rate, I_{ij} is the wheel moment of inertia, and T_{ij} is the output torque of the in-wheel motor.

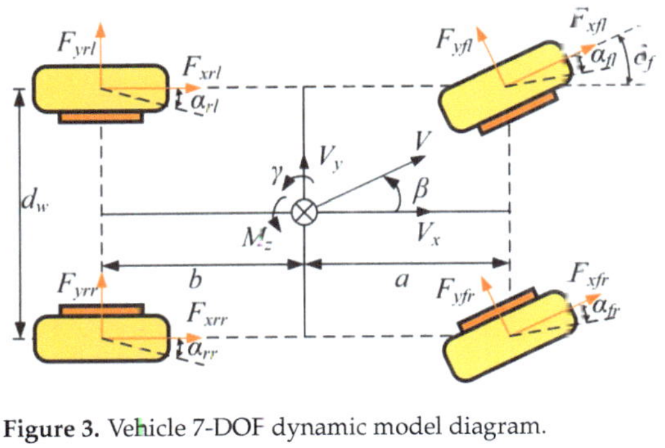

Figure 3. Vehicle 7-DOF dynamic model diagram.

The tire slip angle α_{ij} is expressed as

$$\begin{aligned}\alpha_{fl} &= \arctan\left(\frac{v_y+a\gamma}{v_x-\frac{1}{2}d_w\gamma}\right)-\delta_f,\ \alpha_{rl}=\arctan\left(\frac{v_y-b\gamma}{v_x-\frac{1}{2}d_w\gamma}\right)\\ \alpha_{fr} &= \arctan\left(\frac{v_y+a\gamma}{v_x+\frac{1}{2}d_w\gamma}\right)-\delta_f,\ \alpha_{rr}=\arctan\left(\frac{v_y-b\gamma}{v_x+\frac{1}{2}d_w\gamma}\right)\end{aligned} \tag{17}$$

The tire slip ratio λ_{ij} is calculated by

$$\lambda_{ij} = (v_{ij} - \omega_{\text{ij}} R_\omega)/v_{ij} \tag{18}$$

The expressions of the four-wheel speeds are given as

$$\begin{aligned}v_{fl} &= (v_y + a\gamma)\sin\delta_f + (v_x - \gamma d_w/2)\cos\delta_f,\ v_{rl} = v_x - \gamma d_w/2\\ v_{fr} &= (v_y + a\gamma)\sin\delta_f + (v_x + \gamma d_w/2)\cos\delta_f,\ v_{rr} = v_x + \gamma d_w/2\end{aligned} \tag{19}$$

The tire force F_{zij} is expressed as

$$\begin{aligned}F_{zfl} &= \frac{m}{a+b}\left(\frac{gb}{2} - \frac{a_x h_g}{2} - \frac{a_y h_g b}{d_w}\right)\\ F_{zfr} &= \frac{m}{a+b}\left(\frac{gb}{2} - \frac{a_x h_g}{2} + \frac{a_y h_g b}{d_w}\right)\\ F_{zrl} &= \frac{m}{a+b}\left(\frac{ga}{2} + \frac{a_x h_g}{2} - \frac{a_y h_g a}{d_w}\right)\\ F_{zrr} &= \frac{m}{a+b}\left(\frac{ga}{2} + \frac{a_x h_g}{2} + \frac{a_y h_g a}{d_w}\right)\end{aligned} \tag{20}$$

where h_g is the height of the center of gravity.

As the tire force needs to meet the attachment ellipse, F_{xij} and F_{yij} are expressed as:

$$F_{xij} = F_{x0ij}\psi_{xij}/\psi_{ij} \quad F_{yij} = F_{y0ij}\psi_{xij}/\psi_{ij} \tag{21}$$

where $\psi_{ij} = \sqrt{\psi_{xij}^2 + \psi_{yij}^2}$, $\psi_{xij} = -\frac{\lambda_{ij}}{1+\lambda_{ij}}$, $\psi_{yij} = -\frac{\tan(\alpha_{ij})}{1+\lambda_{ij}}$.$F_{x0ij}$ and F_{y0ij} are calculated by Equation (1).

3. The TVC Algorithm

This section describes the details of the RL-based TVC algorithm, which includes an active safety control layer and an RL-based torque allocation layer.

3.1. Active Safety Control Layer

The role of the active safety control layer is to calculate the total longitudinal force F_x and additional yaw moment M_z necessary to ensure vehicle stability, which will serve as reference values for the torque allocation layer. If the calculated final torque of each wheel by the torque allocation layer can achieve the total longitudinal force and additional yaw moment reference values, the stability of the vehicle can be ensured.

According to Equations (14)–(20), we can obtain the following nonlinear continuous system.

$$\begin{aligned}\dot{x}(t) &= f(x(t), u(t))\\ y(t) &= C \cdot x(t)\end{aligned} \tag{22}$$

where $x(t) = [v_x(t), v_y(t), \gamma(t)]^{\text{T}}$, $y(t) = [v_x(t), \beta(t), \gamma(t)]^{\text{T}}$, $u = [F_x(t), M_z(t)]^{\text{T}}$, $C = \begin{bmatrix}1 & 0 & 0\\ 0 & 1/v_x(t) & 0\\ 0 & 0 & 1\end{bmatrix}$.

Discretizing the Equation (22), we can obtain the difference equation at time k:

$$\begin{aligned} v_x(k+1) &= T_s\Big(\tfrac{1}{m}\Big(F_{xrl}+F_{xrr}+\Big(F_{xfl}+F_{xfr}\Big)\cos\delta_f-\Big(F_{yfl}+F_{yfr}\Big)\sin\delta_f\Big)+v_y(k)\gamma(k)\Big)+v_x(k) \\ v_y(k+1) &= T_s\Big(\tfrac{1}{m}\Big(F_{yrl}+F_{yrr}+\Big(F_{xfl}-F_{xfr}\Big)\sin\delta_f+\Big(F_{yfl}+F_{yfr}\Big)\cos\delta_f\Big)+v_x(k)\gamma(k)\Big)+v_y(k) \\ \gamma(k+1) &= T_s\Big(\tfrac{1}{I_z}\Big(a\Big(F_{yfl}+F_{yfr}\Big)\cos\delta_f-b\Big(F_{yrl}-F_{yrr}\Big)+\tfrac{d_w}{2}\Big(F_{xfl}-F_{xfr}\Big)\sin\delta_f+M_z\Big)\Big)+\gamma(k) \\ y(k) &= C\cdot x(k) \end{aligned} \tag{23}$$

With the active safety control layer defined in Equation (23), we can calculate the M_z and F_x to ensure vehicle safety.

Nonlinear MPC involves the optimization of a nonlinear objective function, which is subject to constraints on the system dynamics and control inputs. The nonlinear MPC controller for the system described in Equation (23) can be designed by formulating an optimal control problem:

$$\min_{X,U}\sum_{k=0}^{Np-1}\left[(Y-Y_{ref})^T W_Q (Y-Y_{ref})+U^T W_R U+\Delta U^T W_S \Delta U\right] \tag{24}$$

$$\text{s.t.}\quad x_0 = x_{in} \tag{25}$$

$$\begin{aligned} x_{k+1} &= f_d(x_k,u_k) \\ y(k) &= C\cdot x(k) \end{aligned} \tag{26}$$

$$x_{\min}\le x_k\le x_{\max} \tag{27}$$

$$u_{\min}\le u_k\le u_{\max} \tag{28}$$

where $Y(k)=\left[y(k),y(k+1),\cdots,y(k+N_p)\right]^{\mathrm{T}}$ and $U(k)=\left[u(k),u(k+1),\cdots,u(k+N_c)\right]^{\mathrm{T}}$ are the N_p step state vector and the N_c step input vector, respectively. $\Delta u = u_k - u_{k-1}$. Additionally, $Y_{ref}(k) = \left[y_{ref}(k), y_{ref}(k+1),\cdots,y_{ref}(k+N_p)\right]^{\mathrm{T}}$ where $y_{ref}(k)=[v_{xd}(k),\beta_d(k),\gamma_d(k)]$ is the reference of the state at step k and f_d is the difference equation in Equation (23). W_Q, W_R, W_S are the weighting matrices.

The goal of optimal control is to minimize the objective function given by Equation (24), subject to constraints such as initial conditions Equation (25), discrete nonlinear system dynamics Equation (26), state constraints Equation (27), and control constraints Equation (28). The objective function Equation (24) comprises three main components, namely, the minimization of state-reference trajectory error, system input, and input variation.

To solve the nonlinear programming problem represented by Equations (23)–(27), the widely used sequential quadratic programming algorithm [24] can be employed. Once the optimal input vector is obtained, the first control input $u(k)$ is then applied to the system, and the process is repeated at the next time step.

3.2. Torque Allocation Layer

Based on Equations (14) and (15), the vehicle longitudinal force F_x and the additional yaw moment M_z are expressed as:

$$F_x = (T_{fl}\cos\delta_f + T_{fr}\cos\delta_f + T_{rl} + T_{rr})/R_\omega \tag{29}$$

$$\begin{aligned} M_z = &\tfrac{1}{R_\omega}((-\tfrac{d_w}{2}\cos\delta_f + a\sin\delta_f)T_{fl} \\ &+(\tfrac{d_w}{2}\cos\delta_f + a\sin\delta_f)T_{fr} - \tfrac{d_w}{2}T_{rl} + \tfrac{d_w}{2}T_{rr}) \end{aligned} \tag{30}$$

To ensure the stability of the vehicle, the four-wheel torque needs to satisfy Equations (29) and (30). The aim of the torque allocation layer is to distribute the four-wheel torque to achieve M_z and F_x. For 4MIDEV, it is a typical over-actuated system with more actuators than degrees of freedom of the system. Therefore, torque allocation is a problem worth researching; it is important to reduce the power consumption of the motor while ensuring safety.

In the torque allocation layer, we considered a combination of safety and power consumption. In particular, the power consumption and safety weights are dynamically adjusted according to the current vehicle state based on RL. To this end, the task of this paper is the torque allocation of the 4MIDEV, which guarantees the safety and power consumption of the EV.

3.2.1. Average Allocation Method

As presented in references [18,25], the most common method of torque allocation is the average allocation method. The average allocation algorithm here serves two purposes. The first is as a base method for comparison. The second is as a heuristic training method for RL in the early stages of training. We have stated this in the revised manuscript. The average allocation methods are shown in the following [26]:

$$\begin{aligned} T_{fl} &= \frac{1}{4}\frac{F_x R_\omega}{\cos\delta_f} + \frac{1}{4}\frac{M_z R_\omega}{-\frac{1}{2}d_w\cos\delta_f + a\sin\delta_f} \\ T_{fr} &= \frac{1}{4}\frac{F_x R_\omega}{\cos\delta_f} + \frac{1}{4}\frac{M_z R_\omega}{\frac{1}{2}d_w\cos\delta_f + a\sin\delta_f} \\ T_{rl} &= \frac{1}{4}F_x R_\omega - \frac{1}{4}\frac{M_z R_\omega}{\frac{1}{2}d_w} \\ T_{rr} &= \frac{1}{4}F_x R_\omega + \frac{1}{4}\frac{M_z R_\omega}{\frac{1}{2}d_w} \end{aligned} \tag{31}$$

3.2.2. RL-Based Torque Allocation Algorithm

This section describes the details of the RL-based torque allocation algorithm. The state plays a crucial role in the RL algorithm, and in this study, we define the state space using four states: the yaw rate of deviation $e_\gamma = \gamma - \gamma_{des}$, the stability indicator ε, velocity v, and velocity deviation e_v. Additionally, the action is defined as four-wheel torque.

$$S = \{e_\gamma, \varepsilon, v, e_v\}, A = \left\{T_{fl}, T_{fr}, T_{rl}, T_{rr}\right\} \tag{32}$$

where the ε is obtained from the phase plane. In vehicle dynamics, the driving stability and instability regions can be depicted using a phase portrait [27]. The ε is defined as follows.

$$\varepsilon = \left|\beta/B_2 + \dot{\beta}B_1/B_2\right| \tag{33}$$

where B_1 and B_2 are the parameters associated with the adhesion coefficient. Their corresponding values are specified in reference [28]. This paper aims to improve the economy of EVs while ensuring safety, so the reward function R is expressed as the following four parts:

Firstly, Equations (29) and (30) ensure that the four-wheel torque satisfies the additional yaw moment M_z and longitudinal force F_x from the active safety control layer. Therefore, to ensure vehicle stability, the torque allocation layer needs to satisfy the following equation:

$$A_s u_c = B_u \tag{34}$$

where $A_s = \frac{1}{R}\begin{bmatrix} \cos\delta_f & \cos\delta_f & 1 & 1 \\ -\frac{l_w}{2}\cos\delta_f + l_f\sin\delta_f & \frac{l_w}{2}\cos\delta_f + l_f\sin\delta_f & -\frac{l_w}{2} & \frac{l_w}{2} \end{bmatrix}$, $B_u = \begin{bmatrix} F_x & M_Z \end{bmatrix}^T$, $u_c = \begin{bmatrix} T_{fl} & T_{fr} & T_{rl} & T_{rr} \end{bmatrix}^T$.

Therefore, we constructed R_{upper} to consider vehicle stability in the torque allocation layer from the perspective of the active safety layer:

$$R_{upper} = -(A_s u_c - B_u)^T W_s (A_s u_c - B_u) \tag{35}$$

Second, the drive system power consumption is minimized by using the reward function Equation (36).

$$R_{economy} = -W_{economy} \sum_{ij=\{fl,fr,rl,rr\}} P_{ij} = -W_{economy} \sum_{ij=\{fl,fr,rl,rr\}} \frac{T_{ij}\omega_{ij}}{\eta(T_{ij}, \omega_{ij})} \quad (36)$$

where $W_{economy}$ is the penalty factors of economy, and ω_{ij} is the speed of motor ij. Additionally, the motor efficiency η is shown in Figure 4.

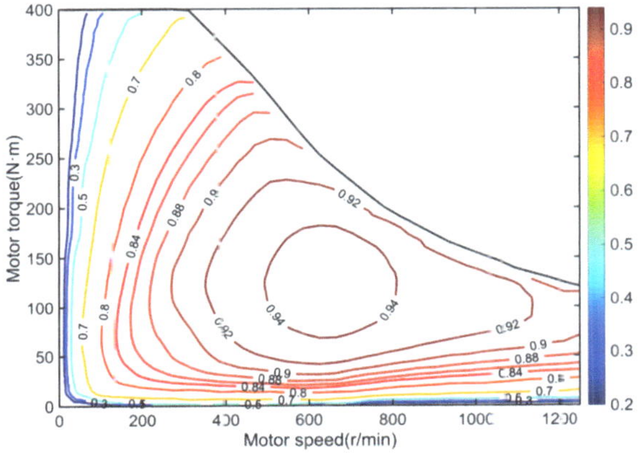

Figure 4. Motor efficiency map.

Lastly, the stability indicators and driver load are considered in the reward function:

$$\begin{aligned} R_{stability} &= -W_{stability}\left((\beta - \beta_d)^2 + (\gamma - \gamma_d)^2\right) \\ R_{driver} &= -W_{driver}\left(\dot{\delta}_{sw}^2 + W_{a\_\delta} \cdot a_x^2\right) \end{aligned} \quad (37)$$

where $W_{stability}$ and W_{driver} are the penalty factors for safety and driver load, respectively.

$$R = R_{upper} + R_{economy} - R_{stability} + R_{driver} \quad (38)$$

The safety of torque allocation consideration consists of two parts: R_{upper} and $R_{stability}$. As safety is the primary concern in vehicle operation, we chose penalty factors to ensure that the reward function satisfies the following relationship:

$$\underbrace{R_{upper} + R_{stability}}_{safety} > R_{economy} > R_{driver} \quad (39)$$

The goal of the RL is to learn a policy that maximizes the expected cumulative reward over time. To encourage exploration and prevent premature convergence, the entropy term is included in the policy π:

$$\pi^* = \underset{\pi}{\operatorname{argmax}} \underset{\tau\sim\pi}{\mathbb{E}}\left[\gamma_f(r_t + \xi\mathcal{H}(\pi(\cdot|s_{t+1})))\right] \quad (40)$$

where ξ is a temperature parameter that determines the level of exploration, and the additional entropy term $\mathcal{H}(\cdot)$ encourages exploration and prevents premature convergence to suboptimal policies.

Heuristic REDQ is an improved algorithm of REDQ. REDQ is a deep RL algorithm that combines ideas from both ensemble methods and double Q-learning to improve stability and sample efficiency [29]. In the REDQ algorithm, the Q function is expressed using the Bellman equation:

$$Q^{\pi}(s_t, a_t) = \mathop{\mathbb{E}}_{a_{t+1}\sim\pi}[r_t + \gamma_f \max_{a_{t+1}} Q^{\pi}(s_{t+1}, a_{t+1}) + \xi\mathcal{H}(\pi(\cdot|s_{t+1}))] \tag{41}$$

where $Q^{\pi}(s_t, a_t)$ is the expected long-term reward of taking action a_t based on policy π in state s_t, $r_t = \mathcal{R}(s_t)$ is the immediate reward based on state s_t, $\gamma_f \in [0,1]$ is the discount factor, s_{t+1} is the next state after taking action a_t in state s_t, and a_{t+1} is the next action.

To improve stability and reduce overestimation, the REDQ algorithm maintains an ensemble of N Q-networks, where each network is represented by a set of weights θ_i, and the target value of Q-function is calculated by randomly selecting M Q-networks from N Q-networks:

$$y^{REDQ} = r_t + \gamma_f \left(\min_{m\in\mathcal{I}_M} \widetilde{Q}_m(s_{t+1}, a_{t+1}, \widetilde{\theta}_t) + \xi\mathcal{H}(\pi(\cdot|s_{t+1})) \right) \tag{42}$$

where the set $\mathcal{I}_M \subseteq \{1, 2, \ldots, N\}$ is M random elements.

The updated rule of evaluating Q-networks' weights is by minimizing the loss:

$$\mathcal{L}_{\theta}(\theta_i) = \mathop{\mathbb{E}}_{(s_t,a_t,r_t,s_{t+1})}\left[\left(y^{REDQ} - Q_{\theta_i}(s_t, a_t)\right)^2\right], i = 1, 2, \ldots N \tag{43}$$

The target Q-networks are updated using a Polyak averaging:

$$\widetilde{\theta}_i \leftarrow \rho\widetilde{\theta}_i + (1-\rho)\theta_i \tag{44}$$

where ρ is a hyperparameter that determines the smoothing factor.

In addition, the parameter ϕ of the policy π_{ϕ} is trained by minimizing the loss:

$$\mathcal{L}_{\phi}(\phi) = \mathop{\mathbb{E}}_{(s_t,a_t,r_t,s_{t+1})}\left[\frac{1}{N}\sum_{i=1}^{N} Q_{\theta_i}(s_t, a_t) + \xi\mathcal{H}(\pi(a_{\phi}(s)|s_t))\right] \tag{45}$$

As the action here is the four-wheeled torque, its action space is four-dimensional, which adds difficulty to the training convergence of the agent. Therefore, a heuristic decay method is introduced here, with the final action being randomly chosen from the set of action strategies $\{a_{OP}, a_{rule}\}$ with distribution $\{1 - P_{rule}, P_{rule}\}$. The probability of selecting action a_{rule} is:

$$P_{rule} = \left(1 - \tanh\left(\frac{n_r}{\tau}\right)\right) \tag{46}$$

where $\tanh(x) = \frac{e^{2x}-1}{e^{2x}+1}$, n_r is the cumulative number of runs, and τ is a constant that denotes the decay speed. The a_{rule} is obtained from Section 4.1, and the a_{OP} is expressed as:

$$a_{OP}(s_t) = \tanh(\mu_{\phi}(s_t) + \sigma_{\phi}(s_t) \odot \varphi)T_{\max} \tag{47}$$

where $\varphi \sim N(0, I)$ is an independent noise, $\mu_{\phi}(s_t)$ is the mean of the stochastic policy that maps the state s_t to an action, and $\sigma_{\phi}(s_t)$ is a random noise sampled from a probability distribution that encourages exploration. $\odot$ is the Hadamard product, and $T_{\max}$ is the maximum motor torque, which is determined by the external characteristics of the motor.

To reduce overestimation and improve stability, REDQ uses an ensemble of Q-networks to estimate the Q-values and takes the minimum of the target Q-values from all the networks. To further improve sample efficiency, REDQ uses an update-to-data ratio denoted by G, which enables control over the number of times the data is reused. Algorithm 2 shows the detailed steps of the heuristic REDQ.

Algorithm 2. Heuristic REDQ algorithm

1. Initialize an ensemble of Q-networks $Q_{\theta}(s_t, a_t)$ with parameters θ_i, Set target parameters
2. Initialize the target Q-networks $\widetilde{Q}_{\widetilde{\theta}}(s_t, a_t)$ with parameters $\widetilde{\theta}_i \leftarrow \theta_i$
3. Initialize the replay buffer
4. **For** each step t **do:**
5. Randomly sample an action a_t from the set of action strategies $\{a_{OP}, a_{rule}\}$ with distribution $\{1 - P_{rule}, P_{rule}\}$
6. Execute the action a_t and observe the next state s_{t+1}, reward r_t
7. Store the experience tuple $\{s_t, a_t, r_t, s_{t+1}\}$ in the replay buffer
8. **for** G updates **do**
9. Sample a mini-batch experiences $\{(s_t, a_t, r_t, s_{t+1})\}$ from replay buffer
10. Randomly select m numbers from the set $\{1, 2, \cdots, N\}$ as a set $\mathcal{I}_M$
11. Based on (42) compute the Q-value estimates y^{REDQ}
12. **for** $i = 1, 2, \cdots, N$ **do**
13. Based on (43), update the parameters θ_i using gradient descent method
14. Based on (44), Update each target Q-network $\widetilde{Q}_{\widetilde{\theta}}(s_t, a_t)$
15. **end for**
16. **end for**
17. **end for**
18. Return the learned Q-network ensemble.

4. Evaluation Indicators and Simulation Results

4.1. Simulation Environment

To validate the effectiveness of the TVC algorithm, a joint simulation of CarSim, Simulink, and Python is used in this paper. The vehicle model in CarSim is based on the Magic Formula tire model. The vehicle model includes four in-wheel motors, which are controlled by the torque allocation algorithm implemented in Simulink. Python is used for training the RL algorithm. As the trained RL algorithm will not impose an additional computational burden on the TVC system, the main computational consumption of this paper lies in the nonlinear MPC. We have verified the real-time problem of the nonlinear MPC controller in the published paper [26], and the implemented nonlinear MPC algorithm by the ACADO toolkit [24], which has demonstrated its numerical efficiency for EV control in previous studies [5,30]. In this paper, the following five representative methods were selected for comparison and validation.

1. RLES. The torque allocation algorithm proposed in this paper. The active safety control layer is a nonlinear MPC controller, and the lower controller is based on a heuristic REDQ deep RL algorithm which integrates considering economy and safety
2. MPC-CO. The torque allocation algorithm proposed in reference [26] which integrates considering economy and safety, where the lower controller is a quadratic planning algorithm.
3. LQR-EQ. The active safety control layer is the LQR controller in reference [31], and the torque allocation layer is a common average allocation method in Section 3.2.1. This controller considers vehicle safety only.
4. w/o control. There is no additional vehicle lateral control; steering is controlled by the driver.

4.2. Performance Indicators

To evaluate the effectiveness of the proposed torque allocation algorithm, several evaluation indicators are employed in this paper. These indicators include:

1. Handling stability

$$\varepsilon_s = -\int_0^t \left(e_\beta^2 + W_{\beta\_\gamma} e_\gamma^2\right) dt \tag{48}$$

where $e_\beta = \beta - \beta_r$, $e_\gamma = \gamma - \gamma_r$, and W_{β_γ} are the weights between e_β^2 and e_γ^2.

2. Driver workload

$$\varepsilon_{driver} = \int_0^t \left(\dot{\delta}_{sw}(\tau)^2 + W_{a\_\delta}(\tau) \cdot a_x(\tau)^2 \right) d\tau \tag{49}$$

where a_x is the longitudinal acceleration, δ_{sw} is the driver steering wheel angle, and W_{a_δ} is the weight between $\dot{\delta}_{sw}^{\ 2}$ and $a_x^{\ 2}$.

3. Motor load

Frequent variations in motor torque may result in thermal saturation, which will increase the wear and tear on the motor.

$$\varepsilon_{motor} = \int_0^t \left(\Delta T_{fl}(\tau)^2 + \Delta T_{fr}(\tau)^2 + \Delta T_{rl}(\tau)^2 + \Delta T_{rr}(\tau)^2 \right) d\tau \tag{50}$$

4. Additional yaw moment

The magnitude of M_z can represent the cost of stability control.

$$\varepsilon_{Mz} = \int_0^t M_z(\tau)^2 d\tau \tag{51}$$

5. Velocity tracking

Vehicle velocity is important for safety assessment, and ensuring vehicle safety while maintaining speed tracking is the goal of stability control.

$$\varepsilon_v = \int_0^t \left(v_{ref}(\tau) - v_x(\tau) \right)^2 d\tau \tag{52}$$

4.3. Training Performance

To ensure the performance of the controller under extreme conditions. We trained the RL agent in the lower controller on the simulation platform described in Section 3.2.2. The training condition is a low adhesion road with an adhesion coefficient of 0.3, and the vehicle velocity was increased from 60 km/h to 100 km/h in 10 s. The RL sampling time is 0.02 s, and the training time per episode is 10 s. The total number of training episodes is 800. The training results are shown in Figure 5.

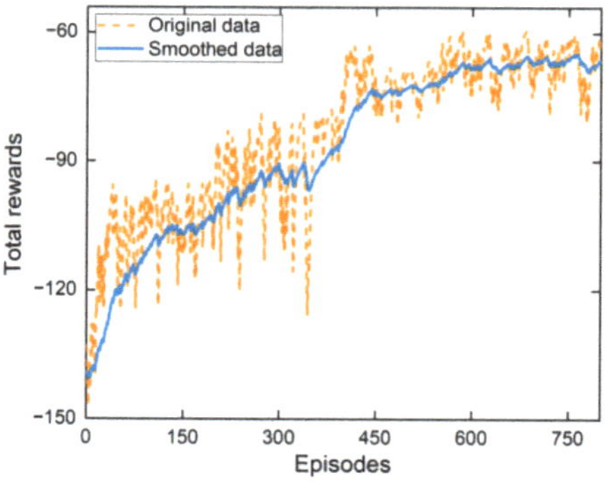

Figure 5. RL algorithm training reward curve.

The training reward curve of RL shows how the agent's cumulative reward changes over the course of the training process. We can see that after about 580 episodes of training, the total rewards of the agent begin to converge at −65. During the early stages of training, the reward curve is erratic, with the agent sometimes achieving high rewards and sometimes achieving low rewards. As the agent learns and the training progresses, the reward curve gradually increases and stabilizes, indicating that the agent is becoming more proficient at the task.

4.4. DLC Maneuver on Slippery Road

The effectiveness of the proposed algorithm is tested under DLC conditions on slippery road surfaces. The longitudinal initial velocity is 72 km/h, and the tire–road friction coefficient is set as 0.3.

The vehicle trajectory in Figure 6a shows that the proposed RLES torque allocation algorithm provides better stability and control during the double line change maneuver compared to the MPC CO and LQR EQ controllers. The vehicle is able to maintain a smooth trajectory and quickly change its direction without large deviation or instability. Additionally, the vehicle under w/o control deviated significantly from the desired path. The yaw rate and side slip angle in Figure 6b,c also show improvements with the proposed RLES algorithm. The yaw rate and side slip angle of the RLES and MPC CO controllers can track the reference value throughout the maneuver, indicating good vehicle handling and control. The yaw rate and side slip angle in the LQR EQ and w/o controller show more oscillations and instability, which may cause the vehicle to lose control. The phase trajectory portrait in Figure 7a also proves this point.

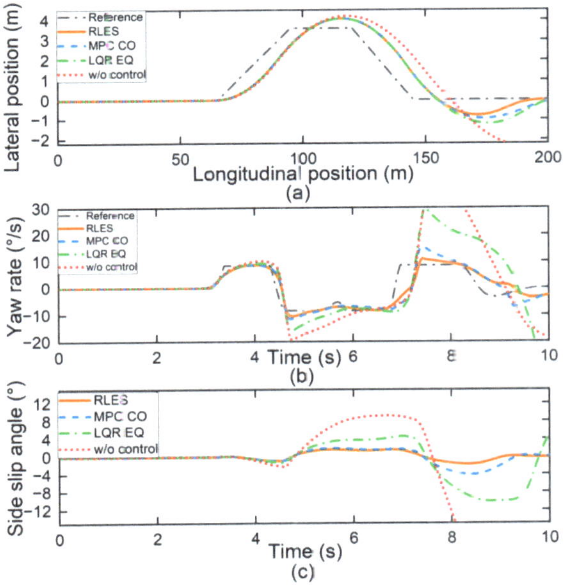

Figure 6. Simulation results on slippery road. (a) Vehicle displacement; (b) yaw rate; (c) side slip angle.

The motor power consumption in Figure 7b also shows improvement with the proposed RLES controller. The motor power consumption is reduced, indicating better fuel economy. The other control strategies show higher motor power consumption, which may result in higher energy consumption and lower fuel economy. Compared to w/o control, RLES, MPC CO, and LQR EQ reduce motor power consumption by 80.2%, 73.4%, and 31.4%, respectively. Table 2 shows the results of the evaluation indicators for the four controllers. The RLES controller shows advantages for all indicators, especially for ε_{driver}, ε_{motor}, and ε_{v}.

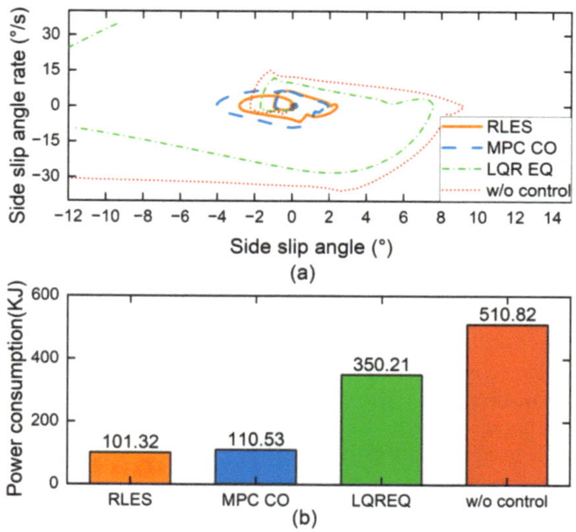

Figure 7. Simulation results on slippery road. (**a**) Phase trajectory portrait; (**b**) motor power consumption.

Table 2. Performance indicators in DLC maneuver on slippery road.

Controller	ε_s		ε_{driver}		ε_{motor}		ε_{Mz}	ε_v
RLES	0.4085	−96%	16.28	−81%	20,820	−93%	735,200	1.1×10^{-5}
MPC CO	0.5640	−94%	19.58	−77%	23,430	−92%	887,200	1.5×10^{-5}
LQR EQ	3.6532	−61%	32.54	−62%	29,680	−90%	892,000	3.5×10^{-4}
w/o control	9.260	-	85.46	-	306,300	-	0	2.9×10^{2}

Overall, the simulation results demonstrate the effectiveness of the torque allocation algorithm in improving vehicle stability, control, and fuel economy during a double-line change maneuver. This is because the RL-based RLES controller can learn the optimal strategy in continuous interaction with the environment and thus output the optimal four-wheel torque according to the current vehicle state.

4.5. DLC Maneuver on Joint Road

To simulate the condition of suddenly encountering an icy road surface at the start of a lane change and analyze the controller performance under extreme conditions, we set the road surface adhesion coefficient as Figure 8. Additionally, the target vehicle speed was 72 km/h, the EV was run in a DLC condition, and the reference trajectory is shown in the reference term in Figure 9a.

The trajectory of a vehicle controlled by RLES has a smaller displacement offset than other controllers as shown in Figure 9a. The vehicle can still complete the DLC maneuver under the w/o controller because of the decrease in vehicle speed, as shown in Figure 10b where the minimum speed of the w/o controller is 51 km/h. The side slip angle of the vehicle controlled by RLES was also well-controlled and remained close to zero, while the LQR EQ and w/o control controllers showed larger deviations from zero, indicating poorer vehicle stability, as shown in Figure 9c. The phase trajectory portrait in Figure 10a shows the relationship between the side slip angle and side slip angle rate, which demonstrates that the proposed RLES torque allocation algorithm provides better stability and control. The phase trajectory of the RLES is within the stability range, while the phase trajectory of the LQR EQ and w/o control controller exceeded the stability boundary and cannot return to the stability point, which may cause the vehicle to lose control.

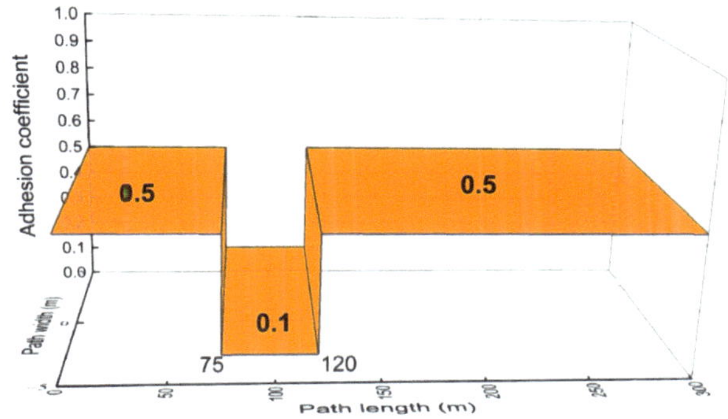

Figure 8. Adhesion coefficient of the joint road.

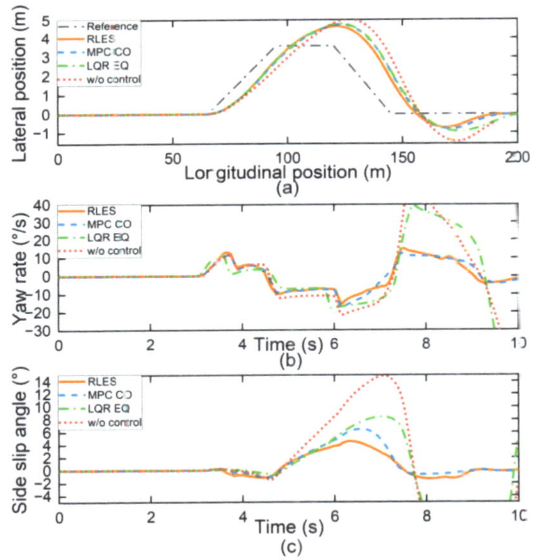

Figure 9. Simulation results under joint road. (**a**) Vehicle displacement; (**b**) yaw rate; (**c**) side slip angle.

Figure 11 shows the stability index and power consumption of the four controllers. It can be seen that both the RLES and MPC CO controllers perform significantly better than the LQR EQ and w/o controller. RLES and MPC CO improve the vehicle economy while ensuring vehicle safety. The performance of RLES is better than the MPC CO controller in terms of stability and power consumption, which indicates that the RLES based on RL training is suitable for different working conditions and can adaptively adopt the optimal control strategy according to the current vehicle state.

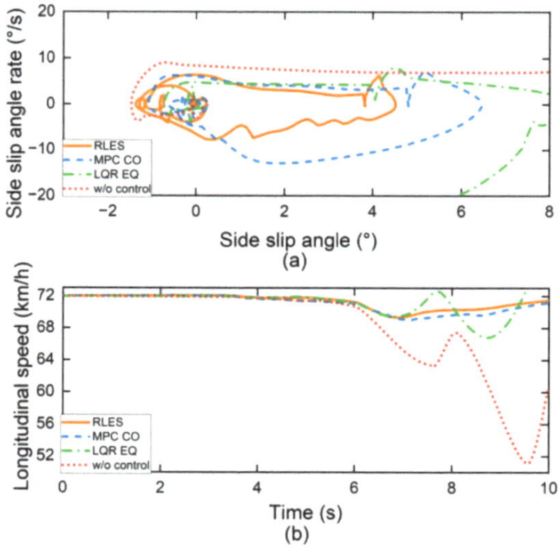

Figure 10. Simulation results under joint road. (**a**) Phase trajectory portrait; (**b**) longitudinal velocity.

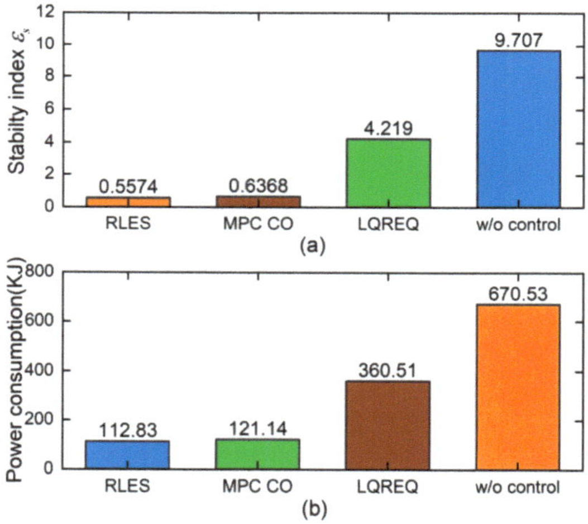

Figure 11. Simulation results under joint road. (**a**) Stability index ε_s; (**b**) motor power consumption.

Table 3 summarizes the performance comparison of the four controllers. The simulation results demonstrate that the proposed RLES controller achieves superior performance in terms of power consumption, driver burden, and vehicle safety, making it a promising solution for the torque allocation of 4MIDEV.

Table 3. Performance indicators in DLC maneuver on joint road.

Controller	ε_s		ε_{driver}		ε_{motor}		ε_{Mz}	ε_v
RLES	0.5574	−94%	16.28	−81%	20,820	−93%	735,200	0.88
MPC CO	0.6368	−93%	19.58	−77%	23,430	−92%	887,200	1.42
LQR EQ	4.2190	−57%	32.54	−62%	29,680	−90%	892,000	2.47
w/o control	9.707	-	85.46	-	306,300	-	0	36.78

4.6. Step Steering Maneuver

To verify the dynamic characteristics of the vehicle, a step steering maneuver was conducted for simulation validation. The vehicle was driving at a speed of 72 km/h on a good road surface with a road adhesion coefficient of 0.75. Within 0.5 s, the steering wheel angle increased from 0 deg to 120 deg and remained unchanged. The simulation results are shown in Figures 12 and 13.

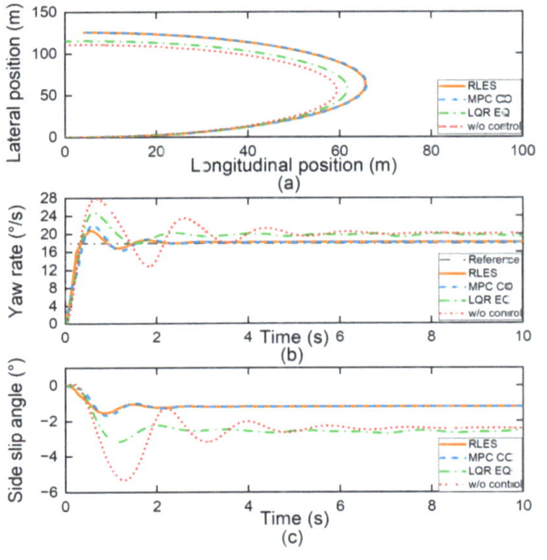

Figure 12. Simulation results under step steering maneuver. (**a**) Vehicle displacement; (**b**) yaw rate; (**c**) side slip angle.

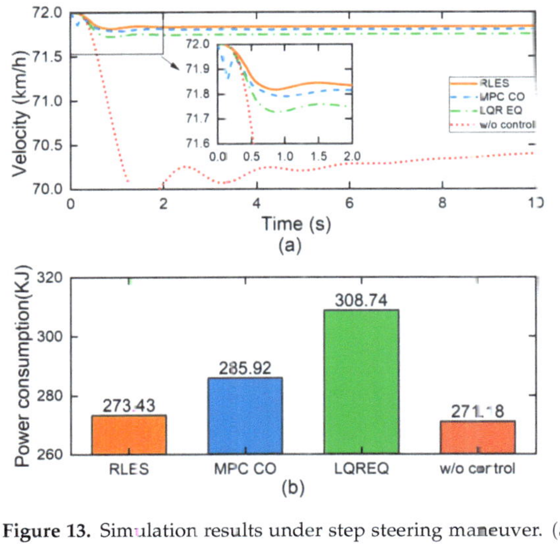

Figure 13. Simulation results under step steering maneuver. (**a**) Longitudinal velocity; (**b**) motor power consumption.

As shown in Figure 12a, the RLES control method shows relative understeering characteristics, which is beneficial for improving driver maneuverability. The yaw rate is shown in Figure 12b, indicating that both the RLES and MPC CO control methods could track the reference value well. Similarly, the side slip angle for the RLES and MPC CO control methods in Figure 12c is better than that of the LQR EQ and w/o control methods, and the RLES control method has the best maneuvering stability. Figure 13a shows the vehicle velocity variation curve in the step steering simulation, and the RLES method has better velocity tracking performance compared to other methods. The vehicle stability is improved under the RLES method while the longitudinal speed tracking performance is guaranteed. As seen in Figure 13b, except for the w/o control method, RLES consumed the least energy during the simulation. The w/o control method consumed the least energy because it does not require torque distribution to generate additional yaw moment by providing differential drive-assisted steering. However, this control method which sacrifices the vehicle stability is actually unsafe.

Table 4 shows the objective evaluation indicators under different controllers. It can be seen from the table that, except for the w/o control method, the RLES controller outperformed the other controllers in terms of vehicle stability, driver burden, and longitudinal speed tracking performance. In conclusion, the controller based on RLES not only ensures vehicle safety but also reduces the energy consumption of the drive system and improves economy under open-loop simulation conditions.

Table 4. Performance indicators in step steering maneuver.

Controller	ε_s		ε_{driver}		ε_{motor}	ε_{Mz}	ε_v
RLES	0.0183	−62%	42.46	−1.7%	9930	1×10^7	0.01895
MPC CO	0.0193	−60%	42.46	−1.7%	9873	1×10^7	0.0273
LQR EQ	0.0274	−43%	43.21	0.1%	4166	170,100	0.04486
w/o control	0.0483	-	43.18	-	132.7	0	2.202

4.7. Driving Cycles

To evaluate the effectiveness of the proposed TVC algorithm in terms of energy savings, a series of tests were conducted under two different driving cycles: New European Driving Cycle (NEDC) and US06. The NEDC case was designed to simulate urban and high-speed driving scenarios, while the US06 driving cycle was designed to replicate a more aggressive driving behavior on the highway. Figure 14a depicts the vehicle velocity allocation for each driving cycle. Note that the driving cycle conditions do not include lateral control, so the performance of the LQR EQ controller is the same as without the control.

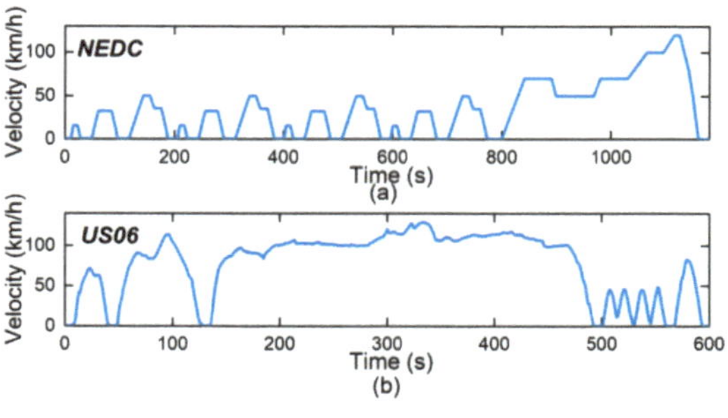

Figure 14. Velocity distribution of driving cycles. (**a**) NEDC; (**b**) US06.

As depicted in Figure 15, the RLES algorithm increases the motor efficiency by 10.9% and reduces the motor power consumption by 11.0% compared to the LQR EQ algorithm in the NEDC cycle. This reduction in power consumption is achieved by optimizing the torque allocation strategy, thus allowing the motor to operate in the high-efficiency range as much as possible. In addition, the RLES also achieves optimal performance in the US60 driving cycle in Figure 15. These results demonstrate the RLES algorithm can adapt to different driving conditions and optimize the torque allocation strategy to achieve better energy efficiency.

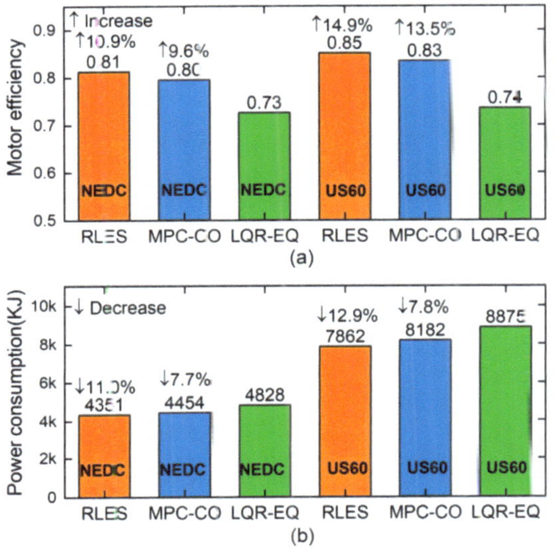

Figure 15. Simulation results under driving cycles. (**a**) Motor efficiency; (**b**) motor power consumption.

5. Conclusions

The 4MIDEV has the feature of independently controllable four-wheel motors. To fully utilize this feature, the paper introduces a novel RL-based TVC algorithm for 4MIDEV that takes into account both economy and safety. The four-wheel tire model is identified using FTO with experimental data, and the active safety control layer utilizes a nonlinear MPC to calculate the required additional yaw moment and longitudinal force. The torque allocation layer employs a heuristic REDQ deep RL algorithm to compute the optimal four-wheel torque. The proposed RLES controller is validated and compared with typical MPC CO, LQR EQ, and w/o controllers under different driving scenarios. The results demonstrate that the proposed RLES enhances vehicle economy and reduces driver workload while ensuring vehicle safety. Future work will focus on designing a TVC algorithm for different driver steering characteristics and conducting real vehicle experiments.

Author Contributions: Conceptualization, H.D.; methodology, H.D.; software, Q.W.; validation, F.L.; formal analysis, Y.Z.; investigation, Y.Z.; writing—original draft preparation, H.D.; writing—review and editing, Y.Z.; supervision, Y.Z. All authors have read and agreed to the published version of the manuscript

Funding: This research was funded by the Postgraduate Research and Practice Innovation Program of Jiangsu Province, grant number KYCX21_0188, the National Natural Science Foundation of China grant number 52272397, 11672127, the Fundamental Research Funds for the Central Universities, grant number NP2022408, and the Army Research and the National Engineering Laboratory of High Mobility anti-riot vehicle technology, grant number HTF B2021 0017.

Institutional Review Board Statement: Not applicable

Informed Consent Statement: Not applicable.

Data Availability Statement: No external data were used in this study.

Acknowledgments: The authors gratefully acknowledge the financial support from the Postgraduate Research and Practice Innovation Program of Jiangsu Province (No. KYCX21_0188).

Conflicts of Interest: The authors declare no conflict of interest.

References

1. Wu, J.; Zhang, J.; Nie, B.; Liu, Y.; He, X. Adaptive Control of PMSM Servo System for Steering-by-Wire System With Disturbances Observation. *IEEE Trans. Transp. Electrif.* **2022**, *8*, 2015–2028. [CrossRef]
2. Wu, J.; Kong, Q.; Yang, K.; Liu, Y.; Cao, D.; Li, Z. Research on the Steering Torque Control for Intelligent Vehicles Co-Driving With the Penalty Factor of Human–Machine Intervention. *IEEE Trans. Syst. Man Cybern. Syst.* **2023**, *53*, 59–70. [CrossRef]
3. Lei, F.; Bai, Y.; Zhu, W.; Liu, J. A novel approach for electric powertrain optimization considering vehicle power performance, energy consumption and ride comfort. *Energy* **2019**, *167*, 1040–1050. [CrossRef]
4. Karki, A.; Phuyal, S.; Tuladhar, D.; Basnet, S.; Shrestha, B.P. Status of Pure Electric Vehicle Power Train Technology and Future Prospects. *Appl. Syst. Innov.* **2020**, *3*, 35. [CrossRef]
5. Dalboni, M.; Tavernini, D.; Montanaro, U.; Soldati, A.; Concari, C.; Dhaens, M.; Sorniotti, A. Nonlinear Model Predictive Control for Integrated Energy-Efficient Torque-Vectoring and Anti-Roll Moment Distribution. *IEEE/ASME Trans. Mechatron.* **2021**, *26*, 1212–1224. [CrossRef]
6. Chatzikomis, C.; Zanchetta, M.; Gruber, P.; Sorniotti, A.; Modic, B.; Motaln, T.; Blagotinsek, L.; Gotovac, G. An energy-efficient torque-vectoring algorithm for electric vehicles with multiple motors. *Mech. Syst. Sig. Process.* **2019**, *128*, 655–673. [CrossRef]
7. Xu, W.; Chen, H.; Zhao, H.; Ren, B. Torque optimization control for electric vehicles with four in-wheel motors equipped with regenerative braking system. *Mechatronics* **2019**, *57*, 95–108. [CrossRef]
8. Hu, X.; Wang, P.; Hu, Y.; Chen, H. A stability-guaranteed and energy-conserving torque distribution strategy for electric vehicles under extreme conditions. *Appl. Energy* **2020**, *259*, 114162. [CrossRef]
9. Ding, S.H.; Liu, L.; Zheng, W.X. Sliding Mode Direct Yaw-Moment Control Design for In-Wheel Electric Vehicles. *IEEE Trans. Ind. Electron.* **2017**, *64*, 6752–6762. [CrossRef]
10. Zhao, B.; Xu, N.; Chen, H.; Guo, K.; Huang, Y. Stability control of electric vehicles with in-wheel motors by considering tire slip energy. *Mech. Syst. Sig. Process.* **2019**, *118*, 340–359. [CrossRef]
11. Zhang, L.; Chen, H.; Huang, Y.; Wang, P.; Guo, K. Human-Centered Torque Vectoring Control for Distributed Drive Electric Vehicle Considering Driving Characteristics. *IEEE Trans. Veh. Technol.* **2021**, *70*, 7386–7399. [CrossRef]
12. Li, Q.; Zhang, J.; Li, L.; Wang, X.; Zhang, B.; Ping, X. Coordination Control of Maneuverability and Stability for Four-Wheel-Independent-Drive EV Considering Tire Sideslip. *IEEE Trans. Transp. Electrif.* **2022**, *8*, 3111–3126. [CrossRef]
13. Deng, H.; Zhao, Y.; Nguyen, A.T.; Huang, C. Fault-Tolerant Predictive Control With Deep-Reinforcement-Learning-Based Torque Distribution for Four In-Wheel Motor Drive Electric Vehicles. *IEEE/ASME Trans. Mechatron.* **2023**, early access. [CrossRef]
14. Aradi, S. Survey of Deep Reinforcement Learning for Motion Planning of Autonomous Vehicles. *IEEE Trans. Intell. Transp. Syst.* **2020**, *23*, 740–759. [CrossRef]
15. Zhu, Y.; Wang, Z.; Chen, C.; Dong, D. Rule-Based Reinforcement Learning for Efficient Robot Navigation With Space Reduction. *IEEE/ASME Trans. Mechatron.* **2022**, *27*, 846–857. [CrossRef]
16. Kiran, B.R.; Sobh, I.; Talpaert, V.; Mannion, P.; Sallab, A.A.A.; Yogamani, S.; Pérez, P. Deep Reinforcement Learning for Autonomous Driving: A Survey. *IEEE Trans. Intell. Transp. Syst.* **2022**, *23*, 4909–4926. [CrossRef]
17. Wei, H.; Zhang, N.; Liang, J.; Ai, Q.; Zhao, W.; Huang, T.; Zhang, Y. Deep reinforcement learning based direct torque control strategy for distributed drive electric vehicles considering active safety and energy saving performance. *Energy* **2022**, *238*, 121725. [CrossRef]
18. Peng, H.; Wang, W.; Xiang, C.; Li, L.; Wang, X. Torque Coordinated Control of Four In-Wheel Motor Independent-Drive Vehicles With Consideration of the Safety and Economy. *IEEE Trans. Veh. Technol.* **2019**, *68*, 9604–9618. [CrossRef]
19. Cabrera, J.A.; Ortiz, A.; Carabias, E.; Simon, A. An Alternative Method to Determine the Magic Tyre Model Parameters Using Genetic Algorithms. *Veh. Syst. Dyn.* **2004**, *41*, 109–127. [CrossRef]
20. Alagappan, A.; Rao, K.V.N.; Kumar, R.K. A comparison of various algorithms to extract Magic Formula tyre model coefficients for vehicle dynamics simulations. *Veh. Syst. Dyn.* **2015**, *53*, 154–178. [CrossRef]
21. Hu, C.; Wang, R.R.; Yan, F.J.; Chen, N. Should the Desired Heading in Path Following of Autonomous Vehicles be the Tangent Direction of the Desired Path? *IEEE Trans. Intell. Transp. Syst.* **2015**, *16*, 3084–3094. [CrossRef]
22. Ji, X.; He, X.; Lv, C.; Liu, Y.; Wu, J. A vehicle stability control strategy with adaptive neural network sliding mode theory based on system uncertainty approximation. *Veh. Syst. Dyn.* **2018**, *56*, 923–946. [CrossRef]
23. Zhang, H.; Liang, J.; Jiang, H.; Cai, Y.; Xu, X. Stability Research of Distributed Drive Electric Vehicle by Adaptive Direct Yaw Moment Control. *IEEE Access* **2019**, *7*, 106225–106237. [CrossRef]
24. Houska, B.; Ferreau, H.J.; Diehl, M. An auto-generated real-time iteration algorithm for nonlinear MPC in the microsecond range. *Automatica* **2011**, *47*, 2279–2285. [CrossRef]
25. Wang, J.; Luo, Z.; Wang, Y.; Yang, B.; Assadian, F. Coordination Control of Differential Drive Assist Steering and Vehicle Stability Control for Four-Wheel-Independent-Drive EV. *IEEE Trans. Veh. Technol.* **2018**, *67*, 11453–11467. [CrossRef]
26. Deng, H.; Zhao, Y.; Feng, S.; Wang, Q.; Zhang, C.; Lin, F. Torque vectoring algorithm based on mechanical elastic electric wheels with consideration of the stability and economy. *Energy* **2021**, *219*, 119643. [CrossRef]
27. Wu, X.; Zhou, B.; Wen, G.; Long, L.; Cui, Q. Intervention criterion and control research for active front steering with consideration of road adhesion. *Veh. Syst. Dyn.* **2018**, *56*, 553–578. [CrossRef]

28. Zhai, L.; Sun, T.M.; Wang, J. Electronic Stability Control Based on Motor Driving and Braking Torque Distribution for a Four In-Wheel Motor Drive Electric Vehicle. *IEEE Trans. Veh. Technol.* **2016**, *65*, 4726–4739. [CrossRef]
29. Chen, X.; Wang, C.; Zhou, Z.; Ross, K. Randomized Ensembled Double Q-Learning: Learning Fast Without a Model. *arXiv* **2021**, arXiv:2101.05982.
30. Parra, A.; Tavernini, D.; Gruber, P.; Sorniotti, A.; Zubizarreta, A.; Perez, J. On Nonlinear Model Predictive Control for Energy-Efficient Torque-Vectoring. *IEEE Trans. Veh. Technol.* **2021**, *70*, 173–188. [CrossRef]
31. Mirzaei, M. A new strategy for minimum usage of external yaw moment in vehicle dynamic control system. *Transp. Res. Part C Emerg. Technol.* **2010**, *18*, 213–224. [CrossRef]

machines

MDPI

Article

An LWPR-Based Method for Intelligent Lower-Limb Prosthesis Control by Learning the Dynamic Model in Real Time

Yi Liu, Honglei An *, Hongxu Ma and Qing Wei

College of Intelligence Science and Technology, National University of Defense Technology, Changsha 410073, China
* Correspondence: eric_hong@nudt.edu.cn

Abstract: A significant number of people in the world suffer from limb losses, while prosthesis is the hopeful way to help the amputees back to normal life. Recently, the most popular control method used in intelligent prosthesis is FSM-IC (finite state machine with impedance control), which requires a significant amount of manual parameter adjustments to achieve a good model compensation in a discrete way. Taking the lower-limb prosthesis as the research object, this paper applies an LWPR (locally weighted projection regression) model to learn the dynamic model of a prosthesis in real time in order to achieve a better model compensation in a continuous way and propose scientific experimental schemes to verify the control method. First, the basic control framework of lower-limb prosthesis is given. Then, the control law is derived on the basis of model building and LWPR's addition. Finally, the proper experimental schemes are designed to carry out the control method effectively in a safe way. The experimental results show that the control law with the LWPR model can greatly improve the tracking performance during the swing phase and obtain rather good compliance during the stance phase. Moreover, the results also indicate that the LWPR model can approximate the dynamic model online. This method is hoped to be extended to more applications and fields.

Keywords: prosthesis; locally weighted projection regression; control law; compliance; tracking performance; real time

Citation: Liu, Y.; An, H.; Ma, H.; Wei, Q. An LWPR-Based Method for Intelligent Lower-Limb Prosthesis Control by Learning the Dynamic Model in Real Time. *Machines* **2023**, *11*, 186. https://doi.org/10.3390/machines11020186

Academic Editors: Jian Wu, Xiangkun He and Guangfei Xu

Received: 14 December 2022
Revised: 26 January 2023
Accepted: 27 January 2023
Published: 30 January 2023

1. Introduction

Amputation of the lower limbs is becoming more and more common due to traffic accidents, diabetes, congenital diseases, tumors [1], etc. Lower-limb losses will lead to limited mobility and psychological burden. In order to restore the mobility and return to normal life, the prosthetic technique is the option of most amputees. Most commercial prostheses are passive because they are cheap and simple in structure. However, walking with a passive prothesis will consume more energy [2] than the able-bodied and the asymmetry may cause secondary damage [3]. Therefore, the research on intelligent prosthesis has become more popular due to its potential in speed switch and adaptation in different terrains.

The general control frameworks of intelligent lower-limb prosthesis are hierarchical into high-level controller, mid-level controller, and low-level controller [4]. The high-level controller aims at human intention recognition including locomotion mode classification, gait parameter determination, and gait phase determination. The low-level controller makes the actuators output the desired torques. The mid-level controller pays attention to intent-to-state conversion and achieves a good interaction between human and prosthesis. For now, a significant amount of control methods have been proposed. Some (e.g., CLME (complementary limb motion estimation) [5] and echo control [6]) collect the kinematic data of the sound leg to complete the auxiliary control of the prosthesis. Model-based methods [7,8] build the dynamic model or human neuromuscular system to generate the joint torques. Phase-based methods introduce continuous gait phase variable to calculate joint trajectory which is followed by the controller. Refs. [9,10] adopt the CCP as a mechanical representation of the gait cycle phase, while [11] extracts proper phase variables from the motion information of the thigh. Ref [12] uses discrete Fourier transform (DFT) to characterize the desired periodic joint trajectories as functions of gait phase. Refs. [13–15] apply FSM–IC (finite state machine with impedance control), which divides a complete gait into several gait stages. In each gait stage, virtual spring damping systems are used to model the impedance of different joints and a significant amount of impedance parameters need to be adjusted for a good model compensation. Furthermore, [16] automatically tunes the impedance parameters by the ADP (approximate dynamic programming) approach. The above control methods mainly consider the tracking performance of joints' trajectories while the performance of compliance is not well considered. To achieve both good compliance and tracking performance for intelligent lower-limb prosthesis, this paper mainly focuses on the mid-level controller to compute the desired torques of knee and ankle.

For an intelligent lower-limb prosthesis, its control goal is improving the tracking performance during the swing phase and compliance during the support phase, which conforms to the human motion mechanism. In order to achieve the above goal, we need to compute the model compensation as a feedforward item by adjusting impedance parameters or modeling. FSM-IC methods adjust impedance parameters and model compensation in a discrete way. Due to uncertain joint friction, model error, and other possible factors, modeling methods can achieve good results in simulation but face the problems of disturbance estimation and real-time model compensation in the real world. Ref. [17] enhances compliance by using the LWPR method [18] to fit the dynamic model. However, the control strategy in [17] is only validated on a simulated biped walker rather than a real human–machine system. Here, we have achieved the migration process from a simulation of the real world by combining the LWPR model and PD controller. The contributions of this paper are listed below: (1) Through theoretical analysis, the control law is obtained and the training method is determined (the LWPR model is only trained in the swing phase); (2) According to the complexity of the model from simple to complex, four experiments are designed to gradually verify the feasibility of the proposed control method and ensure the safety of people; (3) As for prosthesis application, the LWPR model has the ability to rapidly fit the dynamic model in real time, which is input into the controller as a real-time model compensation; (4) Compared with previous works, there are less parameters to be adjusted. (5) The real-time learning of the LWPR model can adapt to the model change.

2. Materials and Methods

The control framework used in this paper is shown in Figure 1, and relevant variables are shown in Figure 2. Under this control framework, [17] completes a simulation experiment and finds a good result. In order to achieve the subjects' walking experiment on the treadmill safely and successfully, there are still significant amounts of analysis and tests to conduct. Each part of the control framework will be introduced below.

2.1. Phase Variable Selection

We install the IMU (inertial measurement unit) on the part where the prosthesis is fixedly connected with the injured thigh of the transfemoral amputee. On the one hand, the motions of the injured thigh can reflect the amputees' motion intention. On the other hand, the sensors are required to be on-board (installed on the prosthesis rather than on the people), which can reduce system complexity and improve convenience. The installed IMU can detect the movement of the thigh in real time with a frequency of 100 Hz (sampling time $t_s = 0.01$ s). The real-time gait phase is obtained through thigh movement [11] when the amputees walk with periodic gaits (e.g., level walking, stair ascent, stair descent, ramp ascent, ramp descent). The real-time gait phase is computed as

$$\varphi(t) = \frac{\mathrm{atan2}[\dot{\theta}_t'(t), \theta_t'(t)] + 180^\circ}{360^\circ} \tag{1}$$

where $\dot{\theta}_t(t)$ and $\theta_t(t)$ are thigh angular velocity and thigh angle, respectively, and $\dot{\theta}_t'(t)$ and $\theta_t'(t)$ are the filtered and normalized thigh angular velocity and thigh angle, respectively [15].

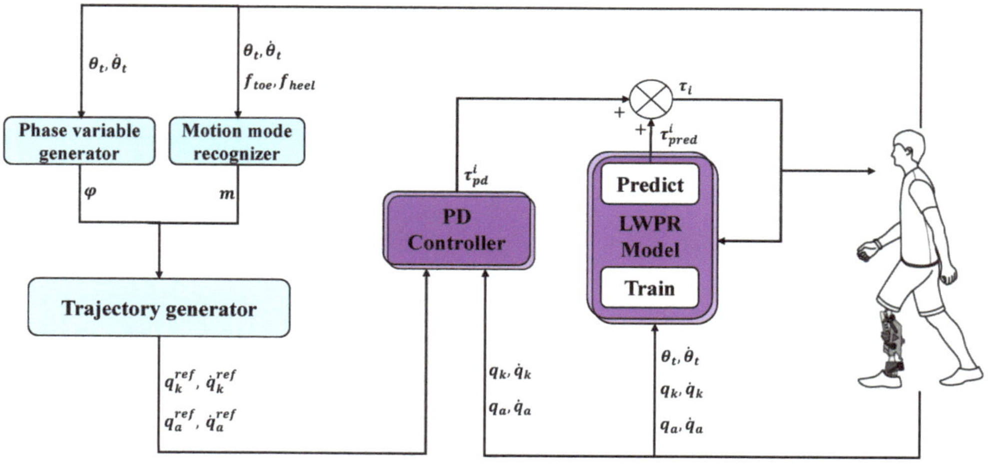

Figure 1. The control framework of intelligent lower-limb prosthesis. The blue parts are the high-level and mid-level controller, which aims at human intention recognition and joint trajectory planning. The purple parts are the main research subjects in this paper.

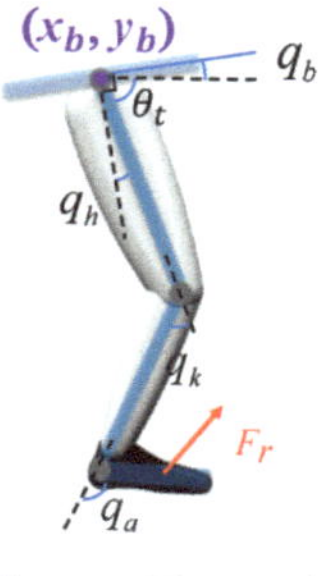

Figure 2. The simplified lower-limb prosthesis model.

2.2. Motion Mode Recognizer

Different motion modes have different joint reference trajectories. This module aims at classifying the locomotion mode according to the sensors' information. However, the focus of this paper is not on the high-level controller. Therefore, we set our locomotion mode as level walking for the convenience of narration.

2.3. Reference Trajectory Generation

For level walking, which corresponds to periodic motion, we use Fourier series to describe the reference trajectory.

$$q_i^{ref}(t) = \sum_{j=0}^{3} A_j \cos[2\pi j \varphi(t) + B_j] \quad (2)$$

where $i \in \{k, a\}$ represents the controlled joint, B_j is the offset term, and A_j is the amplitude of the j-th cosine harmonic component. A_j and B_j are obtained by discrete Fourier transform of the collected joint angle data set off-line.

2.4. Controller Design

We simplify the lower-limb prosthesis to a three-joint manipulator with a floating base (Figure 2). Then, its dynamic equation can be written as

$$u_G = M_G\ddot{x}_G + C_G\dot{x}_G + G_G + J_s^T F_r \tag{3}$$

where $M_G \in R^{6\times 6}$ is the inertia matrix, $C_G \in R^{6\times 6}$ is the centrifugal and Coriolis force term, $G_G \in R^{6\times 1}$ is the gravity term and $J_s^T F_r$ is the projection of ground reaction force (GRF) acting on the foot into forces acting in the degrees of freedom (DOFs). u_G and x_G are defined as

$$u_G \equiv [f_x \quad f_y \quad \tau_b \quad u]^T, x_G \equiv [x_b \quad y_b \quad q_b \quad q]^T \tag{4}$$

where f_x, f_y, and τ_b are forces and torque on the sagittal plane of floating base. x_b, y_b, and q_b are the three DOFs of floating base. $q = [q_h \ q_k \ q_a]^T$ is the joint angle vector and $u = [\tau_h \ \tau_k \ \tau_a]^T$ is the torque input vector. For knee and ankle, their dynamic equations are

$$\begin{bmatrix}\tau_k\\ \tau_a\end{bmatrix} = \begin{bmatrix}M_k\\ M_a\end{bmatrix}\ddot{x}_G + \begin{bmatrix}C_k\\ C_a\end{bmatrix}\dot{x}_G + \begin{bmatrix}G_k\\ G_a\end{bmatrix} + \begin{bmatrix}\tau_d^k\\ \tau_d^a\end{bmatrix} \tag{5}$$

where τ_d^k and τ_d^a are the projection of GRF acting on the foot into forces acting in the knee and ankle.

Here, we intend to use the LWPR model to fit the lower-limb prosthesis model. The LWPR model is of multi-input and single output. For each joint $i \in \{k, a\}$

$$f^i\{x\} + \tau_d^i = \tau_i \tag{6}$$

where x is $(\ddot{q}, \dot{q}, q, q_b, \dot{q}_b, \ddot{q}_b, \dot{x}_b, \dot{y}_b, \ddot{x}_b, \ddot{y}_b)^T$ according to the dynamic model in Table A1 of Appendix A and $f^i(x) = M_i\ddot{x}_G + C_i\dot{x}_G + G_i$. We use LWPR models [2] to approximate f^i

$$\begin{aligned} f^i(x) &= f^i(\ddot{q}, \dot{q}, q, q_b, \dot{q}_b, \ddot{\theta}_t, \dot{x}_b, \dot{y}_b, \ddot{x}_b, \ddot{y}_b) \\ &= \hat{f}^i(\overline{x}) + \varepsilon_i = \sum_{n=1}^{N^i} \omega_n^i \hat{f}_n^i(\overline{x}) / \sum_{n=1}^{N} \omega_n^i + \varepsilon_i \end{aligned} \tag{7}$$

where $\overline{x} = (\dot{\theta}_t, \dot{\theta}_t^l, \theta_t, q_k, \dot{q}_k, \dot{q}_k^l, q_a, \dot{q}_a, \dot{q}_a^l)^T$ is the input vector of the LWPR model (variables in $\overline{x}$ can be obtained from the thigh IMU and joint encoders). When $\overline{x}$ is given, every linear model $\hat{f}_n^i$ calculates a predicted value $\hat{f}_n^i(\overline{x})$, and the normalized weighted mean of all N^i linear models $\hat{f}^i(\overline{x})$ are the output of the LWPR model. $\dot{\theta}_t^l$, $\dot{q}_k^l$, and $\dot{q}_a^l$ are the angular velocity of thigh IMU, knee, and ankle at the last moment. ω_n are Gaussian functions called RFs (receptive fields) which are computed as seen in Equation (8), N is the number of local linear models, and ε is the error between the real dynamic model and the LWPR model.

$$\omega_n^i = \exp\left(-\frac{1}{2}(\overline{x} - c_n^i)^T D_n^i (\overline{x} - c_n^i)\right) \tag{8}$$

where D_n is the distance metric and c_n are the centers of RFs.

Is the given input $\overline{x}$ sufficient for the fitting? We record the coordinates of motion capture markers (Figure 3) [19] and thigh IMU data when the subject walks on the treadmill with a 3 km/h speed and computes related variables by OpenSim [20].

We can obtain the gait phase φ from $\dot{\theta}_t$ and θ_t, and it can be concluded from the plots in Figure 4 that q_h, $\dot{x}_b$, and $\dot{y}_b$ are the functions of gait phase φ when the subject is walking at a constant speed, which means that

$$q_h = g_1(\varphi), \dot{x}_b = g_2(\varphi), \dot{y}_b = g_3(\varphi) \tag{9}$$

Figure 3. The markers of motion capture system.

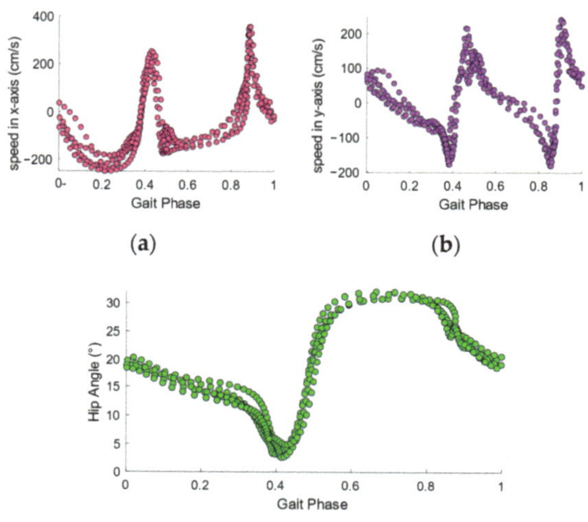

(**a**) (**b**)

(**c**)

Figure 4. (**a**) Scatter plot of $\dot{x}_b$ and gait phase. (**b**) Scatter plot of $\dot{y}_b$ and gait phase. (**c**) Scatter plot of hip joint angle q_h and gait phase. All those points are computed by motion capture data during walking on the treadmill with a speed of 3 km/h.

According to relationships shown in Figure 2 and speed calculation formula, we have

$$q_b = 90 - \theta_t - q_h = 90 - \theta_t - g_1(\varphi), \ddot{q}_k = (\dot{q}_k - \dot{q}_k^l)/t_s, \ddot{q}_a = (\dot{q}_a - \dot{q}_a^l)/t_s \tag{10}$$

where q_b, $\ddot{q}_k$, and $\ddot{q}_a$ can be computed by $\overline{x}$. The derivatives of q_h, $\dot{x}_b$, $\dot{y}_b$, and q_b also are the functions of gait phase φ according to Equation (11).

$$\begin{gathered} \dot{q}_h = \dot{g}_1(\varphi)\dot{\varphi}, \ddot{x}_b = \dot{g}_2(\varphi)\dot{\varphi}, \ddot{y}_b = \dot{g}_3(\varphi)\dot{\varphi} \\ \ddot{q}_h = \ddot{g}_1(\varphi)\dot{\varphi}^2 + \dot{g}_1(\varphi)\ddot{\varphi} \\ \dot{q}_b = -\dot{\theta}_t - \dot{q}_h, \ddot{q}_b = -\ddot{\theta}_t - \ddot{q}_h \end{gathered} \tag{11}$$

where $\ddot{\theta}_t = (\dot{\theta}_t - \dot{\theta}_t^l)/t_s$. When the subject walks with a constant speed, $\dot{\varphi}$ can be regarded as a constant and $\ddot{\varphi} = 0$.

Thus, it is proved that x can be determined by $\overline{x}$ when the subject moves at a constant speed.

The control laws adopted are as follows:

$$\tau_i = \hat{f}^i(x) + K_p^i e_i + K_d^i \dot{e}_i \tag{12}$$

where $e_i = q_i - q_i^{ref}$, and K_p and K_d are the stiffness coefficient and damping coefficient, respectively.

Let us combine Equations (6), (7), and (12):

$$K_p^i e_i + K_d^i \dot{e}_i = \tau_d^i + \varepsilon_i \tag{13}$$

(1) *During swing phase*

When the lower limb is in the swing phase, there is no contact between the foot and the ground, so the GRF is zero. Therefore, it can be assumed that

$$\tau_d^i = 0 \tag{14}$$

Therefore, Equation (13) can be simplified as

$$K_p^i e_i + K_d^i \dot{e}_i = \varepsilon_i \tag{15}$$

The solution of the differential Equation (15) is

$$e_i(t) = C_1^i e^{-\frac{K_p}{K_d}t} + C_2^i \varepsilon_i \tag{16}$$

where $K_p^i, K_d^i > 0$. When $t \to \infty$, $e_i(t) \to C_2^i \varepsilon_i$. The author of [17] shows that the LWPR model can converge to a low error ε with proper model parameters as well as applied and continuous real-time learning. Equation (16) indicates that high stiffness K_p and low damping K_d can improve convergence speed.

(2) *During stance phase*

GRF acts on the prosthesis when the lower limb is in the stance phase, and we have

$$\tau_{ext}^i = \tau_d^i = (J_c^T F_r)_i \tag{17}$$

where τ_{ext} is the external torque on joint. Therefore, we have

$$\tau_{ext}^i + \varepsilon_i = K_p^i e + K_d^i \dot{e} \tag{18}$$

Equation (18) shows that when $\varepsilon_i \to 0$, the joint i is of the dynamic characteristics of the spring damping system. It can help to adjust the relationship between external torque and joint motion, which is called compliance control.

2.5. Learning Policy of LWPR Model

2.5.1. LWPR Model Configuration

A significant number of parameters in the LWPR model need to be adjusted to improve the fitting accuracy and speed. init_D is the initial value of D_k (set initial distance metric $D_k = \text{init\_D} * I_n$ where I_n is n-dimensional identity matrix and n is the number of the LWPR model's input variables); init_alpha is the learning rate of updating D_k; w_gen is a threshold to determine when to create a new receptive filed. More detailed setup and initialization of parameters can be found in the library [21]. Table 1 reports some of the model parameters' values in this paper.

Table 1. Model parameters of LWPR set in the paper.

Parameters	Value (Knee)	Value (Knee)
init_D	50	50
w_gen	0.2	0.2
init_alpha	250	150

2.5.2. Model Learning Policy

According to Equation (7), the LWPR model is intended to fit the internal dynamic model $f^i(x) = M_i\ddot{x}_G + C_i\dot{x}_G + G_i$ of the i joint. The internal dynamic model does not include the external torque (external torque mainly comes from the GRF during walking). Therefore, the LWPR model only trains itself when there are not any external forces while its training input is real-time measurable states of the internal dynamic model and its training output is the real-time motor output torque. Before the LWPR model is added to the control law as a model compensation, it should be trained for a while to approach the internal dynamic model when there are no external forces.

The experiments in this paper all follow the learning policy of the LWPR model above.

3. Results

According to the complexity of the model from simple to complex, four experiments are designed to gradually verify the feasibility of the proposed control method and ensure the safety of people. The single motor (dynamic model of one-dimension) experiment aims at testing the LWPR model's astringency. Considering more complex situations (the centrifugal and Coriolis force term, the gravity term, and higher dimension inputs are added into the dynamic model), a hanging up experiment is intended for further testing the capability of LWPR. After wearing the prosthesis, the real dynamic model will change due to human's involvement. By the leg swinging experiment, the LWPR model is closer to the changed model. When subjects walk on the treadmill, the GRF will appear. The walking experiment is the final test to verify the whole prosthesis control system's performance including tracking performance and compliance.

All the motors used in above four experiments are the same. The motors' reduction ratio is 1:50 and their angle and angular speed are sampled with a frequency of 100 Hz. Except the single motor experiment, the thigh angle and angular velocity in the sagittal plane are recorded with a frequency of 100 Hz.

3.1. Single Motor Experiment

The control method is tested with one motor which is laid flat on the table. In the first 30 s, the PD controller is applied and the LWPR model is only learning while the motor input torque $\tau_{motor} = \tau_{pd}$. The training inputs of the LWPR model are the angle and angular speed of the motor, and the model's training output is τ_{motor}. After 30 s, the LWPR model's output is added into the motor input torque $\tau_{motor} = \tau_{pd} + \tau_{pred}$. We calculate the RMSE to assess the tracking performance.

$$\text{RMSE} = \sqrt{\frac{\sum_{i=1}^{n}(X_{obs,i} - X_{ref,i})^2}{n}} \tag{19}$$

In this paper, $n = 100$. The experiment result is shown in Figure 5. In pure PD mode, the single motor cannot keep up with the sinusoidal curve (in fact, [22] shows that the PID controller cannot track a sine signal with zero error) When the LWPR model is used, the tracking error decreases by about four times and almost reaches a stable value.

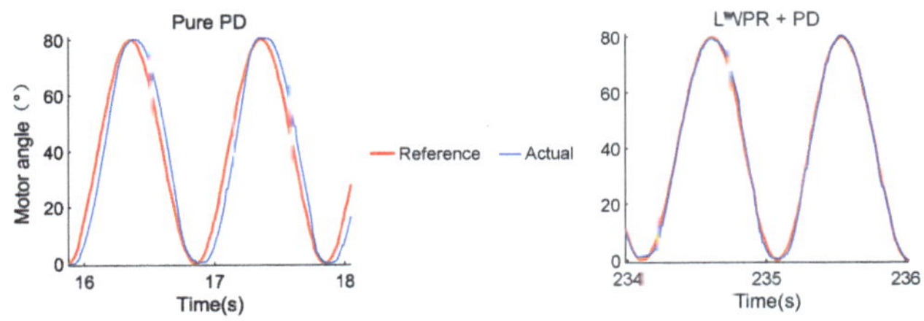

(**a**)

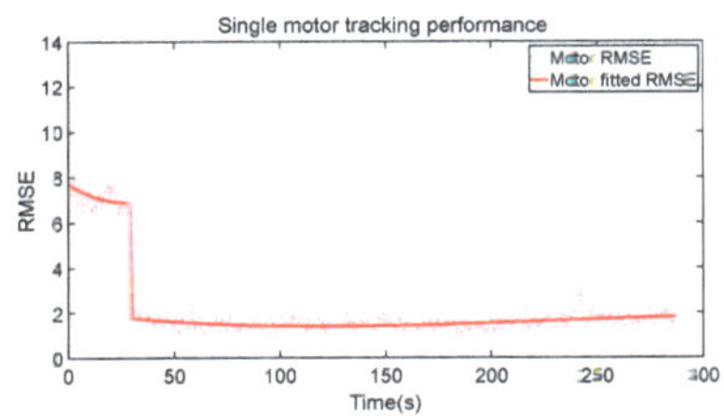

(**b**)

Figure 5. (**a**) Motor's position tracking under different modes. (**b**) Tracking performance of single motor.

3.2. Hanging Up Experiment

The control framework shown in Figure 1 is used in the experiment. We hung the lower-limb prosthesis on a fixed frame (Figure 6). We simulated human thigh movement by swinging the handle to generate phase firstly. Then, we use Fourier series to compute the reference trajectory.

(**a**)

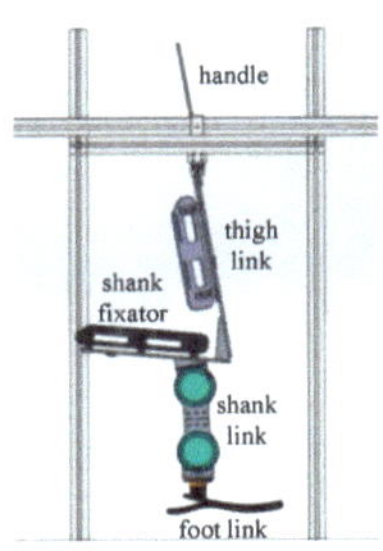

(**b**)

Figure 6. (**a**) Lower-limb prosthesis designed for healthy subjects. (**b**) Mechanical structure diagram of lower-limb prosthesis.

In the first 30 s. Step 1: only the PD controller is applied, while the LWPR model is trained but not applied during this time. The training input of the LWPR model is the vector $\overline{x}$, and the training output is τ_{pd}.

After 30 s. Step 2: we add the predicted value of the LWPR model τ_{pred} to the motor input torques. At this stage, the training input of the LWPR model is the vector $\overline{x}$, and the training output is the $\tau_{pd} + \tau_{pred}$. Then, the value of K_p and K_d will be decreased to K'_p and K'_d.

$$\begin{cases} \tau_i(t) = \tau^i_{pd}(t) = K^i_p e_i(t) + K^i_d \dot{e}_i(t) & 0 \leq t \leq 30 \\ \tau_i(t) = \tau^i_{pred}(t) + \tau^i_{pd}(t) = \hat{f}^i(x(t)) + K'^i_p e_i(t) + K'^i_d \dot{e}_i(t) & t > 30 \end{cases} \tag{20}$$

Figure 7 reports our results. The results show that under LWPR+PD mode, the RMSE of knee angle reduced by five times and the RMSE of ankle angle remain almost unchanged. We hypothesize that the reason for this is that during the experiment, the ankle torque is small and the error of the LWPR model are at the same level, which leads to no improvement of tracking performance.

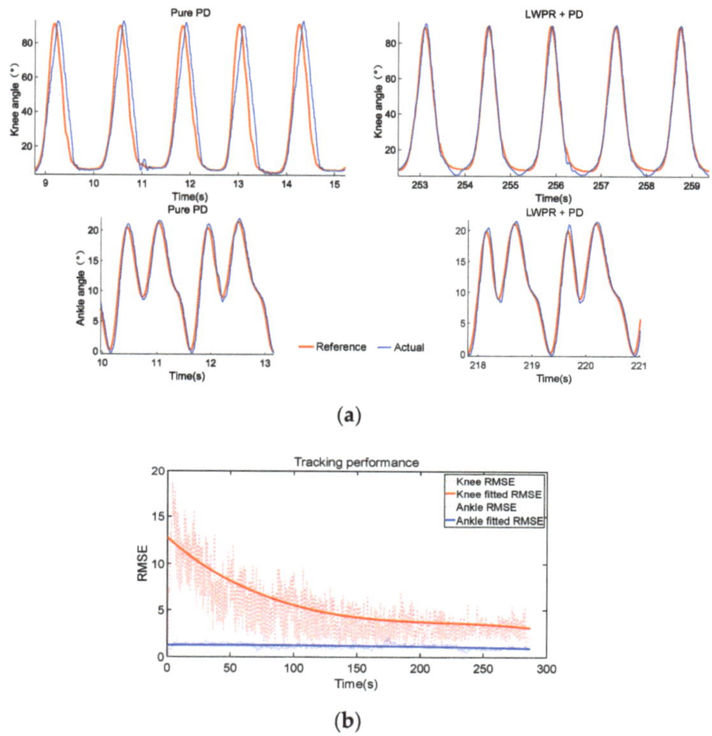

Figure 7. (**a**) Knee and ankle position tracking under different modes. (**b**) Tracking performance of knee and ankle.

In this experiment, K_p and K_d of knee and ankle are, respectively,

$$\begin{aligned} K_p^k &= 6, K_p^{k\prime} = 3; K_d^k = 0.15, K_d^{k\prime} = 0.08; \\ K_p^a &= 4.25, K_p^{a\prime} = 2; K_d^a = 0.12, K_d^{a\prime} = 0.06. \end{aligned} \quad (21)$$

The decrease in the knee's RMSE indicates that the LWPR model provides model compensation but there is still some which is not compensated for. After model compensation, the PD controller can focus more on the error control rather than the uncertain model compensation.

3.3. Leg Swinging Experiment

In this experiment, the subject is required to wear the lower-limb prosthesis, which will change the dynamic model indeed. In order to ensure the subjects' safety, we need to further train our LWPR model to approach the changed model. The experiment is set as follows: The healthy leg stands still and the other leg swings in the air to train the LWPR model for a few minutes (Figure 8a) until the tracking performance is better (Figure 8b). The trained LWPR model in the hanging up experiment is used as the initial model here. Only step 2 in the hanging up experiment is applied in the training process.

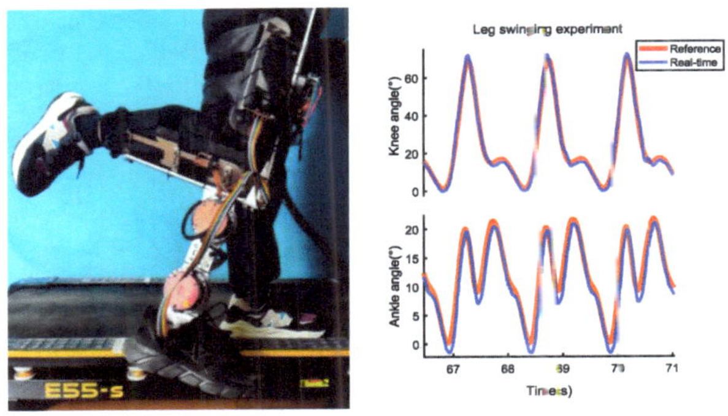

(a) (b)

Figure 8. (**a**) The platform of leg swinging experiment. (**b**) Leg swinging experiment results.

3.4. Walking Experiment

Based on the LWPR model trained in 3.3 as the initial LWPR model, the subjects are required to walk on the treadmill with a comfortable velocity (3 km/h in our experiment). Only step 2 in the hanging up experiment is applied in the training process.

From the experimental results in Figure 9, during the swing phase, we have a rather good tracking performance on knee and ankle position. During the stance phase, the tracking errors of knee and ankle increase because the external torques appear, which proves that the model compensation of LWPR is effective and that the proposed method can bring high compliance to the prosthesis. However, at the beginning of the stance phase, the tracking errors have not been increased immediately, which indicates that there are still errors between the LWPR model and the real dynamic model.

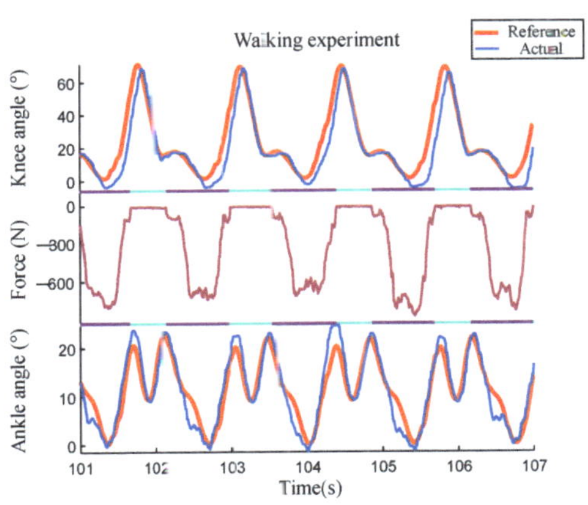

Figure 9. The results of walking experiment on treadmill. Force is the value of GRF and the downward direction is positive. The stance phase is marked with purple horizontal bars while the swing phase is marked with blueish green bars.

According to Equation (18), the smaller ε_i is, the better the compliance and the LWPR model. The good tracking performance and high compliance prove that the LWPR model can fit the real dynamic model in real time.

We capture a one-minute Video S1 of a walking experiment and upload it. According to the subjects' feeling, the impact force between prosthesis and ground is similar to the healthy side, and there is little lag in the swing of the lower-limb prosthesis. Due to the influence of the shank which is on the fixator, the upper body needs to be forward to keep balance. Moreover, the height of the knee joints of both legs is also different due to the difference between healthy people and disabled people. Next, we will improve the mechanical structure and invite the transfemoral amputees to participate in the experiment.

4. Discussion

Through theoretical derivation, the control law is given and the LWPR model is applied. In order to verify the LWPR model's effectiveness (whether model compensation can be achieved) and test the performance of the prosthesis control system, we plan our experiments step by step in a scientific way. The results of the walking experiment show that our method can improve tracking performance during swing phase and bring good compliance during the stance phase, while these two performances are exactly what the lower-limb prosthesis control system is seeking. This paper's contribution can be summarized as follows.

First, this paper proposes the control law of lower-limb prosthesis by theoretical derivation. According to the control law, we need to train the LWPR model only during the swing phase to obtain the performances we want. If we train the LWPR model during the stance phase, the external force torque will be included in the LWPR model and the compliance will not be shown during the stance phase.

Second, a scientific experimental scheme is proposed to initialize the LWPR model. From simulation to experiment is a difficult process, especially when people are involved. A series of experiments seems a little redundant, but they help establish the confidence of the subjects and ensure safety. In human–computer interaction, it is crucial to consider human factors both physically and psychologically.

Finally, good performances in the walking experiment prove that the proposed control method is good in use and that the LWPR model can learn the dynamic model in real-time to achieve a good model compensation.

5. Conclusions

This paper's control method preliminarily shows that the LWPR model is an effective tool for prosthesis application. In high-dimensional space, LWPR greatly improves its computational efficiency by performing linear approximate fitting locally [18]. Due to measurement error and limited measurement information, the fitting error still exists. In the future, the follow points need to be studied further. (1) How to measure the error of the LWPR model and control it? (2) When real dynamic model changes (such as speed changes), how long does it take to for the LWPR model to learn and how to improve learning efficiency? (3) How to use prior knowledge in the LWPR model? (4) Are more sensors necessary? If necessary, which sensors are added? To improve the performance of lower-limb prosthesis, those problems above are what we will study next.

Supplementary Materials: The following supporting information can be downloaded at: https://www.mdpi.com/article/10.3390/machines11020186/s1, Video S1: Walking Experiment.

Author Contributions: Conceptualization, Y.L. and H.A.; methodology, Y.L. and H.A.; software, Y.L.; validation, Y.L., H.A., and H.M.; formal analysis, Y.L.; data curation, Y.L.; writing—original draft preparation, Y.L.; writing—review and editing, H.A., H.M., and Q.W.; visualization, Y.L.; supervision, H.A., H.M., and Q.W.; project administration, H.A.; funding acquisition, H.A., H.M., and Q.W. All authors have read and agreed to the published version of the manuscript.

Funding: This research was funded by the project of the National Key Research and Development Program of China, grant number (2018YFC2001304).

Informed Consent Statement: Not applicable.

Data Availability Statement: Not applicable.

Conflicts of Interest: The authors declare no conflict of interest.

Appendix A

The simplified model of lower-limb prosthesis has been built before. Its dynamic equation is shown in Equation (3). We define

$$
\begin{aligned}
S_1 &= \sin(q_b+q_h), C_1 = \cos(q_b+q_h) \\
S_2 &= \sin(q_k-q_b-q_h), C_2 = \cos(q_k-q_b-q_h) \\
S_3 &= \sin(q_a+q_b+q_h-q_k), C_3 = \cos(q_a+q_b+q_h-q_k)
\end{aligned}
\tag{A1}
$$

The values of parameters in the dynamic equation are given in Table A1. m_b, m_t, m_s, and m_f are the masses of floating base, thigh, shank, and foot, respectively: I_b, I_t, I_s, and I_f are the moment of inertia of floating base, thigh, shank, and foot, respectively; L_t and L_s are the length of thigh and shank, respectively; L_{tc} is the length from hip joint to mass center of thigh; L_{sc} is the length from knee joint to mass center of shank; L_{fc} is the length from ankle joint to mass center of foot, and the definitions of the rest of the variables are shown in Figure 2.

Table A1. The values of parameters in dynamic equation.

Parameters	Value
$M_G(1,1), M_G(2,2)$	$m_b+m_f+m_s+m_t$
$M_G(1,2)$, $M_G(2,1)$	0
$M_G(1,3)$, $M_G(1,4)$, $M_G(3,1)$, $M_G(4,1)$	$m_tL_{tc}C_1 + m_s(L_tC_1+L_{sc}C_2) + m_f(L_tC_1 + L_sC_2 + L_{fc}C_3)$
$M_G(1,5), M_G(5,1)$	$-m_sL_{sc}C_2 - m_f(L_sC_2 + L_{fc}C_3)$
$M_G(1,6), M_G(6,1)$	$m_fL_{fc}C_3$
$M_G(2,3), M_G(2,4)$, $M_G(3,2)$, $M_G(4,2)$	$m_tL_{tc}S_1 + m_s(L_tS_1 - L_{sc}S_2) + m_f(L_tS_1 - L_sS_2 + L_{fc}S_3)$
$M_G(2,5)$, $M_G(5,2)$	$m_sL_{sc}S_2 - m_f(L_sS_2 - L_{fc}S_3)$
$M_G(2,6)$, $M_G(6,2)$	$m_fL_{fc}S_3$
$M_G(3,3)$	$I_b + I_t + I_s + I_f + 2m_fL_{fc}^2 + m_fL_s^2 - m_fL_t^2 + 2m_sL_{sc}^2 + m_sL_t^2 + 2m_tL_{tc}^2 + 2m_fL_sL_{fc}\cos(q_a) + 2m_fL_tL_s\cos(q_k) + 2m_sL_tL_{sc}\cos(q_k) + 2m_fL_tL_{fc}\cos(q_a - q_k)$
$M_G(3,4)$, $M_G(4,3)$, $M_G(4,4)$	$M_G(3,3) - I_b$
$M_G(3,5)$, $M_G(4,5)$, $M_G(5,3)$, $M_G(5,4)$	$-I_s - I_f - 2m_sL_{sc}^2 - m_sL_tL_{sc}\cos(q_k) - 2m_fL_{fc}^2 - m_fL_s^2 - 2m_fL_sL_{fc}\cos(q_a) - m_fL_tL_s\cos(q_k) - m_fL_tL_{fc}\cos(q_a - q_k)$
$M_G(3,6)$, $M_G(4,6)$, $M_G(6,3)$, $M_G(6,4)$	$I_f + 2m_fL_{fc}^2 + m_fL_sL_{fc}\cos(q_a) + m_fL_tL_{fc}\cos(q_a - q_k)$
$M_G(5,5)$	$I_s + I_f + 2m_sL_{sc}^2 + m_fL_s^2 + 2m_fL_{fc}^2 + 2m_fL_sL_{fc}\cos(q_a)$
$M_G(5,6)$, $M_G(6,5)$	$-I_f - 2m_fL_{fc}^2 - m_fL_sL_{fc}\cos(q_a)$
$M_G(6,6)$	$I_f + 2m_fL_{fc}^2$
$C_G(1,1)$, $C_G(1,2)$, $C_G(2,1)$, $C_G(2,2)$, $C_G(3,1)$, $C_G(3,2)$, $C_G(4,1)$, $C_G(4,2)$, $C_G(5,1)$, $C_G(5,2)$, $C_G(6,1)$, $C_G(6,2)$, $C_G(6,6)$	0
$C_G(1,3)$, $C_G(1,4)$	$[m_f(L_{fc}S_3 - L_sS_2) - m_sL_{sc}S_2]\dot{q}_k - m_fL_{fc}S_3\dot{q}_a - [m_f(L_{fc}S_3 - L_sS_2 + L_tS_1) + m_s(L_tS_1 - L_{sc}S_2) + m_tL_{tc}S_1](\dot{q}_h + \dot{q}_b)$
$C_G(1,5)$	$[m_f(L_{fc}S_3 - L_sS_2) - m_sL_{sc}S_2](\dot{q}_h + \dot{q}_b - \dot{q}_k) + m_fL_{fc}S_3\dot{q}_a$
$C_G(1,6)$	$-m_fL_{fc}S_3(\dot{q}_h + \dot{q}_b - \dot{q}_k + \dot{q}_a)$
$C_G(2,3)$, $C_G(2,4)$	$-[m_f(L_{fc}C_3 + L_sC_2) + m_sL_{sc}C_2]\dot{q}_k + m_fL_{fc}C_3\dot{q}_a + [m_f(L_{fc}C_3 + L_sC_2 + L_tC_1) + m_s(L_tC_1 + L_{sc}C_2) + m_tL_{tc}C_1](\dot{q}_h + \dot{q}_b)$
$C_G(2,5)$	$[m_f(L_{fc}C_3+L_sC_2) + m_sL_{sc}C_2](\dot{q}_k - \dot{q}_h - \dot{q}_b) - m_fL_{fc}C_3\dot{q}_a$
$C_G(2,6)$	$m_fL_{fc}C_3(\dot{q}_h+\dot{q}_b - \dot{q}_k + \dot{q}_a)$
$C_G(3,3)$, $C_G(3,4)$, $C_G(4,3)$, $C_G(4,4)$	$[m_fL_sL_{fc}\sin(q_a) + m_fL_tL_{fc}\sin(q_a - q_k)]\dot{q}_a - [m_fL_tL_s\sin(q_k) + m_sL_tL_{sc}\sin(q_k) - m_fL_tL_{fc}\sin(q_a - q_k)]\dot{q}_k$
$C_G(3,5)$, $C_G(4,5)$	$[m_fL_sL_{fc}\sin(q_a) + m_fL_tL_{fc}\sin(q_a - q_k)]\dot{q}_a + [m_fL_tL_s\sin(q_k) + m_sL_tL_{sc}\sin(q_k) - m_fL_tL_{fc}\sin(q_a - q_k)](\dot{q}_k - \dot{q}_h - \dot{q}_b)$
$C_G(3,6)$, $C_G(4,5)$	$-[m_fL_sL_{fc}\sin(q_a) + m_fL_tL_{fc}\sin(q_a - q_k)](\dot{q}_h + \dot{q}_b - \dot{q}_k + \dot{q}_a)$
$C_G(5,3)$, $C_G(5,4)$	$m_fL_sL_{fc}\sin(q_a)\dot{q}_a + [m_fL_tL_s\sin(q_k) + m_sL_tL_{sc}\sin(q_k) - m_fL_tL_{fc}\sin(q_a - q_k)](\dot{q}_h + \dot{q}_b)$

Table A1. *Cont.*

Parameters	Value
$C_G(5,5)$	$-m_f L_s L_{fc} \sin(q_a)\dot{q}_a$
$C_G(5,6)$	$m_f L_s L_{fc} \sin(q_a)(\dot{q}_h + \dot{q}_b - \dot{q}_k + \dot{q}_a)$
$C_G(6,3)$, $C_G(6,4)$	$[m_f L_s L_{fc} \sin(q_a) + m_f L_t L_{fc} \sin(q_a - q_k)](\dot{q}_h + \dot{q}_b) - m_f L_s L_{fc} \sin(q_a)\dot{q}_k$
$C_G(6,5)$	$m_f L_s L_{fc} \sin(q_a)(\dot{q}_h + \dot{q}_b - \dot{q}_k)$
$G_G(1,1)$	0
$G_G(2,1)$	$(m_b + m_f + m_s + m_t)g$
$G_G(3,1)$, $G_G(4,1)$	$(L_t S_1 - L_{sc} S_2) m_s g + (L_t S_1 - L_s S_2 + L_{fc} S_3) m_f g + L_{tc} S_1 m_t g$
$G_G(5,1)$	$(L_{fc} S_3 - L_s S_2) m_f g + L_{sc} S_2 m_s g$
$G_G(6,1)$	$L_{fc} S_3 m_f g$

References

1. Sagawa, Y.; Turcot, K.; Armand, S.; Thevenon, A.; Vuillerme, N.; Watelain, E. Biomechanics and physiological parameters during gait in lower-limb amputees: A systematic review. *Gait Posture* **2011**, *33*, 511–526. [CrossRef] [PubMed]
2. Ettema, S.; Kal, E.; Han, H. Energy cost of walking in people after lower limb amputation: A systematic review and meta-analysis. *Gait Posture* **2020**, *81*, 89–90. [CrossRef]
3. Robert, G.; Kerry, A.; Julie, C.; Jennifer, K.; Mariah, R. Review of secondary physical conditions associated with lower-limb amputation and long-term prosthesis use. *J. Rehabil. Res. Dev.* **2008**, *45*, 15–30.
4. Tucker, M.R.; Olivier, J.; Pagel, A.; Bleuler, H.; Bouri, M.; Lambercy, O.; del R Millán, J.; Riener, R.; Vallery, H.; Gassert, R. Control strategies for active lower extremity prosthetics and orthotics: A review. *J. NeuroEngineering Rehabil.* **2015**, *12*, 1. [CrossRef] [PubMed]
5. Pfeifer, S.; Allery, H.; List, R.J.; Perreault, E.J.; Riener, R. Finding Best Predictors for the Control of Transfemoral Prostheses. 2010. Available online: https://www.automed2010.ethz.ch/tagungsband/sitzung5_beitrag4_pfeifer.pdf (accessed on 26 January 2023).
6. Wang, W.; Li, J.; Li, W.; Sun, L.N. An Echo-Based Gait Phase Determination Method of Lower Limb Prosthesis. *Mechatron. Intell. Mater. III PTS 1-3* **2013**, *706*, 629–634. [CrossRef]
7. Lyu, Y.; Fang, H.; Xu, J.; Jianmin, M.; Qining, W.; Xiaoxu, Z. Dynamic Modeling and Analysis of the Lower Limb Prosthesis with Four-bar Linkage Prosthetic Knee. *Chin. J. Theor. Appl. Mech.* **2020**, *52*, 1157–1173.
8. Thatte, N.; Geyer, H. Toward Balance Recovery with Leg Prostheses Using Neuromuscular Model Control. *IEEE Trans. Biomed. Eng.* **2016**, *63*, 904–913. [CrossRef] [PubMed]
9. Gregg, R.D.; Sensinger, J.W. Towards Biomimetic Virtual Constraint Control of a Powered Prosthetic Leg. *IEEE Trans. Control Syst. Technol.* **2014**, *22*, 246–254. [CrossRef] [PubMed]
10. Gregg, R.D.; Lenzi, T.; Hargrove, L.J.; Sensinger, J.W. Virtual Constraint Control of a Powered Prosthetic Leg: From Simulation to Experiments with Transfemoral Amputees. *IEEE Trans. Robot.* **2014**, *30*, 1455–1471. [CrossRef] [PubMed]
11. Quintero, D.; Lambert, D.J.; Villarreal, D.J.; Gregg, R.D.; Department of Mechanical Engineering. Real-Time continuous gait phase and speed estimation from a single sensor. In Proceedings of the Conference on Control Technology and Applications (CCTA), Maui, HI, USA, 27–30 August 2017.
12. Quintero, D.; Villarreal, D.J.; Lambert, D.J.; Kapp, S.; Gregg, R.D. Continuous-Phase Control of a Powered Knee–Ankle Prosthesis: Amputee Experiments Across Speeds and Inclines. *IEEE Trans. Robot.* **2018**, *34*, 686–701. [CrossRef] [PubMed]
13. Shultz, A.H.; Goldfarb, M. A Unified Controller for Walking on Even and Uneven Terrain with a Powered Ankle Prosthesis. *IEEE Trans. Neural Syst. Rehabil. Eng.* **2018**, *26*, 788–797. [CrossRef] [PubMed]
14. Culver, S.; Bartlett, H.; Shultz, A.; Goldfarb, M. A Stair Ascent and Descent Controller for a Powered Ankle Prosthesis. *IEEE Trans. Neural Syst. Rehabil. Eng.* **2018**, *26*, 993–1002. [CrossRef] [PubMed]
15. Ledoux, E.D.; Goldfarb, M. Control and Evaluation of a Powered Transfemoral Prosthesis for Stair Ascent. *IEEE Trans. Neural Syst. Rehabil. Eng.* **2017**, *25*, 917–924. [CrossRef] [PubMed]
16. Wen, Y.; Si, J.; Brandt, A.; Gao, X.; Huang, H.H. Online Reinforcement Learning Control for the Personalization of a Robotic Knee Prosthesis. *IEEE Trans. Cybern.* **2019**, *50*, 2346–2356. [CrossRef] [PubMed]
17. Heins, S.; Tolu, S.; Ronsse, R. Online Learning of the Dynamical Internal Model of Transfemoral Prosthesis for Enhancing Compliance. *IEEE Robot. Autom. Lett.* **2021**, *6*, 6156–6163. [CrossRef]
18. Vijayakumar, S.; D'Souza, A.; Schaal, S. Incremental Online Learning in High Dimensions. *Neural Comput.* **2005**, *17*, 2602–2634. [CrossRef]
19. Camargo, J.; Ramanathan, A.; Flanagan, W.; Young, A. A comprehensive, open-source dataset of lower limb biomechanics in multiple conditions of stairs, ramps, and level-ground ambulation and transitions. *J. Biomech.* **2021**, *119*, 110320. [CrossRef] [PubMed]
20. Camargo, J.; Ramanathan, A.; Csomay-Shanklin, N.; Young, A. Automated gap-filling for marker-based biomechanical motion capture data. *Comput. Methods Biomech. Biomed. Eng.* **2020**, *23*, 1180–1189. [CrossRef] [PubMed]

21. Klanke, S.; Vijayakumar, S.; Schaal, S. A Library for Locally Weighted Projection Regression. *J. Mach. Learn. Res.* **2008**, *9*, 623–626.
22. Alejandro, G.Y.; Francisco, D.F.; Jesus, D.G.; Oscar, L. Effects of Discretization Methods on the Performance of Resonant Controllers. *IEEE Trans. Power Electron.* **2010**, *25*, 1692–1712.

machines

MDPI

Article

Visual Place Recognition in Changing Environments with Sequence Representations on the Distance-Space Domain

Ioannis Tsampikos Papapetros [1,*], Ioannis Kansizoglou [1], Loukas Bampis [2] and Antonios Gasteratos [1]

[1] Department of Production and Management Engineering, Democritus University of Thrace, Vas. Sophias 12, GR-671 32 Xanthi, Greece; ikansizo@pme.duth.gr (I.K.); agaster@pme.duth.gr (A.G.)
[2] Department of Electrical and Computer Engineering, Democritus University of Thrace, Building B, Kimmeria Campus, GR-671 32 Xanthi, Greece; lbampis@ee.duth.gr
* Correspondence: ipapapet@pme.duth.gr; Tel.: +30-2541-079359

Abstract: Navigating in a perpetually changing world can provide the basis for numerous challenging autonomous robotic applications. With a view to long-term autonomy, visual place recognition (vPR) systems should be able to robustly operate under extreme appearance changes in their environment. Typically, the utilized data representations are heavily influenced by those changes, negatively affecting the vPR performance. In this article, we propose a sequence-based technique that decouples such changes from the similarity estimation procedure. This is achieved by remapping the sequential representation data into the distance-space domain, i.e., a domain in which we solely consider the distances between image instances, and subsequently normalize them. In such a way, perturbations related to different environmental conditions and embedded into the original representation vectors are avoided, therefore the scene recognition efficacy is enhanced. We evaluate our framework under multiple different instances, with results indicating a significant performance improvement over other approaches.

Keywords: visual place recognition; changing environments; sequence matching; localization; navigation

Citation: Papapetros, I.T. ; Kansizoglou, I.; Bampis, L.; Gasteratos, A. Visual Place Recognition in Changing Environments with Sequence Representations on the Distance-Space Domain. *Machines* **2023**, *11*, 558. https://doi.org/10.3390/machines11050558

Academic Editors: Jian Wu, Xiangkun He and Guangfei Xu

Received: 27 March 2023
Revised: 5 May 2023
Accepted: 11 May 2023
Published: 16 May 2023

1. Introduction

The utilization of autonomous mobile robots promises to revolutionize numerous industrial and domestic fields. Navigating safely through the environment is of utmost importance for the majority of those applications. Yet, at the center of every navigation module lies a localization framework, the purpose of which is to determine the platforms' position relative to their environment. In the vast variety of related approaches, visual localization [1] holds a vital role and can offer an edge over the alternatives [2–4], depending on the nature of the underlying problem.

The task of visual place recognition (vPR) describes the process of searching a visual scene into a database of mapped places in order to locate a match, should one exist [5]. Multiple methods have been proposed to address this task, offering adequate results [6]. However, in most cases, their employment in a continuously running scenario significantly limits their performance both in terms of scalability [7] as well as the change in the environment due to the difference in the visual conditions [8] or the existence of dynamic objects [9]. In spite of applications where the observable environment can be assumed to have a constant state, long-term practices should be able to recognize places under severe environmental changes [10–12].

With this article, we propose an unsupervised long-term visual place recognition technique which utilizes sequential observations to reinforce the individual scene representations. We argue that those individual representations are not a direct depiction of the content of the environment, but are heavily affected by the observed visual conditions. The aim of this work is to alleviate the aforementioned condition effect and thus improve the performance of vPR. This is achieved by remapping the image encodings from the descriptor-space to the distance-space and subsequently normalizing the produced signals. In such a way, any existing constant noise between the underlying representations is removed prior to the matching process. Thus, the scene's structure is captured in a similar manner between the database and query observations, alleviating the effects of environmental changes. More specifically, the main contributions of this work can be summarized as follows:

1. Introducing a process for data manipulation in the distance-space domain, with a view to solve the vPR problem;
2. Offering a concept for a descriptor-agnostic procedure for long-term vPR practices, utilizing multiple sequential observations and generalizing place-specific representations into different and unseen environments;
3. Providing insights about the utilization of image sequences for scene representations;
4. Using multiple datasets to evaluate the behavior of the proposed method under different situations, such as velocity and viewpoint fluctuations, varied environments, as well as observed conditions, etc.

In the following section we briefly analyze the related literature, wherein we situate this work. In Section 3, an intuitive explanation of this framework is presented, followed by an extensive description of its implementation details. Subsequently, Section 4 features the method's evaluation procedure, while Section 5 presents the conclusions, a discussion of the key findings and a prompt for possible extensions.

2. Related Work

A vPR framework should be able to efficiently map the observed environment, so that later impressions of a place can be easily identified and retrieved, despite any condition-related changes in its appearance. During navigation, a robot constantly perceives its environment via repetitive discrete sensor readings. Those discrete readings—in this article considered as images—can be used as a source of information relative to the place they were observed. Representing a visual scene with encodings of that information is a fairly popular technique in the vPR challenge [6,7].

Employing local features, i.e., individual patches within the image [13], one can detach the extracted information from the underlying viewpoint-related spatial structure of the scene, enabling the matching of different poses of the same content observations [14–17]. On the other hand, utilizing the spatial relationship of the extracted information on the image plane may limit the viewpoint tolerance but can significantly improve the matching quality of places with major condition-related differences [8,18–20].

While the single-view scene encodings can provide a natively simple vPR method with satisfactory results, augmenting the input data with information from temporally close observations can significantly improve the final outcome. A way of utilizing the sequential information of an observation is to reinforce or weaken the belief that a matched place is a true positive detection. This can be achieved by observing whether or not temporally close observations are matched to the same place or are scattered in the database [11,21,22]. This technique (sequence-based matching) is highly effective, especially in cases prone to perceptual aliasing, and it can be used as an extra layer to filter out uncertain detections.

Furthermore, appending scene information from sequential and discrete observations into a single representation vector, or exploiting the spatial relationship of the visual clues beyond the two-dimensional plane, so as to form the sequence-based representations of a place, can enhance the amount of the extracted information [23–26]. However, merging sequential data without some kind of supervision can introduce noise, degrading the place recognition performance. Thus, several methods have adopted a sequence generation module to filter out noisy visual clues [21,27].

In a similar manner, our method utilizes multiple successive frames to reinforce the perception of the observable scene. Although the lack of a sequence generation module unavoidably introduces some kind of noise, the proposed framework is able to achieve state-of-the-art performance.

3. Method

3.1. Intuition

The purpose of place recognition is to map the environment and speculate whether or not a scene has been seen in the past. To achieve that, each scene has to be somehow abstracted. In this subsection, we will attempt to intuitively analyze the overall pipeline of describing a visual scene in place recognition.

Ideally, we would like to measure the similarity of the scenes based on their content. Unfortunately, the available scene perception methods fail to identify the whole of the underlying information. Furthermore, the process of capturing this information, the condition of the environment, as well as the utilized encoding method can significantly affect the produced representation. Building on that intuition, our method aims to remove the condition effect from the final similarity result of the matching between two scenes. This is achieved by remapping the representation vectors from the description-space into the distance-space, while appropriately representing them, so as for matching assessment to be feasible (Figure 1). In this context, the distance-space is the domain where observations are represented as their distance from a given target. Thus, instead of directly measuring the similarity between scenes, we utilize the relative distance of sequential readings from a common reference. Typically, the calculated relative distance among a specific pair of representation vectors could be generated by a multitude of different vectors, thus decreasing its distinctiveness. In contrast, using sequences of those relative distance values generates highly improbable combinations to be observed randomly. This way, a mutual frame of reference is provided allowing a fair matching. With the reference image belonging into the first 'query' scene, the produced distance vector of the second one can be intuitively perceived as a similar vector displaced by an approximately constant coefficient. This coefficient originated from the difference in environmental conditions between the query frame and the second scene, and it is not present in the first scene's vector. Therefore, a normalization scheme is employed to remove this component, discarding the condition effect prior to the matching procedure.

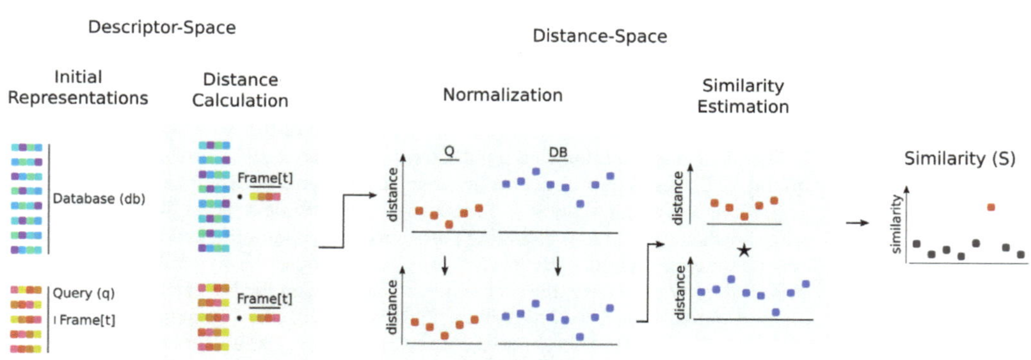

Figure 1. An overview of the proposed method. First, the distances between the representations of the frame of interest and both the query (q) and database (db) sequences, are calculated, forming the new distance-space representations Q and DB. Then, the normalization process takes place, regularizing the produced distance signals to lie within the same range. Finally, the similarity among the frame at timestep t and the database is estimated by sliding the newly formed Q signal over the normalized DB one and computing their cross-correlation (star symbol).

3.2. Implementation

As already mentioned, in order to effectively compare visual scenes, some sort of representation vectors should be generated from the captured images. Considering the marginal effect of the employed representation method in our work, without loss of generality, we assume that the image representation vectors have been already generated.

Let us define a given database (db) as a stream of all seen scenes represented by multidimensional vectors, with l_{db} indicating the length of db. In addition, we consider a query trajectory as a similar stream, from which we form the query sequences (q), by selecting a vector of the image of interest concatenated with its adjacent ones. Given the above, we transform both the database and the query sequences into a one-dimensional time-series. In specific, considering the image of interest at time t, we compute its distance from each image in db as well as each corresponding one in q. Using the following formulas, the new query $Q = [\ldots, Q_i, \ldots], \forall i \in \{i \in \mathbb{N} : t - w \leq i \leq t + w\}$ and database $DB = [\ldots, DB_j, \ldots], \forall j \in \{j \in \mathbb{N} : 0 \leq j \leq l_{db}\}$ signals result from:

$$Q_i = 1 - \frac{\hat{q}_i \cdot \hat{q}_t}{||q_i|| \cdot ||q_t||}, \forall i \in \{i \in \mathbb{N} : t - w \leq i \leq t + w\}, \quad (1)$$

$$DB_j = 1 - \frac{\hat{db}_j \cdot \hat{q}_t}{||db_j|| \cdot ||q_t||}, \forall j \in \{j \in \mathbb{N} : 0 \leq j \leq l_{db}\}, \quad (2)$$

where w represents half the size of the sliding window containing the adjacent images in the query stream. The upper plot in Figure 2 shows a snapshot of the produced one-dimensional signals on a sample subset of the Nordland [12] dataset. It can be observed that the database signal tends to match the query one, displaced by a constant component.

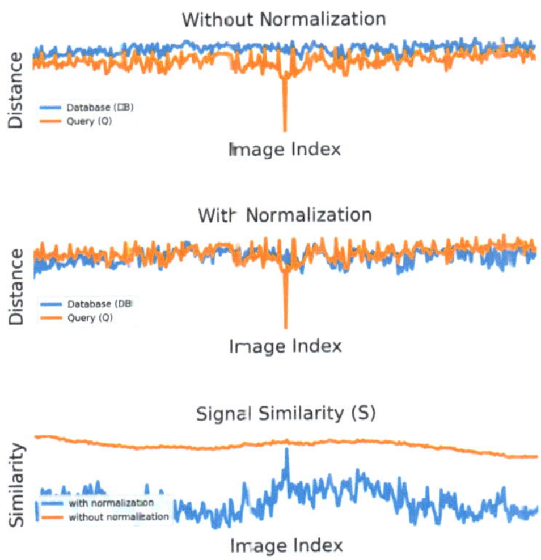

Figure 2. A formed query (Q) representation signal along its database (DB) correspondence on the Nordland dataset [12], before (upper row) and after (middle row) the normalization procedure. Using Q as a sliding window over DB, the cross-correlation results (lower row) of those signal pairs demonstrate that the normalized ones, produce a crisp similarity peak, vigorously indicating a query match in the database.

By utilizing the process of template matching, i.e., calculating the similarity of the template Q over DB, one can deduce meaningful information regarding the origin of the query signal in the database. Although, due to the non-standardized amplitude of those signals, the resulting similarity can be severely affected (Figure 2). To that end, a z-score normalization [28] is applied over both signals, to express the underlying data relative to their mean and standard deviation values, thus eliminating the observed displacement on the amplitude (vertical) axis, as shown in the middle plot of Figure 2. The z-score function is defined as:

$$zs(x) = \frac{(x - \mu)}{\sigma}, \tag{3}$$

where x represents the data, whilst μ and σ are the mean value and the standard deviation of that data, respectively.

Finally, the process of identifying the similarity (S) between the query Q and the database DB signals is performed by calculating the correlation of the normalized Q with the DB representations. In particular:

$$S = zs(DB) \star zs(Q), \tag{4}$$

with $\star$ notating the cross-correlation operation between the two signals. The output of (4), as well as the matching result of the equivalent non-normalized DB and Q signals are shown in the lower plot of Figure 2.

4. Evaluation

A robust place recognition framework should be able to recognize scenes under severe environmental changes. In this section, we assess the behavior of the proposed method in several critical aspects of such a system.

4.1. Datasets

1. *The Nordland dataset:* to evaluate the framework's resilience in recognizing visual scenes under different seasonal conditions, the Nordland dataset [12] was employed. Its content features front-facing images of a 729 km long railway journey, captured in four seasons. Moreover, due to the rail-moving camera, this dataset minimizes the viewpoint variations between each trajectory. For this evaluation, we selected the test partitions [8] of the summer and winter trajectories, each comprising of 3450 synchronized frames.
2. *The Oxford dataset:* in contrast to the one-to-one image correspondence of the Nordland tracks, the Oxford RobotCar dataset [10,29] contains routes from the central Oxford with inconsistent velocities and viewpoints. The data were captured by a car-mounted sensor during different conditions. From the available data, we chose the central view of the front-facing camera of the day and night trajectories (parts 01 of 2015-02-17-14-42-12 and 2014-12-16-18-44-24, respectively). Moreover, due to faulty GPS signal, we discarded the first 2000 frames, resulting in two routes, each one consisting of 4000 images. Within the trajectory, the mean traveled distance between two successive frames is 0.26 m.
3. *The COLD dataset:* unlike the above-mentioned datasets, the COLD collection (sequence 1 of the Freiburg set) [30] comprises of indoor routes, depicting an office-like environment at different time periods. Similar to the Oxford dataset, its trajectories are asynchronous and they exhibit several viewpoint mutations. Within this work, we selected the sunny and night instances, containing 1598 and 1911 images, respectively, with a mean step distance equal to 0.04 m.

From the above three cases, the COLD dataset was not considered during the evaluation of our proposal's behavior over different parameterizations since its respective results were less informative. However, it is considered in order to provide a distinct scenario for our assessment when comparing against the state-of-the-art.

4.2. Metrics

To examine the performance of our approach, we need to quantify and compare the produced results. Among the plethora of available metrics, in this section we utilize precision-recall and their area under curve (AUC) metrics [31].

4.3. The Parameter w

As described in Section 3.2, the length of the query observation depends on the parameter w, which denotes half the window size (in terms of the number of images) around frame t. Figure 3 contains the precision and recall curves we obtained, by varying a similarity threshold over score S for different values of w on the Nordland and Oxford datasets. The employed image representations for this experiment were generated using the NetVLAD method [18]. Whilst changes in w do not seem to significantly affect the framework's performance on the Nordland routes (Figure 3a), higher values tend to produce better results on the Oxford ones (Figure 3b). This can be explained as a consequence of the non-synchronous frames acquisition of the latter dataset among its trajectories. More specifically, comparisons among highly dissimilar places, i.e., distant from time t, introduce noise during the cross-correlation operation in (4), degrading the final outcome. Yet, larger sequences can accumulate additional information about the observed place and thus reinforce the underlying representations, leading to enhanced performance. Furthermore, as mentioned in Section 3.1, representing an observation using a single value offers trivial distinctiveness. On the contrary, aggregating multiple values from a sequence of observations produces a highly distinctive combination that leads to greater matching performance. Indeed, the results confirm this prediction, as in both datasets higher w values tend to produce better results.

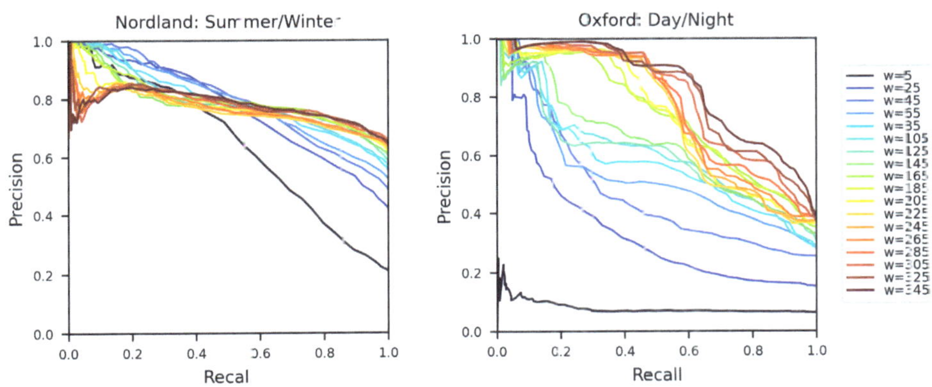

(**a**) (**b**)

Figure 3. The resulting precision and recall curves of our method on the Nordland (**a**) and the Oxford (**b**) datasets, for various values of parameter w. The utilized ground truth radius is $R = 10$ and $R = 40$ for Nordland and Oxford datasets accordingly.

4.4. Different Descriptors

To evaluate the descriptor-agnostic comportment of our framework, we tested a couple of different image encoders. More specifically, we employed the NetVLAD encoder [18], which constitutes a popular well-established network choice for vPR tasks [32,33] and the histogram of oriented gradient (HOG) [34] technique, i.e., a classical unsupervised method utilized by many contemporary systems in the field [17,35,36]. Figure 4 shows the performance of the proposed method, in terms of the change-rate of the AUC score over the raw representations, while feeding different descriptor types as an input. The AUC change-rate is calculated using:

$$cr = \frac{s - r}{r}, \quad (5)$$

with s indicating the AUC score of our method. The values r were obtained by varying a threshold over the similarity metrics among NetVLAD and HOG descriptors, respectively, and computing the corresponding AUC scores. In this manner, the AUC change-rate indicates the performance improvement provided by our technique regardless of the underlined description approach, with higher values of the w parameter consistently performing much better than the baseline.

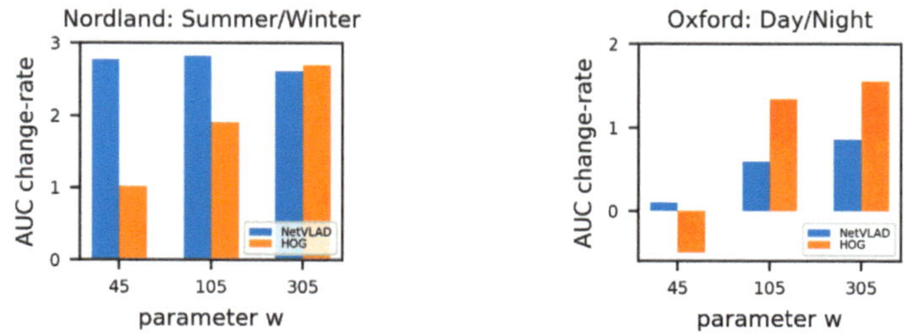

(**a**) (**b**)

Figure 4. Performance evaluation over different representation methods, with the applied metric being the AUC score change-rate of our method over the raw descriptors' one in (**a**) the Nordland and (**b**) the Oxford datasets. The used ground truth radius is $R = 10$ and $R = 40$ respectively.

4.5. Sequence Shuffling

When revisiting a place, images of the environment do not always appear in the same order as its initial observation. In this subsection, we introduce a case study wherein we randomly shuffle the observed images during the query time to emulate such an event. This operation adds noise to the similarity estimation of (4), thus degrading the overall performance. For this case study, we chose to run the experiments on the Nordland dataset due to its one-to-one frame correspondence, as well as the consistent viewpoint of the captured images. Moreover, we fix the parameter w at an average performing query window ($w = 105$). In Figure 5, we plot the AUC score of such experiments, using different percentages of shuffled images from the query sequence. As presented, despite the high percentage of randomly shuffled images within the query observation, the proposed method manages to maintain great results, while outperforming the raw descriptors' score.

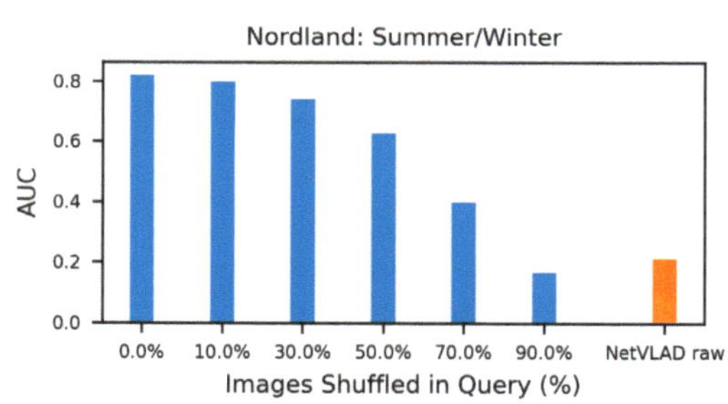

Figure 5. The AUC score of the proposed method, while randomly shuffling a different percentage of the images within the query sequence, alongside the single-image NetVLAD descriptors' result. The applied ground truth radius is $R = 10$. Despite the large percentage of shuffled images, our method manages to maintain great results, outperforming the raw descriptors' score.

4.6. Execution Time

The significance of the place recognition module in the robot navigation pipeline places it under real-time execution constraints. Table 1 shows the mean execution time per image of every part of our system while processing the Nordland dataset, with NetVLAD-generated encodings. As outlined, our method manages to improve the raw NetVLAD descriptors' performance with minimal overhead. The experiments were performed using a Python-based on an 8-core 7-7700K CPU @4.20 GHz and 32 GB of memory.

Table 1. Execution time per image of our method. The data were captured using NetVLAD-generated image representations from the Nordland dataset.

	$\mu \pm \sigma$ **(ms)**
Distance Calculation	17.76 ± 0.55
Normalization	0.28 ± 0.04
Similarity Estimation	0.26 ± 0.08
Total	18.31± 0.68

4.7. Comparison with State-of-the-Art

Whilst the proposed framework demonstrates promising results, it is important to compare them with the performance of similar state-of-the-art techniques, so as to offer an integral understanding regarding its placement within related works in the field. For that purpose, the sequence-based representation method of Delta Descriptors [25] was employed, being the most relevant state-of-the-art approach; alongside the computed raw single-image NetVLAD descriptors, to form the comparison study of this subsection. Figure 6 outlines the benchmark results of the tested methods against the selected datasets, in terms of the precision and recall curves. Similarly, Table 2 exhibits the same results in a single benchmarking value, in terms of AUC score. Note that for the above results, the required place recognition ground-truth was obtained by accepting true-positive matches with the database instance lays within a certain radius R from the query one, in terms of frames for the Nordland and meters for the Oxford and COLD datasets. As shown, the proposed method tends to outperform both the state-of-the-art, as well as the baseline. Moreover, it is worth noting the slightly reduced efficacy relative to the general trend, when using a lower radius for accepting true-positive matches as per the ground-truth. Such a reduction is mainly owed to the multi-frame observations, which are embedded in the used representations, offering a reliable yet imprecise estimate of the correct place in the database. Lastly, the execution time of both methods can be considered minimal, with Delta Descriptors performing slightly slower at 20.47 ms, as reported by deploying the publicly available source codeon the same hardware.

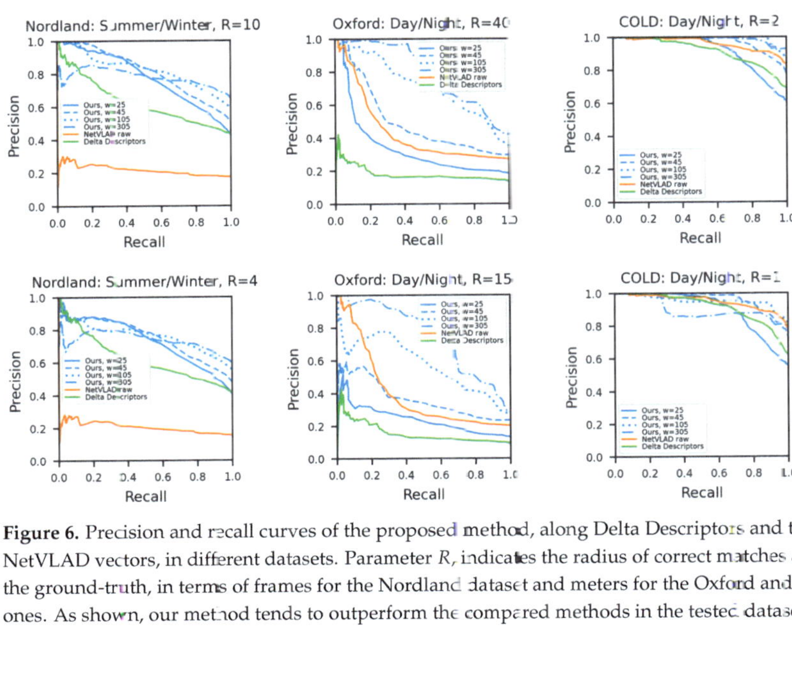

Figure 6. Precision and recall curves of the proposed method, along Delta Descriptors and the raw NetVLAD vectors, in different datasets. Parameter R, indicates the radius of correct matches around the ground-truth, in terms of frames for the Nordland dataset and meters for the Oxford and COLD ones. As shown, our method tends to outperform the compared methods in the tested datasets.

Table 2. The AUC score of the proposed method, along Delta Descriptors and the raw NetVLAD vectors, in different datasets. Parameter R indicates the radius of correct matches around the ground-truth, in terms of frames for the Nordland dataset and meters for the Oxford and COLD ones.

	Nordland $R = 10$	Oxford $R = 40$	COLD $R = 2$	Nordland $R = 4$	Oxford $R = 15$	COLD $R = 1$
Delta Descriptors	0.626	0.166	0.913	0.621	0.147	0.914
NetVLAD raw	0.215	0.438	0.957	0.203	0.401	0.946
Ours $w = 25$	0.767	0.351	0.914	0.725	0.267	0.898
Ours $w = 45$	0.813	0.499	0.969	0.752	0.367	**0.961**
Ours $w = 105$	**0.823**	0.714	**0.990**	**0.760**	0.580	0.951
Ours $w = 305$	0.779	**0.799**	0.988	0.736	**0.751**	0.896

4.8. Query Window as a Time Trace

So far, we have analyzed the performance of the proposed method while placing the query image in the center of the q sequence. In such a layout, the representation benefits by incorporating images from the trajectory without over-extending to frames far from t, thus limiting the drawback of including noise by matching highly dissimilar places, as described in Section 4.3. Yet, in scenarios where the most recent image needs to be utilized for vPR, the described layout is not feasible. For those cases, we modify the query sequence as $Q_{tt} = [\ldots, Q_i, \ldots], \forall i \in \{i \in \mathbb{N} : t - 2w \leq i \leq t\}$ using:

$$Q_i = 1 - \frac{\hat{q}_i \cdot \hat{q}_t}{||q_i|| \cdot ||q_t||}, \forall i \in \{i \in \mathbb{N} : t - w * 2 \leq i \leq t\}. \tag{6}$$

We conducted the same experiments described in Section 4.7 employing the modified query sequence Q_{tt}. Table 3 shows the results of the time trace window experiment. As observed, the AUC score slightly deteriorates in comparison to the results from Table 2, as well as the top-performing window sizes tend to shift to smaller w values. The presented outcome is expected, as a consequence of the abovementioned limitation regarding the added noise in the query sequence.

Table 3. The AUC score of the proposed method, using the time trace window approach, in different datasets. Parameter R indicates the radius of correct matches around the ground-truth, in terms of frames for the Nordland dataset and meters for the Oxford and COLD ones.

	Nordland $R = 10$	Oxford $R = 40$	COLD $R = 2$	Nordland $R = 4$	Oxford $R = 15$	COLD $R = 1$
$w = 25$	0.664	0.415	**0.914**	0.602	0.381	**0.753**
$w = 45$	0.642	**0.618**	0.913	0.590	**0.483**	0.686
$w = 105$	0.735	0.497	0.693	0.687	0.372	0.352
$w = 305$	**0.848**	0.577	0.455	**0.825**	0.225	0.384

5. Conclusions

The autonomous operation of mobile robots in highly mutable environments constitutes a great challenge. Severe appearance changes can significantly limit their navigation ability, thus prohibiting their use in many real-world applications. The article at hand proposed a sequence-based unsupervised vPR technique, capable of operating under severe environmental changes, with the aim to support the navigation procedure of autonomous mobile robots in real-life scenarios. In contrast to the single-frame solutions, utilizing multiple observations to form a single representation significantly enhanced the quality of the produced results. Furthermore, by utilizing the distance-space representation and normalization, our method managed to robustly match scenes across different environmental conditions, without explicit training or parameterization.

Although our method manages to produce superior results along a multitude of environments, observed under challenging conditions, a trade-off between quantity and quality is realized. As described in Section 4.3, the w parameter regulates the number of observations incorporated in the query representation. While higher values of w augment the useful information leading to enhanced matching performance between the query scene and the database, a noise generated from the matching of inconsistent frames amid the associated sequences degrades the final outcome. Furthermore, due to the decreased distinctiveness of the generated distance vectors (Section 3.1), the utilized sequence sizes tend to be quite large. Depending on the specifics of the application, this peculiarity poses a consideration.

As part of future work, we intend to extend our framework with the utilization of an automated sequence generation module. Such a module should suppress the above-mentioned noise, as well as improve the achieved performance for identifying the exact revisited database instance, as explained in Section 4.7.

Author Contributions: Conceptualization, I.T.P. and A.G.; methodology, I.T.P.; software, I.T.P.; validation, I.T.P., I.K. and L.B.; formal analysis, I.T.P. and I.K.; data curation, I.T.P.; writing—original draft preparation, I.T.P.; writing—review and editing, I.K., L.B. and A.G.; visualization, I.T.P.; supervision, L.B.; project administration, A.G.; funding acquisition, A.G. All authors have read and agreed to the published version of the manuscript.

Funding: We acknowledge the support of this work by the project "Study, Design, Development and Implementation of a Holistic System for Upgrading the Quality of Life and Activity of the Elderly" (MIS 5047294) which is implemented under the Action "Support for Regional Excellence", funded by the Operational Programme "Competitiveness, Entrepreneurship and Innovation" (NSRF 2014-2020) and co-financed by Greece and the European Union (European Regional Development Fund).

Data Availability Statement: The above research is based on publicly available datasets.

Conflicts of Interest: The authors declare no conflict of interest.

References

1. Yang, B.; Zhang, Y. Localization and Tracking of Closely-Spaced Human Targets Based on Infrared Sensors. *Infrared Phys. Technol.* **2022**, *123*, 104176. [CrossRef]
2. Cadena, C.; Carlone, L.; Carrillo, H.; Latif, Y.; Scaramuzza, D.; Neira, J.; Reid, I.; Leonard, J.J. Past, Present, and Future of Simultaneous Localization and Mapping: Toward the Robust-Perception Age. *IEEE Trans. Robot.* **2016**, *32*, 1309–1332. [CrossRef]
3. Yin, W.; He, K.; Xu, D.; Luo, Y.; Gong, J. Significant Target Analysis and Detail Preserving Based Infrared and Visible Image Fusion. *Infrared Phys. Technol.* **2022**, *121*, 104041. [CrossRef]
4. Yu, D.; Lin, S.; Lu, X.; Wang, B.; Li, D.; Wang, Y. A Multi-Band Image Synchronous Fusion Method Based on Saliency. *Infrared Phys. Technol.* **2022**, *127*, 104466. [CrossRef]
5. Masone, C.; Caputo, B. A Survey on Deep Visual Place Recognition. *IEEE Access* **2021**, *9*, 19516–19547. [CrossRef]
6. Lowry, S.; Sunderhauf, N.; Newman, P.; Leonard, J.J.; Cox, D.; Corke, P.; Milford, M.J. Visual Place Recognition: A Survey. *IEEE Trans. Robot.* **2016**, *32*, 1–19. [CrossRef]
7. Tsintotas, K.A.; Bampis, L.; Gasteratos, A. The Revisiting Problem in Simultaneous Localization and Mapping: A Survey on Visual Loop Closure Detection. *IEEE Trans. Intell. Transp. Syst.* **2022**, *23*, 1–25. [CrossRef]
8. Olid, D.; Fácil, J.M.; Civera, J. Single-View Place Recognition under Seasonal Changes. In Proceedings of the 2018 IEEE International Conference on Intelligent Robots and Systems, 10th Planning, Perception and Navigation for Intelligent Vehicles Workshop, Madrid, Spain, 1–5 October 2018.
9. Osman, H.; Darwish, N.; Bayoumi, A. PlaceNet: A Multi-Scale Semantic-Aware Model for Visual Loop Closure Detection. *Eng. Appl. Artif. Intell.* **2023**, *119*, 105797. [CrossRef]
10. Maddern, W.; Pascoe, G.; Linegar, C.; Newman, P. 1 Year, 1000 Km: The Oxford RobotCar Dataset. *Int. J. Robot. Res.* **2017**, *36*, 3–15. [CrossRef]
11. Milford, M.J.; Wyeth, G.F. SeqSLAM: Visual Route-Based Navigation for Sunny Summer Days and Stormy Winter Nights. In Proceedings of the 2012 IEEE International Conference on Robotics and Automation, St. Paul, MI, USA, 14–18 May 2012; pp. 1643–1649. [CrossRef]
12. Sunderhauf, N.; Neubert, P.; Protzel, P. Are We There Yet? Challenging SeqSLAM on a 3000 Km Journey Across All Four Seasons. In Proceedings of the ICRA 2013 Workshop on Long-Term Autonomy, Karlsruhe, Germany, 6–10 May 2013; p. 3.
13. Lowe, D.G. Distinctive Image Features from Scale-Invariant Keypoints. *Int. J. Comput. Vis.* **2004**, *60*, 91–110. [CrossRef]
14. Galvez-López, D.; Tardos, J.D. Bags of Binary Words for Fast Place Recognition in Image Sequences. *IEEE Trans. Robot.* **2012**, *28*, 1188–1197. [CrossRef]

15. Sivic, J.; Zisserman, A. Video Google: A Text Retrieval Approach to Object Matching in Videos. In Proceedings of the Ninth IEEE International Conference on Computer Vision, Nice, France, 13–16 October 2003; Volume 2, pp. 1470–1477. [CrossRef]
16. Sünderhauf, N.; Shirazi, S.; Jacobson, A.; Pepperell, E.; Dayoub, F.; Upcroft, B.; Milford, M. Place Recognition with ConvNet Landmarks: Viewpoint-Robust, Condition-Robust, Training-Free. In Proceedings of the Robotics: Science and Systems XI, Rome, Italy, 13–17 July 2015; pp. 1–10. [CrossRef]
17. Zaffar, M.; Ehsan, S.; Milford, M.; McDonald-Maier, K. CoHOG: A Light-Weight, Compute-Efficient, and Training-Free Visual Place Recognition Technique for Changing Environments. *IEEE Robot. Autom. Lett.* **2020**, *5*, 1835–1842. [CrossRef]
18. Arandjelovic, R.; Gronat, P.; Torii, A.; Pajdla, T.; Sivic, J. NetVLAD: CNN Architecture for Weakly Supervised Place Recognition. In Proceedings of the IEEE Conference on Computer Vision and Pattern Recognition, Las Vegas, NV, USA, 26 June–1 July 2016; pp. 5297–5307.
19. Gao, P.; Zhang, H. Long-Term Place Recognition through Worst-case Graph Matching to Integrate Landmark Appearances and Spatial Relationships. In Proceedings of the 2020 IEEE International Conference on Robotics and Automation, Paris, France, 31 May–31 August 2020; pp. 1070–1076. [CrossRef]
20. Khaliq, A.; Ehsan, S.; Chen, Z.; Milford, M.; McDonald-Maier, K. A Holistic Visual Place Recognition Approach Using Lightweight CNNs for Significant ViewPoint and Appearance Changes. *IEEE Trans. Robot.* **2020**, *36*, 561–569. [CrossRef]
21. Papapetros, I.T.; Balaska, V.; Gasteratos, A. Visual Loop-Closure Detection via Prominent Feature Tracking. *J. Intell. Robot. Syst.* **2022**, *104*, 54. [CrossRef]
22. Tsintotas, K.A.; Bampis, L.; Gasteratos, A. Modest-Vocabulary Loop-Closure Detection with Incremental Bag of Tracked Words. *Robot. Auton. Syst.* **2021**, *141*, 103782. [CrossRef]
23. Bampis, L.; Amanatiadis, A.; Gasteratos, A. Fast Loop-Closure Detection Using Visual-Word-Vectors from Image Sequences. *Int. J. Robot. Res.* **2018**, *37*, 62–82. [CrossRef]
24. Diamantas, S.; Dasgupta, P. Optical Flow-Based Place Recognition: Bridging the Gap Between Simulation and Real-World Experiments. In Proceedings of the 2022 IEEE 12th Annual Computing and Communication Workshop and Conference, Virtual, 26–29 January 2022; pp. 619–624. [CrossRef]
25. Garg, S.; Harwood, B.; Anand, G.; Milford, M. Delta Descriptors: Change-Based Place Representation for Robust Visual Localization. *IEEE Robot. Autom. Lett.* **2020**, *5*, 5120–5127. [CrossRef]
26. Tsintotas, K.A.; Bampis, L.; Gasteratos, A. Probabilistic Appearance-Based Place Recognition through Bag of Tracked Words. *IEEE Robot. Autom. Lett.* **2019**, *4*, 1737–1744. [CrossRef]
27. Bampis, L.; Gasteratos, A. Sequence-Based Visual Place Recognition: A Scale-Space Approach for Boundary Detection. *Auton. Robot.* **2021**, *45*, 505–518. [CrossRef]
28. Kreyszig, E. Data Analysis. Probability Theory. In *Advanced Engineering Mathematics*, 10th ed.; Wiley: Hoboken, NJ, USA, 2011; p. 1014.
29. Barnes, D.; Gadd, M.; Murcutt, P.; Newman, P.; Posner, I. The Oxford Radar RobotCar Dataset: A Radar Extension to the Oxford RobotCar Dataset. In Proceedings of the 2020 IEEE International Conference on Robotics and Automation, Paris, France, 31 May–31 August 2020; pp. 6433–6438. [CrossRef]
30. Pronobis, A.; Caputo, B. COLD: The CoSy Localization Database. *Int. J. Robot. Res.* **2009**, *28*, 588–594. [CrossRef]
31. Buckland, M.; Gey, F. The Relationship between Recall and Precision. *J. Am. Soc. Inf. Sci.* **1994**, *45*, 12–19. [CrossRef]
32. Hausler, S.; Garg, S.; Xu, M.; Milford, M.; Fischer, T. Patch-NetVLAD: Multi-Scale Fusion of Locally-Global Descriptors for Place Recognition. In Proceedings of the IEEE/CVF Conference on Computer Vision and Pattern Recognition, Nashville, TN, USA, 20–25 June 2021; pp. 14141–14152.
33. Khaliq, A.; Milford, M.; Garg, S. MultiRes-NetVLAD: Augmenting Place Recognition Training with Low-Resolution Imagery. *IEEE Robot. Autom. Lett.* **2022**, *7*, 3882–3889. [CrossRef]
34. Dalal, N.; Triggs, B. Histograms of Oriented Gradients for Human Detection. In Proceedings of the 2005 IEEE Computer Society Conference on Computer Vision and Pattern Recognition, San Diego, CA, USA, 20–26 June 2005; Volume 1, pp. 886–893. [CrossRef]
35. Lowry, S.; Andreasson, H. Lightweight, Viewpoint-Invariant Visual Place Recognition in Changing Environments. *IEEE Robot. Autom. Lett.* **2018**, *3*, 957–964. [CrossRef]
36. Han, F.; Yang, X.; Deng, Y.; Rentschler, M.; Yang, D.; Zhang, H. SRAL: Shared Representative Appearance Learning for Long-Term Visual Place Recognition. *IEEE Robot. Autom. Lett.* **2017**, *2*, 1172–1179. [CrossRef]

machines

Article

Analytical Sensitivity Analysis of Dynamic Problems with Direct Differentiation of Generalized-α Time Integration

Erich Wehrle [1,2,*] and Veit Gufler [2]

[1] Collins Aerospace, Applied Research and Technology Multidisciplinary Design Optimization Research Group, 80805 Munich, Germany

[2] Faculty of Engineering, Free University of Bozen-Bolzano, Universitätsplatz 1, 39100 Bozen, South Tyrol, Italy; veit.gufler@unibz.it

* Correspondence: erich.wehrle@collins.com

Abstract: In this paper, the direct differentiation of generalized-α time integration is derived, equations are introduced and results are shown. Although generalized-α time integration has found usage, the derivation and the resulting equations for the analytical sensitivity analysis via direct differentiation are missing. Thus, here, the sensitivity equations of generalized-α time integration via direct differentiation are provided. Results with generalized-α are compared with Newmark-β time integration and their sensitivities with numerical sensitivities via forward finite differencing in terms of accuracy and performance. An example is shown for each linear structural dynamics and flexible multibody dynamics.

Keywords: sensitivity analysis; time integration; structural dynamics; flexible multibody dynamics

Citation: Wehrle, E.; Gufler, V. Analytical Sensitivity Analysis of Dynamic Problems with Direct Differentiation of Generalized-α Time Integration. *Machines* **2024**, *12*, 128. https://doi.org/10.3390/machines12020128

Academic Editors: Jian Wu, Xiangkun He and Guangfei Xu

Received: 4 December 2023
Revised: 3 February 2024
Accepted: 5 February 2024
Published: 12 February 2024

1. Introduction

Analytical design sensitivity analysis is developed in this paper for the generalized-α family of time integration. After deriving and introducing design sensitivity using direct differentiation, results are compared with numerical differentiation via forward differencing in terms of accuracy and computational effort.

Efficient sensitivity analysis is the centerpiece of efficient and effective gradient-based design optimization. Other uses of sensitivity analysis include uncertainty analysis and quantification, e.g., the adjoint variable method applied to the sensitivity analysis with respect to uncertain and random variables was developed in [1]. Overviews of design sensitivity analysis are provided by [2–6]. In transient problems, sensitivity analysis must be extended to time integration [7,8], which is handled here by introducing and deriving the equations for the direct differentiation of the generalized-α family of time integration.

In the dynamic systems addressed, the transient response (response in time domain) is of interest and therefore a temporal discretization is applied. The calculation of the system responses, here position, velocity and acceleration, can be carried out with explicit or implicit time integration methods. Implicit time integration allows for larger time steps than explicit methods, which is bounded by the Courant criterion to remain stable [9].

Newmark introduced an implicit integration scheme, now referred to as Newmark-β, in 1959 with two integration parameters γ and β [10]. The Hilber–Hughes–Taylor (HHT) method was introduced in 1977 to improve upon Newmark with the additional integration parameter α_f for numerical damping [11], while the Wood–Bossak–Zienkiewicz (WBZ, also known as simply Bossak or Wood) introduced the parameter α_m [12]. Generalized-α was introduced by Chung and Hulbert in 1993 [13] to generalize all these methods with the integration parameters α_m, α_f, β and γ. These methods can be generalized via a single implementation which is therefore referred to as the generalized-α family of time integrators. In the generalized-α method, for which the family is named, all parameters can be controlled by a single value, namely the spectral radius at infinity ϱ_∞ ranging from zero, being the highest amount of damping, to unity, being none [13]. It should be noted that we differentiate between the variants of the Greek letter rho: ϱ is used here for the spectral radius at infinity, while ρ is used below for density. The mapping of ϱ_∞ to α_m, α_f, β and γ is shown below. The choice of this value can affect the response for a given time step, see, e.g., [14].

Generalized-α time integration has found usage in both nonlinear structural dynamics and multibody dynamics. Multibody dynamics addresses the dynamic response with kinematic constraints. Rigid multibody dynamics considers the bodies rigid, while flexible multibody dynamics allows for the deformation in the bodies of the system. Applications are shown including both linear structural dynamics and flexible multibody dynamics.

This paper is concerned with the sensitivity analysis of dynamic problems, which is used when one or more parameters affect the system responses. The calculation of the system responses is referred to here as primal analysis so that it may be distinguished from the sensitivity analysis. This is of interest for the efficient design optimization via gradient-based optimization algorithms in addition to investigating the sensitivities themselves or for uncertainty analyses. Methods for sensitivity analysis include analytical, numerical, complex step and algorithmic approaches [5].

Analytical sensitivities can be derived either manually (by hand) or via the automatic differentiation [15,16] of the implemented code. The automatic differentiation and manual differentiation of rigid multibody systems are compared in [17]. Although prone to human error and not automatic, the former represents an exact and one of the fastest methods for sensitivity analysis [18]. Manual sensitivity analysis requires high implementation effort, but the resulting code is typically highly reusable and must therefore only be derived and implemented once. Due to its physics- and knowledge-based nature, in addition to its increased transparency, hand differentiation is preferred by the authors.

This study limits itself to analytical sensitivity analysis, which is further divided into direct differentiation (also known as forward sensitivity analysis) and the adjoint variable method (also known as backward sensitivity analysis). Generally, it can be said that direct differentiation is more efficient than the adjoint variable method if the number of design variables is less than the number of functions to differentiate, i.e., in optimization, the sum of the number of objective functions (one in case of single objective optimization) and the number of constraints. Due to its nature, the adjoint variable method must be performed after the time integration of the primal analysis and integrated backwards in time. Thus, the adjoint variable method, in contrast to direct differentiation, needs dedicated solver routines to integrate backwards in time and cannot reuse primal forward time integration [7,8]. Furthermore, for nonlinear analysis, the system matrices must be stored or recalculated when integrating backwards in time. Therefore, direct differentiation is chosen in this work. The reader is referred to the following literature for the adjoint variable method: [19–30].

A number of studies have been carried out for sensitivity analysis in transient structural dynamics and multibody dynamics. Early works on sensitivity analysis with transient structural dynamics include [31], which bridges sensitivities in static and dynamic analysis via the adjoint variable method, including maximum-value objectives and constraints. In [32], sensitivity analysis with the adjoint variable method and direct differentiation is introduced for time-averaged and time-specific constraints. The German-language monograph [33] clearly demonstrates the analytic sensitivity analysis of transient linear dynamics with Newmark-β time integration, providing all necessary equations for implementation, which also includes the sensitivity of the duration of a dynamic operation. The adjoint variable method for transient structural dynamics with Newmark-β time integration is derived by [34]. The adjoint variable method and direct differentiation of nonlinear and transient problems is handled by [35] in which the crucial aspects of the nonlinear solver for sensitivity analysis are introduced.

The sensitivity analysis of multibody systems is often limited to the governing equations and often not extended to the nonlinear solver or time integration. Exceptions to this can be found for direct differentiation with the divide-and-conquer-based modal superposition method in [36]; for direct differentiation and the adjoint variable method with general formulations of differential–algebraic equations in [37]; for direct differentiation and the adjoint variable method of ordered differential equations in [38]; and for the adjoint variable method with the floating frame of reference formulation in [20]. The adjoint variable method of the governing equations in descriptor form and for generic optimization functions is shown by [21] for several time integration methods, including the backward difference formula, Newmark-, Adams–Bashforth–Moulton and diagonally implicit Runge–Kutta. The derivation of the adjoint variable method based on generalized-α time integration and applied to the finite-element-based formulation is provided by [39]. The transient response sensitivity analysis of localized nonlinear behavior using Newmark-β [40].

Higher-order sensitivities can also be of interest, especially in cases of a low number of degrees of freedom and design variables. Second-order sensitivities via the adjoint variable method with the Runge–Kutta–Fehlberg method are shown in [41]. Second-order sensitivities for multibody systems with the adjoint variable method and differential–algebraic equations are shown in [42]. Higher-order sensitivities are used in [43,44] to develop a surrogate model for optimization. This work is limited to first-order sensitivities, as required by gradient-based optimization, although the equations provided can be readily extended to higher-order derivatives.

The scope here is thus the investigation of the direct differentiation of the generalized-α method. A method in this regard is introduced in [45], yet the needed equations are not derived and shown. The present work builds upon and extends the works to generalized-α. The authors developed Newmark-β sensitivities in [7,46], which are in turn based on [33,34]. In this paper, Newmark-β serves as a comparison to the developed method for generalized-α.

This work first generally introduces generalized-α time integration and its equations. Afterwards, the sensitivity equations are introduced for structural dynamics and flexible multibody dynamics, respectively. A numerical example is shown for each and the results are provided for both the primal and sensitivity analyses in comparison with Newmark-β time integration. We close with a conclusion, including a discussion of this work. Furthermore, a detailed derivation of the effective system matrices is provided in the Appendices A and B.

2. Sensitivity Analysis with Generalized-α Time Integration

In this section, we introduce the equations for an acceleration-form (a-form) of generalized-α time integration using a predictor–corrector scheme and extend this for sensitivity analysis. The acceleration-form refers to the formulation, where the acceleration is the primary variable (i.e., the variable for which it is being solved), and contrasts with the displacement-form (d-form) in which the displacement is the primary variable. The predictor–corrector scheme allows for nonlinear terms in the governing equation in which the predicted values are updated in the nonlinear solver, e.g., Newton–Raphson, and then corrected.

In this work, we extend the well-established generalized-α time integration method to include design sensitivity analysis, which is defined by the differentiation with respect to the design variables (but could include other parameters, e.g., uncertain parameters). Sensitivity analysis is carried out here by directly differentiating both the equations of the time integration as well as the governing equations. We consider both linear structural dynamics as well as flexible multibody dynamics, which is intrinsically nonlinear. After introducing the general equations for generalized-α time integration and its sensitivities, the domain-specific governing equations for linear structural dynamics and flexible multibody dynamics are given. Flexible multibody dynamics is shown using the specific approach developed in [7,47,48] with an index-1 approach and the floating frame of reference formulation.

2.1. Generalized-α Time Integration

The method for primal and sensitivity analysis with generalized-α time integration is introduced in the following. This is based on Newmark's equations for the updated position $\underline{q}$ and velocity vectors $\underline{\dot{q}}$ with the calculated value of the acceleration vector $\underline{\ddot{q}}$,

$$\underline{q}_{n+1} = \underbrace{\underline{q}_n + \Delta t\underline{\dot{q}}_n + \left(\frac{1}{2} - \beta\right)\Delta t^2\underline{\ddot{q}}_n}_{\text{predictor } \underline{q}_{\text{pred}}} + \underbrace{\beta\Delta t^2\underline{\ddot{q}}_{n+1}}_{\text{corrector}}, \tag{1}$$

$$\underline{\dot{q}}_{n+1} = \underbrace{\underline{\dot{q}}_n + (1-\gamma)\Delta t\underline{\ddot{q}}_n}_{\text{predictor } \underline{\dot{q}}_{\text{pred}}} + \underbrace{\gamma\Delta t\underline{\ddot{q}}_{n+1}}_{\text{corrector}}, \tag{2}$$

where γ and β are the Newmark integration parameters. Though the simulation is evaluated at the times t_n and t_{n+1}, values at intermediate points are interpolated. The position q and velocity $\dot{q}$ use time $t_{n+1-\alpha_f}$ and the acceleration $\ddot{q}$ uses time $t_{n+1-\alpha_m}$, giving the approximations by linear interpolation (see Figure 1),

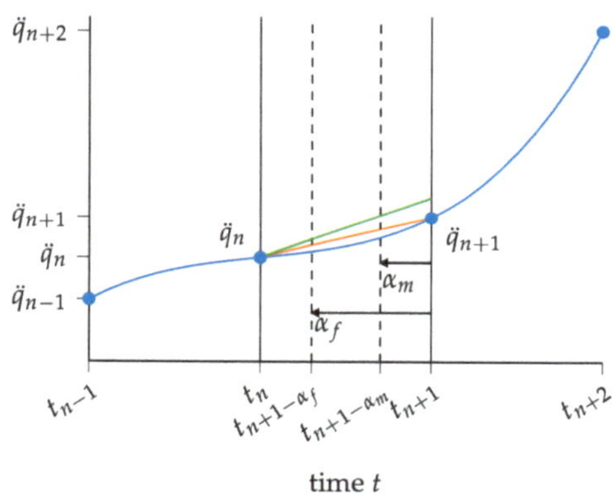

Figure 1. Time integration with — acceleration, — $\gamma = 0.5$, — $\gamma = 0.75$.

$$\underline{q}_{n+1-\alpha_f} = \left(1-\alpha_f\right)\underline{q}_{n+1} + \alpha_f\underline{q}_n, \tag{3}$$

$$\underline{\dot{q}}_{n+1-\alpha_f} = \left(1-\alpha_f\right)\underline{\dot{q}}_{n+1} + \alpha_f\underline{\dot{q}}_n, \tag{4}$$

$$\underline{\ddot{q}}_{n+1-\alpha_m} = (1-\alpha_m)\underline{\ddot{q}}_{n+1} + \alpha_m\underline{\ddot{q}}_n, \tag{5}$$

where α_m and α_f are integration parameters. A predictor–corrector scheme is used in order to accommodate the nonlinear calculation and state-dependency of the system parameters, i.e., the system matrices with respect to position and velocity. The predicted state variables with Newmark's equations are

$$\underline{q}_{\text{pred}} = \underline{q}_n + \Delta t\underline{\dot{q}}_n + \left(\frac{1}{2} - \beta\right)\Delta t^2\underline{\ddot{q}}_n, \tag{6}$$

$$\underline{\dot{q}}_{\text{pred}} = \underline{\dot{q}}_n + (1-\gamma)\Delta t\underline{\ddot{q}}_n, \tag{7}$$

and we calculate $\underline{\ddot{q}}_{n+1}$ using the effective system equations (see below for an explanation of structural dynamics and flexible multibody dynamics, respectively). The predicted values are then corrected with Newmark's equations,

$$\underline{q}_{n+1} = \underline{q}_{\text{pred}} + \beta \Delta t^2 \underline{\ddot{q}}_{n+1}, \tag{8}$$
$$\underline{\dot{q}}_{n+1} = \underline{\dot{q}}_{\text{pred}} + \gamma \Delta t \underline{\ddot{q}}_{n+1}. \tag{9}$$

For the sensitivity analysis within the generalized-α time integration, we use the direct differentiation of the approximations

$$\underline{\underline{\nabla q}}_{n+1-\alpha_f} = \left(1 - \alpha_f\right) \underline{\underline{\nabla q}}_{n-1} + \alpha_f \underline{\underline{\nabla q}}_n, \tag{10}$$
$$\underline{\underline{\nabla \dot{q}}}_{n+1-\alpha_f} = \left(1 - \alpha_f\right) \underline{\underline{\nabla \dot{q}}}_{n-1} + \alpha_f \underline{\underline{\nabla \dot{q}}}_n, \tag{11}$$
$$\underline{\underline{\nabla \ddot{q}}}_{n+1-\alpha_m} = \left(1 - \alpha_m\right) \underline{\underline{\nabla \ddot{q}}}_{n+1} + \alpha_m \underline{\underline{\nabla \ddot{q}}}_n, \tag{12}$$

where the nabla operator ∇ denotes the design sensitivity, i.e., the total derivative with respect to the design variables, e.g., for the sensitivity scalar response with respect to the design variable vector,

$$\underline{\nabla(\cdot)} = \frac{\mathrm{d}(\cdot)}{\mathrm{d}\underline{x}}. \tag{13}$$

The same direct differentiation is applied to the predictors, which are then defined by

$$\underline{\underline{\nabla q}}_{\text{pred}} = \underline{\underline{\nabla q}}_n + \Delta t \underline{\underline{\nabla \dot{q}}}_n + \left(\frac{1}{2} - \beta\right) \Delta t^2 \underline{\underline{\nabla \ddot{q}}}_n, \tag{14}$$
$$\underline{\underline{\nabla \dot{q}}}_{\text{pred}} = \underline{\underline{\nabla \dot{q}}}_n + (1 - \gamma) \Delta t \underline{\underline{\nabla \ddot{q}}}_n. \tag{15}$$

After calculating $\underline{\underline{\nabla \ddot{q}}}_{n+1}$ by use of the effective sensitivity system (see below for explanation of structural dynamics and flexible multibody dynamics, respectively), the following correctors are used:

$$\underline{\underline{\nabla q}}_{n+1} = \underline{\underline{\nabla q}}_{\text{pred}} + \beta \Delta t^2 \underline{\underline{\nabla \ddot{q}}}_{n+1}, \tag{16}$$
$$\underline{\underline{\nabla \dot{q}}}_{n+1} = \underline{\underline{\nabla \dot{q}}}_{\text{pred}} + \gamma \Delta t \underline{\underline{\nabla \ddot{q}}}_{n+1}. \tag{17}$$

For a nonlinear analysis, a Newton solver can be integrated into the solving routine This is explained with flexible multibody systems below in Section 2.3.

The time integration constants of generalized-α are α_m, α_f, β and γ and their choice can lead to a number of established time integration methods, including the Newmark family of integrators, Hilber–Hughes–Taylor and Wood–Bossak–Zienkiewicz. Therefore, this can then be referred to as the α-family of time integrators. The equations introduced above in this section can therefore be generally used within this family. Specifically for the generalized-α time integration used here, the choice of integration constants can be tied to a single user-defined parameter, and the spectral radius at infinity ϱ_∞ [13],

$$\alpha_m = \frac{2\varrho_\infty - 1}{\varrho_\infty + 1}, \tag{18}$$
$$\alpha_f = \frac{\varrho_\infty}{\varrho_\infty + 1}, \tag{19}$$
$$\gamma = \frac{1}{2} - \alpha_m + \alpha_f, \tag{20}$$
$$\beta = \frac{1}{4}\left(1 - \alpha_m + \alpha_f\right)^2. \tag{21}$$

The spectral radius at infinity is sometimes referred to as the (numerical) damping and is limited to a range of $[0, 1]$. The value $\varrho_\infty = 1$ returns the Newmark-β retaining frequencies, which are preserved, while $\varrho_\infty = 0$ annihilates all frequencies above $\frac{\Delta t}{T}$ after one time step in which T is the period of the highest frequency of interest. The results converge to those without numerical damping when the time step goes to zero. Thus, the level of numerical damping introduced to generalized-α time integration depends on both the step size as well as the spectral radius at infinity and the choice is critical for proper results. A parameter study with a fixed-time step and varying the values of the spectral radius at infinity is shown in [14]. As the present study develops a general framework for this methodology, this is not addressed here and the reader is referred to [13,49–51]. In this paper, we use a moderate value of $\varrho_\infty = 0.55$, though the sensitivity analysis methodology developed is general and independent of this parameter value.

2.2. Structural Dynamics

The governing equation for linear structural dynamics is defined by the following second-order equation of motion:

$$\underline{\underline{m}}\,\underline{\ddot{q}}(t) + \underline{\underline{d}}\,\underline{\dot{q}}(t) + \underline{\underline{k}}\,\underline{q}(t) - \underline{F}(t) = \underline{0}, \tag{22}$$

where $\underline{\underline{m}}$ is the mass matrix, $\underline{\underline{d}}$ is the damping matrix, $\underline{\underline{k}}$ is the stiffness matrix and $\underline{F}$ is the external force vector. For linear structural dynamics, the system matrices are constant and independent of position and time. For the solving scheme, we add the residual force vector $\underline{R}(t)$, which is to be zero (to an allowable value) at all times, to the right-hand side. The α-family of integration methods uses intermediate-point approximations between t_n and t_{n+1}, namely $t_{n+1-\alpha_f}$ and $t_{n+1-\alpha_m}$, revealing

$$\underline{R}_{n+1-\alpha_f} = \underline{\underline{m}}\,\underline{\ddot{q}}_{n+1-\alpha_m} + \underline{\underline{d}}\,\underline{\dot{q}}_{n+1-\alpha_f} + \underline{\underline{k}}\,\underline{q}_{n+1-\alpha_f} - \underline{F}_{n+1-\alpha_f}, \tag{23}$$

where the linear approximation of the intermediate external force is

$$\underline{F}_{n+1-\alpha_f} = \left(1-\alpha_f\right)\underline{F}_{n+1} + \alpha_f \underline{F}_n. \tag{24}$$

The effective equation of motion is defined by

$$\underline{\underline{m}}_{\text{eff}}\,\underline{\ddot{q}}_{n+1} = \underline{F}_{\text{eff}}, \tag{25}$$

where

$$\underline{\underline{m}}_{\text{eff}} = (1-\alpha_m)\underline{\underline{m}} + \left(1-\alpha_f\right)\gamma\Delta t\underline{\underline{d}} + \left(1-\alpha_f\right)\beta\Delta t^2\underline{\underline{k}}, \tag{26}$$

$$\begin{aligned}\underline{F}_{\text{eff}} = {} & \left(1-\alpha_f\right)\underline{F}_{n+1} + \alpha_f\underline{F}_n + \\ & - \alpha_m\underline{\underline{m}}\,\underline{\ddot{q}}_n - \underline{\underline{d}}\Big(\left(1-\alpha_f\right)\underline{\dot{q}}_{\text{pred}} + \alpha_f\underline{\dot{q}}_n\Big) - \underline{\underline{k}}\Big(\left(1-\alpha_f\right)\underline{q}_{\text{pred}} + \alpha_f\underline{q}_n\Big).\end{aligned} \tag{27}$$

For the derivation of the effective mass matrix $\underline{\underline{m}}_{\text{eff}}$ and effective force vector $\underline{F}_{\text{eff}}$ for generalized-α method, see Appendix A.1.

The sensitivity analysis is derived using the direct differentiation which is applied to the governing Equation (23). This reveals the sensitivity of the force balance for the α-family of integration methods as

$$\begin{aligned}\underline{\underline{\nabla R}}_{n+1-\alpha_f} = {} & \underline{\underline{\underline{\nabla m}}}\,\underline{\ddot{q}}_{n+1-\alpha_m} + \underline{\underline{m}}\,\underline{\underline{\nabla\ddot{q}}}_{n+1-\alpha_m} + \underline{\underline{\underline{\nabla d}}}\,\underline{\dot{q}}_{n+1-\alpha_f} + \underline{\underline{d}}\,\underline{\underline{\nabla\dot{q}}}_{n+1-\alpha_f} + \\ & + \underline{\underline{\underline{\nabla k}}}\,\underline{q}_{n+1-\alpha_f} + \underline{\underline{k}}\,\underline{\underline{\nabla q}}_{n+1-\alpha_f} - \underline{\underline{\nabla F}}_{n+1-\alpha_f},\end{aligned} \tag{28}$$

where the intermediate external force sensitivity is found with the approximation

$$\underline{\underline{\nabla F}}_{n+1-\alpha_f} = \left(1-\alpha_f\right)\underline{\underline{\nabla F}}_{n+1} + \alpha_f\underline{\underline{\nabla F}}_n. \tag{29}$$

The design sensitivities of the system matrices are three-dimensional matrices and therefore efficient handling is needed.

The effective equation of motion for sensitivities is defined by the differentiation of (25),

$$\underline{\underline{m}}_{\text{eff}} \underline{\underline{\nabla \ddot{q}}}_{n+1} = \underline{\underline{F}}_{\text{pseudo}}, \tag{30}$$

where

$$\underline{\underline{F}}_{\text{pseudo}} = \underline{\underline{\nabla F}}_{\text{eff}} - \underline{\underline{\nabla m}}_{\text{eff}} \ddot{q}_{n+1}, \tag{31}$$

and fully expressed by

$$\begin{aligned} \underline{\underline{F}}_{\text{pseudo}} = & \left(1-\alpha_f\right)\underline{\underline{\nabla F}}_{n+1} + \alpha_f \underline{\underline{\nabla F}}_n - \underline{\underline{\nabla m}}\left((1-\alpha_m)\underline{\ddot{q}}_{n+1} + \alpha_m \underline{\ddot{q}}_n\right) - \alpha_m \underline{\underline{m}}\, \underline{\underline{\nabla \ddot{q}}}_n + \\ & - \underline{\underline{\nabla d}}\left(\left(1-\alpha_f\right)\underline{\dot{q}}_{n+1} + \alpha_f \underline{\dot{q}}_n\right) - \underline{\underline{d}}\left(\left(1-\alpha_f\right)\underline{\underline{\nabla \dot{q}}}_{\text{pred}} + \alpha_f \underline{\underline{\nabla \dot{q}}}_n\right) + \\ & - \underline{\underline{\nabla k}}\left(\left(1-\alpha_f\right)\underline{q}_{n+1} + \alpha_f \underline{q}_n\right) - \underline{\underline{k}}\left(\left(1-\alpha_f\right)\underline{\underline{\nabla q}}_{\text{pred}} + \alpha_f \underline{\underline{\nabla q}}_n\right). \end{aligned} \tag{32}$$

For the derivation of the effective mass matrix $\underline{\underline{m}}_{\text{eff}}$ and effective pseudo force $\underline{\underline{F}}_{\text{pseudo}}$ for the generalized-α method, see Appendices A.1 and A.2.

2.3. Flexible Multibody Dynamics

Flexible multibody dynamics is the analysis of mechanical systems consisting of flexible bodies constrained by kinematic joints modeled with differential–algebraic equations. This work builds upon the previous work of analytical sensitivity analysis with multibody dynamics, including [37,52,53]. In addition to the kinematic constraints with respect to structural dynamics, multibody dynamics also must integrate a nonlinear solver in order to consider the dependencies of the system matrices and vectors on the state variables. This is carried out with Newton steps.

The flexible multibody system is modeled with the floating frame of the reference formulation [54–56]. The generic index-3 differential–algebraic equation for a flexible multibody system with holonomic constraints is written as follows:

$$\underline{\underline{m}}\left(\underline{q},t\right)\underline{\ddot{q}}(t) + \underline{\underline{d}}\left(\underline{q},t\right)\underline{\dot{q}}(t) + \underline{\underline{k}}\left(\underline{q},t\right)\underline{q}(t) + \underline{\underline{J}}_{\Phi}^{T}\left(\underline{q},t\right)\underline{\lambda}(t) = \underline{F}_{\text{ext}}\left(\underline{q},\underline{\dot{q}},t\right) + \underline{F}_{v}\left(\underline{q},\underline{\dot{q}},t\right), \tag{33}$$

$$\underline{\Phi}\left(\underline{q},t\right) = \underline{0}, \tag{34}$$

where $\underline{q}$, $\underline{\dot{q}}$ and $\underline{\ddot{q}}$ are the generalized position, velocity and acceleration vectors; $\underline{\lambda}$ is the vector of Lagrange multipliers of kinematic constraints; the matrices $\underline{\underline{m}}$, $\underline{\underline{d}}$, and $\underline{\underline{k}}$ are the mass, damping and stiffness matrices; $\underline{\underline{J}}_{\Phi}$ is the Jacobian matrix of kinematic constraints; $\underline{F}_{\text{ext}}$ is the vector of external forces; $\underline{F}_{v}$ is the quadratic velocity vector; and $\underline{\Phi}$ is the vector of kinematic constraints at the position level.

Numerical problems related to the solution of index-3 differential–algebraic equations [57] can be avoided by a number of methods including index reduction, coordinate reduction, augmented Lagrangian formulation and preconditioning [58–60]. Although generalized-α time integration is suitable for directly solving index-3 or index-2 differential–algebraic equations, in this work, we address the problem using index reduction leading to an index-1 differential–algebraic equation in which the kinematic constraint function is differentiated twice with respect to time,

$$\underline{\underline{m}}\,\underline{\ddot{q}} + \underline{\underline{d}}\,\underline{\dot{q}} + \underline{\underline{k}}\,\underline{q} + \underline{\underline{J}}_{\Phi}^{T}\underline{\lambda} = \underline{F}_{\text{ext}} + \underline{F}_{v}, \tag{35}$$

$$\underline{\underline{J}}_{\Phi}\,\underline{\ddot{q}} = \underline{F}_{c}, \tag{36}$$

where $\underline{F}_c$ is the right-hand side of acceleration constraints. As index-1 differential–algebraic equations are susceptible to drift in the kinematic constraints, Baumgarte stabilization [61] is used. Kinematic constraint drift is stabilized in this method by adding damping and restoring terms to the acceleration constraint equations that are proportional to the velocity and position constraints. Details of its implementation, including the sensitivity analysis, are described in [14,62].

The generalized-α time integration is based on

$$\left(\underline{\underline{m}}\,\ddot{\underline{q}}\right)_{n+1-\alpha_m} + \left(\underline{\underline{d}}\,\dot{\underline{q}}\right)_{n+1-\alpha_f} + \left(\underline{\underline{k}}\,\underline{q}\right)_{n+1-\alpha_f} + \left(\underline{\underline{\mathrm{J}}}_{\Phi}^{T}\,\underline{\lambda}\right)_{n+1-\alpha_m} = \underline{F}_{\mathrm{ext},n+1-\alpha_f} + \underline{F}_{v,n+1-\alpha_f}, \tag{37}$$

$$\left(\underline{\underline{\mathrm{J}}}_{\Phi}\,\ddot{\underline{q}}\right)_{n+1-\alpha_m} = \underline{F}_{c,n+1-\alpha_f}, \tag{38}$$

with the intermediate approximations of all force terms of the equations of motion. These intermediate approximations are shown in Appendix B.1. Inserting the intermediate approximations of the force terms acting on the flexible multibody system leads to the equations of motion in residual form given by

$$\begin{aligned}\underline{R}_{1,n+1-\alpha_f} = {} & (1-\alpha_m)\underline{\underline{m}}_{n+1}\ddot{\underline{q}}_{n+1} + \alpha_m\underline{\underline{m}}_n\ddot{\underline{q}}_n + \left(1-\alpha_f\right)\underline{\underline{d}}_{n+1}\dot{\underline{q}}_{n+1} + \alpha_f\underline{\underline{d}}_n\dot{\underline{q}}_n + \\ & + \left(1-\alpha_f\right)\underline{\underline{k}}_{n+1}\underline{q}_{n+1} + \alpha_f\underline{\underline{k}}_n\underline{q}_n + (1-\alpha_m)\underline{\underline{\mathrm{J}}}_{\Phi,n+1}^{T}\underline{\lambda}_{n+1} + \alpha_m\underline{\underline{\mathrm{J}}}_{\Phi,n}^{T}\underline{\lambda}_n + \\ & - \left(1-\alpha_f\right)\underline{F}_{\mathrm{ext},n+1} - \alpha_f\underline{F}_{\mathrm{ext},n} - \left(1-\alpha_f\right)\underline{F}_{v,n+1} - \alpha_f\underline{F}_{v,n},\end{aligned} \tag{39}$$

$$\underline{R}_{2,n+1-\alpha_f} = (1-\alpha_m)\underline{\underline{\mathrm{J}}}_{\Phi,n+1}\ddot{\underline{q}}_{n+1} + \alpha_m\underline{\underline{\mathrm{J}}}_{\Phi,n}\ddot{\underline{q}}_n - \left(1-\alpha_f\right)\underline{F}_{c,n+1} - \alpha_f\underline{F}_{c,n}. \tag{40}$$

In the case of a nonlinear mass matrix, as is the case with the floating frame of reference formulation, this weighted formulation of the residual equations are first-order accurate. For a formulation of second-order accuracy, see, e.g., [63], where the dynamic equilibrium is enforced exactly at the time steps of the simulation. The derived time integration and its differentiation with the weighted residual formulation can be extended and applied to further formulations.

The effective equation of motion is defined by

$$\begin{bmatrix} \underline{\underline{m}}_{\mathrm{eff}} & \underline{\underline{\mathrm{J}}}_{\Phi,\mathrm{eff}}^{T} \\ \underline{\underline{\mathrm{J}}}_{\Phi,\mathrm{eff}} & \underline{\underline{0}} \end{bmatrix} \begin{bmatrix} \ddot{\underline{q}}_{n+1} \\ \underline{\lambda}_{n+1} \end{bmatrix} = \begin{bmatrix} \underline{F}_{a,\mathrm{eff}} \\ \underline{F}_{c,\mathrm{eff}} \end{bmatrix}, \tag{41}$$

where

$$\underline{\underline{m}}_{\mathrm{eff}} = (1-\alpha_m)\underline{\underline{m}}_{n+1} + \left(1-\alpha_f\right)\gamma\Delta t\underline{\underline{d}}_{n+1} + \left(1-\alpha_f\right)\beta\Delta t^2\underline{\underline{k}}_{n+1}, \tag{42}$$

$$\underline{\underline{\mathrm{J}}}_{\Phi,\mathrm{eff}} = (1-\alpha_m)\underline{\underline{\mathrm{J}}}_{\Phi,n+1}, \tag{43}$$

$$\begin{aligned}\underline{F}_{a,\mathrm{eff}} = {} & \left(1-\alpha_f\right)\underline{F}_{\mathrm{ext},n+1} + \alpha_f\underline{F}_{\mathrm{ext},n} + \left(1-\alpha_f\right)\underline{F}_{v,n+1} + \alpha_f\underline{F}_{v,n} + \\ & - \alpha_m\,\underline{\underline{m}}_n\,\ddot{\underline{q}}_n - \left(1-\alpha_f\right)\underline{\underline{d}}_{n+1}\dot{\underline{q}}_{\mathrm{pred}} - \alpha_f\underline{\underline{d}}_n\dot{\underline{q}}_n + \\ & - \left(1-\alpha_f\right)\underline{\underline{k}}_{n+1}\underline{q}_{\mathrm{pred}} - \alpha_f\underline{\underline{k}}_n\underline{q}_n - \alpha_m\underline{\underline{\mathrm{J}}}_{\Phi,n}^{T}\underline{\lambda}_n,\end{aligned} \tag{44}$$

$$\underline{F}_{c,\mathrm{eff}} = \left(1-\alpha_f\right)\underline{F}_{c,n+1} + \alpha_f\underline{F}_{c,n} - \alpha_m\underline{\underline{\mathrm{J}}}_{\Phi,n}\ddot{\underline{q}}_n. \tag{45}$$

The derivations of the effective equation of motion and the terms $\underline{\underline{m}}_{\mathrm{eff}}$, $\underline{\underline{\mathrm{J}}}_{\Phi,\mathrm{eff}}$, $\underline{F}_{a,\mathrm{eff}}$ and $\underline{F}_{c,\mathrm{eff}}$ are shown in Appendix B.1.

With flexible multibody dynamics, the system parameters (matrix- and vector-valued for $n_{\mathrm{dof}} > 1$) including mass, damping, stiffness, constraints and forces are generally dependent on state variables. Therefore, a nonlinear solver is needed to solve the system and the Newton–Raphson method is used here of the following form:

$$\begin{bmatrix} \dfrac{\partial\underline{R}_{1,n+1-\alpha_f}}{\partial\ddot{\underline{q}}_{n+1}} & \dfrac{\partial\underline{R}_{1,n+1-\alpha_f}}{\partial\underline{\lambda}_{n+1}} \\ \dfrac{\partial\underline{R}_{2,n+1-\alpha_f}}{\partial\ddot{\underline{q}}_{n+1}} & \dfrac{\partial\underline{R}_{2,n+1-\alpha_f}}{\partial\underline{\lambda}_{n+1}} \end{bmatrix} \begin{bmatrix} \Delta\ddot{\underline{q}}_{n+1} \\ \Delta\underline{\lambda}_{n+1} \end{bmatrix} + \begin{bmatrix} \underline{R}_{1,n+1-\alpha_f} \\ \underline{R}_{2,n+1-\alpha_f} \end{bmatrix} = \underline{0}. \tag{46}$$

With the dependencies shown in (33) and (34), and the terms of the Jacobian matrix (tangent operator) of the residuals with respect to $\underline{\ddot{q}}_{n+1}$ and $\underline{\lambda}_{n+1}$ are given by

$$\begin{aligned}
\frac{\partial \underline{R}_{1,n+1-\alpha_f}}{\partial \underline{\ddot{q}}_{n+1}} = &(1-\alpha_m)\left(\beta\Delta t^2 \frac{\partial \underline{\underline{m}}_{n+1}}{\partial \underline{q}_{n+1}} \underline{\ddot{q}}_{n+1} + \underline{\underline{m}}_{n+1}\right)+ \\
&+\left(1-\alpha_f\right)\left(\beta\Delta t^2 \frac{\partial \underline{\underline{d}}_{n+1}}{\partial \underline{q}_{n+1}} \underline{\dot{q}}_{n+1} + \gamma\Delta t \underline{\underline{d}}_{n+1}\right)+ \\
&+\left(1-\alpha_f\right)\left(\beta\Delta t^2 \frac{\partial \underline{\underline{k}}_{n+1}}{\partial \underline{q}_{n+1}} \underline{q}_{n+1} + \beta\Delta t^2 \underline{\underline{k}}_{n+1}\right)+ \\
&+(1-\alpha_m)\beta\Delta t^2 \frac{\partial \underline{\underline{J}}^T_{\Phi,n+1}}{\partial \underline{q}_{n+1}} \underline{\lambda}_{n+1}+ \\
&-\left(1-\alpha_f\right)\left(\beta\Delta t^2 \frac{\partial \underline{F}_{\mathrm{ext},n+1}}{\partial \underline{q}_{n+1}} + \gamma\Delta t \frac{\partial \underline{F}_{\mathrm{ext},n+1}}{\partial \underline{\dot{q}}_{n+1}}\right)+ \\
&-\left(1-\alpha_f\right)\left(\beta\Delta t^2 \frac{\partial \underline{F}_{v,n+1}}{\partial \underline{q}_{n+1}} + \gamma\Delta t \frac{\partial \underline{F}_{v,n+1}}{\partial \underline{\dot{q}}_{n+1}}\right),
\end{aligned} \tag{47}$$

$$\begin{aligned}
\frac{\partial \underline{R}_{2,n+1-\alpha_f}}{\partial \underline{\ddot{q}}_{n+1}} = &(1-\alpha_m)\left(\beta\Delta t^2 \frac{\partial \underline{\underline{J}}_{\Phi,n+1}}{\partial \underline{q}_{n+1}} \underline{\ddot{q}}_{n+1} + \underline{\underline{J}}_{\Phi,n+1}\right)+ \\
&-\left(1-\alpha_f\right)\left(\beta\Delta t^2 \frac{\partial \underline{F}_{c,n+1}}{\partial \underline{q}_{n+1}} + \gamma\Delta t \frac{\partial \underline{F}_{c,n+1}}{\partial \underline{\dot{q}}_{n+1}}\right),
\end{aligned} \tag{48}$$

$$\frac{\partial \underline{R}_{1,n+1-\alpha_f}}{\partial \underline{\lambda}_{n+1}} = (1-\alpha_m)\underline{\underline{J}}^T_{\Phi,n+1}, \tag{49}$$

$$\frac{\partial \underline{R}_{2,n+1-\alpha_f}}{\partial \underline{\lambda}_{n+1}} = \underline{\underline{0}}. \tag{50}$$

The derivation of these terms is shown in Appendix B.2.

The direct differentiation of the equations of motion (39) and (40) leads to the governing equation of the sensitivity analysis,

$$\begin{aligned}
\underline{\underline{\nabla R}}_{1,n+1-\alpha_f} = &(1-\alpha_m)\underline{\underline{\nabla m}}_{n+1}\underline{\ddot{q}}_{n+1} + \alpha_m \underline{\underline{\nabla m}}_{n}\underline{\ddot{q}}_{n}+ \\
&+(1-\alpha_m)\underline{\underline{m}}_{n+1}\underline{\underline{\nabla \ddot{q}}}_{n+1} + \alpha_m \underline{\underline{m}}_{n}\underline{\underline{\nabla \ddot{q}}}_{n}+ \\
&+\left(1-\alpha_f\right)\underline{\underline{\nabla d}}_{n+1}\underline{\dot{q}}_{n+1} + \alpha_f \underline{\underline{\nabla d}}_{n}\underline{\dot{q}}_{n}+ \\
&+\left(1-\alpha_f\right)\underline{\underline{d}}_{n+1}\underline{\underline{\nabla \dot{q}}}_{n+1} + \alpha_f \underline{\underline{d}}_{n}\underline{\underline{\nabla \dot{q}}}_{n}+ \\
&+\left(1-\alpha_f\right)\underline{\underline{\nabla k}}_{n+1}\underline{q}_{n+1} + \alpha_f \underline{\underline{\nabla k}}_{n}\underline{q}_{n}+ \\
&+\left(1-\alpha_f\right)\underline{\underline{k}}_{n+1}\underline{\underline{\nabla q}}_{n+1} + \alpha_f \underline{\underline{k}}_{n}\underline{\underline{\nabla q}}_{n}+ \\
&+(1-\alpha_m)\underline{\underline{\nabla J}}^T_{\Phi,n+1}\underline{\lambda}_{n+1} + \alpha_m \underline{\underline{\nabla J}}^T_{\Phi,n}\underline{\lambda}_{n}+ \\
&+(1-\alpha_m)\underline{\underline{J}}^T_{\Phi,n+1}\underline{\underline{\nabla \lambda}}_{n+1} + \alpha_m \underline{\underline{J}}^T_{\Phi,n}\underline{\underline{\nabla \lambda}}_{n}+ \\
&-\left(1-\alpha_f\right)\underline{\underline{\nabla F}}_{\mathrm{ext},n+1} - \alpha_f \underline{\underline{\nabla F}}_{\mathrm{ext},n} - \left(1-\alpha_f\right)\underline{\underline{\nabla F}}_{v,n+1} - \alpha_f \underline{\underline{\nabla F}}_{v,n},
\end{aligned} \tag{51}$$

$$\begin{aligned}
\underline{\underline{\nabla R}}_{2,n+1-\alpha_f} = &(1-\alpha_m)\underline{\underline{\nabla J}}_{\Phi,n+1}\underline{\ddot{q}}_{n+1} + \alpha_m \underline{\underline{\nabla J}}_{\Phi,n}\underline{\ddot{q}}_{n} \\
&+(1-\alpha_m)\underline{\underline{J}}_{\Phi,n+1}\underline{\underline{\nabla \ddot{q}}}_{n+1} + \alpha_m \underline{\underline{J}}_{\Phi,n}\underline{\underline{\nabla \ddot{q}}}_{n}+ \\
&-\left(1-\alpha_f\right)\underline{\underline{\nabla F}}_{c,n+1} - \alpha_f \underline{\underline{\nabla F}}_{c,n}.
\end{aligned} \tag{52}$$

The governing sensitivity equation of the effective system is defined by the differentiation of (41),

$$\begin{bmatrix} \underline{\underline{m}}_{\text{eff}} & \underline{\underline{J}}^T_{\Phi,\text{eff}} \\ \underline{\underline{J}}_{\Phi,\text{eff}} & \underline{\underline{0}} \end{bmatrix} \begin{bmatrix} \underline{\underline{\nabla \ddot{q}}}_{n+1} \\ \underline{\underline{\nabla \lambda}}_{n+1} \end{bmatrix} = \begin{bmatrix} \underline{\underline{F}}_{a,\text{pseudo}} \\ \underline{\underline{F}}_{c,\text{pseudo}} \end{bmatrix}, \tag{53}$$

where

$$\begin{bmatrix} \underline{\underline{F}}_{a,\text{pseudo}} \\ \underline{\underline{F}}_{c,\text{pseudo}} \end{bmatrix} = \begin{bmatrix} \underline{\underline{\nabla F}}_{a,\text{eff}} \\ \underline{\underline{\nabla F}}_{c,\text{eff}} \end{bmatrix} - \begin{bmatrix} \underline{\underline{\nabla m}}_{\text{eff}} & \underline{\underline{\nabla J}}^T_{\Phi,\text{eff}} \\ \underline{\underline{\nabla J}}_{\Phi,\text{eff}} & \underline{\underline{0}} \end{bmatrix} \begin{bmatrix} \underline{\ddot{q}}_{n+1} \\ \underline{\lambda}_{n+1} \end{bmatrix}, \tag{54}$$

leading to

$$\begin{aligned} \underline{\underline{F}}_{a,\text{pseudo}} = {} & \left(1-\alpha_f\right)\underline{\underline{\nabla F}}_{\text{ext},n+1} + \alpha_f \underline{\underline{\nabla F}}_{\text{ext},n} + \left(1-\alpha_f\right)\underline{\underline{\nabla F}}_{v,n+1} + \alpha_f \underline{\underline{\nabla F}}_{v,n} + \\ & - (1-\alpha_m)\underline{\underline{\nabla m}}_{n+1}\underline{\ddot{q}}_{n+1} - \alpha_m \underline{\underline{\nabla m}}_{n}\underline{\ddot{q}}_{n} - \alpha_m \underline{\underline{m}}_n \underline{\underline{\nabla \ddot{q}}}_n + \\ & - \left(1-\alpha_f\right)\underline{\underline{\nabla d}}_{n+1}\underline{\dot{q}}_{n+1} - \alpha_f \underline{\underline{\nabla d}}_{n}\underline{\dot{q}}_{n} - \left(1-\alpha_f\right)\underline{\underline{d}}_{n+1}\underline{\underline{\nabla \dot{q}}}_{\text{pred}} - \alpha_f \underline{\underline{d}}_n \underline{\underline{\nabla \dot{q}}}_n + \\ & - \left(1-\alpha_f\right)\underline{\underline{\nabla k}}_{n+1}\underline{q}_{n+1} - \alpha_f \underline{\underline{\nabla k}}_{n}\underline{q}_{n} - \left(1-\alpha_f\right)\underline{\underline{k}}_{n+1}\underline{\underline{\nabla q}}_{\text{pred}} - \alpha_f \underline{\underline{k}}_n \underline{\underline{\nabla q}}_n + \\ & - (1-\alpha_m)\underline{\underline{\nabla J}}^T_{\Phi,n+1}\underline{\lambda}_{n+1} - \alpha_m \underline{\underline{\nabla J}}^T_{\Phi,n}\underline{\lambda}_n - \alpha_m \underline{\underline{J}}^T_{\Phi,n}\underline{\underline{\nabla \lambda}}_n, \end{aligned} \tag{55}$$

$$\begin{aligned} \underline{\underline{F}}_{c,\text{pseudo}} = {} & \left(1-\alpha_f\right)\underline{\underline{\nabla F}}_{c,n+1} + \alpha_f \underline{\underline{\nabla F}}_{c,n} + \\ & - (1-\alpha_m)\underline{\underline{\nabla J}}_{\Phi,n+1}\underline{\ddot{q}}_{n+1} - \alpha_m \underline{\underline{\nabla J}}_{\Phi,n}\underline{\ddot{q}}_n - \alpha_m \underline{\underline{J}}_{\Phi,n}\underline{\underline{\nabla \ddot{q}}}_n . \end{aligned} \tag{56}$$

The derivation of the effective mass matrix and the effective constraint Jacobian matrix is shown in Appendix B.1 and the derivation of the effective pseudo force terms is shown in Appendix B.3.

The Newton–Raphson step for the sensitivity analysis is given by

$$\begin{bmatrix} \frac{\partial \underline{\underline{\nabla R}}_{1,n+1-\alpha_f}}{\partial \underline{\underline{\nabla \ddot{q}}}_{n+1}} & \frac{\partial \underline{\underline{\nabla R}}_{1,n+1-\alpha_f}}{\partial \underline{\underline{\nabla \lambda}}_{n+1}} \\ \frac{\partial \underline{\underline{\nabla R}}_{2,n+1-\alpha_f}}{\partial \underline{\underline{\nabla \ddot{q}}}_{n+1}} & \frac{\partial \underline{\underline{\nabla R}}_{2,n+1-\alpha_f}}{\partial \underline{\underline{\nabla \lambda}}_{n+1}} \end{bmatrix} \begin{bmatrix} \Delta \underline{\underline{\nabla \ddot{q}}}_{n+1} \\ \Delta \underline{\underline{\nabla \lambda}}_{n+1} \end{bmatrix} + \begin{bmatrix} \underline{\underline{\nabla R}}_{1,n+1-\alpha_f} \\ \underline{\underline{\nabla R}}_{2,n+1-\alpha_f} \end{bmatrix} = \underline{\underline{0}}. \tag{57}$$

As shown in [7], the Jacobian of the residual sensitivity with respect to $\underline{\underline{\nabla \ddot{q}}}_{n+1}$ and $\underline{\underline{\nabla \lambda}}_{n+1}$ reduces to

$$\begin{bmatrix} \frac{\partial \underline{\underline{\nabla R}}_{1,n+1-\alpha_f}}{\partial \underline{\underline{\nabla \ddot{q}}}_{n+1}} & \frac{\partial \underline{\underline{\nabla R}}_{1,n+1-\alpha_f}}{\partial \underline{\underline{\nabla \lambda}}_{n+1}} \\ \frac{\partial \underline{\underline{\nabla R}}_{2,n+1-\alpha_f}}{\partial \underline{\underline{\nabla \ddot{q}}}_{n+1}} & \frac{\partial \underline{\underline{\nabla R}}_{2,n+1-\alpha_f}}{\partial \underline{\underline{\nabla \lambda}}_{n+1}} \end{bmatrix} = \begin{bmatrix} \frac{\partial \underline{R}_{1,n+1-\alpha_f}}{\partial \underline{\ddot{q}}_{n+1}} & \frac{\partial \underline{R}_{1,n+1-\alpha_f}}{\partial \underline{\lambda}_{n+1}} \\ \frac{\partial \underline{R}_{2,n+1-\alpha_f}}{\partial \underline{\ddot{q}}_{n+1}} & \frac{\partial \underline{R}_{2,n+1-\alpha_f}}{\partial \underline{\lambda}_{n+1}} \end{bmatrix} \underline{\underline{e}}, \tag{58}$$

which reduces the dimensions of the coefficient matrix from $\left(n_q + n_\Phi\right) \times n_x \times \left(n_q + n_\Phi\right) \times n_x$ to $\left(n_q + n_\Phi\right) \times \left(n_q + n_\Phi\right)$. The solution of the resulting equation,

$$\begin{bmatrix} \frac{\partial \underline{R}_{1,n+1-\alpha_f}}{\partial \underline{\ddot{q}}_{n+1}} & \frac{\partial \underline{R}_{1,n+1-\alpha_f}}{\partial \underline{\lambda}_{n+1}} \\ \frac{\partial \underline{R}_{2,n+1-\alpha_f}}{\partial \underline{\ddot{q}}_{n+1}} & \frac{\partial \underline{R}_{2,n+1-\alpha_f}}{\partial \underline{\lambda}_{n+1}} \end{bmatrix} \begin{bmatrix} \Delta \underline{\underline{\nabla \ddot{q}}}_{n+1} \\ \Delta \underline{\underline{\nabla \lambda}}_{n+1} \end{bmatrix} + \begin{bmatrix} \underline{\underline{\nabla R}}_{1,n+1-\alpha_f} \\ \underline{\underline{\nabla R}}_{2,n+1-\alpha_f} \end{bmatrix} = \underline{\underline{0}}, \tag{59}$$

is found using the same solving scheme for the nonlinear solver of the primal analysis (46). Furthermore, both the primal analysis (46) and sensitivity analysis (59) have the same coefficient matrix, which also allows one to reuse the coefficient matrix of the primal analysis in the sensitivity analysis without the need for further evaluations.

3. Numerical Results

In this section, numerical results are shown comparing the introduced method for the sensitivity analysis for generalized-α time integration (primal analysis introduced in Section 2) with that of Newmark-β. The methods are compared with applications to both structural dynamics as well as flexible multibody dynamics.

3.1. Structural Dynamics

The linear structural dynamic problem used here as a demonstration example is based on [64], which has found usage as a benchmark in [65–71]. This unitless problem is schematically shown in Figure 2. This problem is defined by

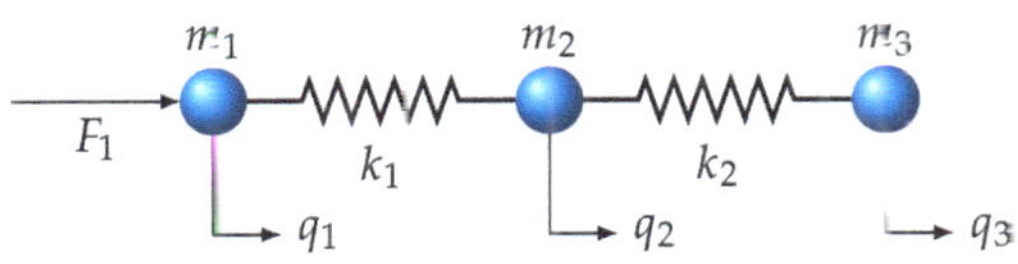

Figure 2. Structural dynamics example.

$$\overbrace{\begin{bmatrix} m_1 & 0 & 0 \\ 0 & m_2 & 0 \\ 0 & 0 & m_3 \end{bmatrix}}^{\underline{\underline{m}}} \overbrace{\begin{bmatrix} \ddot{q}_1 \\ \ddot{q}_2 \\ \ddot{q}_3 \end{bmatrix}}^{\underline{\ddot{q}}} + \overbrace{\begin{bmatrix} k_1 & -k_1 & 0 \\ -k_1 & k_1 + k_2 & -k_2 \\ 0 & -k_2 & k_2 \end{bmatrix}}^{\underline{\underline{k}}} \overbrace{\begin{bmatrix} q_1 \\ q_2 \\ q_3 \end{bmatrix}}^{\underline{q}} - \overbrace{\begin{bmatrix} F_1 \\ 0 \\ 0 \end{bmatrix}}^{\underline{F}} = \overbrace{\begin{bmatrix} 0 \\ 0 \\ 0 \end{bmatrix}}^{\underline{R}}, \tag{60}$$

with the prescribed displacement,

$$q_1 = \sin \omega t, \tag{61}$$

and the unitless values

$$\begin{aligned} m_1 &= 0, \\ m_2 &= 1, \\ m_3 &= 1, \\ k_1 &= 10^7, \\ k_2 &= 1, \\ \omega &= 1.2. \end{aligned}$$

The masses m_2 and m_3 as well as the stiffnesses k_1 and k_2 are considered design variables with respect to which the design sensitivities are calculated. The problem consists of stiff (k_1) and flexible (k_2) parts that are typical of complex, practical structural–mechanical systems [64].

Considering the prescribed displacement, this problem can be reduced to two degrees of freedom,

$$\overbrace{\begin{bmatrix} m_2 & 0 \\ 0 & m_3 \end{bmatrix}}^{\underline{\underline{m}}} \overbrace{\begin{bmatrix} \ddot{q}_2 \\ \ddot{q}_3 \end{bmatrix}}^{\underline{\ddot{q}}} + \overbrace{\begin{bmatrix} k_1 + k_2 & -k_2 \\ -k_2 & k_2 \end{bmatrix}}^{\underline{\underline{k}}} \overbrace{\begin{bmatrix} q_2 \\ q_3 \end{bmatrix}}^{\underline{q}} - \overbrace{\begin{bmatrix} k_1 q_1 \\ 0 \end{bmatrix}}^{\underline{F}} = \overbrace{\begin{bmatrix} 0 \\ 0 \end{bmatrix}}^{\underline{R}}, \tag{62}$$

which we implement in our in-house, Python-based solver framework SIMULI in the time $t = [0, 10]$ with the initial conditions

$$\begin{aligned} q_2 &= 0, \\ q_3 &= 0, \\ \dot{q}_2 &= 0, \\ \dot{q}_3 &= 0. \end{aligned}$$

As reported in [64], unitless time steps of $\Delta t = 0.2618$ are used, giving the following values for the ratio step size to the fundamental periods of the system:

$$\frac{\Delta t}{T_F} = 0.05,$$
$$\frac{\Delta t}{T_1} = 0.041667,$$
$$\frac{\Delta t}{T_2} = 131.76,$$

where T_F is the fundamental period of the forcing function and $T_{1,2}$, for the first and second modes, $T_1 = 6.283$ and $T_2 = 0.001987$.

The results for both Newmark-β and generalized-α time integration methods are shown in Figure 3 and compared with the exact solution. The exact solution is calculated via mode superposition per [72]. The numerical damping parameter of generalized-α time integration is set as $\varrho_\infty = 0.55$. As expected with generalized-α time integration, the unphysical, high-frequent oscillatory behavior of the response can be avoided. As such, the beneficial effects of numerical damping with generalized-α time integration can be clearly seen in the velocity and acceleration of node 2, while maintaining the integrity of the other responses.

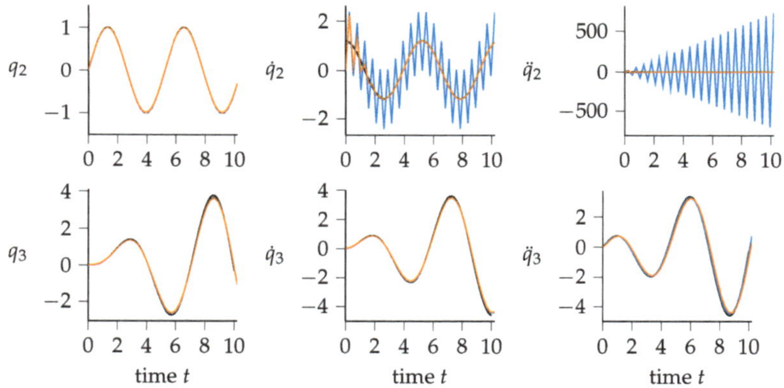

Figure 3. Responses of the structural dynamic example (unitless) calculated with — Newmark-β and — generalized-α compared with — exact solution (undermost line).

The sensitivity analysis developed above is applied and these results can be seen in Figures 4–6. Analytical sensitivity analysis with Newmark-β and generalized-α time integration methods are also compared with their numerical sensitivity implementations via finite differencing using a relative perturbation of 10^{-6}.

In cases where the Newmark-β time integration shows instability or noise, the numerical damping allows for usable sensitivities. In some cases, noise—albeit at a much lower level—persists in the first iterations before being damped out—after approximately 2.5 s (or approximately 10 iterations). Attention is drawn to those sensitivities susceptible to stability issues, namely

- sensitivities of position 2 with respect to stiffness 1 and mass 2 (Figure 4);
- sensitivities of velocity 2 with respect to stiffness 1 and mass 2 (Figure 5);
- sensitivities of acceleration 2 with respect to stiffness 1, stiffness 2 and mass 2 (Figure 6);
- sensitivities of acceleration 3 with respect to stiffness 1 and mass 2 (Figure 6).

Generalized-α time integration is not afflicted by this instability, resulting in sensitivities that can be successfully used in gradient-based algorithms.

Furthermore, generalized-α time integration eliminates the noise in those sensitivities present when using Newmark-β. This is seen with the following sensitivity responses:

- sensitivities of velocity 2 with respect to stiffness 2 (Figure 5);
- sensitivities of velocity 3 with respect to stiffness 1 and mass 2 (Figure 5).

The analytical and numerical results are, in most plots, nearly exact. There are cases, however, where the negative consequences of the numerical instability with Newmark-β are increased. The reader's attention is especially called to

- sensitivity of velocity 2 with respect to stiffness 2 (Figure 5);
- sensitivity of velocity 2 with respect to mass 3 (Figure 5);
- sensitivity of acceleration 2 with respect to stiffness 2 (Figure 6);
- sensitivity of acceleration 2 with respect to mass 3 (Figure 6).

It can be seen that both analytical and numerical sensitivities with generalized-α time integration are not negatively affected by numerical instabilities, as is the case with Newmark-β.

The number of time steps needed for the damping to take effect can be reduced by reducing the numerical damping parameter ϱ_∞ towards zero (i.e., more damping), though also affecting the physical response to a greater degree. As time continues, the instability grows, with Newmark-β time integration and therefore a clear advantage can be seen with generalized-α. It can be clearly seen how gradient-based solvers would struggle to use the gradient information provided with Newmark-β time integration.

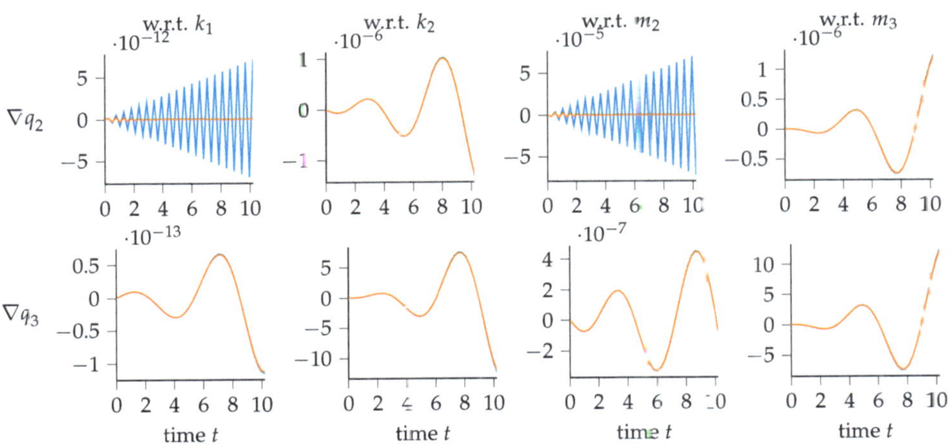

Figure 4. Position sensitivities ∇q of the structural dynamic example with respect to k_1, k_2, m_2 and m_3 (unitless), analytically calculated with — Newmark-β and — generalized-α compared with those calculated numerically via finite differencing with — Newmark-β and — generalized-α (undermost lines). Note: no visible difference between finite differencing and analytical sensitivities.

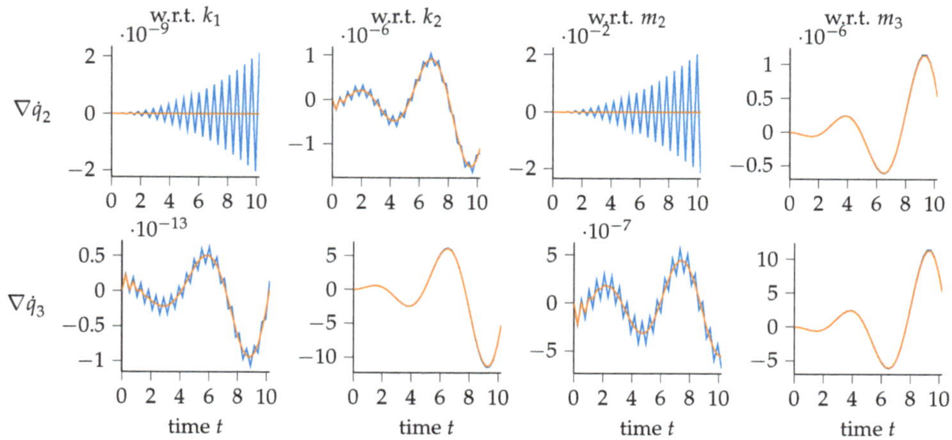

Figure 5. Velocity sensitivities $\nabla \dot{q}$ of the structural dynamic example with respect to k_1, k_2, m_2 and m_3 (unitless), analytically calculated with — Newmark-β and — generalized-α compared with those calculated numerically via finite differencing with — Newmark-β and — generalized-α (undermost lines). Note: no visible difference between finite differencing and analytical sensitivities.

For our modest case of four design variables, we show an acceleration of two, where the speedup S is defined by the ratio of computational time for finite differencing t_{FD} to that of analytical sensitivities t_{ana},

$$S = \frac{t_{\text{FD}}}{t_{\text{ana}}}. \tag{63}$$

The computational time required for analytical sensitivity analysis represents approximately the ideal computational effort twice that of the primal analysis, i.e., analysis without sensitivity analysis. Computation reduction is the same for both Newmark-β and generalized-α time integration methods, as can be seen in Table 1.

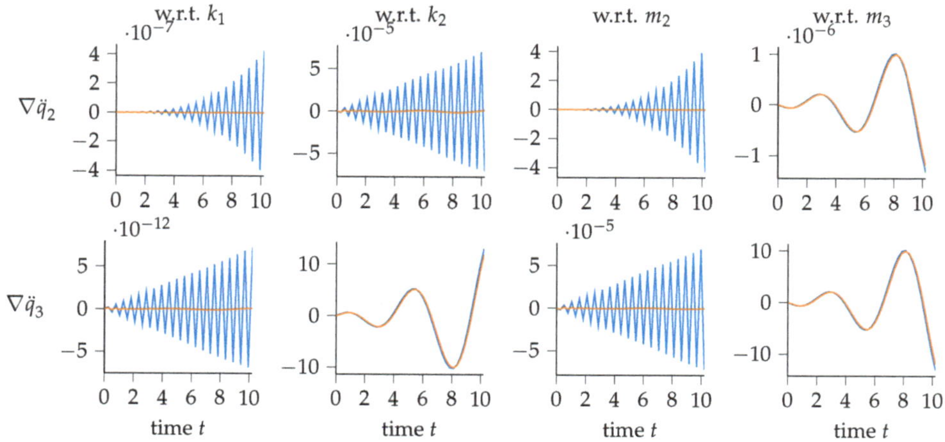

Figure 6. Acceleration sensitivities $\nabla \ddot{q}$ of the structural dynamic example with respect to k_1, k_2, m_2 and m_3 (unitless) analytically calculated with — Newmark-β and — generalized-α compared with those numerically calculated via finite differencing with — Newmark-β and — generalized-α (undermost lines). Note: no visible difference between finite differencing and analytical sensitivities.

Table 1. Computational effort for the structural dynamics example.

Time Integration	Sensitivity Analysis	Calculation Time (s) [1]	Speedup [2]
Newmark-β	None	0.0555	–
Newmark-β	Finite differencing	0.2088	–
Newmark-β	Analytical	0.1023	2.0410
Generalized-α	None	0.0458	–
Generalized-α	Finite differencing	0.2122	–
Generalized-α	Analytical	0.0993	2.1370

[1] Intel Core i5-7200U CPU 2.50 GHz, [2] acceleration of analytical sensitivity analysis with respect to finite differencing.

3.2. Flexible Slider–Crank Mechanism

The slider–crank mechanism, as shown in Figure 7, is modeled with flexible multibody dynamics using the floating frame of reference formulation. The mechanism consists of four bodies that are connected by three revolute joints and one prismatic joint, modeled as ideal joints without friction. The ground and the slider are considered to be rigid, while the crank and the rod are considered to be flexible and are modeled with Euler–Bernoulli beams. Figure 8 shows the flexible multibody system of the mechanism. The inertial reference frame coinciding with the reference frame of the ground is shown in blue. The body reference frames of the crank, the rod and the slider are shown in red. The undeformed bodies of the crank and the rod are shown in gray and the deformed bodies are shown in black, where the points represent the nodes of the finite elements for the geometric discretization. The coordinates that are investigated in detail for the primal analysis and the sensitivity analysis are shown in green. The dimensions of the bodies are listed in Table 2. The bodies are made of steel and Table 3 shows the material properties. A linear elastic material model with isotropic behavior is applied. In this example, the SI-derived MPa unit system is used with length in millimeters (mm), mass in tonnes (t), time in second (s), force in Newtons (N), Angle in radian (rad) and stress in megapascals (MPa).

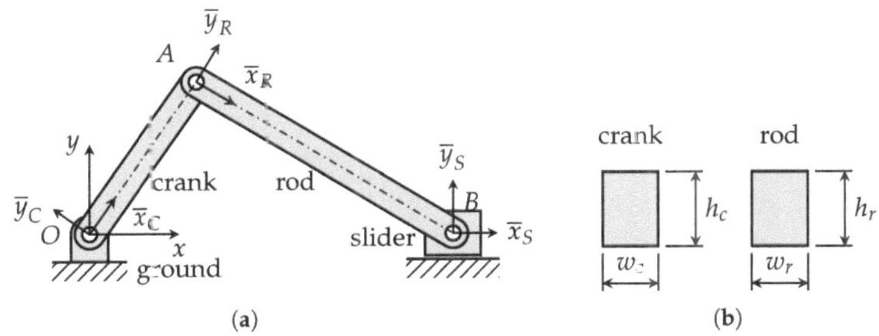

Figure 7. Slider–crank mechanism: (**a**) schematic diagram; and (**b**) cross-sections.

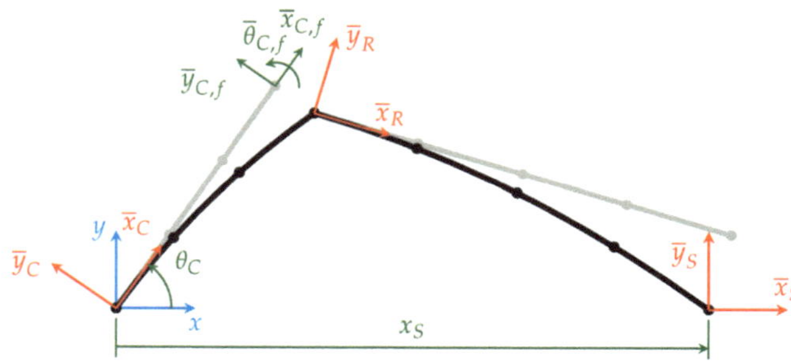

Figure 8. Flexible multibody system of the slider–crank mechanism with the undeformed bodies in gray, the deformed bodies in black, the inertial reference frame in blue, the body reference frames in red and the coordinates that are investigated in green.

Table 2. Geometry of the slider–crank mechanism.

Property	Symbol	Crank	Rod	Slider	Units
Width	w	20	20	30	mm
Height	h	30	30	30	mm
Length	ℓ	120	180	40	mm

Table 3. Material properties.

Property	Symbol	Value	Units
Density	ρ	7.85×10^{-9}	$\frac{\text{t}}{\text{mm}^3}$
Elastic modulus	E	210 000	MPa
Poisson ratio	ν	0.3	–

The generalized coordinates of a flexible multibody system modeled with the floating frame of the reference formulation and Euler–Bernoulli beams are given by

$$\underline{q} = \begin{bmatrix} \underline{r}^T & \theta & \overline{\underline{q}}_f^T \end{bmatrix}^T, \tag{64}$$

where $\underline{r}$ is the position of the body reference frame, θ is the orientation of the body reference frame and $\overline{\underline{q}}_f$ are the nodal displacements in the body coordinates of each node on the flexible body. Since no damping is considered, the equations of motion of a flexible body are given by

$$\underline{\underline{m}}\,\ddot{\underline{q}} + \underline{\underline{k}}\,\underline{q} + \underline{\underline{J}}_\Phi^T\,\underline{\lambda} = \underline{F}_{\text{ext}} + \underline{F}_v, \tag{65}$$

$$\underline{\underline{J}}_\Phi\,\ddot{\underline{q}} = \underline{F}_c, \tag{66}$$

where the system matrices of the flexible bodies are given by

$$\underline{\underline{m}} = \begin{bmatrix} \underline{\underline{m}}_{tt} & \underline{\underline{m}}_{tr} & \underline{\underline{m}}_{tf} \\ & \underline{\underline{m}}_{rr} & \underline{\underline{m}}_{rf} \\ \text{sym.} & & \underline{\underline{m}}_{ff} \end{bmatrix}, \tag{67}$$

$$\underline{\underline{k}} = \begin{bmatrix} 0 & 0 & 0 \\ & 0 & 0 \\ \text{sym.} & & \underline{\underline{k}}_{ff} \end{bmatrix}, \tag{68}$$

and the system vectors are given by

$$\underline{F}_{ext} = \begin{bmatrix} \underline{F}_{ext,t} \\ \underline{F}_{ext,r} \\ \underline{F}_{ext,f} \end{bmatrix}, \tag{69}$$

$$\underline{F}_{v} = \begin{bmatrix} \underline{F}_{v,t} \\ \underline{F}_{v,r} \\ \underline{F}_{v,f} \end{bmatrix}, \tag{70}$$

with the index denoting translational terms t, rotational terms r and flexible terms f.

For the dynamic simulation of the mechanism, the above-mentioned solver framework SIMULI is used. The simulation is performed in the time interval $t = [0\,\text{s}, 0.025\,\text{s}]$. For both generalized-α and Newmark-β time integration, a time step of $\Delta t = 1 \times 10^{-4}$ s is chosen for comparison. The resulting values for the ratio step size to fundamental periods of the system for the first ten non-zero eigenfrequencies are found in Figure 9. The values for the ratio step size to fundamental periods range from 0.2119 s for first non-zero eigenfrequency to 3.9919 s for the tenth non-zero eigenfrequency and to 17.1157 s for the twentieth non-zero eigenfrequency (not pictured in the plot).

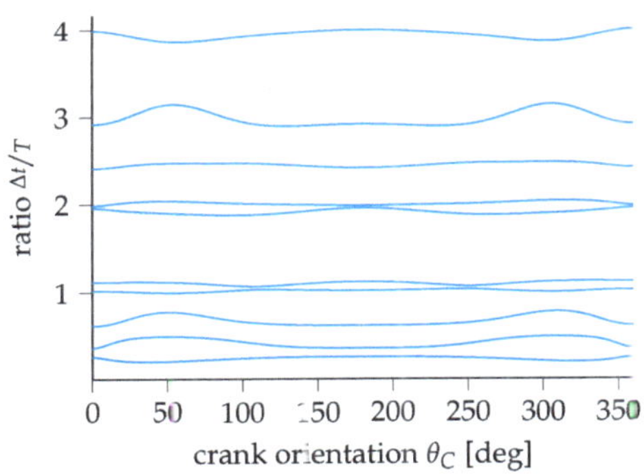

Figure 9. Ratio time step to fundamental periods for the first ten non-zero eigenfrequencies of the slider–crank mechanism.

For the initial conditions, the angular position and velocity of the crank are defined with

$$\theta_C = 0\,\text{rad},$$

$$\dot{\theta}_C = 150\pi\,\frac{\text{rad}}{\text{s}},$$

and the initial conditions of the other bodies are derived through the kinematic analysis of the system. The dimensions of the cross-sections of the crank h_C, w_C and the rod h_R, w_R are used as design variables in the sensitivity analysis.

The results of the simulation of the coordinates that are highlighted in Figure 8 are shown in Figure 10. With Newmark-β, vibrations at high frequencies are observed. With generalized-α, vibrations are observed in the beginning of the simulation that are numerically damped with ongoing simulation time. In addition to the generalized coordinates, the maximum stresses on the upper and lower fiber of the beam elements of the flexible bodies including the crank and the rod are computed. The Kreisselmeier–Steinhauser function [73,74] is used to approximate the maximum stress value of the bodies at each time step with a differentiable function and the results are shown in Figure 11. With Newmark-β, the vibrating behavior of the system causes higher flexible deformation and therefore unrealistic high stress in the components. The numerical damping with generalized-α leads to lower maximum stress values and a more realistic stress distribution over time.

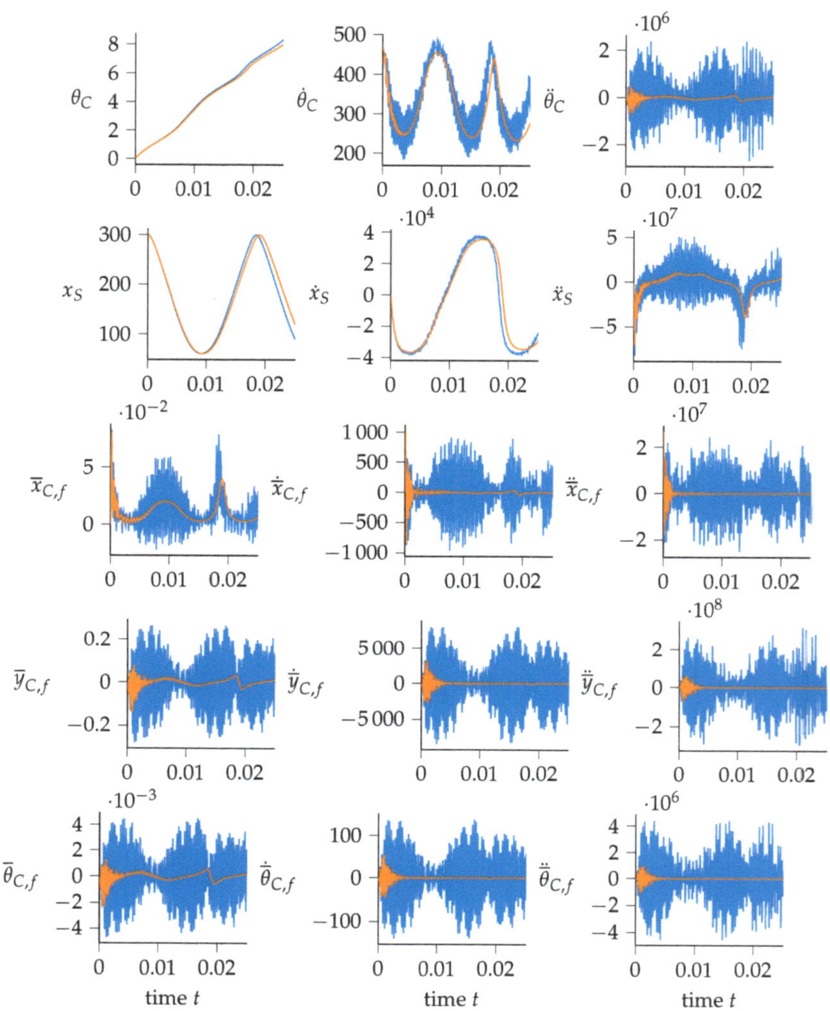

Figure 10. Responses of the slider–crank mechanism (values in MPa units, i.e., s, mm, $\frac{mm}{s}$, $\frac{mm}{s^2}$ for translations and rad, $\frac{rad}{s}$, $\frac{rad}{s^2}$ for rotations) calculated with — Newmark-β and — generalized-α.

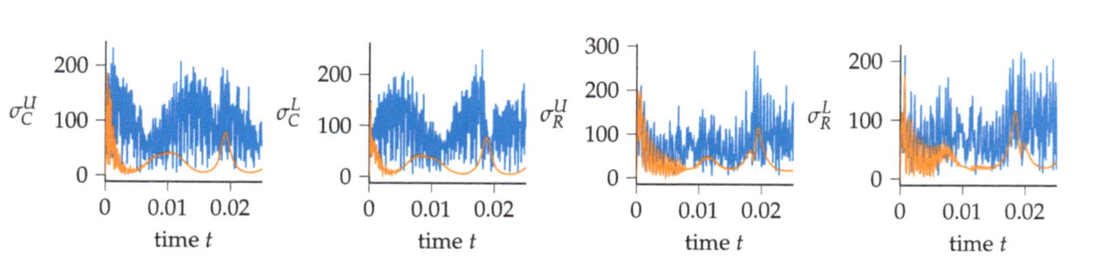

Figure 11. Maximum stress in MPa on the upper and lower fiber of the crank and the rod over time in s calculated with — Newmark-β and — generalized-α.

The results of the sensitivity analysis of the coordinates that are highlighted in Figure 8 are shown in Figures 12–14. The analyzed coordinates include the rotation of the floating frame of the crank θ_C, the translation in the x direction of the floating frame of the slider x_S and the three flexible coordinates of the node in A on the crank $\overline{x}_{C,f}$, $\overline{y}_{C,f}$ and $\overline{\theta}_{C,f}$. The design variables used for the sensitivity analysis are the cross-sectional dimensions of the crank h_C, w_C and the rod h_R, w_R.

Similarly to the above, with Newmark-β, the results show a vibrating behavior that would lead to significant problems when using the sensitivities in gradient-based design optimization. With generalized-α, the numerical damping leads to usable sensitivity values after a few time steps. Figure 12 shows that the position sensitivities of the position and orientation coordinates of the floating frames are less noisy compared to the flexible coordinates.

Position sensitivities that suffer little from noise are limited to the position and orientation coordinates of the floating frames with respect to several design variables and include the following:

- sensitivities of θ_C with respect to w_C and w_R;
- sensitivities of x_S with respect to h_C, w_C, and w_R.

Position sensitivities with significant noise include the position and orientation coordinates of the floating frames with respect to a few specific design variables as well as the flexible coordinates with respect to all design variables. The list of noisy responses of position sensitivities is given in the following:

- sensitivities of θ_C with respect to h_C and h_R;
- sensitivities of x_S with respect to h_R;
- sensitivities of $\overline{x}_{C,f}$ with respect to h_C w_C, h_R and w_R;
- sensitivities of $\overline{y}_{C,f}$ with respect to h_C; w_C, h_R and w_R;
- sensitivities of $\overline{\theta}_{C,f}$ with respect to h_C, w_C, h_R and w_R.

The sensitivities of the velocities as shown in Figure 13 and the sensitivities of the accelerations as shown in Figure 14 have noisy responses with Newmark that are damped with generalized-α for all investigated coordinates and all used design variables. The same behavior also continues with the sensitivity values for the maximum stresses in the crank and the rod shown in Figure 15.

To check the correctness of the results, the sensitivity values with the introduced method are compared to numerical sensitivity analysis with forward differencing. The results are in good agreement, with only small numerical differences. In fact, there are no visible differences in Figures 12–15 with the finite differencing values lying under the respective analytical values.

Table 4 shows a comparison of the computational effort with different methods. The computational effort with generalized-α is about 30 % lower compared to the computational effort with Newmark-β. The main reason for that is the continuing vibrating behavior with Newmark-β requires more iteration steps of the nonlinear solver. The comparison of the sensitivity methods with numerical finite differencing and analytical direct differentiation shows the high efficiency of the analytical method. With the four shown design variables, as shown in (63), an acceleration of about $S \approx 4$ of the analytical method compared to the numerical method is obtained. The additional computational effort of analytical sensitivity analysis is just a fraction of the computational effort of the primal analysis. This is because of the simplification shown in (59), which also allows one to reuse the Jacobian matrix of the residual built in the primal analysis for the sensitivity analysis.

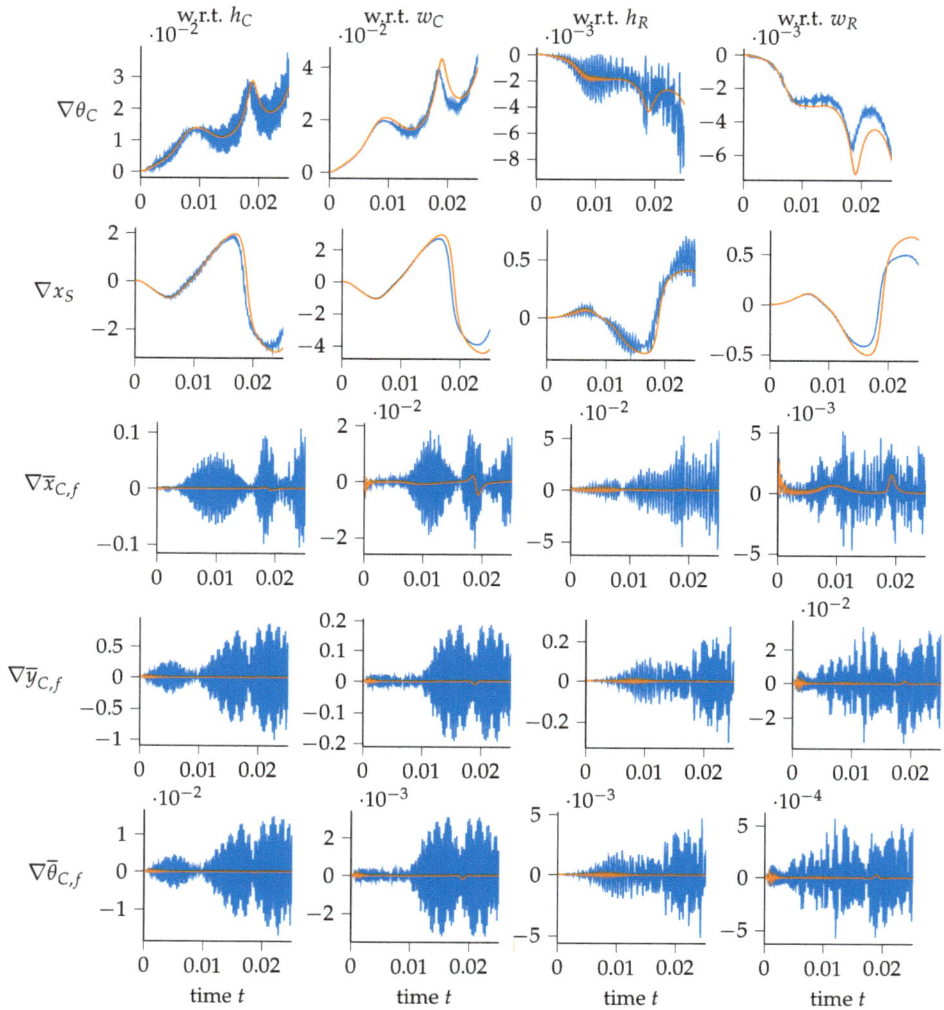

Figure 12. Position sensitivities ∇q of the slider–crank mechanism with respect to h_C, w_C, h_R and w_R (values in MPa system of units, i.e., s, mm, $\frac{\text{mm}}{\text{s}}$, $\frac{\text{mm}}{\text{s}^2}$ for translations and rad, $\frac{\text{rad}}{\text{s}}$, $\frac{\text{rad}}{\text{s}^2}$ for rotations) calculated with — Newmark-β and — generalized-α compared with those numerically calculated via finite differencing with — Newmark-β and — generalized-α (undermost lines). Note: there is no visible difference between finite differencing and analytical sensitivities.

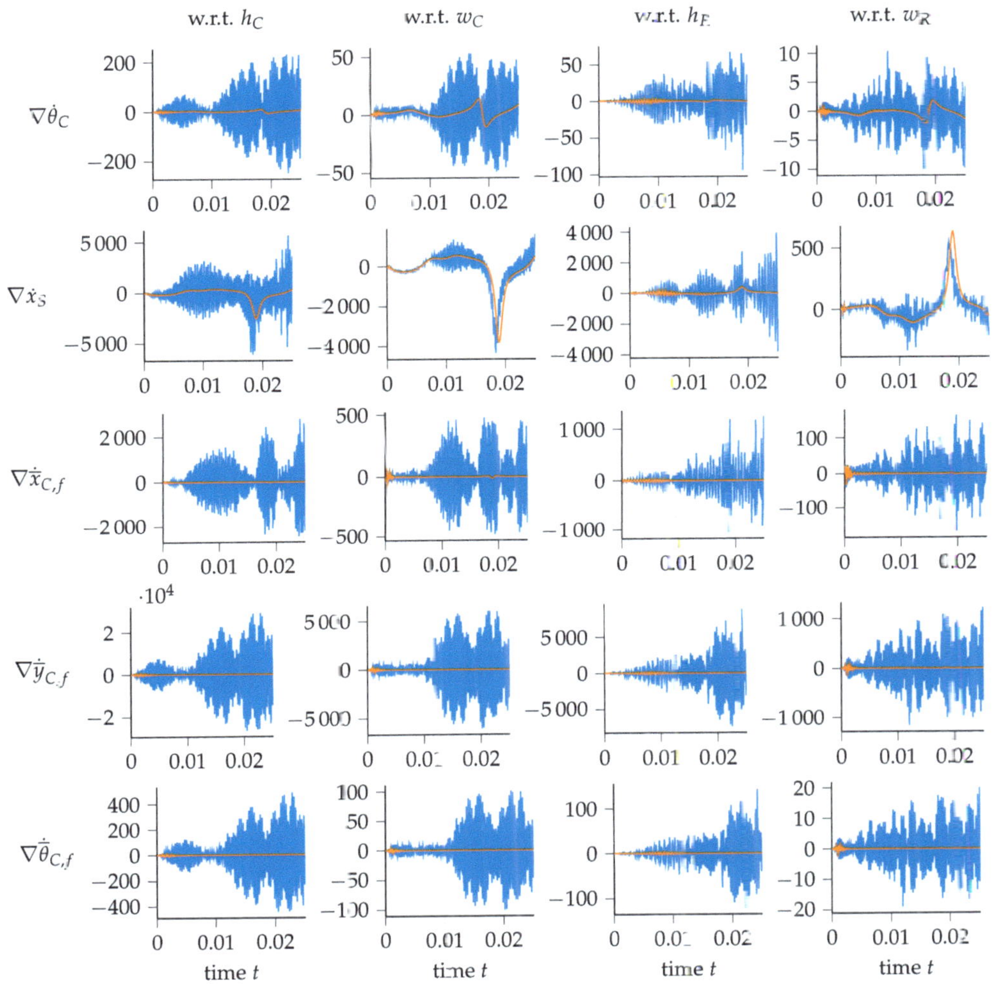

Figure 13. Velocity sensitivities $\nabla\dot{q}$ of the slider–crank mechanism with respect to h_C, w_C, h_R and w_R (values in MPa system of units, i.e., s, mm, $\frac{\text{mm}}{\text{s}}$, $\frac{\text{mm}}{\text{s}^2}$ for translations and rad, $\frac{\text{rad}}{\text{s}}$, $\frac{\text{rad}}{\text{s}^2}$ for rotations) calculated with — Newmark-β and — generalized-α compared with those calculated numerically via finite differencing with — Newmark-β and — generalized-α (undermost lines). Note: no visible difference between finite differencing and analytical sensitivities.

Table 4. Computational effort for the slider–crank mechanism.

Time Integration	Sensitivity Analysis	Calculation Time [mm:ss] [1]	Speedup [2]
Newmark-β	None	03:44	–
Newmark-β	Finite differencing	18:12	–
Newmark-β	Analytical	04:29	4.0595
Generalized-α	None	02:42	–
Generalized-α	Finite differencing	13:22	–
Generalized-α	Analytical	03:30	3.8190

[1] Intel Core i7-8700 CPU 3.20 GHz, [2] speedup of analytical sensitivity analysis with respect to finite differencing.

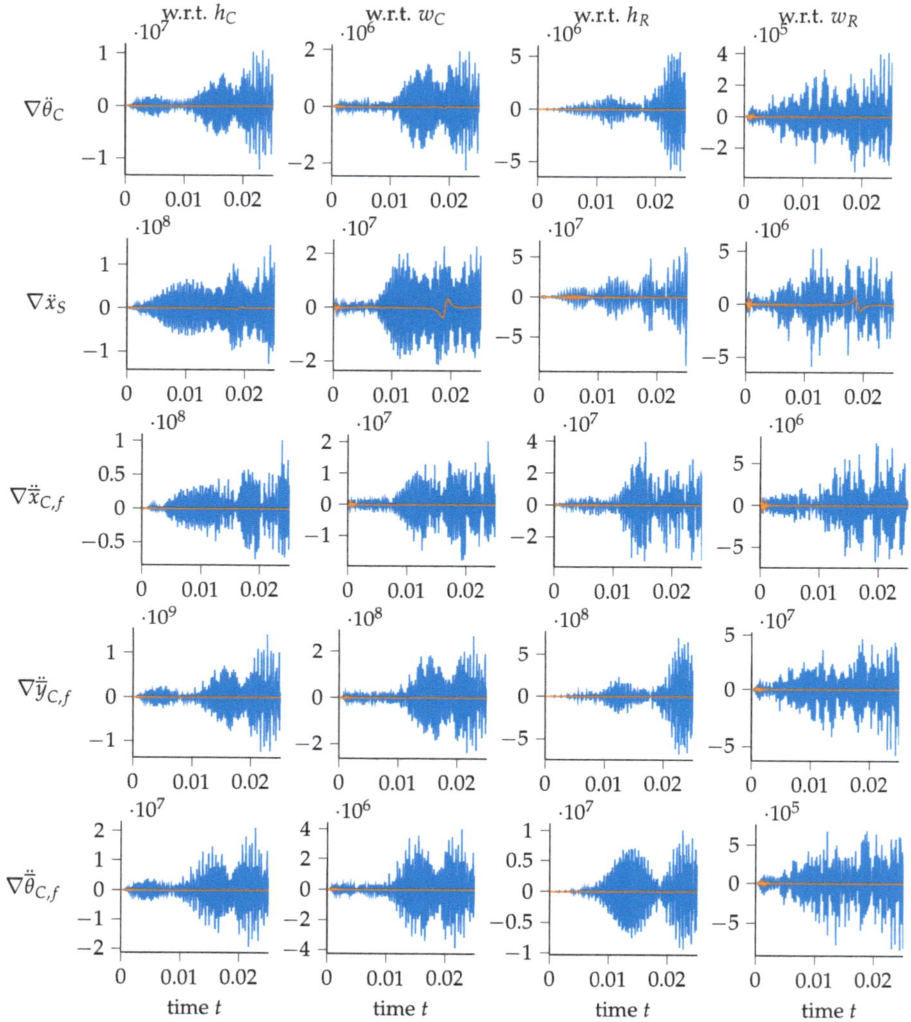

Figure 14. Acceleration sensitivities $\nabla\ddot{q}$ of the slider–crank mechanism with respect to h_C, w_C, h_R and w_R (values in MPa system of units, i.e., s, mm, $\frac{\text{mm}}{\text{s}}$, $\frac{\text{mm}}{\text{s}^2}$ for translations and rad, $\frac{\text{rad}}{\text{s}}$, $\frac{\text{rad}}{\text{s}^2}$ for rotations) calculated with — Newmark-β and — generalized-α compared with those calculated numerically via finite differencing with — Newmark-β and — generalized-α (undermost lines). Note: there is no visible difference between finite differencing and analytical sensitivities.

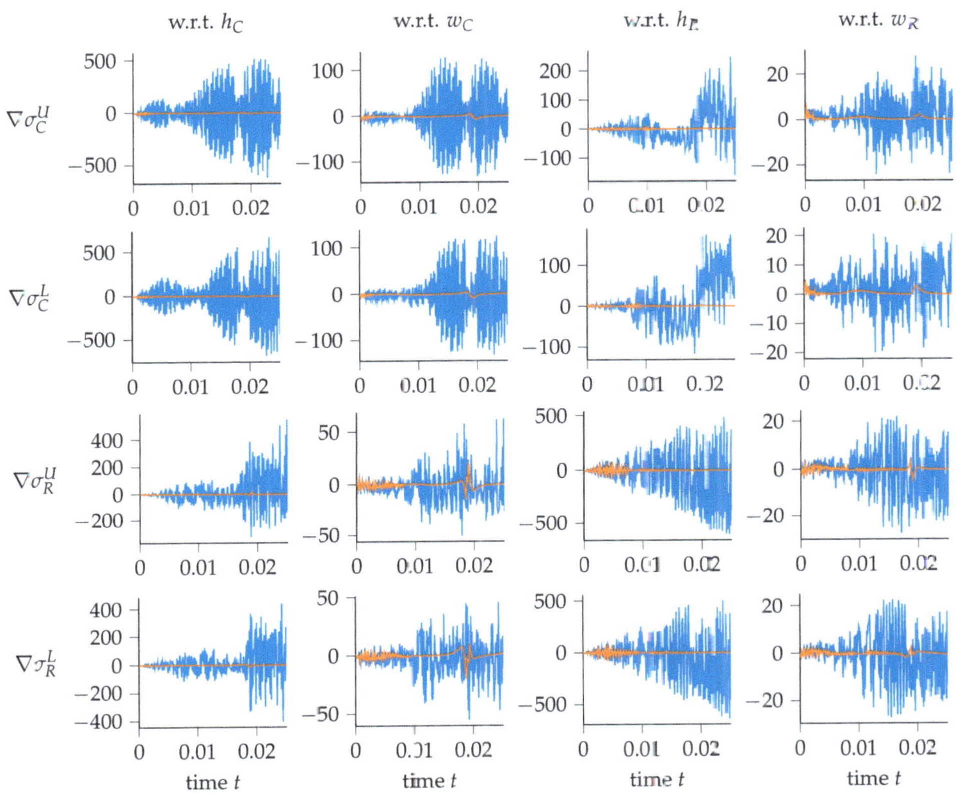

Figure 15. Maximum stress sensitivities on the upper and lower fibers of the crank and the rod (values in MPa system of units, i.e., s, mm, $\frac{\text{mm}}{\text{s}}$, $\frac{\text{mm}}{\text{s}^2}$ for translations and rad, $\frac{\text{rad}}{\text{s}}$, $\frac{\text{rad}}{\text{s}^2}$ for rotations) calculated with — Newmark-β and — generalized-α compared with those calculated numerically via finite differencing with — Newmark-β and — generalized-α (undermost lines). Note: there is no visible difference between finite differencing and analytical sensitivities.

4. Conclusions

This work develops the analytical sensitivity analysis for the α-family of time integration, which has a single implementation for several common time integration schemes. This therefore has consequences for gradient-based optimization frameworks for dynamic systems. Analytical sensitivity analysis is crucial for efficient design optimization, parameter fitting and uncertainty analysis. In this work, the direct differentiation of the generalized-α time integration is derived and implemented for linear structural dynamics and flexible multibody dynamics, the latter of which requires a nonlinear solution routine. All equations needed for implementation are provided and extendable to further governing equations. The beneficial effects of the step-size-dependent numerical damping provided by the generalized-α method can clearly be seen in the comparison of the results with Newmark-β. The numerical noise of the responses and sensitivities can lead to convergence problems when using gradient-based optimization algorithms. This can be avoided when using the generalized-α time integration and its sensitivities as developed herein.

If the functions of interest are nonsmooth, discontinuous and bifurcated, non-gradient-based methods can be successfully used, e.g., design cases with crash analysis. Such cases can be solved with gradient-free optimization algorithms (e.g., [75–77]) and approximation methods (e.g., [78–80]).

The system responses with respect to the design variables are indeed often smooth, continuous and unbifurcated and thus enabling gradient-based methods where efficient and accurate sensitivity analysis is critical for both effectiveness and efficiency. This is shown above with linear structural dynamics and flexible multibody dynamics. The method and equations developed herein are of general nature and are extendable to further formulations and applicable to larger problems, both of larger number of degrees of freedom and number of design variables. Future investigations are needed on these large-scale problems to assess the acceleration and include a comparison with the adjoint variable method.

With the above study and introduction of the direct differentiation of the α-family of time integrators, the authors hope to provide the readers as design engineers a tool in the pursuit of optimal structures and mechanisms under dynamic loading.

Author Contributions: Conceptualization, E.W. and V.G.; methodology, E.W. and V.G.; software, E.W. and V.G.; validation, E.W. and V.G.; formal analysis, E.W. and V.G.; investigation, E.W. and V.G.; resources, E.W. and V.G.; data curation, E.W. and V.G.; writing—original draft preparation, E.W. and V.G.; writing—review and editing, E.W. and V.G.; visualization, E.W. and V.G.; supervision, E.W.; project administration, E.W.; funding acquisition, E.W. All authors have read and agreed to the published version of the manuscript.

Funding: This work is supported by the project CRC 2017 TN2091 doloMULTI Design of Lightweight Optimized structures and systems under MULTIdisciplinary considerations through integration of MULTIbody dynamics in a MULTIphysics framework, funded by the Free University of Bozen-Bolzano.

Data Availability Statement: Data available upon request.

Conflicts of Interest: The authors declare no conflicts of interest. The numerical studies by the first author were carried out during his previous position at the Free University of Bozen-Bolzano.

Appendix A. Derivation of Effective Equations of Motion for Structural Dynamics

Appendix A.1. Primal Analysis

The base for the α-family of integration methods is the equation of motion which for linear structural dynamics which is given by

$$\underline{\underline{m}}\,\underline{\ddot{q}}_{n+1-\alpha_m} + \underline{\underline{d}}\,\underline{\dot{q}}_{n+1-\alpha_f} + \underline{\underline{k}}\,\underline{q}_{n+1-\alpha_f} = \underline{F}_{n+1-\alpha_f}, \tag{A1}$$

and the intermediate values shown in Equations (3)–(5). Inserting the intermediate values into the equation of motion leads to

$$\underline{\underline{m}}\Big((1-\alpha_m)\underline{\ddot{q}}_{n+1} + \alpha_m\underline{\ddot{q}}_n\Big) + \underline{\underline{d}}\Big(\Big(1-\alpha_f\Big)\underline{\dot{q}}_{n+1} + \alpha_f\underline{\dot{q}}_n\Big) + \underline{\underline{k}}\Big(\Big(1-\alpha_f\Big)\underline{q}_{n+1} + \alpha_f\underline{q}_n\Big) \;=\; \Big(1-\alpha_f\Big)\underline{F}_{n+1} + \alpha_f\underline{F}_n. \tag{A2}$$

With Newmark's equation, (1) and (2), this becomes

$$\begin{aligned}\underline{\underline{m}}\Big((1-\alpha_m)\underline{\ddot{q}}_{n+1} + \alpha_m\underline{\ddot{q}}_n\Big) + \underline{\underline{d}}\Big(\Big(1-\alpha_f\Big)\Big(\underline{\dot{q}}_{\text{pred}} + \gamma\Delta t\underline{\ddot{q}}_{n+1}\Big) + \alpha_f\underline{\dot{q}}_n\Big) + \\ + \underline{\underline{k}}\Big(\Big(1-\alpha_f\Big)\Big(\underline{q}_{\text{pred}} + \beta\Delta t^2\underline{\ddot{q}}_{n+1}\Big) + \alpha_f\underline{q}_n\Big) = \Big(1-\alpha_f\Big)\underline{F}_{n+1} + \alpha_f\underline{F}_n.\end{aligned} \tag{A3}$$

This can be rearranged in the following form

$$\begin{aligned}&\Big((1-\alpha_m)\underline{\underline{m}} + \Big(1-\alpha_f\Big)\gamma\Delta t\underline{\underline{d}} + \Big(1-\alpha_f\Big)\beta\Delta t^2\underline{\underline{k}}\Big)\underline{\ddot{q}}_{n+1} = \\ &\Big(1-\alpha_f\Big)\underline{F}_{n+1} + \alpha_f\underline{F}_n - \alpha_m\underline{\underline{m}}\,\underline{\ddot{q}}_n - \underline{\underline{d}}\Big(\Big(1-\alpha_f\Big)\underline{\dot{q}}_{\text{pred}} + \alpha_f\underline{\dot{q}}_n\Big) - \underline{\underline{k}}\Big(\Big(1-\alpha_f\Big)\underline{q}_{\text{pred}} + \alpha_f\underline{q}_n\Big).\end{aligned} \tag{A4}$$

Therefore, the effective equation of motion is defined by

$$\underline{\underline{m}}_{\text{eff}}\,\underline{\ddot{q}}_{n+1} = \underline{F}_{\text{eff}}, \tag{A5}$$

where

$$\underline{\underline{m}}_{\text{eff}} = (1-\alpha_m)\underline{\underline{m}} + \left(1-\alpha_f\right)\gamma\Delta t\underline{\underline{d}} + \left(1-\alpha_f\right)\beta\Delta t^2\underline{\underline{k}}, \tag{A6}$$

$$\underline{F}_{\text{eff}} = \left(1-\alpha_f\right)\underline{F}_{n+1} + \alpha_f\underline{F}_n - \alpha_m\underline{\underline{m}}\,\underline{\ddot{q}}_n - \underline{\underline{d}}\left(\left(1-\alpha_f\right)\underline{\dot{q}}_{\text{pred}} + \alpha_f\underline{\dot{q}}_n\right) - \underline{\underline{k}}\left(\left(1-\alpha_f\right)\underline{q}_{\text{pred}} + \alpha_f\underline{q}_n\right). \tag{A7}$$

Appendix A.2. Sensitivity Analysis

The governing equation for sensitivities is the first derivative of the equation of motion,

$$\underline{\underline{\nabla m}}\,\underline{\ddot{q}}_{n+1-\alpha_m} + \underline{\underline{m}}\,\underline{\underline{\nabla \ddot{q}}}_{n+1-\alpha_m} + \underline{\underline{\nabla d}}\,\underline{\dot{q}}_{n+1-\alpha_f} + \underline{\underline{d}}\,\underline{\underline{\nabla \dot{q}}}_{n+1-\alpha_f} + \underline{\underline{\nabla k}}\,\underline{q}_{n+1-\alpha_f} + \underline{\underline{k}}\,\underline{\underline{\nabla q}}_{n+1-\alpha_f} = \underline{\underline{\nabla F}}_{n+1-\alpha_f}, \tag{A8}$$

where we introduce the intermediate sensitivity values (10)–(12),

$$\begin{aligned}
&\underline{\underline{\nabla m}}\left((1-\alpha_m)\underline{\ddot{q}}_{n+1} + \alpha_m\underline{\ddot{q}}_n\right) + \underline{\underline{m}}\left((1-\alpha_m)\underline{\underline{\nabla \ddot{q}}}_{n+1} + \alpha_m\underline{\underline{\nabla \ddot{q}}}_n\right) + \\
&+\underline{\underline{\nabla d}}\left(\left(1-\alpha_f\right)\underline{\dot{q}}_{n+1} + \alpha_f\underline{\dot{q}}_n\right) + \underline{\underline{d}}\left(\left(1-\alpha_f\right)\underline{\underline{\nabla \dot{q}}}_{n+1} + \alpha_f\underline{\underline{\nabla \dot{q}}}_n\right) + \\
&+\underline{\underline{\nabla k}}\left(\left(1-\alpha_f\right)\underline{q}_{n+1} + \alpha_f\underline{q}_n\right) + \underline{\underline{k}}\left(\left(1-\alpha_f\right)\underline{\underline{\nabla q}}_{n-1} + \alpha_f\underline{\underline{\nabla q}}_n\right) = \\
&\qquad\left(1-\alpha_f\right)\underline{\underline{\nabla F}}_{n+1} + \alpha_f\underline{\underline{\nabla F}}_n,
\end{aligned} \tag{A9}$$

and which we update with Newmark's sensitivity Equations (16) and (17)

$$\begin{aligned}
&\underline{\underline{\nabla m}}\left((1-\alpha_m)\underline{\ddot{q}}_{n+1} + \alpha_m\underline{\ddot{q}}_n\right) + \underline{\underline{m}}\left((1-\alpha_m)\underline{\underline{\nabla \ddot{q}}}_{n+1} + \alpha_m\underline{\underline{\nabla \ddot{q}}}_n\right) + \\
&+\underline{\underline{\nabla d}}\left(\left(1-\alpha_f\right)\underline{\dot{q}}_{n+1} + \alpha_f\underline{\dot{q}}_n\right) + \underline{\underline{d}}\left(\left(1-\alpha_f\right)\left(\underline{\underline{\nabla \dot{q}}}_{\text{pred}} + \gamma\Delta t\underline{\underline{\nabla \ddot{q}}}_{n+1}\right) + \alpha_f\underline{\underline{\nabla \dot{q}}}_n\right) + \\
&+\underline{\underline{\nabla k}}\left(\left(1-\alpha_f\right)\underline{q}_{n+1} + \alpha_f\underline{q}_n\right) + \underline{\underline{k}}\left(\left(1-\alpha_f\right)\left(\underline{\underline{\nabla q}}_{\text{pred}} + \beta\Delta t^2\underline{\underline{\nabla \ddot{q}}}_{n+1}\right) + \alpha_f\underline{\underline{\nabla q}}_n\right) = \\
&\qquad\left(1-\alpha_f\right)\underline{\underline{\nabla F}}_{n+1} + \alpha_f\underline{\underline{\nabla F}}_n.
\end{aligned} \tag{A10}$$

This equation can be rearranged to the following form,

$$\begin{aligned}
&\left((1-\alpha_m)\underline{\underline{m}} + \left(1-\alpha_f\right)\gamma\Delta t\underline{\underline{d}} + \left(1-\alpha_f\right)\beta\Delta t^2\underline{\underline{k}}\right)\underline{\underline{\nabla \ddot{q}}}_{n+1} = \\
&\quad\left(1-\alpha_f\right)\underline{\underline{\nabla F}}_{n+1} + \alpha_f\underline{\underline{\nabla F}}_n - \underline{\underline{\nabla m}}\left((1-\alpha_m)\underline{\ddot{q}}_{n+1} + \alpha_m\underline{\ddot{q}}_n\right) - \alpha_m\underline{\underline{m}}\,\underline{\underline{\nabla \ddot{q}}}_n + \\
&\quad-\underline{\underline{\nabla d}}\left(\left(1-\alpha_f\right)\underline{\dot{q}}_{n+1} + \alpha_f\underline{\dot{q}}_n\right) - \underline{\underline{d}}\left(\left(1-\alpha_f\right)\underline{\underline{\nabla \dot{q}}}_{\text{pred}} + \alpha_f\underline{\underline{\nabla \dot{q}}}_n\right) + \\
&\quad-\underline{\underline{\nabla k}}\left(\left(1-\alpha_f\right)\underline{q}_{n+1} + \alpha_f\underline{q}_n\right) - \underline{\underline{k}}\left(\left(1-\alpha_f\right)\underline{\underline{\nabla q}}_{\text{pred}} + \alpha_f\underline{\underline{\nabla q}}_n\right).
\end{aligned} \tag{A11}$$

Therefore, the effective equation of motion for sensitivities is defined by

$$\underline{\underline{m}}_{\text{eff}}\,\underline{\underline{\nabla \ddot{q}}}_{n+1} = \underline{\underline{F}}_{\text{pseudo}}, \tag{A12}$$

where $\underline{\underline{m}}_{\text{eff}}$ is given by Equation (A6) and the pseudo load is defined by

$$\begin{aligned}
\underline{\underline{F}}_{\text{pseudo}} = &\left(1-\alpha_f\right)\underline{\underline{\nabla F}}_{n+1} + \alpha_f\underline{\underline{\nabla F}}_n - \underline{\underline{\nabla m}}\left((1-\alpha_m)\underline{\ddot{q}}_{n+1} + \alpha_m\underline{\ddot{q}}_n\right) - \alpha_m\underline{\underline{m}}\,\underline{\underline{\nabla \ddot{q}}}_n + \\
&-\underline{\underline{\nabla d}}\left(\left(1-\alpha_f\right)\underline{\dot{q}}_{n+1} + \alpha_f\underline{\dot{q}}_n\right) - \underline{\underline{d}}\left(\left(1-\alpha_f\right)\underline{\underline{\nabla \dot{q}}}_{\text{pred}} + \alpha_f\underline{\underline{\nabla \dot{q}}}_n\right) + \\
&-\underline{\underline{\nabla k}}\left(\left(1-\alpha_f\right)\underline{q}_{n+1} + \alpha_f\underline{q}_n\right) - \underline{\underline{k}}\left(\left(1-\alpha_f\right)\underline{\underline{\nabla q}}_{\text{pred}} + \alpha_f\underline{\underline{\nabla q}}_n\right).
\end{aligned} \tag{A13}$$

Appendix B. Derivation of Effective Equations of Motion for Flexible Multibody Dynamics

Appendix B.1. Primal Analysis

The base of the generalized-α method for index-1 differential algebraic equations of a flexible multibody system is given by

$$\left(\underline{\underline{m}}\,\ddot{\underline{q}}\right)_{n+1-\alpha_m} + \left(\underline{\underline{d}}\,\dot{\underline{q}}\right)_{n+1-\alpha_f} + \left(\underline{\underline{k}}\,\underline{q}\right)_{n+1-\alpha_f} + \left(\underline{\underline{\mathrm{J}}}_{\Phi}^{T}\,\underline{\lambda}\right)_{n+1-\alpha_m} = \underline{F}_{\mathrm{ext},n+1-\alpha_f} + \underline{F}_{v,n+1-\alpha_f}, \tag{A14}$$

$$\left(\underline{\underline{\mathrm{J}}}_{\Phi}\,\ddot{\underline{q}}\right)_{n+1-\alpha_m} = \underline{F}_{c,n+1-\alpha_f}. \tag{A15}$$

The intermediate point approximations of the force terms acting on the flexible multibody system are given by

$$\left(\underline{\underline{m}}\,\ddot{\underline{q}}\right)_{n+1-\alpha_m} = (1-\alpha_m)\underline{\underline{m}}_{n+1}\ddot{\underline{q}}_{n+1} + \alpha_m\underline{\underline{m}}_n\ddot{\underline{q}}_n, \tag{A16}$$

$$\left(\underline{\underline{d}}\,\dot{\underline{q}}\right)_{n+1-\alpha_f} = \left(1-\alpha_f\right)\underline{\underline{d}}_{n+1}\dot{\underline{q}}_{n+1} + \alpha_f\underline{\underline{d}}_n\dot{\underline{q}}_n, \tag{A17}$$

$$\left(\underline{\underline{k}}\,\underline{q}\right)_{n+1-\alpha_f} = \left(1-\alpha_f\right)\underline{\underline{k}}_{n+1}\underline{q}_{n+1} + \alpha_f\underline{\underline{k}}_n\underline{q}_n, \tag{A18}$$

$$\left(\underline{\underline{\mathrm{J}}}_{\Phi}^{T}\,\underline{\lambda}\right)_{n+1-\alpha_m} = (1-\alpha_m)\underline{\underline{\mathrm{J}}}_{\Phi,n+1}^{T}\underline{\lambda}_{n+1} + \alpha_m\underline{\underline{\mathrm{J}}}_{\Phi,n}^{T}\underline{\lambda}_n, \tag{A19}$$

$$\left(\underline{\underline{\mathrm{J}}}_{\Phi}\,\ddot{\underline{q}}\right)_{n+1-\alpha_m} = (1-\alpha_m)\underline{\underline{\mathrm{J}}}_{\Phi,n+1}\ddot{\underline{q}}_{n+1} + \alpha_m\underline{\underline{\mathrm{J}}}_{\Phi,n}\ddot{\underline{q}}_n, \tag{A20}$$

$$\underline{F}_{\mathrm{ext},n+1-\alpha_f} = \left(1-\alpha_f\right)\underline{F}_{\mathrm{ext},n+1} + \alpha_f\underline{F}_{\mathrm{ext},n}, \tag{A21}$$

$$\underline{F}_{v,n+1-\alpha_f} = \left(1-\alpha_f\right)\underline{F}_{v,n+1} + \alpha_f\underline{F}_{v,n}, \tag{A22}$$

$$\underline{F}_{c,n+1-\alpha_f} = \left(1-\alpha_f\right)\underline{F}_{c,n+1} + \alpha_f\underline{F}_{c,n}. \tag{A23}$$

Applying these linear approximations of the generalized-α method to the equations of motion leads to

$$\begin{aligned}
&(1-\alpha_m)\underline{\underline{m}}_{n+1}\ddot{\underline{q}}_{n+1} + \alpha_m\underline{\underline{m}}_n\ddot{\underline{q}}_n + \\
&+\left(1-\alpha_f\right)\underline{\underline{d}}_{n+1}\dot{\underline{q}}_{n+1} + \alpha_f\underline{\underline{d}}_n\dot{\underline{q}}_n + \\
&+\left(1-\alpha_f\right)\underline{\underline{k}}_{n+1}\underline{q}_{n+1} + \alpha_f\underline{\underline{k}}_n\underline{q}_n + \\
&+(1-\alpha_m)\underline{\underline{\mathrm{J}}}_{\Phi,n+1}^{T}\underline{\lambda}_{n+1} + \alpha_m\underline{\underline{\mathrm{J}}}_{\Phi,n}^{T}\underline{\lambda}_n = \left(1-\alpha_f\right)\underline{F}_{\mathrm{ext},n+1} + \alpha_f\underline{F}_{\mathrm{ext},n} + \\
&\qquad + \left(1-\alpha_f\right)\underline{F}_{v,n+1} + \alpha_f\underline{F}_{v,n},
\end{aligned} \tag{A24}$$

$$(1-\alpha_m)\underline{\underline{\mathrm{J}}}_{\Phi,n+1}\ddot{\underline{q}}_{n+1} + \alpha_m\underline{\underline{\mathrm{J}}}_{\Phi,n}\ddot{\underline{q}}_n = \left(1-\alpha_f\right)\underline{F}_{c,n+1} + \alpha_f\underline{F}_{c,n}. \tag{A25}$$

Inserting Newmark's Equations (1) and (2) gives

$$\begin{aligned}
&(1-\alpha_m)\underline{\underline{m}}_{n+1}\ddot{\underline{q}}_{n+1} + \alpha_m\underline{\underline{m}}_n\ddot{\underline{q}}_n + \\
&+\left(1-\alpha_f\right)\underline{\underline{d}}_{n+1}\left(\dot{\underline{q}}_{\mathrm{pred}} + \gamma\Delta t\ddot{\underline{q}}_{n+1}\right) + \alpha_f\underline{\underline{d}}_n\dot{\underline{q}}_n + \\
&\left(1-\alpha_f\right)\underline{\underline{k}}_{n+1}\left(\underline{q}_{\mathrm{pred}} + \beta\Delta t^2\ddot{\underline{q}}_{n+1}\right) + \alpha_f\underline{\underline{k}}_n\underline{q}_n + \\
&(1-\alpha_m)\underline{\underline{\mathrm{J}}}_{\Phi,n+1}^{T}\underline{\lambda}_{n+1} + \alpha_m\underline{\underline{\mathrm{J}}}_{\Phi,n}^{T}\underline{\lambda}_n = \left(1-\alpha_f\right)\underline{F}_{\mathrm{ext},n+1} + \alpha_f\underline{F}_{\mathrm{ext},n} + \\
&\qquad + \left(1-\alpha_f\right)\underline{F}_{v,n+1} + \alpha_f\underline{F}_{v,n},
\end{aligned} \tag{A26}$$

$$(1-\alpha_m)\underline{\underline{\mathrm{J}}}_{\Phi,n+1}\ddot{\underline{q}}_{n+1} + \alpha_m\underline{\underline{\mathrm{J}}}_{\Phi,n}\ddot{\underline{q}}_n = \left(1-\alpha_f\right)\underline{F}_{c,n+1} + \alpha_f\underline{F}_{c,n}. \tag{A27}$$

which can be rewritten to

$$\left((1-\alpha_m)\underline{\underline{m}}+\left(1-\alpha_f\right)\gamma\Delta t\underline{\underline{d}}+\left(1-\alpha_f\right)\beta\Delta t^2\underline{\underline{k}}\right)\underline{\ddot{q}}_{n+1}+$$
$$+(1-\alpha_m)\underline{\underline{J}}_{\Phi,n+1}^T\underline{\lambda}_{n+1}=\left(1-\alpha_f\right)\underline{F}_{\text{ext},n+1}+\alpha_f\underline{F}_{\text{ext},n}+$$
$$+\left(1-\alpha_f\right)\underline{F}_{v,n+1}+\alpha_f\underline{F}_{v,n}-\alpha_m\underline{\underline{m}}_n\underline{\ddot{q}}_n+$$
$$-\left(1-\alpha_f\right)\underline{\underline{d}}_{n+1}\underline{\dot{q}}_{\text{pred}}-\alpha_f\underline{\underline{d}}_n\underline{\dot{q}}_n+$$
$$-\left(1-\alpha_f\right)\underline{\underline{k}}_{n+1}\underline{q}_{\text{pred}}-\alpha_f\underline{\underline{k}}_n\underline{q}_n+$$
$$-\alpha_m\underline{\underline{J}}_{\Phi,n}^T\underline{\lambda}_n, \tag{A28}$$

$$(1-\alpha_m)\underline{\underline{J}}_{\Phi,n+1}\underline{\ddot{q}}_{n+1}=\left(1-\alpha_f\right)\underline{F}_{c,n+1}+\alpha_f\underline{F}_{c,n}-\alpha_m\underline{\underline{J}}_{\Phi,n}\underline{\ddot{q}}_n. \tag{A29}$$

The equations of motion of the effective system are given by

$$\begin{bmatrix}\underline{\underline{m}}_{\text{eff}} & \underline{\underline{J}}_{\Phi,\text{eff}}^T\\ \underline{\underline{J}}_{\Phi,\text{eff}} & \underline{\underline{0}}\end{bmatrix}\begin{bmatrix}\underline{\ddot{q}}_{n+1}\\ \underline{\lambda}_{n+1}\end{bmatrix}=\begin{bmatrix}\underline{F}_{a,\text{eff}}\\ \underline{F}_{c,\text{eff}}\end{bmatrix}, \tag{A30}$$

where

$$\underline{\underline{m}}_{\text{eff}}=(1-\alpha_m)\underline{\underline{m}}_{n+1}+\left(1-\alpha_f\right)\gamma\Delta t\underline{\underline{d}}_{n+1}+\left(1-\alpha_f\right)\beta\Delta t^2\underline{\underline{k}}_{n+1}, \tag{A31}$$

$$\underline{\underline{J}}_{\Phi,\text{eff}}=(1-\alpha_m)\underline{\underline{J}}_{\Phi,n+1}, \tag{A32}$$

$$\underline{F}_{a,\text{eff}}=\left(1-\alpha_f\right)\underline{F}_{\text{ext},n+1}+\alpha_f\underline{F}_{\text{ext},n}+\left(1-\alpha_f\right)\underline{F}_{v,n+1}+\alpha_f\underline{F}_{v,n}-\alpha_m\underline{\underline{m}}_n\underline{\ddot{q}}_n+$$
$$-\left(1-\alpha_f\right)\underline{\underline{d}}_{n+1}\underline{\dot{q}}_{\text{pred}}-\alpha_f\underline{\underline{d}}_n\underline{\dot{q}}_n-\left(1-\alpha_f\right)\underline{\underline{k}}_{n+1}\underline{q}_{\text{pred}}-\alpha_f\underline{\underline{k}}_n\underline{q}_n-\alpha_m\underline{\underline{J}}_{\Phi,n}^T\underline{\lambda}_n, \tag{A33}$$

$$\underline{F}_{c,\text{eff}}=\left(1-\alpha_f\right)\underline{F}_{c,n+1}+\alpha_f\underline{F}_{c,n}-\alpha_m\underline{\underline{J}}_{\Phi,n}\underline{\ddot{q}}_n. \tag{A34}$$

Appendix B.2. Nonlinear Solver

The base for the nonlinear solver are the residual equations of the flexible multibody system, given by

$$\underline{R}_{1,n+1-\alpha_f}=(1-\alpha_m)\underline{\underline{m}}_{n+1}\underline{\ddot{q}}_{n+1}+\alpha_m\underline{\underline{m}}_n\underline{\ddot{q}}_n+\left(1-\alpha_f\right)\underline{\underline{d}}_{n+1}\underline{\dot{q}}_{n+1}+\alpha_f\underline{\underline{d}}_n\underline{\dot{q}}_n+$$
$$+\left(1-\alpha_f\right)\underline{\underline{k}}_{n+1}\underline{q}_{n+1}+\alpha_f\underline{\underline{k}}_n\underline{q}_n+(1-\alpha_m)\underline{\underline{J}}_{\Phi,n+1}^T\underline{\lambda}_{n+1}+\alpha_m\underline{\underline{J}}_{\Phi,n}^T\underline{\lambda}_n+$$
$$-\left(1-\alpha_f\right)\underline{F}_{\text{ext},n+1}-\alpha_f\underline{F}_{\text{ext},n}-\left(1-\alpha_f\right)\underline{F}_{v,n+1}-\alpha_f\underline{F}_{v,n}, \tag{A35}$$

$$\underline{R}_{2,n+1-\alpha_f}=(1-\alpha_m)\underline{\underline{J}}_{\Phi,n+1}\underline{\ddot{q}}_{n+1}+\alpha_m\underline{\underline{J}}_{\Phi,n}\underline{\ddot{q}}_n-\left(1-\alpha_f\right)\underline{F}_{c,n+1}-\alpha_f\underline{F}_{c,n}. \tag{A36}$$

The Newton–Raphson method is used here

$$\begin{bmatrix}\dfrac{\partial\underline{R}_{1,n+1-\alpha_f}}{\partial\underline{\ddot{q}}_{n+1}} & \dfrac{\partial\underline{R}_{1,n+1-\alpha_f}}{\partial\underline{\lambda}_{n+1}}\\ \dfrac{\partial\underline{R}_{2,n+1-\alpha_f}}{\partial\underline{\ddot{q}}_{n+1}} & \dfrac{\partial\underline{R}_{2,n+1-\alpha_f}}{\partial\underline{\lambda}_{n+1}}\end{bmatrix}\begin{bmatrix}\Delta\underline{\ddot{q}}_{n+1}\\ \Delta\underline{\lambda}_{n+1}\end{bmatrix}+\begin{bmatrix}\underline{R}_{1,n+1-\alpha_f}\\ \underline{R}_{2,n+1-\alpha_f}\end{bmatrix}=\underline{0}, \tag{A37}$$

where the terms of the Jacobian matrix are given by

$$
\begin{aligned}
\frac{\partial \underline{R}_{1,n+1-\alpha_f}}{\partial \ddot{\underline{q}}_{n+1}} = {} & (1-\alpha_m)\frac{\partial \underline{\underline{m}}_{n+1}}{\partial \ddot{\underline{q}}_{n+1}}\ddot{\underline{q}}_{n+1} + \alpha_m \frac{\partial \underline{\underline{m}}_{n}}{\partial \ddot{\underline{q}}_{n+1}}\ddot{\underline{q}}_{n} + (1-\alpha_m)\underline{\underline{m}}_{n+1}\frac{\partial \ddot{\underline{q}}_{n+1}}{\partial \ddot{\underline{q}}_{n+1}} + \alpha_m \underline{\underline{m}}_{n}\frac{\partial \ddot{\underline{q}}_{n}}{\partial \ddot{\underline{q}}_{n+1}} + \\
& + \left(1-\alpha_f\right)\frac{\partial \underline{\underline{d}}_{n+1}}{\partial \ddot{\underline{q}}_{n+1}}\dot{\underline{q}}_{n+1} + \alpha_f \frac{\partial \underline{\underline{d}}_{n}}{\partial \ddot{\underline{q}}_{n+1}}\dot{\underline{q}}_{n} + \left(1-\alpha_f\right)\underline{\underline{d}}_{n+1}\frac{\partial \dot{\underline{q}}_{n+1}}{\partial \ddot{\underline{q}}_{n+1}} + \alpha_f \underline{\underline{d}}_{n}\frac{\partial \dot{\underline{q}}_{n}}{\partial \ddot{\underline{q}}_{n+1}} + \\
& + \left(1-\alpha_f\right)\frac{\partial \underline{\underline{k}}_{n+1}}{\partial \ddot{\underline{q}}_{n+1}}\underline{q}_{n+1} + \alpha_f \frac{\partial \underline{\underline{k}}_{n}}{\partial \ddot{\underline{q}}_{n+1}}\underline{q}_{n} + \left(1-\alpha_f\right)\underline{\underline{k}}_{n+1}\frac{\partial \underline{q}_{n+1}}{\partial \ddot{\underline{q}}_{n+1}} + \alpha_f \underline{\underline{k}}_{n}\frac{\partial \underline{q}_{n}}{\partial \ddot{\underline{q}}_{n+1}} + \\
& + (1-\alpha_m)\frac{\partial \underline{\underline{J}}_{\Phi,n+1}^{T}}{\partial \ddot{\underline{q}}_{n+1}}\underline{\lambda}_{n+1} + \alpha_m \frac{\partial \underline{\underline{J}}_{\Phi,n}^{T}}{\partial \ddot{\underline{q}}_{n+1}}\underline{\lambda}_{n} + (1-\alpha_m)\underline{\underline{J}}_{\Phi,n+1}^{T}\frac{\partial \underline{\lambda}_{n+1}}{\partial \ddot{\underline{q}}_{n+1}} + \alpha_m \underline{\underline{J}}_{\Phi,n}^{T}\frac{\partial \underline{\lambda}_{n}}{\partial \ddot{\underline{q}}_{n+1}} + \\
& - \left(1-\alpha_f\right)\frac{\partial \underline{F}_{\mathrm{ext},n+1}}{\partial \ddot{\underline{q}}_{n+1}} - \alpha_f \frac{\partial \underline{F}_{\mathrm{ext},n}}{\partial \ddot{\underline{q}}_{n+1}} - \left(1-\alpha_f\right)\frac{\partial \underline{F}_{v,n+1}}{\partial \ddot{\underline{q}}_{n+1}} - \alpha_f \frac{\partial \underline{F}_{v,n}}{\partial \ddot{\underline{q}}_{n+1}},
\end{aligned} \tag{A38}
$$

$$
\begin{aligned}
\frac{\partial \underline{R}_{2,n+1-\alpha_f}}{\partial \ddot{\underline{q}}_{n+1}} = {} & (1-\alpha_m)\frac{\partial \underline{\underline{J}}_{\Phi,n+1}}{\partial \ddot{\underline{q}}_{n+1}}\ddot{\underline{q}}_{n+1} + \alpha_m \frac{\partial \underline{\underline{J}}_{\Phi,n}}{\partial \ddot{\underline{q}}_{n+1}}\ddot{\underline{q}}_{n} + (1-\alpha_m)\underline{\underline{J}}_{\Phi,n+1}\frac{\partial \ddot{\underline{q}}_{n+1}}{\partial \ddot{\underline{q}}_{n+1}} + \alpha_m \underline{\underline{J}}_{\Phi,n}\frac{\partial \ddot{\underline{q}}_{n}}{\partial \ddot{\underline{q}}_{n+1}} + \\
& - \left(1-\alpha_f\right)\frac{\partial \underline{F}_{c,n+1}}{\partial \ddot{\underline{q}}_{n+1}} - \alpha_f \frac{\partial \underline{F}_{c,n}}{\partial \ddot{\underline{q}}_{n+1}},
\end{aligned} \tag{A39}
$$

$$
\begin{aligned}
\frac{\partial \underline{R}_{1,n+1-\alpha_f}}{\partial \underline{\lambda}_{n+1}} = {} & (1-\alpha_m)\frac{\partial \underline{\underline{m}}_{n+1}}{\partial \underline{\lambda}_{n+1}}\ddot{\underline{q}}_{n+1} + \alpha_m \frac{\partial \underline{\underline{m}}_{n}}{\partial \underline{\lambda}_{n+1}}\ddot{\underline{q}}_{n} + (1-\alpha_m)\underline{\underline{m}}_{n+1}\frac{\partial \ddot{\underline{q}}_{n+1}}{\partial \underline{\lambda}_{n+1}} + \alpha_m \underline{\underline{m}}_{n}\frac{\partial \ddot{\underline{q}}_{n}}{\partial \underline{\lambda}_{n+1}} + \\
& + \left(1-\alpha_f\right)\frac{\partial \underline{\underline{d}}_{n+1}}{\partial \underline{\lambda}_{n+1}}\dot{\underline{q}}_{n+1} + \alpha_f \frac{\partial \underline{\underline{d}}_{n}}{\partial \underline{\lambda}_{n+1}}\dot{\underline{q}}_{n} + \left(1-\alpha_f\right)\underline{\underline{d}}_{n+1}\frac{\partial \dot{\underline{q}}_{n+1}}{\partial \underline{\lambda}_{n+1}} + \alpha_f \underline{\underline{d}}_{n}\frac{\partial \dot{\underline{q}}_{n}}{\partial \underline{\lambda}_{n+1}} + \\
& + \left(1-\alpha_f\right)\frac{\partial \underline{\underline{k}}_{n+1}}{\partial \underline{\lambda}_{n+1}}\underline{q}_{n+1} + \alpha_f \frac{\partial \underline{\underline{k}}_{n}}{\partial \underline{\lambda}_{n+1}}\underline{q}_{n} + \left(1-\alpha_f\right)\underline{\underline{k}}_{n+1}\frac{\partial \underline{q}_{n+1}}{\partial \underline{\lambda}_{n+1}} + \alpha_f \underline{\underline{k}}_{n}\frac{\partial \underline{q}_{n}}{\partial \underline{\lambda}_{n+1}} + \\
& + (1-\alpha_m)\frac{\partial \underline{\underline{J}}_{\Phi,n+1}^{T}}{\partial \underline{\lambda}_{n+1}}\underline{\lambda}_{n+1} + \alpha_m \frac{\partial \underline{\underline{J}}_{\Phi,n}^{T}}{\partial \underline{\lambda}_{n+1}}\underline{\lambda}_{n} + (1-\alpha_m)\underline{\underline{J}}_{\Phi,n+1}^{T}\frac{\partial \underline{\lambda}_{n+1}}{\partial \underline{\lambda}_{n+1}} + \alpha_m \underline{\underline{J}}_{\Phi,n}^{T}\frac{\partial \underline{\lambda}_{n}}{\partial \underline{\lambda}_{n+1}} + \\
& - \left(1-\alpha_f\right)\frac{\partial \underline{F}_{\mathrm{ext},n+1}}{\partial \underline{\lambda}_{n+1}} - \alpha_f \frac{\partial \underline{F}_{\mathrm{ext},n}}{\partial \underline{\lambda}_{n+1}} - \left(1-\alpha_f\right)\frac{\partial \underline{F}_{v,n+1}}{\partial \underline{\lambda}_{n+1}} - \alpha_f \frac{\partial \underline{F}_{v,n}}{\partial \underline{\lambda}_{n+1}},
\end{aligned} \tag{A40}
$$

$$
\begin{aligned}
\frac{\partial \underline{R}_{2,n+1-\alpha_f}}{\partial \underline{\lambda}_{n+1}} = {} & (1-\alpha_m)\frac{\partial \underline{\underline{J}}_{\Phi,n+1}}{\partial \underline{\lambda}_{n+1}}\ddot{\underline{q}}_{n+1} + \alpha_m \frac{\partial \underline{\underline{J}}_{\Phi,n}}{\partial \underline{\lambda}_{n+1}}\ddot{\underline{q}}_{n} + (1-\alpha_m)\underline{\underline{J}}_{\Phi,n+1}\frac{\partial \ddot{\underline{q}}_{n+1}}{\partial \underline{\lambda}_{n+1}} + \alpha_m \underline{\underline{J}}_{\Phi,n}\frac{\partial \ddot{\underline{q}}_{n}}{\partial \underline{\lambda}_{n+1}} + \\
& - \left(1-\alpha_f\right)\frac{\partial \underline{F}_{c,n+1}}{\partial \underline{\lambda}_{n+1}} - \alpha_f \frac{\partial \underline{F}_{c,n}}{\partial \underline{\lambda}_{n+1}}.
\end{aligned} \tag{A41}
$$

The derivatives of the system parameters of the previous time step with respect to the state variables of the next time step disappear,

$$
\frac{\partial \underline{\bigcirc}_{n}}{\partial \underline{\bullet}_{n+1}} = \underline{\underline{0}}, \qquad \frac{\partial \underline{\underline{\bigcirc}}_{n}}{\partial \underline{\bullet}_{n+1}} = \underline{\underline{\underline{0}}}. \tag{A42}
$$

The dependencies of the system parameters are shown in (33) and (34), and applying the chain rule of differentiation leads to

$$\frac{\partial \underline{\underline{m}}}{\partial \underline{\ddot{q}}} = \frac{\partial \underline{\underline{m}}}{\partial \underline{q}} \frac{\partial \underline{q}}{\partial \underline{\ddot{q}}}, \tag{A43}$$

$$\frac{\partial \underline{\underline{d}}}{\partial \underline{\ddot{q}}} = \frac{\partial \underline{\underline{d}}}{\partial \underline{q}} \frac{\partial \underline{q}}{\partial \underline{\ddot{q}}}, \tag{A44}$$

$$\frac{\partial \underline{\underline{k}}}{\partial \underline{\ddot{q}}} = \frac{\partial \underline{\underline{k}}}{\partial \underline{q}} \frac{\partial \underline{q}}{\partial \underline{\ddot{q}}}, \tag{A45}$$

$$\frac{\partial \underline{\underline{J}}_{\Phi}}{\partial \underline{\ddot{q}}} = \frac{\partial \underline{\underline{J}}_{\Phi}}{\partial \underline{q}} \frac{\partial \underline{q}}{\partial \underline{\ddot{q}}}, \tag{A46}$$

$$\frac{\partial \underline{F}_{\text{ext}}}{\partial \underline{\ddot{q}}} = \frac{\partial \underline{F}_{\text{ext}}}{\partial \underline{q}} \frac{\partial \underline{q}}{\partial \underline{\ddot{q}}} + \frac{\partial \underline{F}_{\text{ext}}}{\partial \underline{\dot{q}}} \frac{\partial \underline{\dot{q}}}{\partial \underline{\ddot{q}}}, \tag{A47}$$

$$\frac{\partial \underline{F}_{v}}{\partial \underline{\ddot{q}}} = \frac{\partial \underline{F}_{v}}{\partial \underline{q}} \frac{\partial \underline{q}}{\partial \underline{\ddot{q}}} - \frac{\partial \underline{F}_{v}}{\partial \underline{\dot{q}}} \frac{\partial \underline{\dot{q}}}{\partial \underline{\ddot{q}}}, \tag{A48}$$

$$\frac{\partial \underline{\underline{\bigcirc}}}{\partial \underline{\lambda}} = \underline{\underline{0}}. \tag{A49}$$

$$\frac{\partial \underline{\underline{\bigcirc}}}{\partial \underline{\lambda}} = \underline{\underline{0}}. \tag{A50}$$

The derivatives of the state variables with respect to $\underline{\ddot{q}}_{n+1}$ and $\underline{\lambda}_{n+1}$ are given by

$$\begin{aligned}
\frac{\partial \underline{q}_{n+1}}{\partial \underline{\ddot{q}}_{n+1}} &= \beta \Delta t^2 \underline{\underline{e}}, & \frac{\partial \underline{q}_{n+1}}{\partial \underline{\lambda}_{n+1}} &= \underline{\underline{0}}, \\
\frac{\partial \underline{\dot{q}}_{n+1}}{\partial \underline{\ddot{q}}_{n+1}} &= \gamma \Delta t \underline{\underline{e}}, & \frac{\partial \underline{\dot{q}}_{n+1}}{\partial \underline{\lambda}_{n+1}} &= \underline{\underline{0}}, \\
\frac{\partial \underline{\ddot{q}}_{n+1}}{\partial \underline{\ddot{q}}_{n+1}} &= \underline{\underline{e}}, & \frac{\partial \underline{\ddot{q}}_{n+1}}{\partial \underline{\lambda}_{n+1}} &= \underline{\underline{0}}, \\
\frac{\partial \underline{\lambda}_{n+1}}{\partial \underline{\ddot{q}}_{n+1}} &= \underline{\underline{0}}, & \frac{\partial \underline{\lambda}_{n+1}}{\partial \underline{\lambda}_{n+1}} &= \underline{\underline{e}}.
\end{aligned} \tag{A51}$$

Applying Equation (A42), Equations (A43)–(A51) lead to the terms of the Jacobian matrix given by

$$\begin{aligned}
\frac{\partial \underline{R}_{1,n+1-\alpha_f}}{\partial \underline{\ddot{q}}_{n+1}} = {} & (1-\alpha_m)\left(\beta \Delta t^2 \frac{\partial \underline{\underline{m}}_{n+1}}{\partial \underline{q}_{n-1}} \underline{\ddot{q}}_{n+1} + \underline{\underline{m}}_{n+1}\right) + \\
& + \left(1-\alpha_f\right)\left(\beta \Delta t^2 \frac{\partial \underline{\underline{d}}_{n+1}}{\partial \underline{q}_{n+1}} \underline{\dot{q}}_{n+1} + \gamma \Delta t \underline{\underline{d}}_{n+1}\right) + \\
& + \left(1-\alpha_f\right)\left(\beta \Delta t^2 \frac{\partial \underline{\underline{k}}_{n+1}}{\partial \underline{q}_{n+1}} \underline{q}_{n-1} + \beta \Delta t^2 \underline{\underline{k}}_{n+1}\right) + \\
& + (1-\alpha_m)\beta \Delta t^2 \frac{\partial \underline{\underline{J}}^T_{\Phi,n+1}}{\partial \underline{q}_{n+1}} \underline{\lambda}_{n+1} + \\
& - \left(1-\alpha_f\right)\left(\beta \Delta t^2 \frac{\partial \underline{F}_{\text{ext},n+1}}{\partial \underline{a}_{n+1}} + \gamma \Delta t \frac{\partial \underline{F}_{\text{ext},n+1}}{\partial \underline{\dot{q}}_{n+1}}\right) + \\
& - \left(1-\alpha_f\right)\left(\beta \Delta t^2 \frac{\partial \underline{F}_{v,n+1}}{\partial \underline{q}_{n+1}} + \gamma \Delta t \frac{\partial \underline{F}_{v,n+1}}{\partial \underline{\dot{q}}_{n+1}}\right),
\end{aligned} \tag{A52}$$

$$\frac{\partial \underline{R}_{2,n+1-\alpha_f}}{\partial \underline{\ddot{q}}_{n+1}} = (1-\alpha_m)\left(\beta\Delta t^2 \frac{\partial \underline{\underline{J}}_{\Phi,n+1}}{\partial \underline{q}_{n+1}} \underline{\ddot{q}}_{n+1} + \underline{\underline{J}}_{\Phi,n+1}\right) + \\ -\left(1-\alpha_f\right)\left(\beta\Delta t^2 \frac{\partial \underline{F}_{c,n+1}}{\partial \underline{q}_{n+1}} + \gamma\Delta t \frac{\partial \underline{F}_{c,n+1}}{\partial \underline{\dot{q}}_{n+1}}\right), \tag{A53}$$

$$\frac{\partial \underline{R}_{1,n+1-\alpha_f}}{\partial \underline{\lambda}_{n+1}} = (1-\alpha_m)\underline{\underline{J}}^T_{\Phi,n+1}, \tag{A54}$$

$$\frac{\partial \underline{R}_{2,n+1-\alpha_f}}{\partial \underline{\lambda}_{n+1}} = \underline{\underline{0}}. \tag{A55}$$

Appendix B.3. Sensitivity Analysis

The derivative of the equations of motion leads to the governing equation of the sensitivities given by

$$\begin{aligned}
&(1-\alpha_m)\underline{\underline{\underline{\nabla m}}}_{n+1}\underline{\ddot{q}}_{n+1} + \alpha_m \underline{\underline{\underline{\nabla m}}}_{n}\underline{\ddot{q}}_{n} + \\
&+(1-\alpha_m)\underline{\underline{m}}_{n+1}\underline{\underline{\nabla \ddot{q}}}_{n+1} + \alpha_m \underline{\underline{m}}_{n}\underline{\underline{\nabla \ddot{q}}}_{n} + \\
&+\left(1-\alpha_f\right)\underline{\underline{\underline{\nabla d}}}_{n+1}\underline{\dot{q}}_{n+1} + \alpha_f \underline{\underline{\underline{\nabla d}}}_{n}\underline{\dot{q}}_{n} + \\
&+\left(1-\alpha_f\right)\underline{\underline{d}}_{n+1}\underline{\underline{\nabla \dot{q}}}_{n+1} + \alpha_f \underline{\underline{d}}_{n}\underline{\underline{\nabla \dot{q}}}_{n} + \\
&+\left(1-\alpha_f\right)\underline{\underline{\underline{\nabla k}}}_{n+1}\underline{q}_{n+1} + \alpha_f \underline{\underline{\underline{\nabla k}}}_{n}\underline{q}_{n} + \\
&+\left(1-\alpha_f\right)\underline{\underline{k}}_{n+1}\underline{\underline{\nabla q}}_{n+1} + \alpha_f \underline{\underline{k}}_{n}\underline{\underline{\nabla q}}_{n} + \\
&+(1-\alpha_m)\underline{\underline{\underline{\nabla J}}}^T_{\Phi,n+1}\underline{\lambda}_{n+1} + \alpha_m \underline{\underline{\underline{\nabla J}}}^T_{\Phi,n}\underline{\lambda}_{n} + \\
&+(1-\alpha_m)\underline{\underline{J}}^T_{\Phi,n+1}\underline{\underline{\nabla \lambda}}_{n+1} + \alpha_m \underline{\underline{J}}^T_{\Phi,n}\underline{\underline{\nabla \lambda}}_{n} = \left(1-\alpha_f\right)\underline{\underline{\nabla F}}_{\text{ext},n+1} + \alpha_f \underline{\underline{\nabla F}}_{\text{ext},n} + \\
&\qquad + \left(1-\alpha_f\right)\underline{\underline{\nabla F}}_{v,n+1} + \alpha_f \underline{\underline{\nabla F}}_{v,n},
\end{aligned} \tag{A56}$$

and

$$\begin{aligned}
&(1-\alpha_m)\underline{\underline{\underline{\nabla J}}}_{\Phi,n+1}\underline{\ddot{q}}_{n+1} + \alpha_m \underline{\underline{\underline{\nabla J}}}_{\Phi,n}\underline{\ddot{q}}_{n} + \\
&+(1-\alpha_m)\underline{\underline{J}}_{\Phi,n+1}\underline{\underline{\nabla \ddot{q}}}_{n+1} + \alpha_m \underline{\underline{J}}_{\Phi,n}\underline{\underline{\nabla \ddot{q}}}_{n} = \left(1-\alpha_f\right)\underline{\underline{\nabla F}}_{c,n+1} + \alpha_f \underline{\underline{\nabla F}}_{c,n}.
\end{aligned} \tag{A57}$$

Including Newmark's sensitivity, Equations (16) and (17) lead to

$$
\begin{aligned}
&(1-\alpha_m)\underline{\underline{\underline{\nabla m}}}_{n+1}\underline{\ddot{q}}_{n+1}+\alpha_m\underline{\underline{\underline{\nabla m}}}_n\underline{\ddot{q}}_n+\\
&+(1-\alpha_m)\underline{\underline{m}}_{n+1}\underline{\underline{\nabla\ddot{q}}}_{n+1}+\alpha_m\underline{\underline{m}}_n\underline{\underline{\nabla\ddot{q}}}_n+\\
&+\left(1-\alpha_f\right)\underline{\underline{\underline{\nabla d}}}_{n+1}\underline{\dot{q}}_{n+1}+\alpha_f\underline{\underline{\underline{\nabla d}}}_n\underline{\dot{q}}_n+\\
&+\left(1-\alpha_f\right)\underline{\underline{d}}_{n+1}\left(\underline{\underline{\nabla\dot{q}}}_{\text{pred}}+\gamma\Delta t\underline{\underline{\nabla\ddot{q}}}_{n+1}\right)+\alpha_f\underline{\underline{d}}_n\underline{\underline{\nabla\dot{q}}}_n+\\
&+\left(1-\alpha_f\right)\underline{\underline{\underline{\nabla k}}}_{n+1}\underline{q}_{n+1}+\alpha_f\underline{\underline{\underline{\nabla k}}}_n\underline{q}_n+\\
&+\left(1-\alpha_f\right)\underline{\underline{k}}_{n+1}\left(\underline{\underline{\nabla q}}_{\text{pred}}+\beta\Delta t^2\underline{\underline{\nabla\ddot{q}}}_{n+1}\right)+\alpha_f\underline{\underline{k}}_n\underline{\underline{\nabla q}}_n+\\
&+(1-\alpha_m)\underline{\underline{\underline{\nabla \mathrm{J}}}}^T_{\Phi,n+1}\underline{\lambda}_{n+1}+\alpha_m\underline{\underline{\underline{\nabla \mathrm{J}}}}^T_{\Phi,n}\underline{\lambda}_n+\\
&+(1-\alpha_m)\underline{\underline{\mathrm{J}}}^T_{\Phi,n+1}\underline{\underline{\nabla\lambda}}_{n+1}+\alpha_m\underline{\underline{\mathrm{J}}}^T_{\Phi,n}\underline{\underline{\nabla\lambda}}_n=\left(1-\alpha_f\right)\underline{\underline{\nabla F}}_{\text{ext},n+1}+\alpha_f\underline{\underline{\nabla F}}_{\text{ext},n}+\\
&\qquad+\left(1-\alpha_f\right)\underline{\underline{\nabla F}}_{v,n+1}+\alpha_f\underline{\underline{\nabla F}}_{v,n},
\end{aligned}
\tag{A58}
$$

and

$$
\begin{aligned}
&(1-\alpha_m)\underline{\underline{\underline{\nabla \mathrm{J}}}}_{\Phi,n+1}\underline{\ddot{q}}_{n+1}+\alpha_m\underline{\underline{\underline{\nabla \mathrm{J}}}}_{\Phi,n}\underline{\ddot{q}}_n+\\
&+(1-\alpha_m)\underline{\underline{\mathrm{J}}}_{\Phi,n+1}\underline{\underline{\nabla\ddot{q}}}_{n+1}+\alpha_m\underline{\underline{\mathrm{J}}}_{\Phi,n}\underline{\underline{\nabla\ddot{q}}}_n=\left(1-\alpha_f\right)\underline{\underline{\nabla F}}_{c,n+1}+\alpha_f\underline{\underline{\nabla F}}_{c,n}.
\end{aligned}
\tag{A59}
$$

which can be rewritten to

$$
\begin{aligned}
&\left((1-\alpha_m)\underline{\underline{m}}+\left(1-\alpha_f\right)\gamma\Delta t\underline{\underline{d}}+\left(1-\alpha_f\right)\beta\Delta t^2\underline{\underline{k}}\right)\underline{\underline{\nabla\ddot{q}}}_{n+1}+\\
&+(1-\alpha_m)\underline{\underline{\mathrm{J}}}^T_{\Phi,n+1}\underline{\underline{\nabla\lambda}}_{n+1}=\left(1-\alpha_f\right)\underline{\underline{\nabla F}}_{\text{ext},n+1}+\alpha_f\underline{\underline{\nabla F}}_{\text{ext},n}+\\
&\qquad+\left(1-\alpha_f\right)\underline{\underline{\nabla F}}_{v,n+1}+\alpha_f\underline{\underline{\nabla F}}_{v,n}+\\
&\qquad-(1-\alpha_m)\underline{\underline{\underline{\nabla m}}}_{n+1}\underline{\ddot{q}}_{n+1}+\\
&\qquad-\alpha_m\underline{\underline{\underline{\nabla m}}}_n\underline{\ddot{q}}_n-\alpha_m\underline{\underline{m}}_n\underline{\underline{\nabla\ddot{q}}}_n+\\
&\qquad-\left(1-\alpha_f\right)\underline{\underline{\underline{\nabla d}}}_{n+1}\underline{\dot{q}}_{n+1}-\alpha_f\underline{\underline{\underline{\nabla d}}}_n\underline{\dot{q}}_n+\\
&\qquad-\left(1-\alpha_f\right)\underline{\underline{d}}_{n+1}\underline{\underline{\nabla\dot{q}}}_{\text{pred}}-\alpha_f\underline{\underline{d}}_n\underline{\underline{\nabla\dot{q}}}_n+\\
&\qquad-\left(1-\alpha_f\right)\underline{\underline{\underline{\nabla k}}}_{n+1}\underline{q}_{n+1}-\alpha_f\underline{\underline{\underline{\nabla k}}}_n\underline{q}_n+\\
&\qquad-\left(1-\alpha_f\right)\underline{\underline{k}}_{n+1}\underline{\underline{\nabla q}}_{\text{pred}}-\alpha_f\underline{\underline{k}}_n\underline{\underline{\nabla q}}_n+\\
&\qquad-(1-\alpha_m)\underline{\underline{\underline{\nabla \mathrm{J}}}}^T_{\Phi,n+1}\underline{\lambda}_{n+1}+\\
&\qquad-\alpha_m\underline{\underline{\underline{\nabla \mathrm{J}}}}^T_{\Phi,n}\underline{\lambda}_n-\alpha_m\underline{\underline{\mathrm{J}}}^T_{\Phi,n}\underline{\underline{\nabla\lambda}}_n,
\end{aligned}
\tag{A60}
$$

and

$$
\begin{aligned}
&(1-\alpha_m)\underline{\underline{\mathrm{J}}}_{\Phi,n+1}\underline{\underline{\nabla\ddot{q}}}_{n+1}=\left(1-\alpha_f\right)\underline{\underline{\nabla F}}_{c,n+1}+\alpha_f\underline{\underline{\nabla F}}_{c,n}+\\
&\qquad-(1-\alpha_m)\underline{\underline{\underline{\nabla \mathrm{J}}}}_{\Phi,n+1}\underline{\ddot{q}}_{n+1}-\alpha_m\underline{\underline{\underline{\nabla \mathrm{J}}}}_{\Phi,n}\underline{\ddot{q}}_n-\alpha_m\underline{\underline{\mathrm{J}}}_{\Phi,n}\underline{\underline{\nabla\ddot{q}}}_n.
\end{aligned}
\tag{A61}
$$

The governing equation of the sensitivity analysis of the effective system is given by

$$\begin{bmatrix} \underline{\underline{m}}_{\text{eff}} & \underline{\underline{\mathrm{J}}}^{T}_{\Phi,\text{eff}} \\ \underline{\underline{\mathrm{J}}}_{\Phi,\text{eff}} & \underline{\underline{0}} \end{bmatrix} \begin{bmatrix} \underline{\underline{\nabla \ddot{q}}}_{n+1} \\ \underline{\underline{\nabla \lambda}}_{n+1} \end{bmatrix} = \begin{bmatrix} \underline{\underline{F}}_{a,\text{pseudo}} \\ \underline{\underline{F}}_{c,\text{pseudo}} \end{bmatrix}, \tag{A62}$$

where

$$\begin{aligned} \underline{\underline{F}}_{a,\text{pseudo}} =& \left(1-\alpha_f\right)\underline{\underline{\nabla F}}_{\text{ext},n+1} + \alpha_f \underline{\underline{\nabla F}}_{\text{ext},n} + \left(1-\alpha_f\right)\underline{\underline{\nabla F}}_{v,n+1} + \alpha_f \underline{\underline{\nabla F}}_{v,n} + \\ & - (1-\alpha_m)\underline{\underline{\underline{\nabla m}}}_{n+1}\underline{\ddot{q}}_{n+1} - \alpha_m \underline{\underline{\underline{\nabla m}}}_{n}\underline{\ddot{q}}_{n} - \alpha_m \underline{\underline{m}}_n \underline{\underline{\nabla \ddot{q}}}_n + \\ & - \left(1-\alpha_f\right)\underline{\underline{\underline{\nabla d}}}_{n+1}\underline{\dot{q}}_{n+1} - \alpha_f \underline{\underline{\underline{\nabla d}}}_{n}\underline{\dot{q}}_{n} - \left(1-\alpha_f\right)\underline{\underline{d}}_{n+1}\underline{\underline{\nabla \dot{q}}}_{\text{pred}} - \alpha_f \underline{\underline{d}}_n \underline{\underline{\nabla \dot{q}}}_n + \\ & - \left(1-\alpha_f\right)\underline{\underline{\underline{\nabla k}}}_{n+1}\underline{q}_{n+1} - \alpha_f \underline{\underline{\underline{\nabla k}}}_{n}\underline{q}_{n} - \left(1-\alpha_f\right)\underline{\underline{k}}_{n+1}\underline{\underline{\nabla q}}_{\text{pred}} - \alpha_f \underline{\underline{k}}_n \underline{\underline{\nabla q}}_n + \\ & - (1-\alpha_m)\underline{\underline{\underline{\nabla \mathrm{J}}}}^{T}_{\Phi,n+1}\underline{\lambda}_{n+1} - \alpha_m \underline{\underline{\underline{\nabla \mathrm{J}}}}^{T}_{\Phi,n}\underline{\lambda}_{n} - \alpha_m \underline{\underline{\mathrm{J}}}^{T}_{\Phi,n}\underline{\underline{\nabla \lambda}}_n, \end{aligned} \tag{A63}$$

$$\begin{aligned} \underline{\underline{F}}_{c,\text{pseudo}} =& \left(1-\alpha_f\right)\underline{\underline{\nabla F}}_{c,n+1} + \alpha_f \underline{\underline{\nabla F}}_{c,n} + \\ & - (1-\alpha_m)\underline{\underline{\underline{\nabla \mathrm{J}}}}_{\Phi,n+1}\underline{\ddot{q}}_{n+1} - \alpha_m \underline{\underline{\underline{\nabla \mathrm{J}}}}_{\Phi,n}\underline{\ddot{q}}_{n} - \alpha_m \underline{\underline{\mathrm{J}}}_{\Phi,n}\underline{\underline{\nabla \ddot{q}}}_n. \end{aligned} \tag{A64}$$

References

1. Hien, T.D.; Kleiber, M. Stochastic design sensitivity in structural dynamics. *Int. J. Numer. Methods Eng.* **1991**, *32*, 1247–1265. [CrossRef]
2. Cho, S.; Choi, K.K. Design sensitivity analysis and optimization of non-linear transient dynamics. Part I–sizing design. *Int. J. Numer. Methods Eng.* **2000**, *48*, 351–373. [CrossRef]
3. Cho, S.; Choi, K.K. Design sensitivity analysis and optimization of non-linear transient dynamics. Part II–configuration design. *Int. J. Numer. Methods Eng.* **2000**, *48*, 375–399. [CrossRef]
4. van Keulen, F.; Haftka, R.; Kim, N. Review of options for structural design sensitivity analysis—Part 1: Linear systems. *Comput. Methods Appl. Mech. Eng.* **2005**, *194*, 3213–3243. [CrossRef]
5. Martins, J.R.R.A.; Hwang, J.T. Review and unification of methods for computing derivatives of multidisciplinary computational models. *AIAA J.* **2013**, *51*, 2582–2599. [CrossRef]
6. Tortorelli, D.A.; Michaleris, P. Design sensitivity analysis: Overview and review. *Inverse Probl. Eng.* **1994**, *1*, 71–105. [CrossRef]
7. Wehrle, E.; Gufler, V. Lightweight engineering design of nonlinear dynamic systems with gradient-based structural design optimization. In *Proceedings of the Munich Symposium on Lightweight Design 2020*; Springer: Berlin/Heidelberg, Germany, 2021; pp. 44–57. [CrossRef]
8. Wehrle, E. Optimal lightweight engineering via a three-block solver scheme for mechanical analysis. In *Optimal Design and Control of Multibody Systems*; Nachbagauer, K., Held, A., Eds.; Springer: Berlin/Heidelberg, Germany, 2024; Volume 42, pp. 16–29. [CrossRef]
9. Courant, R.; Friedrichs, K.; Lewy, H. Über die partiellen Differenzengleichungen der mathematischen Physik. *Math. Ann.* **1928**, *100*, 32–74. [CrossRef]
10. Newmark, N.M. A method of computation for structural dynamics. *J. Eng. Mech.* **1959**, *85*, 67–94. [CrossRef]
11. Hilber, H.M.; Hughes, T.J.R.; Taylor, R.L. Improved numerical dissipation for time integration algorithms in structural dynamics. *Earthq. Eng. Struct. Dyn.* **1977**, *5*, 283–292. [CrossRef]
12. Wood, W.L.; Bossak, M.; Zienkiewicz, O.C. An alpha modification of Newmark's method. *Int. J. Numer. Methods Eng.* **1980**, *15*, 1562–1566. [CrossRef]
13. Chung, J.; Hulbert, G.M. A time integration algorithm for structural dynamics with improved numerical dissipation: The generalized-α method. *J. Appl. Mech.* **1993**, *60*, 371–375. [CrossRef]
14. Gufler, V.; Wehrle, E.; Vidoni, R. Sensitivity analysis of flexible multibody dynamics with generalized-α time integration and Baumgarte stabilization: A study on numerical stability. In *Mechanisms and Machine Science*; Springer International Publishing: Berlin/Heidelberg, Germany, 2022; pp. 147–155. [CrossRef]
15. Griewank, A.; Walther, A. *Evaluating Derivatives: Principles and Techniques of Algorithmic Differentiation*, 2nd ed.; SIAM: Philadelphia, PA, USA, 2008.
16. Naumann, U. *The Art of Differentiating Computer Programs: An Introduction to Algorithmic Differentiation*; SIAM: Philadelphia, PA, USA, 2012.
17. Callejo, A.; Dopico, D. Direct sensitivity analysis of multibody systems: A vehicle dynamics benchmark. *J. Comput. Nonlinear Dyn.* **2019**, *14*, 021004. [CrossRef]

18. Margossian, C.C. A review of automatic differentiation and its efficient implementation. *WIREs Data Min. Knowl. Discov.* **2019**, *9*, e1305. [CrossRef]
19. Pi, T.; Zhang, Y.; Chen, L. First order sensitivity analysis of flexible multibody systems using absolute nodal coordinate formulation. *Multibody Syst. Dyn.* **2012**, *27*, 153–171. [CrossRef]
20. Held, A.; Knüfer, S.; Seifried, R. Structural sensitivity analysis of flexible multibody systems modeled with the floating frame of reference approach using the adjoint variable method. *Multibody Syst. Dyn.* **2016**, *40*, 287–302. [CrossRef]
21. Boopathy, K.; Kennedy, G. Adjoint-based derivative evaluation methods for flexible multibody systems with rotorcraft applications. In Proceedings of the 55th AIAA Aerospace Sciences Meeting, Grapevine, TX, USA, 9–13 January 2017; American Institute of Aeronautics and Astronautics: Reston, VA, USA, 2017. [CrossRef]
22. Martins, J.R.R.A.; Kennedy, G.J. Enabling large-scale multidisciplinary design optimization through adjoint sensitivity analysis. *Struct. Multidiscip. Optim.* **2021**, *64*, 2959–2974. [CrossRef]
23. Nachbagauer, K.; Oberpeilsteiner, S.; Sherif, K.; Steiner, W. The use of the adjoint method for solving typical optimization problems in multibody dynamics. *J. Comput. Nonlinear Dyn.* **2015**, *10*, 061011. [CrossRef]
24. Lauß, T.; Oberpeilsteiner, S.; Steiner, W.; Nachbagauer, K. The discrete adjoint method for parameter identification in multibody system dynamics. *Multibody Syst. Dyn.* **2017**, *42*, 397–410. [CrossRef]
25. Boopathy, K.; Kennedy, G.J. Parallel finite element framework for rotorcraft multibody dynamics and discrete adjoint sensitivities. *Aiaa J.* **2019**, *57*, 3159–3172. [CrossRef]
26. Boopathy, K. Adjoint Based Design Optimization of Systems with Time Dependent Physics and Probabilistically Modeled Uncertainties. Ph.D. Thesis, Georgia Institute of Technology, Atlanta, GA, USA, 2020.
27. Ebrahimi, M.; Butscher, A.; Cheong, H.; Iorio, F. Design optimization of dynamic flexible multibody systems using the discrete adjoint variable method. *Comput. Struct.* **2019**, *213*, 82–99. [CrossRef]
28. Nejat, A.A.; Moghadasi, A.; Held, A. Adjoint sensitivity analysis of flexible multibody systems in differential-algebraic form. *Comput. Struct.* **2020**, *228*, 106–148. [CrossRef]
29. Held, A. On design sensitivities in the structural analysis and optimization of flexible multibody systems. *Multibody Syst. Dyn.* **2022**, *54*, 53–74. [CrossRef]
30. Solano, D.; Sarojini, D.; Rajaram, D.; Mavris, D.N. Adjoint-based analysis and optimization of beam-like structures subjected to dynamic loads. *Struct. Multidiscip. Optim.* **2022**, *65*, 52. [CrossRef]
31. Haug, E.J.; Arora, J.S. Design sensitivity analysis of elastic mechanical systems. *Comput. Methods Appl. Mech. Eng.* **1978**, *15*, 35–62. [CrossRef]
32. Hsieh, C.C.; Arora, J.S. Structural design sensitivity analysis with general boundary conditions: Dynamic problem. *Int. J. Numer. Methods Eng.* **1985**, *21*, 267–283. [CrossRef]
33. Baier, H.; Seeßelberg, C.; Specht, B. *Optimierung in der Strukturmechanik*; Vieweg: Braunschweig/Wiesbaden, Germany 1994. [CrossRef]
34. Trier, S.; Marthinsen, A.; Sivertsen, O. Design sensitivities by the adjoint variable method in nonlinear structural dynamics. In Proceedings of the SIMS Simulation Conference, Trondheim, Norway, 11–13 June 1996.
35. Michaleris, P.; Tortorelli, D.A.; Vidal, C.A. Tangent operators and design sensitivity formulations for transient non-linear coupled problems with applications to elastoplasticity. *Int. J. Numer. Methods Eng.* **1994**, *37*, 2471–2499. [CrossRef]
36. Bhalerao, K.D.; Poursina, M.; Anderson, K.S. An efficient direct differentiation approach for sensitivity analysis of flexible multibody systems. *Multibody Syst. Dyn.* **2010**, *23*, 121–140. [CrossRef]
37. Zhu, Y. Sensitivity Analysis and Optimization of Multibody Systems. Ph.D. Thesis, Virginia Polytechnic Institute and State University, Blacksburg, VA, USA, 2014.
38. Dopico, D.; Zhu, Y.; Sandu, A.; Sandu, C. Direct and adjoint sensitivity analysis of ordinary differential equation multibody formulations. *J. Comput. Nonlinear Dyn.* **2014**, *10*, 011012. [CrossRef]
39. Callejo, A.; Sonneville, V.; Bauchau, O. Discrete adjoint method for the sensitivity analysis of flexible multibody systems. *J. Comput. Nonlinear Dyn.* **2018**, *14*, 021001. [CrossRef]
40. Cao, Z.; Yao, J.; Jia, Z.; Liang, D. Transient response sensitivity analysis of localized nonlinear structure using direct differentiation method. *Machines* **2022**, *10*, 1039. [CrossRef]
41. Haug, E.J.; Ehle, P.E. Second-order design sensitivity analysis of mechanical system dynamics. *Int. J. Numer. Methods Eng.* **1982**, *18*, 1699–1717. [CrossRef]
42. Ding, J.Y.; Pan, Z.K.; Chen, L.Q. Second order adjoint sensitivity analysis of multibody systems described by differential-algebraic equations. *Multibody Syst. Dyn.* **2007**, *18*, 599–617. [CrossRef]
43. Belotti, R.; Palomba, I.; Wehrle, E.; Vidoni, R. An approximation-based design optimization approach to eigenfrequency assignment for flexible multibody systems. *Appl. Sci.* **2021**, *11*, 11558. [CrossRef]
44. Palomba, I.; Vidoni, R. Flexible-link multibody system eigenvalue analysis parameterized with respect to rigid-body motion. *Appl. Sci.* **2019**, *9*, 5156. [CrossRef]
45. Brüls, O.; Lemaire, E.; Duysinx, P.; Eberhard, P., Multibody dynamics. In *Multibody Dynamics: Computational Methods and Applications*, Arczewski, K., Blajer, W., Fraczek, J., Wojtyra, M., Eds.; Springer: Berlin/Heidelberg, Germany, 2011, Chapter Optimization of Multibody Systems and Their Structural Components, pp. 49–68. [CrossRef]

46. Gufler, V.; Wehrle, E.; Vidoni, R. Multiphysical design optimization of multibody systems: Application to a Tyrolean weir cleaning mechanism. In Proceedings of the 3rd International Conference of IFToMM Italy, Naples, Italy, 9–11 September 2020. [CrossRef]
47. Gufler, V.; Wehrle, E.; Achleitner, J.; Vidoni, R. A semi-analytical approach to sensitivity analysis with flexible multibody dynamics of a morphing forward wing section. *Multibody Syst. Dyn.* **2023**, *58*, 1–20. [CrossRef]
48. Gufler, V.; Zwölfer, A.; Wehrle, E. Analytical derivatives of flexible multibody dynamics with the floating frame of reference formulation. *Multibody Syst. Dyn.* **2022**, *60*, 257–288. [CrossRef]
49. Brüls, O.; Golinval, J. The generalized-α method in mechatronic applications. *ZAMM* **2006**, *86*, 748–758. [CrossRef]
50. Jansen, K.E.; Whiting, C.H.; Hulbert, G.M. A generalized-α method for integrating the filtered Navier–Stokes equations with a stabilized finite element method. *Comput. Methods Appl. Mech. Eng.* **2000**, *190*, 305–319. [CrossRef]
51. Köbis, M.A.; Arnold, M. Convergence of generalized-α time integration for nonlinear systems with stiff potential forces. *Multibody Syst. Dyn.* **2015**, *37*, 107–125. [CrossRef]
52. Bestle, D. *Analyse und Optimierung von Mehrkörpersystemen*; Springer: Berlin/Heidelberg, Germany, 1994. [CrossRef]
53. Haug, E.J.; Arora, J.S. *Applied Optimal Design: Mechanical and Structural Systems*; John Wiley & Sons: Hoboken, NJ, USA, 1979.
54. Gufler, V.; Wehrle, E.; Zwölfer, A. A review of flexible multibody dynamics for gradient-based design optimization. *Multibody Syst. Dyn.* **2021**, *53*, 379–409. [CrossRef]
55. Shabana, A.A. Flexible multibody dynamics: Review of past and recent developments. *Multibody Syst. Dyn.* **1997**, *1*, 189–222. [CrossRef]
56. Shabana, A.A. *Dynamics of Multibody Systems*, 3rd ed.; Cambridge University Press: Cambridge, UK, 2013. [CrossRef]
57. Petzold, L.; Lötstedt, P. Numerical solution of nonlinear differential equations with algebraic constraints II: Practical implications. *SIAM J. Sci. Stat. Comput.* **1986**, *7*, 720–733. [CrossRef]
58. Bottasso, C.L.; Bauchau, O.A.; Cardona, A. Time-step-size-independent conditioning and sensitivity to perturbations in the numerical solution of index three differential algebraic equations. *SIAM J. Sci. Comput.* **2007**, *29*, 397–414. [CrossRef]
59. Bauchau, O.A. *Flexible Multibody Dynamics*; Springer: Berlin/Heidelberg, Germany, 2011. [CrossRef]
60. Dopico, D.; González, F.; Luaces, A.; Saura, M.; García-Vallejo, D. Direct sensitivity analysis of multibody systems with holonomic and nonholonomic constraints via an index-3 augmented Lagrangian formulation with projections. *Nonlinear Dyn.* **2018**, *93*, 2039–2056. [CrossRef]
61. Baumgarte, J. Stabilization of constraints and integrals of motion in dynamical systems. *Comput. Methods Appl. Mech. Eng.* **1972**, *1*, 1–16. [CrossRef]
62. Gufler, V.; Wehrle, E.; Vidoni, R. Analytical sensitivity analysis of flexible multibody dynamics with index-1 differential-algebraic equations and Baumgarte stabilization. *Int. J. Mech. Control* **2023**, *24*, 3–14.
63. Arnold, M.; Brüls, O. Convergence of the generalized-α scheme for constrained mechanical systems. *Multibody Syst. Dyn.* **2007**, *18*, 185–202. [CrossRef]
64. Bathe, K.J.; Noh, G. Insight into an implicit time integration scheme for structural dynamics. *Comput. Struct.* **2012**, *98–99*, 1–6. [CrossRef]
65. Huang, C.; Fu, M. A composite collocation method with low-period elongation for structural dynamics problems. *Comput. Struct.* **2018**, *195*, 74–84. [CrossRef]
66. Kadapa, C.; Dettmer, W.; Perić, D. On the advantages of using the first-order generalised-alpha scheme for structural dynamic problems. *Comput. Struct.* **2017**, *193*, 226–238. [CrossRef]
67. Malakiyeh, M.M.; Shojaee, S.; Javaran, S.H. Development of a direct time integration method based on Bezier curve and 5th-order Bernstein basis function. *Comput. Struct.* **2018**, *194*, 15–31. [CrossRef]
68. Namadchi, A.H.; Jandaghi, E.; Alamatian, J. A new model-dependent time integration scheme with effective numerical damping for dynamic analysis. *Eng. Comput.* **2020**, *37*, 2543–2558. [CrossRef]
69. Noh, G.; Bathe, K.J. The Bathe time integration method with controllable spectral radius: The ρ_∞-Bathe method. *Comput. Struct.* **2019**, *212*, 299–310. [CrossRef]
70. Shojaee, S.; Rostami, S.; Abbasi, A. An unconditionally stable implicit time integration algorithm: Modified quartic B-spline method. *Comput. Struct.* **2015**, *153*, 98–111. [CrossRef]
71. Wen, W.; Wei, K.; Lei, H.; Duan, S.; Fang, D. A novel sub-step composite implicit time integration scheme for structural dynamics. *Comput. Struct.* **2017**, *182*, 176–186. [CrossRef]
72. Bathe, K.J. *Finite Element Procedures*; Prentice Hall: Englewood Cliffs, NJ, USA, 1996; p. 1037.
73. Kreisselmeier, G.; Steinhauser, R. Systematic control design by optimizing a vector performance index. In Proceedings of the International Federation of Active Controls Symposium on Computer-Aided Design of Control Systems, Zürich, Switzerland, 29–31 August 1979. [CrossRef]
74. Martins, J.R.R.A.; Poon, N.M.K. On structural optimization using constraint aggregation. In Proceedings of the 6th World Congress on Structural and Multidisciplinary Optimization, Rio de Janeiro, Brazi, 30 May–3 June 2005.
75. Duddeck, F. Multidisciplinary optimization of car bodies. *Struct. Multidiscip. Optim.* **2008**, *35*, 375–389. [CrossRef]
76. Kurtaran, H.; Eskandarian, A.; Marzougui, D.; Bedewi, N.E. Crashworthiness design optimization using successive response surface approximations. *Comput. Mech.* **2002**, *29*, 409–421. [CrossRef]
77. Langer, H. Extended Evolutionary Algorithms for Multiobjective and Discrete Design Optimization of Structure. Ph.D. Dissertation, Lehrstuhl für Leichtbau, Technische Universität München, Munich, Germany, 2005.

78. Boursier Niutta, C.; Wehrle, E.J.; Duddeck, F.; Belingardi, G. Surrogate modeling in design optimization of structures with discontinuous responses: A new approach for ill-posed problems in crashworthiness design. *Struct. Multidiscip. Optim.* **2018**, *57*, 1857–1869. [CrossRef]
79. Forrester, A.I.J.; Sábester, A.; Keane, A.J. *Engineering Design via Surrogate Modelling: A Practical Guide*; Wiley: Hoboken, NJ, USA, 2008; p. 227.
80. Xu, Q.; Wehrle, E.; Baier, H. Knowledge-based surrogate modeling in engineering design optimization. In *Surrogate-Based Modeling and Optimization*; Koziel, S., Leifsson, L., Eds.; Springer: Berlin/Heidelberg, Germany, 2013; pp. 313–336. [CrossRef]

machines

MDPI

Article

Kinematic Models and the Performance Level Index of a Picking-and-Placing Hybrid Robot

Qi Zou [1], Dan Zhang [2,*] and Guanyu Huang [3]

1 Lassonde School of Engineering, York University, Toronto, ON M3J 1P3, Canada; qizou@yorku.ca
2 Department of Mechanical Engineering, The Hong Kong Polytechnic University, Hung Hom, Kowloon, Hong Kong
3 Intelligent Robotics Research Center, Zhejiang Lab, Hangzhou 311100, China; huanggy@zhejianglab.com
* Correspondence: dan.zhang@polyu.edu.hk

Abstract: The mobile platform of the parallel robot designed for picking and placing operations is usually equipped with one or two extra degree(s) of freedom to enable flexible grasping orientations. However, additional motors indicate extra loads for the moving platform, and the total payload performance shrinks. This paper proposes a spatial picking-and-placing manipulator, in which one actuator that is supposed to be installed on the mobile platform is placed far away from the mobile platform. The platform has a large workspace along one direction. The comprehensive analytical inverse and forward kinematic solutions of this robot are derived. The reachable workspace of the parallel manipulator module is then explored. The novel performance level index is designed to normalize the performance index and demonstrate the performance rank for any pose. A mathematical proof is provided for this novel index. The manipulability index is taken as an example to examine the level indicator. A multi-objective optimization is implemented to pursue optimal performance; then, the initial design and optimized results are compared in detail. A sample trajectory is provided to verify the correctness of the kinematic mathematical model of the parallel mechanism.

Keywords: hybrid robot; kinematic analysis; performance level index; optimum design

Citation: Zou, Q.; Zhang, D.; Huang, G. Kinematic Models and the Performance Level Index of a Picking-and-Placing Hybrid Robot. *Machines* **2023**, *11*, 979. https://doi.org/10.3390/machines11100979

Academic Editors: Jian Wu, Xiangkun He and Guangfei Xu

Received: 9 October 2023
Revised: 20 October 2023
Accepted: 20 October 2023
Published: 23 October 2023

1. Introduction

The products in picking and placing applications may have various shapes, sizes, weights, surface stiffness, positions and orientations, and complex stacking manners [1–3]. Hence, grippers are required to possess large reachable and dexterous workspaces in order to have flexible capabilities. Another essential condition is high speed and acceleration to improve operation efficiency, which makes this kind of task more complicated [4,5].

Apparently, a parallel-structure-based robot can conduct pick-and-place operations according to its intrinsic characteristics [6]. The Delta robot invented by Reymond Clavel [7] was equipped with 3-R(S-S)$_2$ parallel structure (R is the rotary pair, and S indicates a spherical joint). It possessed a three-dimensional (3D) translational workspace and was widely employed in various picking-and-placing operations [8–10]. Belzile et al. [11] explored a 2-CRRH (C and H indicate, respectively, cylindrical and helical joints) parallel architecture for pick-and-place operations. The moving platform could achieve Schönflies motion in space with a premium limitless rotational capability. The X4 robot 4-R(S-S)$_2$R was outlined in [12]. This robot could generate 3T1R (three translations and one rotation). An exhaustive study on the kinematic analysis was expressed, based upon which the optimal design was established. Furthermore, its dynamic performance evaluations and power consumption analysis were conducted in [13,14]. Rahul et al. [15] designed a two-degree-of-freedom (DOF) planar parallel robot to automatically load/unload paper pot seedlings in agriculture applications. This robot was equipped with five revolute joints, and most linkages were manufactured via 3D printing. The same kind of five-bar linkage mechanism was also mentioned in [16]. It was upgraded and named DexTAR, which exhibited high accuracy. Its four groups of operation modes, interference areas, and singularity loci were determined comprehensively. Similarly to H4 and I4R parallel robots, the Par4 parallel architecture with symmetrical features and an articulated mobile platform was introduced in [17]. This novel design could avoid the shortcomings of these previous robots and still possess high speed and acceleration. Briot and Bonev [18] developed a spatial parallel robot, Pantopteron, with pure sliding movements aimed at pick-and-place tasks. Its three DOFs were completely decoupled. It could realize high speeds and large reachable workspace due to the unique pantograph linkage mechanism installed in each kinematic branch.

Designing a picking-and-placing parallel robot with extraordinary strokes along one axis is still challenging, not to mention the requirement of a large workspace in all directions. Considering the different robot structures of serial and parallel robots, a serial–parallel hybrid robot may be an option for meeting the requirements of picking-and-placing operations.

In addition, the performance evaluation of parallel architecture is essential in order to make full use of its intrinsic advantages. A generalized articulated parallel manipulator $(RRP_aR)_2$-R-$(RRP_aRR)_2$ (P_a denotes parallelogram linkage) was established in [19]. Beyond the kinematic mathematic models and kinematic performance evaluations, a special workspace that was far from singularity configurations was selected as the operational workspace. The parallel diamond structure was reported in [20]. The rotary movement of the mobile platform was constrained due to the parallelogram mechanism in each kinematic limb. Detailed kinematic performance analyses, including workspace and dexterity analyses, were conducted. During the designing phase, the force transmission property was considered to be more suitable for battery assembly lines. Its global kinematic constraints and dynamic performance were further evaluated in [21]. A solar tracker based on the U/PRU/PUS (P and U represent prismatic and universal joints, respectively) parallel robot was designed in [22]. The two-axis rotational workspace/area of this device was computed and employed as the optimization target in dimensional synthesis. In the work of [23], the maximum orientation capacity and the global transmission and constraint index for parallel manipulators were deduced. Both were examined in a group of parallel robots: 3-UCR and 3-RCU (C indicates cylindrical joints). The novel dynamic transfer performance index and dynamic transfer balance performance index were defined in [24]. Better performances could be obtained using a larger former index and a smaller latter index. These indices were further employed as optimization objects in the trajectory planning of a hybrid robot. An end-effector transmission index and branch transmission index were proposed in [25]. The authors further brought forward a special local transmission index based on orthogonal degrees. This index was evaluated on a planar 2-DOF parallel robot with various driving modes. To avoid the singularity problem of the Jacobian matrix, the kinetostatic conditioning index was proposed in [26,27]. The designed index was not dependent on the configuration of the parallel robot. Patel and Sobh [28] investigated and concluded the widely used performance indices of parallel mechanisms. In this research study, thirty-six indices were sorted with respect to three methods: the local or global index, the kinematic or dynamic index or neither, and the intrinsic or extrinsic index. Brinker et al. [29] conducted an exhaustive study on kinematic performance indices for parallel structures. Both the advantages and shortcomings of the condition number, manipulability index, and motion/force transmission index were discussed in detail. The solutions for overcoming the limitations of Jacobian-matrix-based indices were provided.

These various indices can provide a more comprehensive understanding with respect to evaluating serial/parallel/hybrid robots. A concern arises with respect to how one can deal with many different categories of performance indices, kinematic/dynamic indices, and performance indices with various ranges. It is also difficult to clarify that one pose of a robot is superior to one specific pose of another robot relative to a given performance measure. The performance ranking of any pose within a reachable workspace is unknown.

Considering the above two main problems, a novel hybrid robot with a large workspace along one direction is proposed in this research study. A novel index is explored to normalize any performance index and provide performance ratings for each position/orientation within the workspace. The article is arranged as follows: the structural design of this robot is presented in Section 2, and it is followed by Section 3, which formulates the analytical kinematic relations. The workspace's analysis and novel level index are investigated in Section 4. Multi-objective linkage optimization and performance comparisons are conducted in Section 5. A case study is shown in Section 6, and the conclusions are provided in Section 7.

2. Mechanical Design

A hybrid manipulator employed for the picking-and-placing operation is proposed, as shown in Figure 1a. This hybrid structure comprises a planar parallel mechanism [30] (shown in Figure 1b), a belt drive system, a rack and pinion module, and a gripper. The coordinate system is located at the bottom left corner, as illustrated in Figure 1b. Based on the right-hand rule, the Z-axis is perpendicular to the XOY plane and points outward.

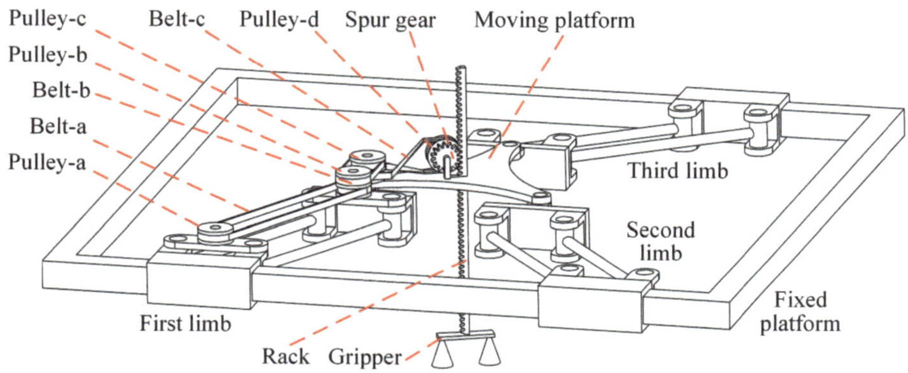

(**a**)

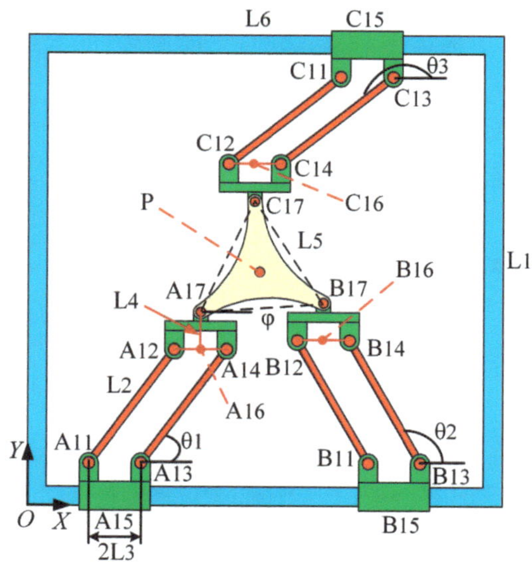

(**b**)

Figure 1. Schematic diagram of the industrial robot. (**a**) The 4-DOF hybrid manipulator; (**b**) the planar parallel mechanism.

This parallel robot includes one fixed platform and a mobile platform connected by three independent kinematic chains. Each chain comprises one active prismatic joint, one parallelogram joint ((R-R)$_2$ linkage), and a revolute joint. All sliding joints move along the X direction, and this linear motion can be realized using a lead screw motion system. The parallelogram joints are placed on the XOY plane. In each kinematic branch, one side of the parallelogram joint is installed on the driving sliding joint, and another side is attached to the moving platform with a rotary joint. The axes of the three rotational joints connected to the mobile platform are all parallel relative to the Z-axis.

Points A_{16}, B_{16}, and C_{16} are the midpoints of $A_{12}A_{14}$, $B_{12}B_{14}$, and $C_{12}C_{14}$, respectively. Pulley-a is placed at the midpoint of $A_{11}A_{13}$. Pulley-b and pulley-c both comprise two pulleys. They are separately mounted on points A_{16} and A_{17}. The shafts of the aforementioned three pulleys are colinear with the Z-axis. Pulley-d is installed on the moving platform, and its axis is perpendicular to the Z-axis. Pulley-a drives pulley-b through belt-a. Pulley-b and pulley-c are linked by belt-b. Pulley-c and pulley-d are connected by belt-c.

The spur gear is fixed to pulley-d, and they share the same rotational axis. The rack matching the spur gear can slide along the Z direction with respect to the mobile platform. The rack is installed vertically at the center of the mobile platform. The gripper is fixed at the bottom end of the rack. The movement of pulley-a manipulates the translation of the gripper.

Although a motor directly installed on the mobile platform can drive the spur gear, this is avoided in this research study. One of the reasons is that a compact moving platform can be enabled. A remote actuator would make the moving platform geometrically more compact. Another concern is the inertial property. The mobile platform with three movements requires a larger mass in order to attach the actuator with the large mass feature. In this case, the inertia features of the mobile platform and the actuator are enlarged. The high-speed performance of the parallel robot module is affected. The requirements for the other three sliding actuators are improved under the same payload conditions. Therefore, this actuator for Z-direction translation is installed far from the mobile platform. The corresponding transmission parts are designed using lightweight structures and materials.

This planar parallel architecture can achieve two translations along the X-axis and Y-axis and one rotational movement about the Z-axis. The belt drive system and the gear rack system enable the gripper to translate along the Z direction. In summary, the entire hybrid robot has four DOFs.

3. Kinematic Solutions

3.1. Inverse Kinematics of the Parallel Robot

The detailed linkage dimensions of the parallel manipulator with three identical chains are also illustrated in Figure 1b. The fixed platform is a rectangle with a length of L_1 and width of L_6. $|A_{11}A_{12}| = |A_{13}A_{14}| = L_2$.$|A_{11}A_{13}| = |A_{12}A_{14}| = 2L_3$. $|A_{16}A_{17}| = L_4$. The corresponding linkages of the three chains have the same dimensions. Mobile platform $A_{17}B_{17}C_{17}$ forms a virtual equilateral triangle with an edge of L_5, and its center is point P. According to the features of this triangle, $|A_{17}P| = |B_{17}P| = |C_{17}P| = \sqrt{3}L_5/3$. The positions along the X direction of active sliding joints A_{15}, B_{15}, and C_{15} are x_1, x_2, and x_3, respectively. The system input of this parallel manipulator is defined as $Q = [x_1, x_2, x_3]^{\mathrm{T}}$. The sloping angles between the positive X-axis and linkages $A_{13}A_{14}$, $B_{13}B_{14}$, and $C_{13}C_{14}$ are, respectively, denoted as θ_1, θ_2, and θ_3. The angle between the virtual edge $A_{17}B_{17}$ and the positive X-axis is φ. The position of point P in the provided coordinate system is (x, y). Thereafter, the pose of the mobile platform is given as $\mathbf{X} = [x, y, \varphi]^{\mathrm{T}}$.

The inverse kinematic problem can be computed with the given $\mathbf{X}$. For the first kinematic chain, the vector-loop equation is obtained as follows:

$$\boldsymbol{OA}_{15} + \boldsymbol{A}_{15}\boldsymbol{A}_{16} + \boldsymbol{A}_{16}\boldsymbol{A}_{17} + \boldsymbol{A}_{17}\boldsymbol{P} = \boldsymbol{OP}. \tag{1}$$

Equation (1) can be expressed along two orthogonal directions:

$$\begin{cases} X\ axis: x_1 + L_2\cos\theta_1 + \sqrt{3}L_5/3\cos(\varphi+\pi/6) = x \\ Y\ axis:\ L_2\sin\theta_1 + L_4 + \sqrt{3}L_5/3\sin(\varphi+\pi/6) = y. \end{cases} \tag{2}$$

From Equation (2), unknown parameter θ_1 should be placed on one side while the others are on the other side to eliminate θ_1. Hence, Equation (2) is rephrased as

$$\begin{cases} L_2\cos\theta_1 = x - \sqrt{3}L_5/3\cos(\varphi+\pi/6) - x_1 \\ L_2\sin\theta_1 = y - \sqrt{3}L_5/3\sin(\varphi+\pi/6) - L_4. \end{cases} \tag{3}$$

Adding the squares of both equations in Equation (3) leads to

$$L_2^2 = \left[x - \sqrt{3}L_5/3\cos(\varphi+\pi/6) - x_1\right]^2 + \left[y - \sqrt{3}L_5/3\sin(\varphi+\pi/6) - L_4\right]^2. \tag{4}$$

The position of the active prismatic joint A_{15} is then solved as

$$x_1 = \left[x - \sqrt{3}L_5/3\cos(\varphi+\pi/6)\right] \pm \sqrt{L_2^2 - \left[y - \sqrt{3}L_5/3\sin(\varphi+\pi/6) - L_4\right]^2}. \tag{5}$$

Similar calculation steps for the first branch will be employed for the remaining supporting branches. The closed-loop equations for the other supporting branches are calculated as

$$\boldsymbol{OB}_{15} + \boldsymbol{B}_{15}\boldsymbol{B}_{16} + \boldsymbol{B}_{16}\boldsymbol{B}_{17} + \boldsymbol{B}_{17}\boldsymbol{P} = \boldsymbol{OP} \tag{6}$$

$$\boldsymbol{OC}_{15} + \boldsymbol{C}_{15}\boldsymbol{C}_{16} + \boldsymbol{C}_{16}\boldsymbol{C}_{17} + \boldsymbol{C}_{17}\boldsymbol{P} = \boldsymbol{OP}. \tag{7}$$

Both Equations (6) and (7) are described in two orthogonal directions, and passive parameters θ_2 and θ_3 are placed on one side, as arranged below.

$$\begin{cases} L_2\cos\theta_2 = x - \sqrt{3}L_5/3\cos(\varphi+5\pi/6) - x_2 \\ L_2\sin\theta_2 = y - \sqrt{3}L_5/3\sin(\varphi+5\pi/6) - L_4 \end{cases} \tag{8}$$

$$\begin{cases} L_2\cos\theta_3 = x - \sqrt{3}L_5/3\cos(\varphi+3\pi/2) - x_3 \\ L_2\sin\theta_3 = y - \sqrt{3}L_5/3\sin(\varphi+3\pi/2) - L_1 + L_4. \end{cases} \tag{9}$$

Summing the squares of both equations in Equations (8) and (9), the following expressions can be deduced, respectively.

$$L_2^2 = \left[x - \sqrt{3}L_5/3\cos(\varphi+5\pi/6) - x_2\right]^2 + \left[y - \sqrt{3}L_5/3\sin(\varphi+5\pi/6) - L_4\right]^2 \tag{10}$$

$$L_2^2 = \left[x - \sqrt{3}L_5/3\cos(\varphi+3\pi/2) - x_3\right]^2 + \left[y - \sqrt{3}L_5/3\sin(\varphi+3\pi/2) - L_1 + L_4\right]^2 . \lim_{x\to\infty} \tag{11}$$

The positions of remaining actuation joints B_{15} and C_{15} can be formulated separately and are listed below.

$$x_2 = \left[x - \sqrt{3}L_5/3\cos(\varphi+5\pi/6)\right] \pm \sqrt{L_2^2 - \left[y - \sqrt{3}L_5/3\sin(\varphi+5\pi/6) - L_4\right]^2} \tag{12}$$

$$x_3 = \left[x - \sqrt{3}L_5/3\cos(\varphi+3\pi/2)\right] \pm \sqrt{L_2^2 - \left[y - \sqrt{3}L_5/3\sin(\varphi+3\pi/2) - L_1 + L_4\right]^2}. \tag{13}$$

The analytical solution of the inverse kinematic model is concluded in Equations (5), (12) and (13).

3.2. Direct Kinematics of the Parallel Robot

The forward kinematic problem can be calculated using the predefined *Q*. Equations (4), (10) and (11) can be deduced in the following form:

$$x^2 + y^2 + e_{11}x + e_{12}y + e_{13} = 0 \tag{14}$$

$$x^2 + y^2 + e_{21}x + e_{22}y + e_{23} = 0 \tag{15}$$

$$x^2 + y^2 + e_{31}x + e_{32}y + e_{33} = 0, \tag{16}$$

where

$$\begin{cases} e_{11} = -2\left[\sqrt{3}L_5/3\cos(\varphi+\pi/6)+x_1\right] \\ e_{12} = -2\left[\sqrt{3}L_5/3\sin(\varphi+\pi/6)+L_4\right] \\ e_{13} = \left[\sqrt{3}L_5/3\cos(\varphi+\pi/6)+x_1\right]^2 + \left[\sqrt{3}L_5/3\sin(\varphi+\pi/6)+L_4\right]^2 - L_2^2 \end{cases}$$

$$\begin{cases} e_{21} = -2\left[\sqrt{3}L_5/3\cos(\varphi+5\pi/6)+x_2\right] \\ e_{22} = -2\left[\sqrt{3}L_5/3\sin(\varphi+5\pi/6)+L_4\right] \\ e_{23} = \left[\sqrt{3}L_5/3\cos(\varphi+5\pi/6)+x_2\right]^2 + \left[\sqrt{3}L_5/3\sin(\varphi+5\pi/6)+L_4\right]^2 - L_2^2 \end{cases}$$

$$\begin{cases} e_{31} = -2\left[\sqrt{3}L_5/3\cos(\varphi+3\pi/2)+x_3\right] \\ e_{32} = -2\left[\sqrt{3}L_5/3\sin(\varphi+3\pi/2)+L_1-L_4\right] \\ e_{33} = \left[\sqrt{3}L_5/3\cos(\varphi+3\pi/2)+x_3\right]^2 + \left[\sqrt{3}L_5/3\sin(\varphi+3\pi/2)+L_1-L_4\right]^2 - L_2^2. \end{cases}$$

Subtracting Equation (15) from Equation (14) results in the following:

$$e_{41}x + e_{42}y + e_{43} = 0, \tag{17}$$

where $e_{4i} = e_{1i} - e_{2i} \quad i = 1,2,3$.

Subtracting Equation (16) from Equation (14) generates the following:

$$e_{51}x + e_{52}y + e_{53} = 0, \tag{18}$$

where $e_{5i} = e_{1i} - e_{3i} \quad i = 1,2,3$.

Parameters x and y can be computed by solving Equations (17) and (18):

$$x = \frac{e_{62}}{e_{61}},\ y = \frac{-e_{63}}{e_{61}}, \tag{19}$$

where

$$\begin{cases} e_{61} = e_{41}e_{52} - e_{42}e_{51} \\ e_{62} = e_{42}e_{53} - e_{43}e_{52} \\ e_{63} = e_{41}e_{53} - e_{43}e_{51} \end{cases}$$

Combining Equations (14) and (19) yields

$$e_{62}^2 + e_{63}^2 + e_{11}e_{61}e_{62} - e_{12}e_{61}e_{63} + e_{13}e_{61}^2 = 0. \tag{20}$$

Hence, variable φ can be computed since it is the only unknown parameter in Equation (20). The tangent half-angle formula can be employed for Equation (20), which is transferred as an eighth-degree polynomial. The polynomial denotes that eight poses at most are obtained with the predefined $\boldsymbol{Q}$.

3.3. Kinematic Model of Z-Displacement

The initial configuration of the vertical (along the Z direction) translation system is illustrated in Figure 2. Figure 2 is depicted in a two-dimensional plane to clearly state all parameters. The dimensions of this system are described, and the radius of pulley-a is R_1. The radii of the two pulleys within pulley-b are R_2 and R_3. The radii of the two pulleys of pulley-c are R_4 and R_5. The radius of pulley-d is R_6. The pitch diameter of the pinion is defined as $2R_7$. The rack has a length of L_7. In this original assembly configuration, the midpoint of the rack is placed at point P of the moving platform. The bottom point of the rack that is attached to the end-effector is represented as E. It is noteworthy that the x- and y-components of the position of point E are the same as that of point P. The only unknown parameter is the z-component of the displacement of point E.

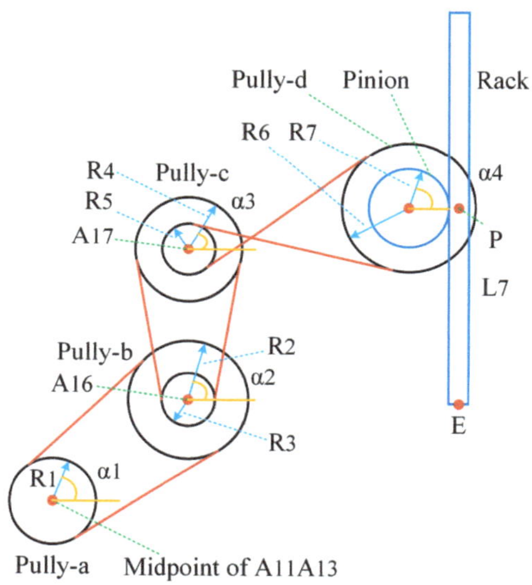

Figure 2. Initial configuration of the Z-translation system.

Once driving pulley-a has a rotation angle of α_1(shown in Figure 2), the rotary angles for pulley-b, pulley-c, and pulley-d are separately denoted as α_2, α_3, and α_4. All angles are in radians, and the reference line is the X-axis. In accordance with the belt mission, the following expressions can be obtained:

$$\begin{cases} S_1 = \alpha_1 R_1 = \alpha_2 R_2 \\ S_2 = \alpha_2 R_3 = \alpha_3 R_4 \\ S_3 = \alpha_3 R_5 = \alpha_4 R_6, \end{cases} \tag{21}$$

where S_1, S_2, and S_3 denote the displacements of belt-a, belt-b, and belt-c, respectively.

Based on rack–pinion transmission, the height of the rack can be calculated as follows:

$$z_{rack} = -\alpha_4 R_7. \tag{22}$$

The position of point E is then derived as

$$z_E = z_{rack} - L_7/2. \tag{23}$$

Combining Equations (21) and (22), the forward kinematic mathematical solution shown in Equation (23) is further expressed as

$$z_E = -\frac{R_1R_3R_5R_7}{R_2R_4R_6}\alpha_1 - L_7/2. \quad (24)$$

The inverse kinematic analytical model of this transmission system can also be formulated using Equation (24), as indicated below.

$$\alpha_1 = -\frac{R_2R_4R_6}{R_1R_3R_5R_7}(z_E + L_7/2). \quad (25)$$

4. Kinematic Performance Assessments

From Figure 1a, the full rotation of pulley-a allows the rack to move along the Z direction, and the corresponding translational range of the four-DOF mechanism is merely dependent on the length of the rack. It is evident that a longer rack enables a larger workspace. It is difficult to achieve such a long stroke along one direction for a traditional parallel mechanism with the same scale. Henceforth, the remaining calculations and discussions are only for the parallel module: the 3-PP$_a$R parallel mechanism.

4.1. Workspace Analysis

The reachable workspace is one of the most essential indices for a parallel manipulator. For introductive purposes, the linkage dimensions of this parallel manipulator are selected as L_1 = 150 mm, L_2 = 120 mm, L_3 = 20 mm, L_4 = 0, L_5 = 50 mm, and L_6 = 400 mm. The first factor that affects the workspace's volume is the strokes of the driving joints. In this mechanism, the strokes of three actuation joints are provided as $L_3 \le x_1, x_2, x_3 \le L_6 - L_3$. In accordance with the current manipulator configuration in Figure 1b, the $\pm$ symbols in Equations (5), (12) and (13) are chosen as $-$, $+$, and $+$, respectively. According to the geometric relations, passive angle θ_i is computed as

$$\begin{cases} \theta_1 = \sin^{-1}\left\{\left[y - \sqrt{3}L_5/3\sin(\varphi + \pi/6) - L_4\right]/L_2\right\} \\ \theta_2 = \sin^{-1}\left\{\left[y - \sqrt{3}L_5/3\sin(\varphi + 5\pi/6) - L_4\right]/L_2\right\} \\ \theta_3 = \sin^{-1}\left\{\left[y - \sqrt{3}L_5/3\sin(\varphi + 3\pi/2) - L_1 + L_4\right]/L_2\right\}. \end{cases} \quad (26)$$

The driven revolute joints are also constrained as

$$\begin{cases} [i\pi/2 - 17\pi/36] \le \theta_i \le [i\pi/2 - \pi/36] \quad i = 1,2,3 \\ -\pi/3 \le \varphi \le 2\pi/3. \end{cases} \quad (27)$$

To solve the reachable workspace, the position range of mobile platform $A_{17}B_{17}C_{17}$ is predefined as $2L_3 \le x \le L_6 - 2L_3$ and $2L_3 \le y \le L_1 - L_5$. The spatial search methodology is employed to obtain the reachable workspace. The flow diagram of this approach is demonstrated in Figure 3. According to the above conditions, the parameters in this flowchart are listed as x_{min} = $2L_3$, x_{max} = L_6 −$2L_3$, x_{step} = 4 mm, y_{min} = $2L_3$, y_{max} = L_1 − L_5, y_{step} = 1 mm, $\varphi_{min} = -\pi/3$ rad, $\varphi_{max} = 2\pi/3$ rad, and $\varphi_{step} = \pi/18$ rad. In the provided $\mathbf{X}$ range, each possible position is employed to compute the corresponding $\mathbf{Q}$ based on the inverse kinematic model introduced in Section 3.1. This position will be feasible once it meets all aforementioned constraints.

The reachable workspace of this planar parallel mechanism is computed and shown in Figure 4. Figure 4a–c are the standard views while Figure 4d presents the 3D view. The irregular shape of this workspace can be revealed by the standard views, selected layers, and 3D view. In accordance with Figure 4, when $\varphi = 0$, the workspace is the smallest, and the workspace increases when orientation φ is larger or smaller. The reachable workspace is irregular and asymmetric, which causes difficulty during applications.

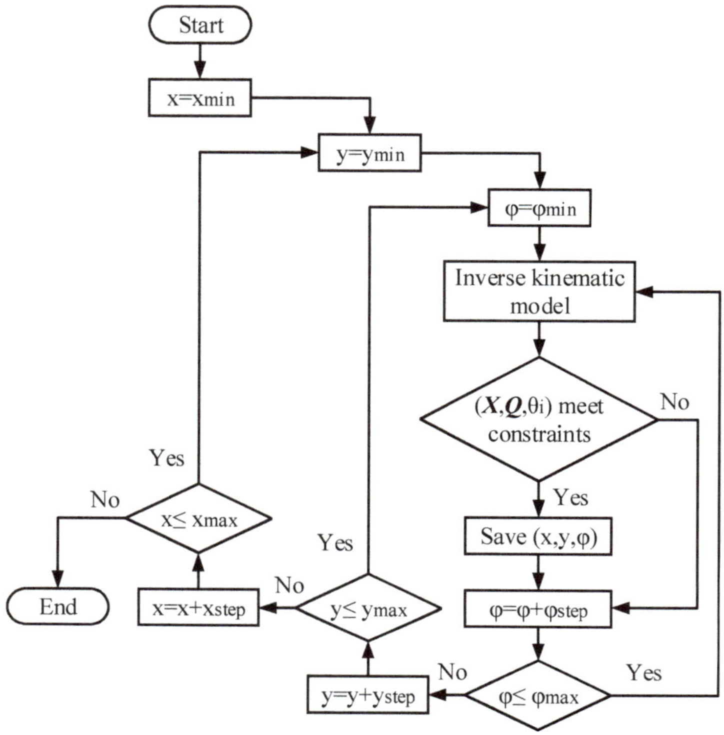

Figure 3. Flow diagram for the reachable workspace.

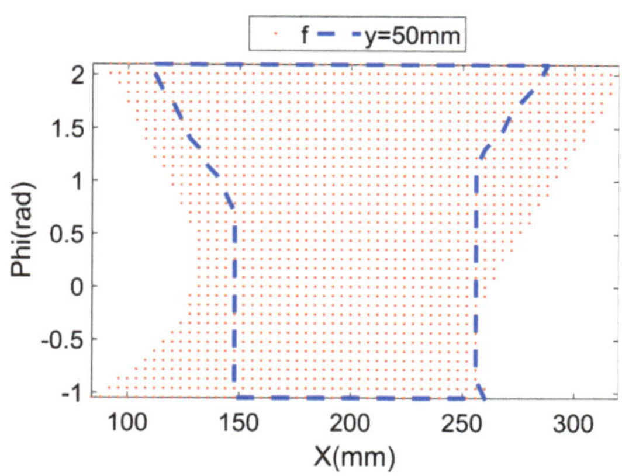

(a)

Figure 4. *Cont.*

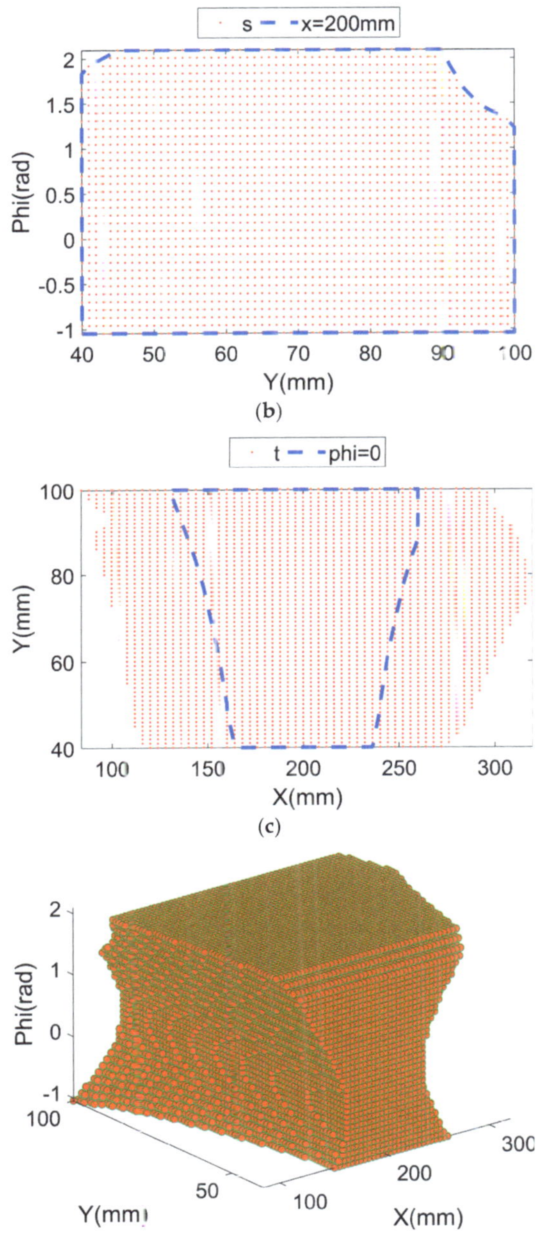

Figure 4. Reachable workspace and specific layers (f, s, and t, respectively, represent front, side, and top views): (**a**) front view; (**b**) side view; (**c**) top view; (**d**) 3D view.

4.2. Level Index

There are various kinds of kinematic and dynamic indices that measure the performance of parallel manipulators. They are generally no less than 0 in the entire workspace of the parallel structure. In this section, the novel level index is proposed for evaluating the performance rank of any position/orientation with respect to the overall performance of the workspace. The level index can be computed as follows:

$$\text{Level Index} = \frac{\mu(\text{pose})}{\overline{\mu} + \mu(\text{pose})}, \tag{28}$$

where $\mu(\text{pose})$ is the performance of any pose of the moving platform, and $\overline{\mu}$ represents the mean value of this kind of performance in the workspace.

Although this level index does not have physical meaning, there are three main advantages of this novel index.

(i) Based on the definition, the level index ranges from 0 to 1, and it is a monotonically increasing function with respect to $\mu(\text{pose})$. In other words, if one performance index (it is assumed that larger values denote higher performances) is chosen, the mean level index is constant, and the pose with higher performances will possess a larger level index. The level index can maintain the features of the original index.

We provide the proof here. There are two positions for points a and b. Their corresponding level indices of one specific indicator are $\text{LI}(a) = \mu(a)/[\overline{\mu} + \mu(a)]$ (LI is the abbreviation for the level index) and $\text{LI}(b) = \mu(b)/[\overline{\mu} + \mu(b)]$, respectively. Parameters $\mu(a)$, $\mu(b)$, $\overline{\mu}$, $\text{LI}(a)$, and $\text{LI}(b)$ are all positive values. Hence, the following expression can be generated.

$$\begin{aligned} \text{LI}(b) - \text{LI}(a) &= \frac{\mu(b)}{\overline{\mu}+\mu(b)} - \frac{\mu(a)}{\overline{\mu}+\mu(a)} \\ &= \frac{\mu(b)[\overline{\mu}+\mu(a)]-\mu(a)[\overline{\mu}+\mu(b)]}{[\overline{\mu}+\mu(a)][\overline{\mu}+\mu(b)]} \\ &= \frac{\overline{\mu}[\mu(b)-\mu(a)]}{[\overline{\mu}+\mu(a)][\overline{\mu}+\mu(b)]}. \end{aligned} \tag{29}$$

Equation (29) has an identical positive or negative symbol with respect to $\mu(b) - \mu(a)$. Therefore, the level index can retain the original index distributions.

It is noteworthy that the performance index with positive values is suitable for the level index. The performance index with negative numbers can also be arranged (e.g., absolute operation and multiplicative inverse operation) in order to adopt this concept.

(ii) The usage of the level index can also assist in shrinking/normalizing the range of the performance index. For example, the local condition index (LCI) of a parallel mechanism generally ranges from 1 to $+\infty$. The corresponding level index is limited between 0 and 1. This process will make depictions in figures easier, as the longer axis for the performance index will be replaced.

(iii) The level index provides the performance rating of the parallel mechanism. When comparing two poses from two parallel mechanisms, the same performances with respect to these two poses are not sufficient. Their ratings in the corresponding manipulators are unknown. It is assumed that there are two parallel manipulators with respect to the provided LCI information. Relative to the first parallel manipulator, the mean LCI is 4, and the LCI of point D_1 is 5. Relative to the second parallel mechanism, the mean LCI is 10, while the LCI of point D_2 is 6. On the one hand, point D_1 has a better LCI than that of point D_2. On the other hand, the LCI of point D_1 is larger than the mean value of the first mechanism. The LCI of point D_2 is lower than the mean value of the second mechanism. Considering this aspect, point D_1 is worse than point D_2. This conclusion can be verified via the local condition level index, which can be deduced using Equation (28). The local condition level indices of these two points are 5/9 and 3/8, respectively, which indicate their ranks in the corresponding mechanisms.

The manipulability index of the mechanism shown in Figure 1b is employed to examine this concept. Taking the derivatives of Equations (4), (10), and (11) with respect to time and rearranging the expression, the following is obtained:

$$J_Q \cdot [x_1, x_2, x_3]^{\text{T}} = J_X \cdot [x, y, \varphi]^{\text{T}}, \tag{30}$$

where

$$J_Q = diag(M_1, M_3, M_5)$$

$$J_X = \begin{bmatrix} M_1 & M_2 & M_1\sqrt{3}L_5/3\sin(\varphi+\pi/6) - M_2\sqrt{3}L_5/3\cos(\varphi+\pi/6) \\ M_3 & M_4 & M_3\sqrt{3}L_5/3\sin(\varphi+5\pi/6) - M_4\sqrt{3}L_5/3\cos(\varphi+5\pi/6) \\ M_5 & M_6 & M_5\sqrt{3}L_5/3\sin(\varphi+3\pi/2) - M_6\sqrt{3}L_5/3\cos(\varphi+3\pi/2) \end{bmatrix}$$

$$M_1 = x - \sqrt{3}L_5/3\cos(\varphi+\pi/6) - x_1$$

$$M_2 = y - \sqrt{3}L_5/3\sin(\varphi+\pi/6) - L_4$$

$$M_3 = x - \sqrt{3}L_5/3\cos(\varphi+5\pi/6) - x_2$$

$$M_4 = y - \sqrt{3}L_5/3\sin(\varphi+5\pi/6) - L_4$$

$$M_5 = x - \sqrt{3}L_5/3\cos(\varphi+3\pi/2) - x_3$$

$$M_6 = y - \sqrt{3}L_5/3\sin(\varphi+3\pi/2) - L_1 + L_4.$$

Equation (30) can be further expressed as follows

$$\dot{Q} = J\dot{X} = (J_Q^{-1}J_X)\dot{X}, \tag{31}$$

where J is the Jacobian matrix.

The limitation of $\left\|\dot{Q}\right\|$ results in the constraints of $\dot{Q}^{\mathrm{T}}\dot{Q} = \dot{X}^{\mathrm{T}}J^{\mathrm{T}}J\dot{X}$. Since term $J^{\mathrm{T}}J$ plays a key role in this equation, the manipulability [31] of this planar parallel manipulator can be defined as

$$M(J) = \sqrt{\det(J^{\mathrm{T}}J)}. \tag{32}$$

Since this planar parallel manipulator has translational and rotational movements, it is essential to compute this index separately to avoid the non-homogeneous unit [32]. Equation (31) is divided into the sum of two terms:

$$\dot{Q} = J_V[x, y]^{\mathrm{T}} + J_W[\varphi]^{\mathrm{T}}, \tag{33}$$

where J_V and J_W possess, respectively, the first two columns and the last column of Jacobian matrix J.

In accordance with the definition in Equation (32), the translational manipulability index (TMI) and the rotational manipulability index (RMI) are generated as follows:

$$\begin{cases} M(J_V) = \sqrt{\det(J_V^{\mathrm{T}}J_V)} \\ M(J_W) = \sqrt{\det(J_W^{\mathrm{T}}J_W)}. \end{cases} \tag{34}$$

The linkages' dimensions are already provided in Section 4.1. Both TMI and RMI can be computed and are not associated with position x according to the numerical calculations provided by the software. TMI-y and RMI-φ are plotted in Figure 5. From Figure 5a, TMI decreases at the beginning and then improves as the moving platform moves along the positive Y-axis. The TMI layouts of $\varphi = 0$ or $\pi/6$ are similar. The TMI ranges from 1.24 to 15.4756 in the entire workspace, and the average value is 2.755. Three sets of (x, y) values (unit: mm) are selected to obtain Figure 5b, (180, 50), (200, 60), and (220, 70). From Figure 5b, the RMI distributions of all these scenarios seem cyclic before the first 2/3 period, while the first case indicates an upward trend, and the other two cases maintain this tendency in the following 1/3 period. In the overall workspace, the smallest and largest RMIs are 26.4522 and 244.619, respectively. The mean RMI is 52.7392.

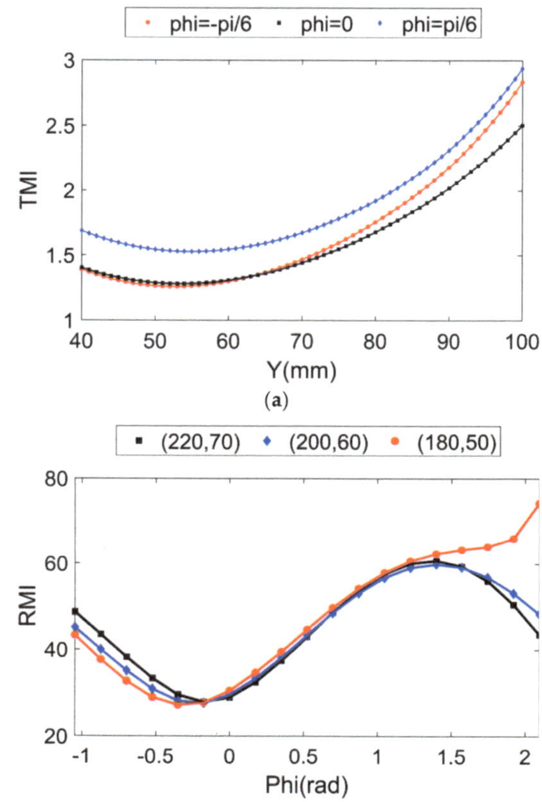

Figure 5. Distributions of two manipulability indices: (**a**) translational manipulability index under constant orientations; (**b**) rotational manipulability index under constant positions.

Based on the definition in Equation (28), the manipulability level index (MLI), translational manipulability level index (TMLI), and rotational manipulability level index (RMLI) are formulated as follows:

$$\begin{cases} \text{MLI} = \frac{M(J)}{\overline{M}(J)+M(J)} \\ \text{TMLI} = \frac{M(J_V)}{\overline{M}(J_V)+M(J_V)} \\ \text{RMLI} = \frac{M(J_W)}{\overline{M}(J_W)+M(J_W)}, \end{cases} \tag{35}$$

where $\overline{M}(J)$, $\overline{M}(J_V)$, and $\overline{M}(J_W)$ represent the global manipulability index (GMI), global translational manipulability index (GTMI), and global rotational manipulability index (GRMI), respectively. Both analytical and discrete expressions for these indices are denoted as follows:

$$\begin{cases} \overline{\text{index}} = \frac{\int_W (\text{index}) dW}{\int_W dW} \\ \overline{\text{index}} = \frac{1}{n}\sum_{i=1}^{n} (\text{index}), \end{cases} \tag{36}$$

where W denotes the workspace, and n indicates the sum of all sample poses in the workspace.

The corresponding distributions for TMLI and RMLI are depicted in Figure 6, and their trends are exactly the same as those shown in Figure 5. The scales for both TMLI and RMLI are extremely narrow. In the entire workspace, the TMLI ranges from 0.3104 to 0.8489, and the mean value is 0.4703. The RMLI ranges from 0.3340 to 0.3226, and the average value is 0.4853.

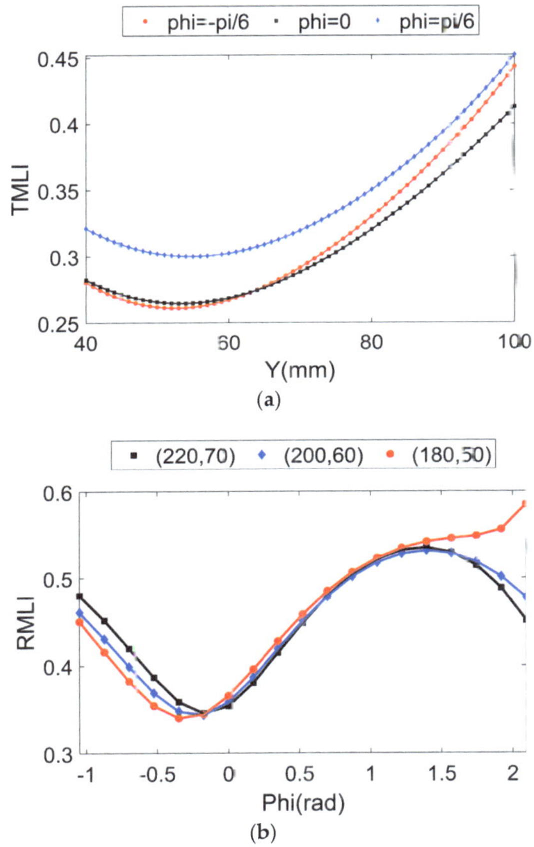

Figure 6. Layouts of two manipulability level indices: (**a**) translational manipulability level index under fixed orientations; (**b**) rotational manipulability level index under fixed positions.

5. Dimensional Synthesis

The proper sets of linkage dimensions have significant impacts on the performance distributions and overall characteristics of a parallel mechanism. The employment of linkage optimization is beneficial to search for optimal performances under the pre-described conditions.

Taking into consideration that the hybrid structure serves as a picking-and-placing robot, the reachable workspace volume is one of the most important indicators. TMI and RMI are also beneficial for the selection of parallel mechanism candidates. Hence, three optimization objectives are described as follows:

$$\begin{cases} \text{Objective 1}: \frac{1}{\text{Workspace Volume}} \\ \text{Objective 2}: \frac{1}{\text{GTMI}} \\ \text{Objective 3}: \frac{1}{\text{GRMI}}, \end{cases} \quad (37)$$

where the volume denotes the amount of discrete feasible configurations.

The commonly employed non-dominated sorting genetic algorithm II [33] was chosen in this case. There are 20 individuals in each generation. The maximum generation is set as 60 to guarantee an efficient and global searching process. The ranges of the linkage lengths are provided below (unit: mm).

$$\begin{cases} 100 \leq L_1 \leq 200 & 80 \leq L_2 \leq 140 & 10 \leq L_3 \leq 40 \\ 0 \leq L_4 \leq 20 & 10 \leq L_5 \leq 60 & 100 \leq L_6 \leq 500. \end{cases} \tag{38}$$

The corresponding constraints of the parallel structure are listed in Section 4.1. After optimization, the trends of the workspace volume, GTMI, and GRMI in each generation are illustrated in Figure 7. As shown in Figure 7, the workspace volume, GTMI, and GRMI exhibit significant increases with respect to generations based on the searched linkage dimensions.

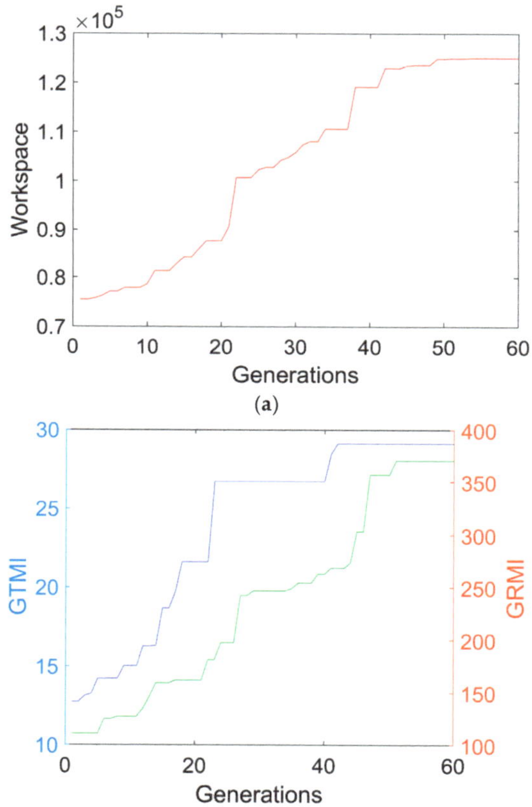

Figure 7. Performances in the optimization process: (**a**) workspace volume; (**b**) GTMI and GRMI.

The performances of all individuals in the final generation (Pareto front) are represented in Figure 8. Considering that the workspace is important in picking-and-placing operations, the individual (shown as a blue pentagram in Figure 8) is selected relative to the largest workspace volume. The other individuals are not chosen because the workspace shrinks sharply, although GTMI and GRMI exhibit improvements. The linkage lengths are provided as L_1 = 1493.9428 mm, L_2 = 114.2580 mm, L_3 = 10.0368 mm, L_4 = 8.0307 mm,

L_5 = 12.9197 mm, and L_6 = 499.5674 mm. This set of linkages are further rounded as L_1 = 150 mm, L_2 = 114 mm, L_3 = 10 mm, L_4 = 8 mm, L_5 = 13 mm, and L_6 = 500 mm. These three optimization objectives are generated and compared with those of the initial linkages' lengths (Section 4.1). The corresponding results are shown in Table 1. The corresponding performance ratio is calculated as the ratio between those of the initial linkages and the chosen linkages. From Table 1, the optimized set of linkage has a significantly larger workspace and greater GTMI, with the sacrifice of GRMI.

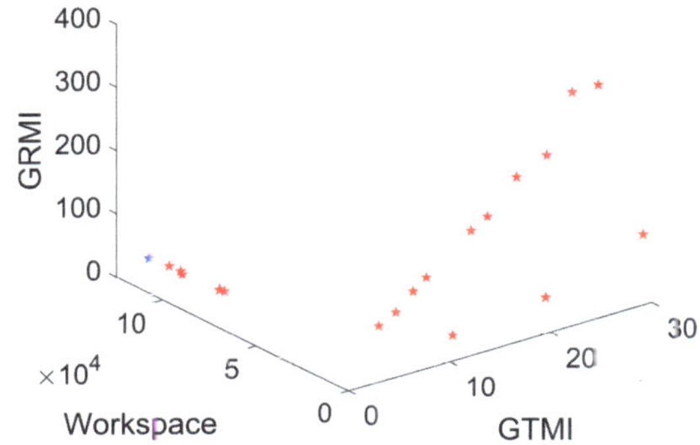

Figure 8. The performance of the last generation (the blue pentagram is the selected performance).

Table 1. Performance comparisons between two sets of linkages.

	Workspace Volume	GTMI	GRMI
Initial linkages	71,428	2.7550	52.7392
Optimized linkages	124,745	3.2327	14.5590
Performance ratio	174.64%	117.34%	27.61%

The reachable workspace of the optimized parallel mechanism is observed in Figure 9. It is evident that the optimized workspace is larger than the initial workspace (Figure 4) in each direction. Compared to the workspace in Figure 4, the improved workspace (Figure 9) is more evenly distributed and is much closer to a regular shape, which ensures that this mechanism is more suitable for industrial applications.

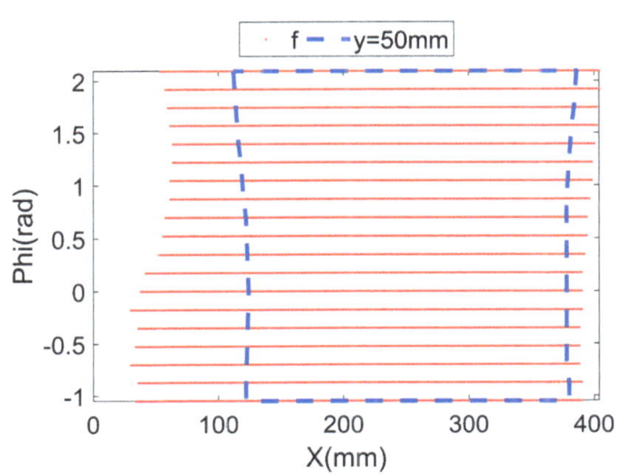

(**a**)

Figure 9. *Cont.*

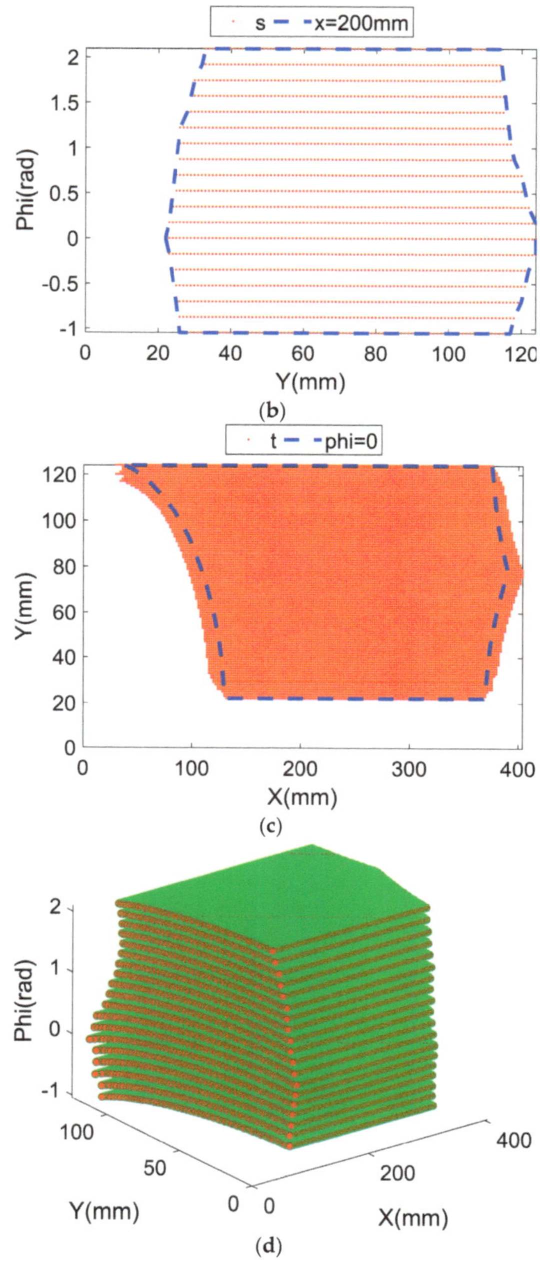

Figure 9. Diagrams for reachable workspace and layers (f, s, and t, respectively, denote front, side, and top views): (**a**) front view; (**b**) side view; (**c**) top view; (**d**) 3D view.

The TMI of the final parallel mechanism can be observed in Figure 10a. Figures 5a and 10a have a similar trend while the latter case has a larger value in each Y-value with respect to a given rotational angle. The RMI of the optimized rods is also illustrated in

Figure 10b, which exhibits lower performances than that in Figure 5b for each φ value with a predefined position.

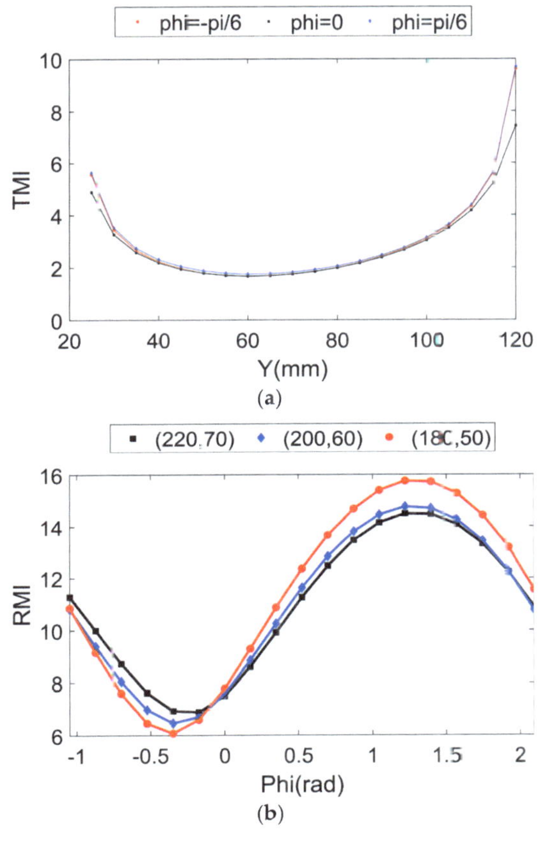

Figure 10. Performance layouts of the optimized parallel structure: (**a**) translational manipulability index under constant orientations; (**b**) rotational manipulability index under constant positions.

6. Case Study

A simulation is provided to testify the analytical inverse kinematic model and Jacobian matrix. The dimension set in Section 4.1 is employed to construct the three-dimensional prototype in MSC ADAMS [34]. In the initial configuration, the position of the moving platform is (200, 68) mm and $\varphi = 0$. Since this parallel robot enables two translations and one rotation, the displacement of this mobile platform is defined based on the current pose, as listed below:

$$x = \begin{cases} 1 - \cos(1000t) & 0 < t \le 0.002\pi \\ 0 & 0.002\pi < t \le 0.009 \end{cases}$$
$$y = \begin{cases} \sin(1000t) & 0 < t \le 0.002\pi \\ 0 & 0.002\pi < t \le 0.009 \end{cases} \tag{39}$$
$$\varphi = \begin{cases} 0 & 0 < t \le 0.002\pi \\ \mathrm{step}(t, 0.002\pi, 0, 0.008, \pi/50) & 0.002\pi < t \le 0.008 \\ \mathrm{step}(t, 0.008, \pi/50, 0.0009, -\pi/50) & 0.008 < t \le 0.009, \end{cases}$$

where t indicates the time, and the unit is in seconds. The linear and angular displacement units are mm and rad, respectively.

From Equation (39), the mobile platform travels on a two-dimensional circular path with a constant orientation from 0 to 0.002π seconds; then, it travels along a smooth curve without translational movements in the remaining period. It is worth noticing that the step function in MSC ADAMS is an approximation step function via the third-degree polynomial.

The positions of three prismatic sliders are measured after the simulation process in MSC ADAMS. The mathematical inverse kinematic model in Section 3.1 is employed to calculate the corresponding analytical positions of these active joints. Figure 11a–c illustrate two sets of results, and the corresponding trajectories of these sliding joints are the same. The positions of the three joints return to their corresponding starting positions during the entire period. The position of the first joint presents an increasing trend from 0 to 0.0027 s and from 0.0058 s to 0.008 s. It decreases from 0.0027 s to 0.0058 s and from 0.008 s to 0.009 s. The position of the second joint decreases from 0 to 0.0005 s and then increases significantly until 0.0036 s. It then exhibits a downward trend until 0.0063 s. The positions are at around 332.4 mm for the rest of this trajectory. The position of the last joint continually increases, with the exception of two other periods: from 0.0027 s to 0.0059 s and from 0.0063 s to 0.008 s.

Furthermore, the linear velocities of these driving joints are measured in MSC ADAMS. The Jacobian matrix expressed in Equation (31) is utilized to compute the analytical velocities of actuators. Both the simulation and analytical results are demonstrated in Figure 11d–f. The two groups of linear velocities match with one another. According to the position trends in Figure 11a–c, the velocities of the first and third joints are similar to two cycles of sine functions while the demarcation point occurs when the time is 0.0062 s; in contrast, the velocities of the second joint possess an additional half cycle of the sine function.

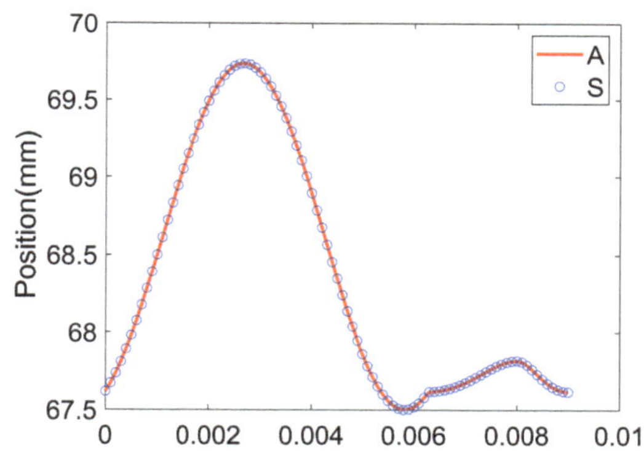

(a)

Figure 11. *Cont.*

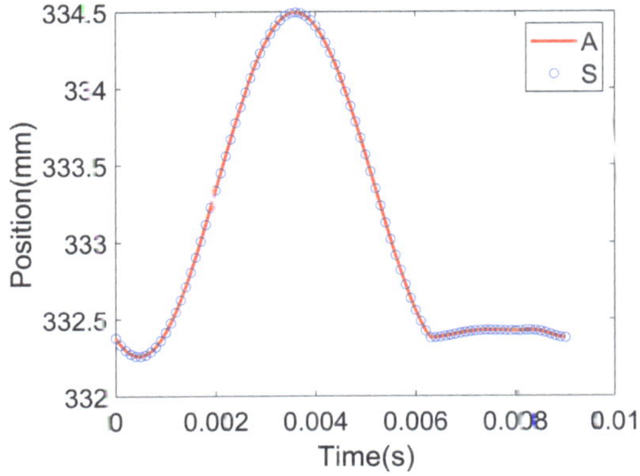

(b)

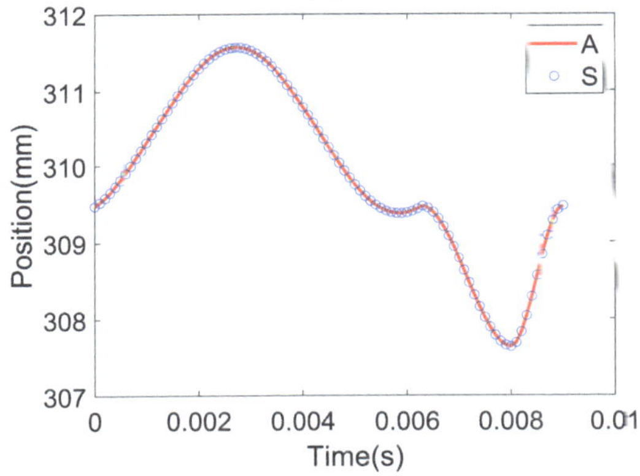

(c)

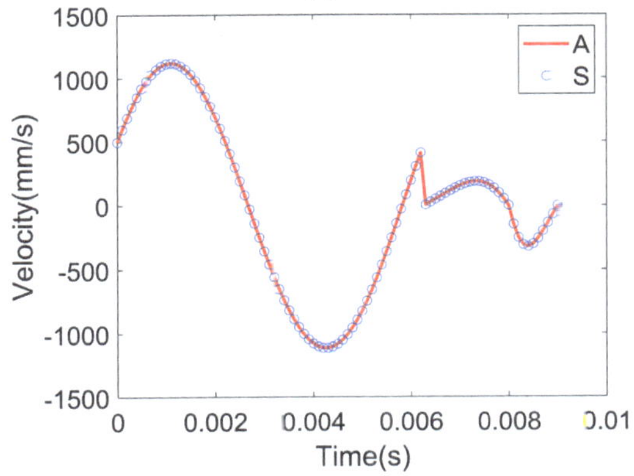

(d)

Figure 11. *Cont.*

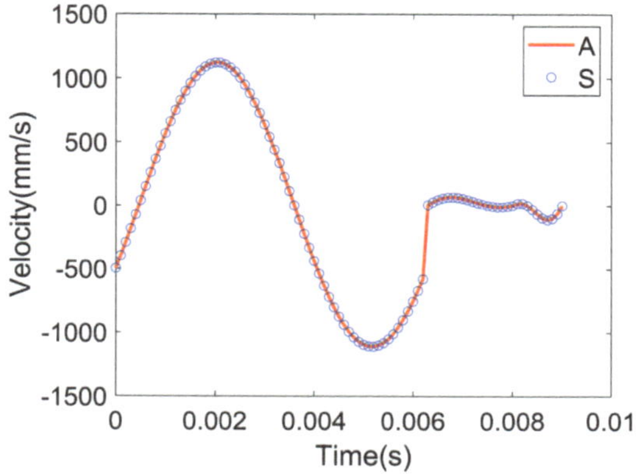

(e)

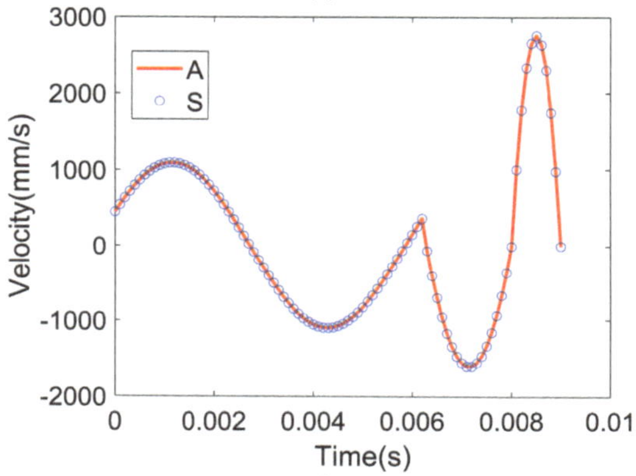

(f)

Figure 11. Comparison between the analytic and simulation results (A—analytics; S—simulation): (**a**) position of the first slider; (**b**) position of the second slider; (**c**) position of the third slider; (**d**) velocity of the first slider; (**e**) velocity of the second slider; (**f**) velocity of the third slider.

The root mean square errors (RMSEs) of the positions and velocities are derived for quantitative analyses, as listed in Table 2. According to Table 2, the discrepancies of each variable are minute compared with the target trajectory. The obtained velocities of the two methodologies demonstrate higher matching performances than that of positions. The errors come from the rounded values of measurements and computations in the software. The correctness of the analytical inverse kinematic model and the Jacobian matrix is proved using this case. It is worth noticing that the simulation results can outperform some intelligent algorithm-based inverse kinematic models (although the robot structures are not the same). For example, the RMSEs of three deep learning models (multi-layer perceptron, deep LSTM (long short-term memory), and GRU (gated recurrent unit) networks) of the Delta robot are 0.0635, 0.0694, and 0.1009, respectively [35]. The inverse kinematic model employing the optimized neural network of the Stanford robot arm can achieve an RMSE of 9.5588×10^{-4} [36].

Table 2. RMSEs of two approaches.

Motor	First Motor	Second Motor	Third Motor
RMSE of position	1.03×10^{-4}	8.973×10^{-4}	5.862×10^{-4}
RMSE of velocity	1.342×10^{-4}	1.555×10^{-4}	1.805×10^{-4}

7. Discussion

This paper investigated a picking-and-placing hybrid robot based on the planar parallel mechanism. The parallel module provided two translations and one rotation, while belt and rack-and-pinion transmissions enabled extra translation with a large stroke. The actuator for this additional movement is not directly mounted on the moving platform via the unique transmission module, although it exhibits a serial–parallel hybrid structure. The detailed kinematic models for the parallel module and the remaining parts were formulated, which indicated that these two groups of motions are decoupled. More specifically, the translations along the X-axis and Y-axis and the rotation about the Z-axis were determined using the parallel module. The translation along the Z direction was commanded by the belt transmission and rack-and-pinion system.

An approach was proposed to explore the reachable workspace of the parallel manipulator. The novel level index was presented to unify the performance indicator and measure the performance rank in the entire reachable workspace. The manipulability index of the planar parallel architecture was employed to testify to its effectiveness. The dimensional synthesis considering the three targets was provided. In the search process, the workspace volume, GTMI, and GRMI exhibited evident improvements. The detailed performances of the optimal set of linkage dimensions were verified and compared with those of the pre-defined set of linkage lengths. To examine the inverse kinematic solution and the velocity relationship between the actuators and moving platform, a simulation was conducted in MSC ADAMS, and the results matched the analytical solutions.

In future studies, the dynamic characteristics of this proposed 3T1R hybrid robot will be studied, and the level index will be employed for other indices to evaluate industrial robots.

Author Contributions: This work was supervised by D.Z. Under D.Z., Q.Z., D.Z. and G.H. conducted the study together. All authors have read and agreed to the published version of the manuscript.

Funding: This research was funded by the Research Institute for Advanced Manufacturing at the Hong Kong Polytechnic University. The third author would like to acknowledge the financial support provided by the National Natural Science Foundation of China (NSFC) (Grant No. 52305039).

Data Availability Statement: Data sharing is not applicable to this article.

Acknowledgments: The authors would like to thank the financial support from the Research Institute for Advanced Manufacturing at the Hong Kong Polytechnic University. The third author would like to acknowledge the financial support provided by the National Natural Science Foundation of China (NSFC) (Grant No. 52305039).

Conflicts of Interest: The authors declare no conflict of interest.

References

1. Zheng, S.R.; Park, T.; Hoang, M.C.; Go, G.; Kim, C.S.; Park, J.O.; Choi, E.; Hong, A. Ascidian-inspired soft robots that can crawl, tumble, and pick-and-place objects. *IEEE Robot. Autom. Lett.* **2021**, *6*, 1722–1728. [CrossRef]
2. Ghadiri, M.N.; Shavarani, S.M.; Güden, H.; Barenji, R.V. Process sequencing for a pick-and-place robot in a real-life flexible robotic cell. *Int. J. Adv. Manuf. Technol.* **2019**, *103*, 3613–3627. [CrossRef]
3. Scalera, L.; Boscariol, P.; Carabin, G.; Vidoni, R.; Gasparetto, A. Enhancing energy efficiency of a 4-DOF parallel robot through task-related analysis. *Machines* **2020**, *8*, 10. [CrossRef]
4. Liu, X.J.; Han, G.; Xie, F.G.; Meng, Q.Z. A novel acceleration capacity index based on motion/force transmissibility for high-speed parallel robots. *Mech. Mach. Theory* **2018**, *126*, 155–170. [CrossRef]
5. Li, Y.H.; Huang, T.; Chetwynd, D.G. An approach for smooth trajectory planning of high-speed pick-and-place parallel robots using quintic B-splines. *Mech. Mach. Theory* **2018**, *126*, 479–490. [CrossRef]

6. Zhao, C.; Wang, K.; Zhao, H.F.; Guo, H.W.; Liu, R.Q. Kinematics, dynamics and experiments of n (3RRIS) reconfigurable series–parallel manipulators for capturing space noncooperative targets. *J. Mech. Robot.* **2022**, *14*, 061002. [CrossRef]
7. Clavel, R. Device for the Movement and Positioning of an Element in Space. U.S. Patent 4,976,582, 11 December 1990.
8. Wu, M.K.; Mei, J.P.; Zhao, Y.Q.; Niu, W.T. Vibration reduction of delta robot based on trajectory planning. *Mech. Mach. Theory* **2020**, *153*, 104004. [CrossRef]
9. Carabin, G.; Scalera, L.; Wongratanaphisan, T.; Vidoni, R. An energy-efficient approach for 3D printing with a Linear Delta Robot equipped with optimal springs. *Robot. Comput. Integr. Manuf.* **2021**, *67*, 102045. [CrossRef]
10. Shen, H.P.; Meng, Q.M.; Li, J.; Deng, J.M.; Wu, G.L. Kinematic sensitivity, parameter identification and calibration of a non-fully symmetric parallel Delta robot. *Mech. Mach. Theory* **2021**, *161*, 104311. [CrossRef]
11. Belzile, B.; Eskandary, P.K.; Angeles, J. Workspace determination and feedback control of a pick-and-place parallel robot: Analysis and experiments. *IEEE Robot. Autom. Lett.* **2019**, *5*, 40–47. [CrossRef]
12. Staicu, S.; Shao, Z.F.; Zhang, Z.K.; Tang, X.Q.; Wang, L.P. Kinematic analysis of the X4 translational–rotational parallel robot. *Int. J. Adv. Robot. Syst.* **2018**, *15*, 1729881418803849. [CrossRef]
13. Mo, J.; Shao, Z.F.; Guan, L.W.; Xie, F.G.; Tang, X.Q. Dynamic performance analysis of the X4 high-speed pick-and-place parallel robot. *Robot. Comput. Integr. Manuf.* **2017**, *46*, 48–57. [CrossRef]
14. Han, G.; Xie, F.G.; Liu, X.J. Evaluation of the power consumption of a high-speed parallel robot. *Front. Mech. Eng.* **2018**, *13*, 167–178. [CrossRef]
15. Rahul, K.; Raheman, H.; Paradkar, V. Design and development of a 5R 2DOF parallel robot arm for handling paper pot seedlings in a vegetable transplanter. *Comput. Electron. Agric.* **2019**, *166*, 105014. [CrossRef]
16. Bourbonnais, F.; Bigras, P.; Bonev, I.A. Minimum-time trajectory planning and control of a pick-and-place five-bar parallel robot. *IEEE/ASME Trans. Mechatron.* **2014**, *20*, 740–749. [CrossRef]
17. Nabat, V.; Rodriguez, M.O.; Company, O.; Krut, S.; Pierrot, F. Par4: Very high speed parallel robot for pick-and-place. In Proceedings of the 2005 IEEE/RSJ International Conference on Intelligent Robots and Systems, Edmonton, AB, Canada, 2–6 August 2005; pp. 553–558.
18. Briot, S.; Bonev, I.A. Pantopteron: A new fully decoupled 3DOF translational parallel robot for pick-and-place applications. *J. Mech. Robot.* **2009**, *1*, 021001. [CrossRef]
19. Meng, Q.Z.; Xie, F.G.; Liu, X.J. Conceptual design and kinematic analysis of a novel parallel robot for high-speed pick-and-place operations. *Front. Mech. Eng.* **2018**, *13*, 211–224. [CrossRef]
20. Huang, T.; Li, Z.X.; Li, M.; Chetwynd, D.G.; Gosselin, C.M. Conceptual design and dimensional synthesis of a novel 2-DOF translational parallel robot for pick-and-place operations. *J. Mech. Des.* **2004**, *126*, 449–455. [CrossRef]
21. Huang, T.; Liu, S.T.; Mei, J.P.; Chetwynd, D.G. Optimal design of a 2-DOF pick-and-place parallel robot using dynamic performance indices and angular constraints. *Mech. Mach. Theory* **2013**, *70*, 246–253. [CrossRef]
22. Du, X.Q.; Li, Y.C.; Wang, P.C.; Ma, Z.H.; Li, D.W.; Wu, C.Y. Design and optimization of solar tracker with U-PRU-PUS parallel mechanism. *Mech. Mach. Theory* **2021**, *155*, 104107. [CrossRef]
23. Chong, Z.H.; Xie, F.G.; Liu, X.J.; Wang, J.S.; Niu, H.F. Design of the parallel mechanism for a hybrid mobile robot in wind turbine blades polishing. *Robot. Comput. Integr. Manuf.* **2020**, *61*, 101857. [CrossRef]
24. Li, Y.B.; Wang, L.; Chen, B.; Wang, Z.S.; Sun, P.; Zheng, H.; Xu, T.T.; Qin, S.Y. Optimization of dynamic load distribution of a serial-parallel hybrid humanoid arm. *Mech. Mach. Theory* **2020**, *149*, 103792. [CrossRef]
25. Wang, L.P.; Zhang, Z.K.; Shao, Z.F.; Tang, X.Q. Analysis and optimization of a novel planar 5R parallel mechanism with variable actuation modes. *Robot. Comput. Integr. Manuf.* **2019**, *56*, 178–190. [CrossRef]
26. Angeles, J. *Fundamentals of Robotic Mechanical Systems: Theory, Methods, and Algorithms*; Springer: Cham, Switzerland, 2007.
27. Wang, Y.J.; Belzile, B.; Angeles, J.; Li, Q.C. Kinematic analysis and optimum design of a novel 2PUR-2RPU parallel robot. *Mech. Mach. Theory* **2019**, *139*, 407–423. [CrossRef]
28. Patel, S.; Sobh, T. Manipulator performance measures-a comprehensive literature survey. *J. Intell. Robot. Syst.* **2015**, *77*, 547–570. [CrossRef]
29. Brinker, J.; Corves, B.; Takeda, Y. Kinematic performance evaluation of high-speed Delta parallel robots based on motion/force transmission indices. *Mech. Mach. Theory* **2018**, *125*, 111–125. [CrossRef]
30. Zou, Q. Type Synthesis and Performance Optimization of Parallel Manipulators. Ph.D. Dissertation, York University, Toronto, ON, Canada, 2022.
31. Cardou, P.; Bouchard, S.; Gosselin, C. Kinematic-sensitivity indices for dimensionally nonhomogeneous Jacobian matrices. *IEEE Trans. Robot.* **2010**, *26*, 166–173. [CrossRef]
32. Khan, S.; Andersson, K.; Wikander, J. Jacobian matrix normalization—A comparison of different approaches in the context of multi-objective optimization of 6-DOF haptic devices. *J. Intell. Robot.* **2015**, *79*, 87–100. [CrossRef]
33. Shentu, S.Z.; Xie, F.G.; Liu, X.J.; Gong, Z. Motion control and trajectory planning for obstacle avoidance of the mobile parallel robot driven by three tracked vehicles. *Robotica* **2021**, *39*, 1037–1050. [CrossRef]
34. Niu, J.Y.; Wang, H.B.; Shi, H.M.; Pop, N.; Li, D.; Li, S.S.; Wu, S.Z. Study on structural modeling and kinematics analysis of a novel wheel-legged rescue robot. *Int. J. Adv. Robot. Syst.* **2018**, *15*, 1729881417752758. [CrossRef]

35. Toquica, J.S.; Oliveira, P.S.; Souza, W.S.R.; Motta, J.M.S.T.; Borges, D.L. An analytical and a Deep Learning model for solving the inverse kinematic problem of an industrial parallel robot. *Comput. Ind. Eng.* **2021**, *151*, 106682. [CrossRef]
36. KöKer, R. A genetic algorithm approach to a neural-network-based inverse kinematics solution of robotic manipulators based on error minimization. *Inf. Sci.* **2013**, *222*, 528–543. [CrossRef]

Disclaimer/Publisher's Note: The statements, opinions and data contained in all publications are solely those of the individual author(s) and contributor(s) and not of MDPI and/or the editor(s). MDPI and/or the editor(s) disclaim responsibility for any injury to people or property resulting from any ideas, methods, instructions or products referred to in the content.

machines

MDPI

Article

Synthesis of $\mathcal{H}_\infty$ Control for Descriptor Hybrid Systems with Actuator Saturation

Chan-eun Park

School of Electronics Engineering, Kyungpook National University, Daegu 41566, Republic of Korea; chaneun@knu.ac.kr

Abstract: This paper addresses a mode-dependent state-feedback $\mathcal{H}_\infty$ control for stochastic descriptor hybrid systems, considering both the absence and presence of actuator saturation. Firstly, the necessary and sufficient conditions for the stochastic admissibility criterion with $\mathcal{H}_\infty$ performance γ of the closed-loop system are proposed. Given the proposed non-convex condition, the author reformulates it into linear matrix inequalities (LMIs). Then, to extend the result to the systems with actuator saturation, the actuator-saturated control input is expressed as a linear combination of a given state-feedback control input and a virtual control input that always remains under the saturation level. To verify this expression, the set invariant condition is also suggested by using the singular mode-dependent Lyapunov function candidate. Therefore, the conditions for the existence of both the mode-dependent state-feedback $\mathcal{H}_\infty$ control and the ellipsoidal shape invariant sets are successfully derived in terms of LMIs. Two numerical examples demonstrate the effectiveness of the proposed method by solving optimization problems subject to the proposed LMIs that minimize $\mathcal{H}_\infty$ performance γ and maximize the invariant set, respectively.

Keywords: descriptor system; hybrid system; stochastic system; $\mathcal{H}_\infty$ control; input saturation; set invariant

Citation: Park, C.-e. Synthesis of $\mathcal{H}_\infty$ Control for Descriptor Hybrid Systems with Actuator Saturation. *Machines* **2024**, *12*, 38. https://doi.org/10.3390/machines12010038

Academic Editors: Jian Wu, Xiangkun He and Guangfei Xu

Received: 26 November 2023
Revised: 25 December 2023
Accepted: 1 January 2024
Published: 5 January 2024

1. Introduction

In the field of control theory, researchers have focused on the analysis of system stability and the design of controllers using state-space equations [1–3]. Linear systems, being the most fundamental form, have been extensively investigated due to the ease of obtaining numerical solutions to problems [4–6]. Hence, researchers have sought to represent real-world systems through variations of linear systems. One well-known example is the descriptor system, also referred to as a generalized state-space system. The descriptor system is characterized by having only some parts of the state vector described by differential equations, while the remaining components are determined by algebraic equations based on the interrelations of the state vector [7–9]. In practical applications, large-scale systems or grid systems typically exhibit the characteristics of descriptor systems. Therefore, a power system model can be considered as one of the well-known examples of descriptor systems [10]. To represent the differential and algebraic equations of the system state in a single form, a square matrix of order n is utilized, where n represents the length of the state vector. This square matrix is used to identify the part of the states having the differential equations. Therefore, its rank is equal to the number of differential equations in the state vector, which is always smaller than n. While the advantage of expressing both dynamic and static characteristics of the system in a single form exists, the presence of a singular matrix introduces challenges in system analysis, necessitating additional considerations compared to regular systems.

On the other hand, hybrid systems have also garnered significant attention over the past few decades. Hybrid systems represent systems undergoing changes in both continuous and discrete time properties. An example is the stochastic hybrid systems or stochastic jump system, representing cases where continuous-time systems experience sudden discrete changes in system parameters due to stochastic processes [11–13]. Systems possessing the characteristics of both stochastic hybrid systems and descriptor systems are known as stochastic descriptor hybrid systems (SDHSs). Due to the advantage of SDHS that can express both abrupt changes on the descriptor systems, it can be used to express various phenomena such as DC motor systems undergoing random load changes and grid systems with network structures [14–16]. For the analysis of SDHSs in the field of control theory, studies on deriving stochastically admissible conditions and researches on controller and filter design have progressed over the past several decades. The authors of the paper [17] presented the stochastic admissibility conditions for SDHS in strict linear matrix inequalities (LMIs). In the context of such research, results on controllers and filters for continuous-time SDHS also exist [18–20]. Among them, Refs. [18,19] proposed necessary and sufficient conditions for the existence of state feedback controllers and dynamic output feedback controllers for SDHSs.

On the other hand, as the presence of disturbances in the real world is inevitable, $\mathcal{H}_\infty$ controllers and $\mathcal{H}_\infty$ filters have been extensively researched both theoretically and practically [21,22]. In its theoretical approach, finding the optimal $\mathcal{H}_\infty$ control or $\mathcal{H}_\infty$ filter has been one of the attractive topics [23]. In the view of optimal controls, LMIs have been widely employed due to their ease in finding optimal solutions. In the case of $\mathcal{H}_\infty$ control for SDHSs, research usually started from the stochastic admissibility criteria with $\mathcal{H}_\infty$ performance γ [24–26]. This criteria is commonly referred to as the bounded real lemma if it holds both necessary and sufficient conditions [27]. The bounded real lemma defines an upper bound on the ratio of the norm between the desired output and the disturbance, referred to as $\mathcal{H}_\infty$ performance γ, and aims to minimize this value since it can minimize the worst-case impact of disturbances. Generally, the desired output depends on both the system state and external disturbance. For linear SDHSs, [25] first presented necessary and sufficient conditions for the bounded real lemma of SDHS with disturbance-affected desired output in LMI form. Previous studies have mainly dealt with optimal $\mathcal{H}_\infty$ control and $\mathcal{H}_\infty$ filter for SDHSs with disturbance-unaffected desired output [26,28] or provided only sufficient conditions for the existence of $\mathcal{H}_\infty$ controls in cases with disturbance-affected desired output [26,29]. This implies that there is still a room for improvement in the $\mathcal{H}_\infty$ control for SDHS with a general desired output, serving as one of the motivations for this study.

Another motivation for this study is the need to investigate $\mathcal{H}_\infty$ control for SDHSs with input saturation. In practical situations, the actuator in every control system has its limits, which result in input saturation [30]. It is known that the input saturation can lead to performance degradation or even instability in the system. To ensure the stable operation of a control system under input saturation, it is necessary to design controllers that guarantee stability in the presence of saturation phenomena. The input saturation in hybrid systems [31,32] or in descriptor systems [33,34] has been addressed through various studies. Recent research for SDHSs with input saturation is covered in the papers [35,36]. However, to the best of the author's knowledge, no prior research has addressed the combined aspects of $\mathcal{H}_\infty$ control and actuator saturation for SDHSs. Therefore, this serves as an additional motivation for this study.

This paper addresses the synthesis problem of $\mathcal{H}_\infty$ control for SDHSs both absence and presence of actuator saturation. First, the author assumes that only the states governed by differential equations, i.e., those with dynamics, are considered controllable. Thus, a structure for differentiable state-feedback $\mathcal{H}_\infty$ control is proposed. Then, by utilizing the closed-loop system with the proposed control, the stochastic admissibility criterion with $\mathcal{H}_\infty$ performance γ is derived. As the proposed criteria is a non-convex formula, it is a challenge to directly find the solutions. Therefore, the equivalent condition is suggested in terms of LMIs. Then, this paper extends its focus to SDHSs with actuator saturation. By introducing a virtual control input, structured similarly to the proposed control and remaining within the saturation level, the closed-loop system is successfully reformulated as a linear SDHS even in the presence of the actuator saturation. Since an assumption about the range of states for this expression is required, a set invariant condition is also examined. By accounting for the structure of the state-feedback control, the ellipsoidal shape of the set invariant is obtained, with dimensions matching the number of components corresponding to states with differential equations. Since both $\mathcal{H}_\infty$ control and actuator saturation phenomena are considered, the results can address two optimization problems: (1) Finding the optimal $\mathcal{H}_\infty$ performance γ, and (2) Identifying the largest invariant set, representing the set of initial states ensuring stochastic convergence to zero. The effectiveness of the proposed approach is demonstrated through two numerical examples, illustrating the optimization results for both scenarios.

The notations used in this paper are standard. For a vector x or matrix X, the superscript T denotes its transpose. For symmetric matrices X and Y, the notation $X \leq (<) Y$ signifies that $Y - X$ is semi-positive (positive) definite. For any square matrix X, the symbol $sym(X) \triangleq X + X^T$. The matrix I denotes the identity matrix with appropriate dimensions, and I_r represents the identity matrix with dimensions $r \times r$. For matrix X, the notation $[X]_{ij}$ specifies the (i,j)-th component. Similarly, for vector x, the notation $[x]_i$ denotes the i-th component. The vector e_i indicates a unit vector with a single nonzero element at the i-th position, i.e., $[e_i]_i = 1,\ [e_i]_k = 0\ \forall k \neq i$. For symmetric matrices, the symbol $(*)$ serves as an ellipsis for terms induced by symmetry.

2. Problem Statements

Consider the following stochastic descriptor hybrid systems (SDHSs):

$$E\dot{x}(t) = A(\theta_t)x(t) + B(\theta_t)u(t) + F(\theta_t)w(t), \tag{1}$$

$$z(t) = C(\theta_t)x(t) + D(\theta_t)u(t) + G(\theta_t)w(t), \tag{2}$$

where the notations $x(t) \in \mathcal{R}^{n_x}$, $u(t) \in \mathcal{R}^{n_u}$, $z(t) \in \mathcal{R}^{n_z}$, $w(t) \in \mathcal{R}^{n_w}$ denote the system state, control input, desired output, and external disturbance, respectively. The matrix $E \in \mathcal{R}^{n_x \times n_x}$ is a square matrix whose rank is smaller than its dimension, i.e., $rank(E) \triangleq r < n_x$. The notation θ_t denotes a continuous-time Markov process defined on a probability space with outcomes in a finite set $N_+ = \{1, 2, \cdots, N\}$. The mode transition rate of the Markov process from mode i to mode j is defined as π_{ij}. Subsequently, the mode transition probability from mode i at time t and mode j at time $t + \Delta t$ are defined as follows:

$$Pr\{\theta_{t+\Delta t} = j | \theta_t = i\} \triangleq \begin{cases} \pi_{ij}\Delta t + o(\Delta t) & \text{if } i \neq j \\ 1 + \pi_{ii}\Delta + o(\Delta t) & \text{if } i = j \end{cases}, \tag{3}$$

where $\Delta t > 0$ and $o(\Delta t)$ denotes little-o of Δt such that $lim_{\Delta t \to 0}\left(\frac{o(\Delta t)}{\Delta t}\right) = 0$. The transition rate matrix $\Pi \in \mathcal{R}^{N \times N}$ can be defined as $[\Pi]_{i,j} = \pi_{ij}$, where $\sum_{j \in N_+} \pi_{ij} = 0$, $\pi_{ij} \geq 0$ for $i \neq j$ and $\pi_{ii} < 0$. To simplify the notations, the mode-dependent matrices at $\theta_t = i$ will be represented by using subscript i, i.e.,

$$\begin{bmatrix} A_i & B_i & F_i \\ C_i & D_i & G_i \end{bmatrix} \triangleq \begin{bmatrix} A(\theta_t = i) & B(\theta_t = i) & F(\theta_t = i) \\ C(\theta_t = i) & D(\theta_t = i) & G(\theta_t = i) \end{bmatrix}. \tag{4}$$

Also, to prevent issues arising from the singularity of matrix E, let us define of full-column matrices E_L, $E_R \in \mathcal{R}^{n_x \times r}$, R, $S^T \in \mathcal{R}^{(n_x - r) \times n_x}$ which hold the following properties regarding the singular matrix E:

$$E_L^T E_R = E\,, RE = 0,\ ES = 0. \tag{5}$$

Then, by using the matrices in (5), we will use the following lemma.

Lemma 1 ([37]). *For a symmetric matrix $P \in \mathcal{R}^{n_x \times n_x}$ which satisfies $E_L^T P E_L > 0$, and of full-rank matrix $Q \in \mathcal{R}^{(n_x-r)\times(n_x-r)}$, the term $(PE + R^T QS^T) \in \mathcal{R}^{n_x \times n_x}$ is of full-rank, and its inversion can be expressed as follows:*

$$(PE + R^T QS^T)^{-1} = \bar{P}E^T + S\bar{Q}R, \tag{6}$$

where $\bar{P}$ and $\bar{Q}$ defined as

$$E_R^T \bar{P} E_R = (E_L^T P E_L)^{-1},\ \bar{Q} = (S^T S)^{-1} Q^{-1} (RR^T)^{-1}. \tag{7}$$

The objective of this study is to analyze the SDHS with disturbances and synthesize a state-feedback $\mathcal{H}_\infty$ control that is robust to disturbances and actuator saturation. Therefore, the following definition and lemmas are employed in the next section to analyze the SDHS with disturbances.

Definition 1 ([17]).

(i) *The continuous-time SDHS (1) with $u(t) = 0$, $w(t) = 0$ is called to be regular if the term $det(sE - A_i)$ is not identically zero for all $i \in \mathcal{N}_+$.*

(ii) *The continuous-time SDHS (1) with $u(t) = 0$, $w(t) = 0$ is called to be impulse-free if $deg(det(sE - A_i)) = rank(E)$ for all $i \in \mathcal{N}_+$.*

(iii) *The continuous-time SDHS (1) with $u(t) = 0$, $w(t) = 0$ is called to be stochastically stable if there exists a scalar $M(x(0), r(t)) > 0$ for all $x(0) \in \mathcal{R}^n$, $r(0) \in \mathcal{N}_+$ such that*

$$\mathcal{E}\left\{ \int_0^\infty ||x(t)||^2 dt \,\middle|\, x(0), r(0) \right\} \leq M(x(0), r(0)), \tag{8}$$

where $\mathcal{E}\{\cdot\}$ denotes the expectation.

(iv) *The continuous-time SDHS (1) with $u(t) = 0$, $w(t) = 0$ is called to be stochastically admissible if it is regular, impulse-free, and stochastically stable.*

Definition 2 ([38]). *The SDHS (1) and (2) with $u(t) = 0$ can be called stochastically admissible with $\mathcal{H}_\infty$ performance γ if the system holds the following two conditions:*

(i) *At $w(t) = 0$, the SDHS (1) and (2) with $u(t) = 0$ is stochastically admissible.*

(ii) *At $x(0) = 0$, the SDHS (1) and (2) with $u(t) = 0$ holds the following inequality.*

$$||T(s)||_\infty \triangleq sup_{\theta(0)\in N_+} sup_{0 \neq w(t) \in \mathcal{L}_2^+} \frac{||z(t)||_2}{||w(t)||_2} < \gamma, \tag{9}$$

where the notation sup means supremum.

Lemma 2 ([25]). *The SDHS (1) and (2) with $u(t) = 0$ is stochastically admissible with $\mathcal{H}_\infty$ performance γ if and only if there exist the symmetric matrices $P_i \in \mathcal{R}^{n_x \times n_x}$, $W_i \in \mathcal{R}^{(n_x-r)\times n_w}$ and of full-rank matrices $Q_i \in \mathcal{R}^{(n_x-r)\times(n_x-r)}$ such that for all $i \in \mathcal{N}_+$*

$$0 < E_L^T P_i E_L, \tag{10}$$

$$0 > \begin{bmatrix} sym\{A_i^T (P_i E + R^T Q_i S^T)\} + \sum_{j\in N_+} \pi_{ij} E^T P_j E & (*) & (*) \\ F_i^T (P_i E + R^T Q_i S^T) + W_i^T R A_i & -\gamma^2 I + sym\{F_i^T R^T W_i\} & (*) \\ C_i & G_i & -I \end{bmatrix} \tag{11}$$

To synthesize a mode-dependent state-feedback $\mathcal{H}_\infty$ control for SDHSs, let us contemplate the following structure:

$$u(t) = K_i E x(t), \tag{12}$$

where K_iE is a mode-dependent control gain to be determined. Then the closed-loop system (1) and (2) with the control input (12) is defined as

$$E\dot{x}(t) = (A_i + B_iK_iE)x(t) + F_iw(t), \quad (13)$$
$$z(t) = (C_i + D_iK_iE)x(t) + G_iw(t). \quad (14)$$

This paper serves two main objectives. Firstly, it aims to determine the control gains K_iE that satisfy the stochastic admissibility criterion with $\mathcal{H}_\infty$ performance γ for the closed-loop system (13) and (14). Secondly, the focus is on finding control gains K_iE that are still valid under the actuator saturation phenomena in the system (1) and (2). When the SDHS (1) and (2) has actuator saturation, it can be represented as follows:

$$E\dot{x}(t) = A_ix(t) + B_i\rho(u(t)) + F_iw(t), \quad (15)$$
$$z(t) = C_ix(t) + D_i\rho(u(t)) + G_iw(t). \quad (16)$$

The symbol $\rho(\cdot)$ denotes the saturation operator such that

$$[\rho(u(t))]_k \triangleq sign([u(t)]_k)min([|u(t)|]_k, \mu), \quad (17)$$

where $\mu > 0$ is a saturation level. Although saturation is a common phenomenon, it induces nonlinearity even when the input signal $u(t)$ maintains linearity. To address this issue, the subsequent representation will prove to be beneficial.

Lemma 3 ([39]). *For any state $x(t) \in \mathcal{L}(H)$, the saturated control input $\rho(Kx(t))$ belongs the following convex-hull:*

$$\rho(Kx(t)) \in \mathbf{Co}\{(M_kK + M_k^- H)x(t),\ k \in [0, 2^{n_u} - 1]\}. \quad (18)$$

The set $\mathcal{L}(H)$ is a set of states where every component of the vector Hx is less than the saturation level, i.e., $\mathcal{L}(H) \triangleq \{x||e_l^THx| \leq \mu, l \in [1, n_u]\}$. The notation **Co** *denotes the convex hull, and the matrix $M_k \in \mathcal{R}^{n_u \times n_u}$ denotes the diagonal matrix whose diagonal elements have all possible combinations of* 1 *and* 0 *and $M_k^- \triangleq I - M_k$. For example, when $n_u = 2$, the following matrices will be used:*

$$M_0 = \begin{bmatrix} 0 & 0 \\ 0 & 0 \end{bmatrix},\ M_0^- = \begin{bmatrix} 1 & 0 \\ 0 & 1 \end{bmatrix}, M_1 = \begin{bmatrix} 1 & 0 \\ 0 & 0 \end{bmatrix},\ M_1^- = \begin{bmatrix} 0 & 0 \\ 0 & 1 \end{bmatrix},$$
$$M_2 = \begin{bmatrix} 0 & 0 \\ 0 & 1 \end{bmatrix},\ M_2^- = \begin{bmatrix} 1 & 0 \\ 0 & 0 \end{bmatrix}, M_3 = \begin{bmatrix} 1 & 0 \\ 0 & 1 \end{bmatrix},\ M_3^- = \begin{bmatrix} 0 & 0 \\ 0 & 0 \end{bmatrix}.$$

With the help of Lemma 3, the term $\rho(K_iEx(t))$ in (15) and (16) can be expressed as following form for the states belonging to the set $\mathcal{L}(H_iE)$:

$$\rho(K_iEx) = \sum_{k=1}^{2_u^n} \zeta_k\{M_kK_iEx(t) + M_k^- H_iEx(t)\}, \quad (19)$$

where the convex parameter ζ_k holds the following property:

$$\sum_{k=1}^{2^{n_u}} \zeta_k = 1,\ 0 \leq \zeta_k \leq 1,\ \forall k \in [1, 2^{n_u}]. \quad (20)$$

Utilizing the aforementioned lemmas, the following section will present two theorems aimed at determining the control gains K_iE under conditions of both absence and presence of actuator saturation.

3. Main Result

In this section, the conditions for the existence of control gains for the closed-loop system to be stochastically admissible with $\mathcal{H}_\infty$ performance γ will be presented. Firstly, by applying Lemma 2 to the closed-loop system (13) and (14), the stochastic admissibility with $\mathcal{H}_\infty$ performance γ of the closed-loop system (13) and (14) is ensured if and only if there exist the solutions P_i, Q_i, W_i and K_iE such that for all $i \in N_+$

$$0 < E_L^T P_i E_L, \tag{21}$$

$$0 > \begin{bmatrix} sym\{(A_i + B_iK_iE)^T\Lambda_i\} + \sum_{j\in N_+}\pi_{ij}E^TP_jE & (*) & (*) \\ F_i^T\Lambda_i + W_i^TRA_i & -\gamma^2 I + sym\{F_i^TR^TW_i\} & (*) \\ C_i + D_iK_iE & G_i & -I \end{bmatrix}, \tag{22}$$

$$\Lambda_i \triangleq P_iE + R^TQ_iS^T. \tag{23}$$

However, finding the solution for (21) and (22) is challenging due to the variable coupled term $K_i^T\Lambda_i$ in (22). To address this challenge, the following theorem presents an equivalent condition for (21) and (22) in terms of strict linear matrix inequalities (LMIs).

Theorem 1. *The assurance of the existence of solutions P_i, Q_i, W_i and K_iE for the conditions* (21) *and* (22)*, representing the stochastic admissibility criterion with $\mathcal{H}_\infty$ performance γ for the closed-loop system* (13) *and* (14)*, is established if and only if there symmetric exist matrices $\bar{P}_i \in \mathcal{R}^{n_x\times n_x}$, of full-rank matrices $\bar{Q}_i \in \mathcal{R}^{(n_x-r)\times(n_x-r)}$, $\bar{W}_i \in \mathcal{R}^{(n_x-r)\times n_w}$, $\bar{K}_i \in \mathcal{R}^{n_u\times n_x}$ for all $i \in N_+$, satisfying the following LMIs:*

$$0 < E_R^T\bar{P}_iE_R, \tag{24}$$

$$0 > \begin{bmatrix} sym\{A_i\bar{\Lambda}_i + B_i\bar{K}_iE^T\} + \pi_{ii}E^T\bar{P}_iE & (*) & (*) & (*) \\ F_i^T - \bar{W}_i^TS^TA_i^T & -\gamma^2 I & (*) & (*) \\ C_i\bar{\Lambda}_i + D_i\bar{K}_iE^T & G_i - C_iS\bar{W}_i & -I & (*) \\ \mathcal{X}_i & 0 & 0 & \mathcal{Y}_i \end{bmatrix}, \tag{25}$$

$$\bar{\Lambda}_i \triangleq \bar{P}_iE^T + S\bar{Q}_iR, \tag{26}$$

$$\mathcal{X}_i \triangleq [\sqrt{\pi_{ij}}E\bar{P}_iE_R]^T_{j\in N_+/\{i\}}, \tag{27}$$

$$\mathcal{Y}_i \triangleq -diag[E_R^T\bar{P}_jE_R]_{j\in N_+/\{i\}}. \tag{28}$$

Proof. To show the if and only if condition, two parts of proof are provided.
(Sufficient proof) To show the existence of the solutions for (21) and (22) from the solutions of the proposed LMIs (24) and (25), let us define the inversion of Λ_i in (23) by using Lemma 1:

$$(\bar{P}_iE^T + S\bar{Q}_iR) = (P_iE + R^TQ_iS^T)^{-1} \triangleq \bar{\Lambda}_i, \tag{29}$$

where $\bar{P}_i$ and $\bar{Q}_i$ satisfy the following conditions:

$$(E_L^TP_iE_L)^{-1} = E_R^T\bar{P}_iE_R, \tag{30}$$

$$\bar{Q}_i = (S^TS)^{-1}Q_i^{-1}(RR^T)^{-1}. \tag{31}$$

Then the condition (21) is equivalent to (24) through the relation (30). Next, to reformulate the condition (22) as (25), we will employ the following full-rank matrix by referring the existing work [25]:

$$\mathcal{T}_i \triangleq \begin{bmatrix} \bar{\Lambda}_i & -\bar{\Lambda}_i R^T W_i & 0 \\ 0 & I & 0 \\ 0 & 0 & I \end{bmatrix}. \quad (32)$$

Then applying the congruence transform to (22) using the matrix $\mathcal{T}_i$ in (32) earns the following inequality:

$$0 > \begin{bmatrix} sym\{A_i\bar{\Lambda}_i + B_iK_iE\bar{\Lambda}_i\} + \sum_{j\in N_+} \pi_{ij}\bar{\Lambda}_i^T E^T P_j E\bar{\Lambda}_i & (*) & (*) \\ F_i^T - W_i^T R(A_i\bar{\Lambda}_i + B_iK_iE\bar{\Lambda}_i)^T & -\gamma^2 I & (*) \\ C_i\bar{\Lambda}_i + D_iK_iE\bar{\Lambda}_i & G_i - (C_i\bar{\Lambda}_i + D_iK_iE\bar{\Lambda}_i)R^T W_i & -I \end{bmatrix}. \quad (33)$$

Taking into account the properties $RE = 0$ and $ES = 0$, the condition (33) transforms into the proposed condition (25) by putting

$$\bar{K}_i \triangleq K_iE\bar{P}_i, \quad (34)$$

$$\bar{Q}_i RR^T W_i \triangleq \bar{W}_i, \quad (35)$$

and applying Schur complement. This means that the existence of the solutions of the proposed LMIs (24) and (25) guarantees the solutions for the non-convex criteria, i.e., we can find at least one set of solutions P_i, Q_i, W_i, K_iE for (21) and (22) using the solutions of LMIs (24) and (25):

$$P_i = (E_L E_L^T)^{-1} E_L (E_R^T \bar{P}_i E_R)^{-1} E_L^T (E_L E_L^T)^{-1}, \quad (36)$$

$$Q_i = (RR^T)^{-1} \bar{Q}_i^{-1} (S^T S)^{-1}, \quad (37)$$

$$W_i = Q_i (S^T S) \bar{W}_i, \quad (38)$$

$$K_i E = \bar{K}_i E^T (P_i E + R^T Q_i S^T) \quad (39)$$

This completes the sufficient proof.
(Necessary proof) In this part, we have to show the existence of the solutions for the proposed LMIs (24) and (25) by using the solutions of the non-convex inequalities (21) and (22). Let us set the following matrices:

$$\tilde{P}_i = (E_R E_R^T)^{-1} E_R (E_L^T P_i E_L)^{-1} E_R^T (E_R E_R^T)^{-1}, \quad (40)$$

$$\tilde{Q}_i = (S^T S)^{-1} Q_i^{-1} (RR^T)^{-1}. \quad (41)$$

Then we can use the following relation

$$\tilde{\Lambda}_i = (P_i E + R^T Q_i S^T)^{-1} = \tilde{P}_i E^T + S\tilde{Q}_i R, \quad (42)$$

and also define the full-rank matrix

$$\tilde{T}_i = \begin{bmatrix} \tilde{\Lambda}_i & -\tilde{\Lambda}_i R^T W_i & 0 \\ 0 & I & 0 \\ 0 & 0 & I \end{bmatrix}. \quad (43)$$

Then, by applying congruence transformation to (21) and (22) by $\tilde{\Lambda}_i$ and $\tilde{T}_i$, respectively, we can obtain the following equivalent conditions:

$$0 < E_R^T \tilde{P}_i E_R, \tag{44}$$

$$0 > \begin{bmatrix} sym\{A_i\tilde{\Lambda}_i + B_iK_iE\tilde{\Lambda}_i\} + \sum_{j\in\mathcal{N}_+} \pi_{ij}\tilde{\Lambda}_i^T E^T P_j E\tilde{\Lambda}_i & (*) & (*) \\ F_i^T - W_i^T R(A_i\tilde{\Lambda}_i + B_iK_iE\tilde{\Lambda}_i)^T & -\gamma^2 I & (*) \\ C_i\tilde{\Lambda}_i + D_iK_iE\tilde{\Lambda}_i & G_i - (C_i\tilde{\Lambda}_i + D_iK_iE\tilde{\Lambda}_i)R^T W_i & -I \end{bmatrix}. \tag{45}$$

Then the goal is to show the existence of the solutions of the proposed LMIs (24) and (25) by using the solutions of the inequality (44) and (45). To find the solutions of (24) and (25), let us reformulate (25) into the following form using Schur complement:

$$0 > \begin{bmatrix} sym\{A_i\bar{\Lambda}_i + B_i\bar{K}_iE^T\} + \sum_{j\in\mathcal{N}_+} \pi_{ij}E^T\bar{P}_jE & (*) & (*) \\ F_i^T - \bar{W}_i^T S^T A_i^T & -\gamma^2 I & (*) \\ C_i\bar{\Lambda}_i + D_i\bar{K}_iE^T & G_i - C_iS\bar{W}_i & -I \end{bmatrix}. \tag{46}$$

By using the properties $RE = 0$ and $ES = 0$, we can construct the following zero constraints using the given matrices P_i, Q_i, K_iE and W_i:

$$0 = \begin{bmatrix} 0 & (2,1)^T & (D_iK_iES\tilde{Q}_iR)^T \\ (2,1) & 0 & (3,2)^T \\ D_iK_iES\tilde{Q}_iR & (3,2) & 0 \end{bmatrix}, \tag{47}$$

$$(2,1) = -W_i^T RE\tilde{P}_i^T (A_i + B_iE)^T - W_i^T RR^T \tilde{Q}_i^T (ES)^T K_i^T B_i^T, \tag{48}$$

$$(3,2) = -(C_i + D_iK_iE)\tilde{P}_i(RE)^T W_i - D_iK_i(ES)\tilde{Q}_iRR^T W_i, \tag{49}$$

where $\tilde{P}_i$ and $\tilde{Q}_i$ are defined in (40)-(41). By inserting the zero constraint (47) into the inequality (46), and putting the solutions as

$$\bar{P}_i = \tilde{P}_i,\ \bar{Q}_i = \tilde{Q}_i,\ \bar{W}_i = \tilde{Q}_iRR^T W_i,\ \bar{K}_i = K_iE\tilde{P}_i, \tag{50}$$

the condition (46) which is equivalent to (24), and the condition (25) conclude to (45) and (44), respectively. Since the solutions of (44) and (45) always exist in this proof, it is clear that the existence of the solutions of the proposed LMIs (24) and (25) is always guaranteed. This completes the necessary proof. □

Remark 1. *The control gain K_iE in the mode-dependent state-feedback $\mathcal{H}_\infty$ control* (12) *can be determined through the following relation:*

$$K_iE = \bar{K}_iE^T(\bar{P}_iE^T + S\bar{Q}_iR)^{-1}, \tag{51}$$

where $\bar{P}_i$, $\bar{K}_i$, $\bar{Q}_i$ are the solutions of Theorem 1.

Remark 2. *The synthesis problem of $\mathcal{H}_\infty$ control for SDHSs has been considered for several decades. However, before the introduction of the new bounded real lemma for SDHSs with disturbance-affected output* (2) *in Lemma 2, the existing research focused on establishing the sufficient conditions of controllers or exclusively examined scenarios with disturbance-unaffected desired output, i.e., $w(t) = 0$ in* (2)*. Hence, it is noteworthy to emphasize that Theorem 1 provides the necessary and sufficient condition of the controller* (12)*, ensuring the stochastic admissibility of the closed-loop system with the proposed controller* (12) *under disturbance-affected output.*

The next topic involves deriving the condition to determine control gains considering actuator saturation. Therefore, let us define the following closed-loop system with saturated control input $\rho(K_iEx(t))$:

$$E\dot{x}(t) = A_ix(t) + B_i\rho(K_iEx(t)) + F_iw(t), \tag{52}$$

$$z(t) = C_ix(t) + D_i\rho(K_iEx(t)) + G_iw(t). \tag{53}$$

By utilizing the formula (19) which is an alternative representation of the saturated input, the closed-loop system (52) and (53) can be expressed as follows:

$$E\dot{x}(t) = \sum_{k=1}^{2^{n_u}} \zeta_k \bar{A}_{i,k} x(t) + F_i w(t), \tag{54}$$

$$z(t) = \sum_{k=1}^{2^{n_u}} \zeta_k \bar{C}_{i,k} x(t) + G_i w(t), \tag{55}$$

$$\bar{A}_{i,k} \triangleq A_i + B_i(M_k K_i + M_k^- H_i)E, \tag{56}$$

$$\bar{C}_{i,k} \triangleq C_i + D_i(M_k K_i + M_k^- H_i)E \tag{57}$$

since $\sum_{k=1}^{2^{n_u}} \zeta_k = 1$. By applying Lemma 2 to the closed-loop system (54) and (55), the criteria for stochastic admissibility with $\mathcal{H}_\infty$ performance γ for the closed-system (54) and (55) is obtained as follows: for all $i \in N_+$, $k \in [1, 2^{n_u}]$

$$0 < E_L^T P_i E_L, \tag{58}$$

$$0 > \begin{bmatrix} sym\{\bar{A}_{i,k}^T \Lambda_i\} + \sum_{j \in N_+} \pi_{ij} E^T P_j E & (*) & (*) \\ F_i^T \Lambda_i & -\gamma^2 I + sym\{W_i^T R F_i\} & (*) \\ \bar{C}_{i,k} & G_i & -I \end{bmatrix}, \tag{59}$$

where Λ_i is defined in (23). This representation is valid only for the states within the set $\mathcal{L}(H_i E)$. To ensure that the range of states belongs to the set $\mathcal{L}(H_i E)$, we need to consider a set-invariant condition for the set $\mathcal{L}(H_i E)$. Before deriving it, let us define an ellipsoid using the condition (58):

$$\mathcal{E}(E^T P_i E) \triangleq \{x(t) \in \mathcal{R}^{n_x} | x^T(t) E^T P_i E x(t) \leq 1\} \tag{60}$$

Utilizing the ellipsoid, the set invariant condition for the $\mathcal{L}(H_i E)$ and the equivalent condition of (58) and (59) are provided in the following theorem.

Theorem 2. *For all states $x(t) \in \mathcal{E}(E^T P_i E)$ in* (60)*, the conditions* (58) *and* (59) *are feasible if and only if there exist symmetric matrices $\bar{P}_i \in \mathcal{R}^{n_x \times n_x}$, non-singular matrices $\bar{Q}_i \in \mathcal{R}^{(n_x - r)\times(n_x - r)}$, matrices $\bar{W}_i \in \mathcal{R}^{(n_x - r)\times n_w}$, $\bar{K}_i, \bar{H}_i \in \mathcal{R}^{n_u \times n_x}$ such that for all $i \in N_+$, $l \in [1, n_u]$ and $k \in [1, 2^{n_u}]$*

$$0 < \begin{bmatrix} E_R^T \bar{P}_i E_R & (*) \\ e_l^T \bar{H}_i E_R & \mu^2 I \end{bmatrix}, \tag{61}$$

$$0 > \begin{bmatrix} (1,1) & (*) & (*) & (*) \\ F_i^T - \bar{W}_i^T S^T A_i^T & -\gamma^2 I & (*) & (*) \\ C_i \bar{\Lambda}_i + D_i(M_k \bar{K}_i E + M_k^- \bar{H}_i) E^T & G_i - C_i S \bar{W}_i & -I & (*) \\ \mathcal{X}_i & 0 & 0 & \mathcal{Y}_i \end{bmatrix}, \tag{62}$$

$$(1,1) \triangleq sym\{A_i \bar{\Lambda}_i + B_i(M_k \bar{K}_{1i} E^T + M_k^- \bar{H}_i E^T)\} + \pi_{ii} E^T \bar{P}_i E, \tag{63}$$

where $\mathcal{X}_i$ and $\mathcal{Y}_i$ are defined in (27) *and* (28)*.*

Proof. Firstly, let us establish the set invariant condition for the set $\mathcal{L}(K_i E)$. If the ellipsoid (60) is within the linear region $\mathcal{L}(H_i E)$, the expression for the saturated input (19) is valid for states within the ellipsoid. Therefore, we can derive the following set invariant condition: for all $l \in [1, n_u]$,

$$\mu^{-2} x^T(t) E^T K_i^T e_l e_l^T K_i E x(t) \leq x^T(t) E^T P_i E x(t), \tag{64}$$

which is equivalent to

$$\mu^{-2}E^T K_i^T e_l e_l^T K_i E \leq E^T P_i E. \tag{65}$$

The condition (65) is equivalent to the following inequality:

$$\mu^{-2}\bar{\Lambda}_i^T E^T K_i^T e_l e_l^T K_i E\bar{\Lambda}_i \leq \bar{\Lambda}_i^T E^T P_i E\bar{\Lambda}_i, \tag{66}$$

where $\bar{\Lambda}_i$ is defined in (26). By utilizing the property $ES = 0$, the condition (66) concludes to the following inequality:

$$\mu^{-2}E_L E_R^T \bar{H}_i^T e_l e_l^T \bar{H}_i E_R E_L^T < E_L E_R^T \bar{P}_i E_R E_L^T, \tag{67}$$

by putting $\bar{H} \triangleq H_i E_L E_R^T \bar{P}_i$. Applying Schur complement to (67) leads to the proposed condition (61), considering the full-column rank E_L.

Secondly, we can derive the equivalent condition for (59) by applying the congruence transformation using $\mathcal{T}_i$ in (32):

$$0 > \begin{bmatrix} (1,1) & (*) & (*) \\ F_i^T - W_i^T R(A_i\bar{\Lambda}_i)^T & -\gamma^2 I & (*) \\ C_i\bar{\Lambda}_i + D_i(M_k K_i + M_k^- H_i)E\bar{\Lambda}_i & G_i - (C_i\bar{\Lambda}_i)R^T W_i & -I \end{bmatrix}. \tag{68}$$

$$(1,1) \triangleq sym\{A_i\bar{\Lambda}_i + B_i(M_k K_i + M_k^- H_i)E\bar{\Lambda}_i\} + \sum_{j\in N_+} \pi_{ij} E\bar{P}_i E^T P_j \bar{P}_i E^T. \tag{69}$$

Similar to the proof of Theorem 1, the condition (68) leads to (62) by defining

$$\bar{K}_i \triangleq K_i E\bar{P}_i,\ \bar{H}_i \triangleq H_i E\bar{P}_i,\ \bar{Q}_i RR^T W_i \triangleq \bar{W}_i. \tag{70}$$

This completes the proof. □

Remark 3. *The control gain $K_i E$ that renders the closed-loop system with actuator saturation (54) and (55) stochastically admissible with $\mathcal{H}_\infty$ performance γ can be constructed using the solutions from Theorem 2, and the formula remains the same as in (51).*

Remark 4. *One of the key contributions of this study is addressing both input saturation and disturbance, whereas the previous work [25] has focused only on disturbance. Lemma 3 enables the representation of a system with input saturation as a linear system with respect to the control input and virtual input. Since the $\mathcal{H}_\infty$ performance could be used as an optimization object, deriving the closed-loop system in a linear form is essential, and Lemma 3 facilitates this process. Additionally, deriving the set invariant condition to verify the linear expression is also crucial. For the invariant condition, we define the ellipsoidal region using the Lyapunov function candidate in (60) for SDHSs. This justifies our choice of employing the differentiable state-feedback control (12), whereas the previous work [25] has used the full state-feedback control such as $u(k) = K_i x(t)$.*

Remark 5. *To achieve a less conservative result in terms of $\mathcal{H}_\infty$ performance, the minimal γ can be determined by solving an optimization problem that minimizes γ^2 while satisfying LMIs suggested in Theorem 1 or Theorem 2.*

Remark 6. *With the aid of Lemma 3, the saturated input $\rho(K_i Ex(t))$ can be expressed as a linear combination of two state feedback controls: $K_i Ex(t)$ and $H_i Ex(t)$. This enables us to consider the closed-loop system as a linear system even when subjected to actuator saturation. However, this alternative representation is only valid for states belonging to the given set $\mathcal{L}(H_i E)$. Therefore, the set-invariant condition is proposed in (61). It implies that the SDHSs with only the initial states within the invariant set $\mathcal{E}(E^T P_i E)$ in (60) can stochastically converge to zero. Therefore,* *maximizing the area of the invariant set is an essential issue. The largest invariant set can be found by solving an optimization problem that maximizes α subject to:*

$$(i)\ LMIs\ \ (61)\ and\ (62), \tag{71}$$

$$(ii)\ \alpha x_0^w \in \bigcap_{i=1}^{N} \mathcal{E}(E_L^T P_i E_L),\ \forall w \in [1, g], \tag{72}$$

where x_0^w is the component to express the region of initial states, i.e., $x(0) \in X_0 \in \mathbf{Co}\{x_0^w,\ w \in [1, g]\}$, $x_0^w \in \mathcal{R}^{n-r}$. The condition (72) is equivalent to

$$1 \geq \alpha^2 (x_0^w)^T E_L^T P_i E_L x_0^w = \alpha^2 (x_0^w)^T (E_R^T \bar{P}_i E_R)^{-1} x_0^w, \tag{73}$$

and it can be expressed as the following LMIs after applying the Schur complement:

$$0 \geq \begin{bmatrix} \bar{\alpha} & x_0^w \\ x_0^w & E_R^T \bar{P}_i E_R \end{bmatrix},\ \forall w \in [1, g],\ i \in N_+, \tag{74}$$

where $\bar{\alpha} \overset{\triangle}{=} \alpha^{-2}$.

4. Numerical Example

In this section, two numerical examples are considered to demonstrate the effectiveness of the proposed mode-dependent state-feedback $\mathcal{H}_\infty$ control with actuator saturation.

Example 1. *In this example, control gains with their minimal $\mathcal{H}_\infty$ performance will be found. To obtain them, we need to solve LMIs in Theorem 2 by minimizing γ^2. Consider the following SDHS with input saturation* (15) *and* (16) *with following parameters:*

$$E = \begin{bmatrix} 1 & 0.5 & 0 \\ 0 & 1 & 0 \\ 0 & 0 & 0 \end{bmatrix}, A_1 = \begin{bmatrix} -0.7 & -0.3 & 0 \\ 0.7 & -0.7 & -0.5 \\ 0.1 & 0 & 1.2 \end{bmatrix}, A_2 = \begin{bmatrix} -0.5 & 1.8 & 1.3 \\ -0.2 & -2.1 & -0.1 \\ 1.2 & 2.5 & -1 \end{bmatrix},$$

$$B_1 = \begin{bmatrix} 0 & 0.2 \\ 0.1 & 0 \\ 0 & 0.1 \end{bmatrix}, B_2 = \begin{bmatrix} 2.1 & 0.2 \\ 0.6 & 0 \\ -0.1 & 0 \end{bmatrix}, C_1 = \begin{bmatrix} 1 & 0 & 1 \\ 0 & 1 & 0 \end{bmatrix}, C_2 = \begin{bmatrix} 0 & 1 & 1 \\ 1 & 0 & 0 \end{bmatrix},$$

$$D_1 = \begin{bmatrix} 0.7 & 1 \\ 0 & 0 \end{bmatrix}, D_2 = \begin{bmatrix} 1 & 0.9 \\ 0 & 0 \end{bmatrix}, F_1 = \begin{bmatrix} 1 & 0 \\ 0 & 2 \\ 0 & 0 \end{bmatrix}, F_2 = \begin{bmatrix} -0.1 & 1 \\ 0 & 1 \\ 0.1 & 0 \end{bmatrix},$$

$$G_1 = \begin{bmatrix} 0.3 & 0.1 \\ 0 & 0.1 \end{bmatrix}, G_2 = \begin{bmatrix} 0 & 0 \\ 0 & 0.2 \end{bmatrix},\ \mu = 3.$$

The mode transition rate matrix is defined as

$$\Pi = \begin{bmatrix} -1.4 & 1.4 \\ 1.1 & -1.1 \end{bmatrix},$$

and the matrices E_L, E_R, R and S for the matrix E are defined as follows:

$$E_L = \begin{bmatrix} 1 & 0.5 \\ 0 & 1 \\ 0 & 0 \end{bmatrix}, E_R = \begin{bmatrix} 1 & 0 \\ 0 & 1 \\ 0 & 0 \end{bmatrix} R = \begin{bmatrix} 0 \\ 0 \\ 1 \end{bmatrix}^T, S = \begin{bmatrix} 0 \\ 0 \\ 1 \end{bmatrix}.$$

Figure 1 shows the state trajectories for the unforced case, i.e., $u(t) = 0$. It indicates that the SDHS with the given system parameters is unstable. To render the closed-loop system stochastically admissible, control input derived from Theorem 2 is applied. In this example, we solve an optimization *problem that minimizes γ^2 in Theorem 2, and the following control gains with minimal $\gamma = 1.4525$ are obtained:*

$$K_1 E = \begin{bmatrix} -0.1743 & 0.1546 & 0 \\ -0.9417 & -0.1837 & 0 \end{bmatrix}, \tag{75}$$

$$K_2 E = \begin{bmatrix} -1.9441 & -2.1215 & 0 \\ 0.2144 & -1.9336 & 0 \end{bmatrix}, \tag{76}$$

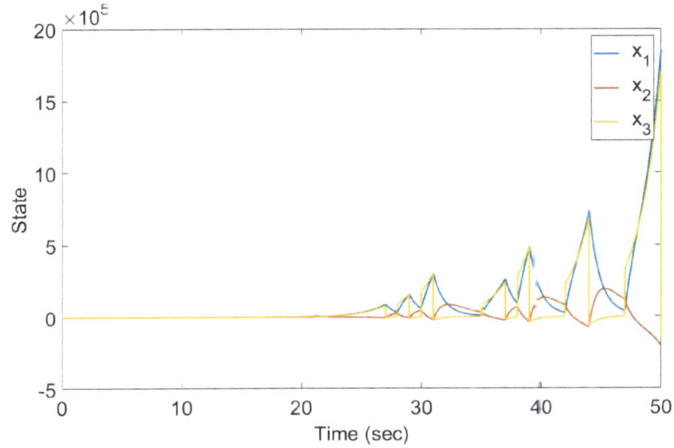

Figure 1. The unforced response of systems in Example 1.

The state trajectories of the closed-loop system are depicted in Figure 2. The initial state $x(0) = [5, -5, -6.5]^T$ is considered, and mode evolution and disturbance are also illustrated in Figure 2. Since a saturation level of 3 is considered, control input cannot exceed this limit.

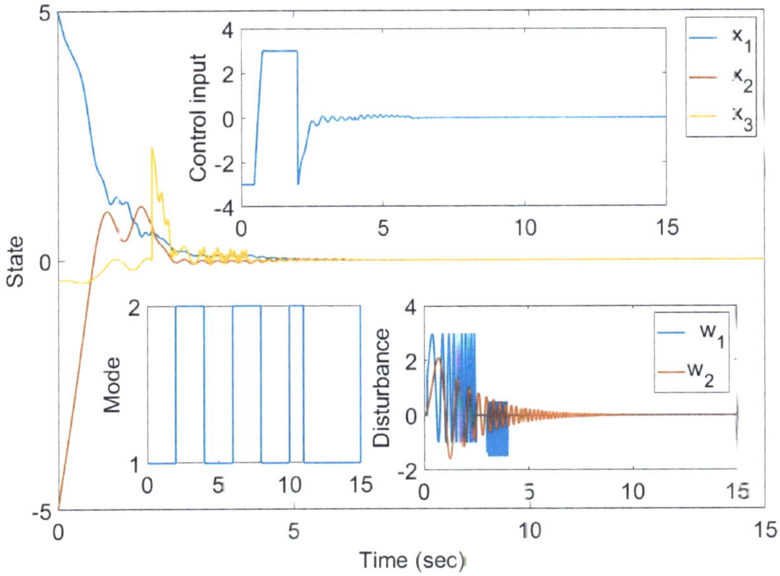

Figure 2. The state trajectories, mode evolution and control input of the closed-loop system in Example 1.

Example 2. *In this example, control gains with their maximal invariant set will be found. To obtain them, we need to solve LMIs in Theorem 2 with additional condition* (74) *by minimizing* $\bar{\alpha}$. *Let us use the same system parameters as in Example 1. By solving the LMIs via the optimization problem, the solutions of* $\bar{P}_i$, $i \in [1, 2]$ *are obtained. Then, using the relation* $E_L^T P_i E_L = (E_R^T \bar{P}_i E_R)^{-1}$, *we can determine the largest invariant set. Figure 3 shows the region of attraction obtained from the ellipsoids* $\mathcal{E}(E_L^T P_i E_L)$, *and the state trajectories starting from the boundary of the ellipsoids successfully converge to zero. Figure 4 shows the state trajectories over time, the control input and the mode evolution. In this simulation, the disturbance is considered to be 1/10th of the scale of Example 1.*

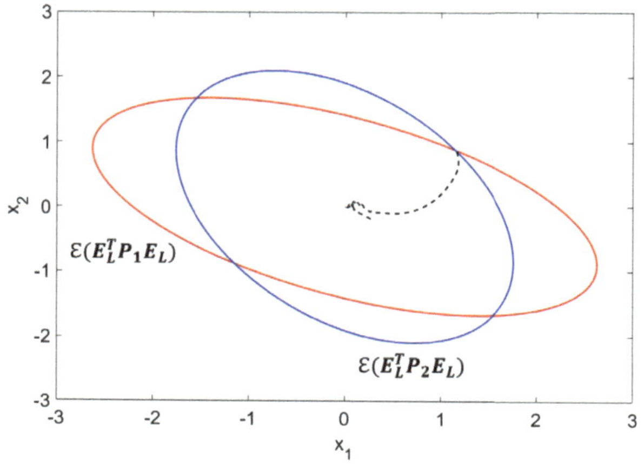

Figure 3. The ellipsoid $\mathcal{E}(E_L^T P_i E_L)$ and the state trajectory starting from the boundary of the ellipsoid.

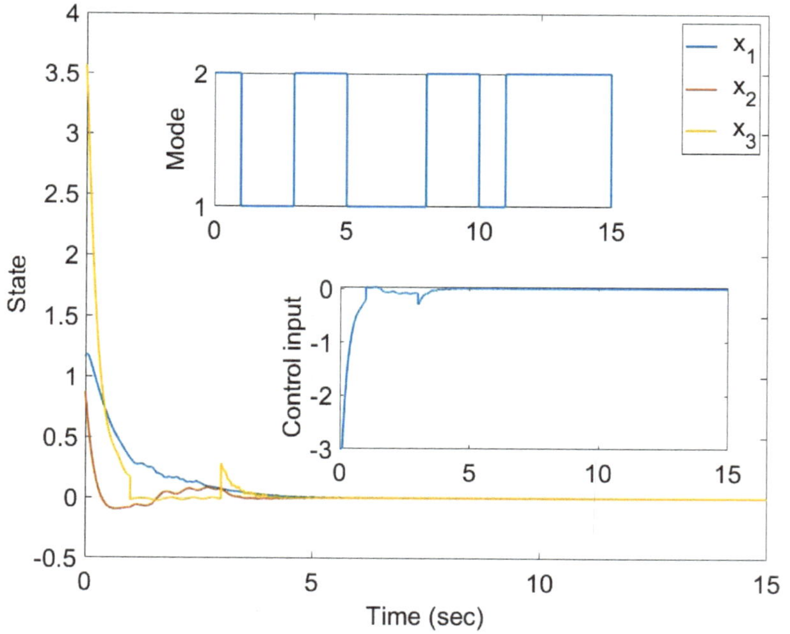

Figure 4. The state trajectories over time, the control input and the mode evolution of Example 2.

5. Conclusions

This paper considered the mode-dependent state-feedback $\mathcal{H}_\infty$ control for SDHSs, considering both the absence and presence of actuator saturation. Firstly, we established the necessary and sufficient condition for the stochastic admissibility criterion with $\mathcal{H}_\infty$ performance γ of the closed-loop system using the proposed non-saturated control input. Since the proposed condition was expressed as a non-convex formula, we reformulated it into the LMIs. Next, we extended our result to the closed-loop system with actuator saturation, expressing it as SDHSs with linear state-feedback control inputs through the introduction of a virtual control input that always remains under the saturation level. To verify this expression, the set invariant condition was also considered. By utilizing the singular mode-dependent Lyapunov function candidate, we suggested the ellipsoidal shape of the invariant set and provided a method to determine the largest invariant set. The key motivation of this study compared to the existing work was to derive the synthesis criterion for SDHS with both input saturation and disturbance *in terms of LMIs*. It implies that the proposed method can address an optimization problem subject to the both largest invariant sets and optimal $\mathcal{H}_\infty$ performance. Example 1 and Example 2 showed the results of optimization problems that minimize $\mathcal{H}_\infty$ performance, and maximize the region of the invariant set, respectively. Our future work involves extending the results to stochastic hybrid descriptor systems with semi-Markov processes or applying them to practical systems, such as power grid systems.

Funding: This research was supported by Kyungpook National University Research Fund, 2022.

Informed Consent Statement: Informed consent was obtained from all subjects involved in the study.

Data Availability Statement: Data sharing is not applicable to this article as no new data were generated or analyzed in the current study.

Conflicts of Interest: The authors declare no conflict of interest.

References

1. Chow, S.M.; Ho, M.h.R.; Hamaker, E.L.; Dolan, C.V. Equivalence and differences between structural equation modeling and state-space modeling techniques. *Struct. Equ. Model.* **2010**, *17*, 303–332. [CrossRef]
2. Aoki, M. *State Space Modeling of Time Series*; Springer Science & Business Media: Berlin/Heidelberg, Germany, 2013.
3. Kailath, T. *Linear Systems*; Prentice-Hall: Englewood Cliffs, NJ, USA, 1980; Volume 156.
4. Hespanha, J.P. *Linear Systems Theory*; Princeton University Press: Princeton, NJ, USA, 2018.
5. Hendricks, E.; Jannerup, O.; Sørensen, P.H. *Linear Systems Control: Deterministic and Stochastic Methods*; Springer Berlin/Heidelberg, Germany, 2008.
6. Sun, Z. *Switched Linear Systems: Control and Design*; Springer Science & Business Media: Berlin/Heidelberg, Germany, 2006.
7. Takaba, K.; Morihira, N.; Katayama, T. A generalized Lyapunov theorem for descriptor system. *Syst. Control. Lett.* **1995**, *24*, 49–51. [CrossRef]
8. Fridman, E.; Shaked, U. A descriptor system approach to $\mathcal{H}_\infty$ control of linear time-delay systems. *IEEE Trans. Autom. Control.* **2002**, *47*, 253–270. [CrossRef]
9. Zhu, Y.; Zhong, Z.; Basin, M.V.; Zhou, D. A descriptor system approach to stability and stabilization of discrete-time switched PWA systems. *IEEE Trans. Autom. Control.* **2018**, *63*, 3456–3463. [CrossRef]
10. Nugroho, S.A.; Taha, A.F. Load-and Renewable-Following Control of Linearization-Free Differential Algebraic Equation Power System Models. *IEEE Trans. Control. Syst. Technol.* **2023** . [CrossRef]
11. Shi, P.; Li, F. A survey on Markovian jump systems: Modeling and design. *Int. J. Control. Autom. Syst.* **2015**, *13*, 1–16. [CrossRef]
12. Fei, Z.; Gao, H.; Shi, P. New results on stabilization of Markovian jump systems with time delay. *Automatica* **2009**, *45*, 2300–2306. [CrossRef]
13. Zhao, P.; Kang, Y.; Zhao, Y.B. A brief tutorial and survey on markovian jump systems: Stability and control. *IEEE Syst. Man, Cybern. Mag.* **2019**, *5*, 37-C3. [CrossRef]
14. Wang, G.; Zhang, Q.; Yan, X. *Analysis and Design of Singular Markovian Jump Systems*; Springer: Berlin/Heidelberg, Germany, 2015.
15. Li, F.; Du, C.; Yang, C.; Gui, W. Passivity-based asynchronous sliding mode control for delayed singular Markovian jump systems. *IEEE Trans. Autom. Control.* **2017**, *63*, 2715–2721. [CrossRef]
16. Yao, X.; Park, J.H.; Wu, L.; Guo, L. Disturbance-observer-based composite hierarchical antidisturbance control for singular Markovian jump systems. *IEEE Trans. Autom. Control.* **2018**, *64*, 2875–2882. [CrossRef]

17. Xia, Y.; Boukas, E.K.; Shi, P.; Zhang, J. Stability and stabilization of continuous-time singular hybrid systems. *Automatica* **2009**, *45*, 1504–1509. [CrossRef]
18. Kao, Y.; Xie, J.; Wang, C. Stabilization of singular Markovian jump systems with generally uncertain transition rates. *IEEE Trans. Autom. Control.* **2014**, *59*, 2604–2610. [CrossRef]
19. Kwon, N.K.; Park, I.S.; Park, P.; Park, C. Dynamic output-feedback control for singular Markovian jump system: LMI approach. *IEEE Trans. Autom. Control.* **2017**, *62*, 5396–5400. [CrossRef]
20. Park, I.S.; Kwon, N.K.; Park, P. Dynamic output-feedback control for singular Markovian jump systems with partly unknown transition rates. *Nonlinear Dyn.* **2019**, *95*, 3149–3160. [CrossRef]
21. Gahinet, P.; Apkarian, P. A linear matrix inequality approach to $\mathcal{H}_\infty$ control. *Int. J. Robust Nonlinear Control.* **1994**, *4*, 421–448. [CrossRef]
22. Lee, B.K.; Chen, Y.H.; Chen, B.S. Robust $\mathcal{H}_\infty$ power control for cdma cellular communication systems. *IEEE Trans. Signal Process.* **2006**, *54*, 3947–3956. [CrossRef]
23. Başar, T.; Bernhard, P. *$\mathcal{H}_\infty$ Optimal Control and Related Minimax Design Problems: A Dynamic Game Approach*; Springer Science & Business Media: Berlin/Heidelberg, Germany, 2008.
24. Zhou, J.; Lai, H.; Liu, C.; Men, B. $\mathcal{H}_\infty$ controller design for singular Markov jump systems with time-delays. *Proc. Inst. Mech. Eng. Part I J. Syst. Control. Eng.* **2022**, *236*, 1566–1577. [CrossRef]
25. Park, C.e.; Kwon, N.K.; Park, P. New bounded real lemma for singular Markovian jump systems: Application to $\mathcal{H}_\infty$ control. *Int. J. Robust Nonlinear Control.* **2021**, *31*, 907–919. [CrossRef]
26. Park, C.e.; Kwon, N.K.; Park, I.S.; Park, P. $\mathcal{H}_\infty$ filtering for singular Markovian jump systems with partly unknown transition rates. *Automatica* **2019**, *109*, 108528. [CrossRef]
27. Wang, H.S.; Yung, C.F.; Chang, F.R. Bounded real lemma and $\mathcal{H}_\infty$ control for descriptor systems. *IEE Proc.-Control. Theory Appl.* **1998**, *145*, 316–322. [CrossRef]
28. Wu, Z.; Su, H.; Chu, J. $\mathcal{H}_\infty$ filtering for singular Markovian jump systems with time delay. *Int. J. Robust Nonlinear Control. IFAC-Affil. J.* **2010**, *20*, 939–957. [CrossRef]
29. Kwon, N.K.; Park, I.S.; Park, P. $\mathcal{H}_\infty$ control for singular Markovian jump systems with incomplete knowledge of transition probabilities. *Appl. Math. Comput.* **2017**, *295*, 126–135. [CrossRef]
30. Hu, T.; Lin, Z. *Control Systems with Actuator Saturation: Analysis and Design*; Springer Science & Business Media: Berlin/Heidelberg, Germany, 2001.
31. Shi, P.; Yin, Y.; Liu, F.; Zhang, J. Robust control on saturated Markov jump systems with missing information. *Inf. Sci.* **2014**, *265*, 123–138. [CrossRef]
32. Liu, H.; Boukas, E.K.; Sun, F.; Ho, D.W. Controller design for Markov jumping systems subject to actuator saturation. *Automatica* **2006**, *42*, 459–465. [CrossRef]
33. Lin, Z.; Lv, L. Set invariance conditions for singular linear systems subject to actuator saturation. *IEEE Trans. Autom. Control.* **2007**, *52*, 2351–2355. [CrossRef]
34. Chen, Y.; Xu, T.; Zeng, C.; Zhou, Z.; Zhang, Q. Simultaneous stabilization for uncertain descriptor systems with input saturation. *Int. J. Robust Nonlinear Control.* **2012**, *22*, 1938–1951. [CrossRef]
35. Ma, Y.; Jia, X.; Zhang, Q. Robust observer-based finite-time $\mathcal{H}_\infty$ control for discrete-time singular Markovian jumping system with time delay and actuator saturation. *Nonlinear Anal. Hybrid Syst.* **2018**, *28*, 1–22. [CrossRef]
36. Park, C.e.; Kwon, N.K.; Park, P. Output-feedback control for singular Markovian jump systems with input saturation. *Nonlinear Dyn.* **2018**, *93*, 1231–1240. [CrossRef]
37. Uezato, E.; Ikeda, M. Strict LMI conditions for stability, robust stabilization, and $\mathcal{H}_\infty$ control of descriptor systems. In Proceedings of the Proceedings of the 38th IEEE Conference on Decision and Control (Cat. No. 99CH36304), Phoenix, AZ, USA, 7–10 December 1999 ; IEEE: Piscataway, NJ, USA, 1999; Volume 4, pp. 4092–4097.
38. Feng, Z.; Shi, P. Two equivalent sets: Application to singular systems. *Automatica* **2017**, *77*, 198–205. [CrossRef]
39. Cao, Y.Y.; Lin, Z.; Shamash, Y. Set invariance analysis and gain-scheduling control for LPV systems subject to actuator saturation. *Syst. Control. Lett.* **2002**, *46*, 137–151. [CrossRef]

machines

Article

EHB Gear-Drive Symmetric Dead-Zone Finite-Time Adaptive Control

Shuai Wang, Qinghua Cao, Fukuo Ma and Jian Wu *

School of Automotive Engineering, Liaocheng University, Liaocheng 252000, China; ws15854393910@163.com (S.W.); 18353643398@163.com (Q.C.); 15552307021@163.com (F.M.)
* Correspondence: wujian@lcu.edu.cn

Abstract: Intelligent driving vehicles require more accurate and stable braking control. Electro-hydraulic braking (EHB) systems can better adapt to the development of autonomous driving technology. The gear transmission system plays a crucial role in EHB deceleration and torque increase mechanisms. However, its dead-zone nonlinearity poses challenges for EHB control. To address the position-control problem in the EHB gear transmission system, we propose a finite-time adaptive control method for the symmetrical dead zone. This approach combines adaptive control theory with finite-time control theory and designs parameter-updating laws for the unknown parameters in the system. Boundary estimates are introduced into the parameter-update laws and control laws to compensate for unknown disturbances. By adjusting the relevant parameters, the convergence rate can be improved, ensuring that errors converge within a specified range within a limited time. After modifying the parameter-updating laws and control laws, all closed-loop signals remain bounded. Finally, we validate the proposed control strategy through simulation and hardware-in-the-loop (HIL) testing. The results demonstrate that the control strategy developed in this study achieves high tracking accuracy and stability even in the presence of dead zones, unknown parameters, and unknown interferences in the EHB gear-drive servo system.

Keywords: adaptive control; finite-time control; symmetric dead-zone nonlinearity; gear transmission servosystems

Citation: Wang, S.; Cao, Q.; Ma, F.; Wu, J. EHB Gear-Drive Symmetric Dead-Zone Finite-Time Adaptive Control. *Machines* **2023**, *11*, 1002. https://doi.org/10.3390/machines11111002

Academic Editor: Domenico Mundo

Received: 13 September 2023
Revised: 19 October 2023
Accepted: 26 October 2023
Published: 30 October 2023

1. Introduction

With the development of intelligent vehicles and autonomous driving technology, braking systems are playing an increasingly important role. In order to improve the application of automatic driving technology in vehicles, a more reliable braking system has become a research hotspot for countless researchers [1–3]. In this context, EHB has gradually replaced traditional braking methods and been applied and promoted in intelligent vehicles due to its advantages such as simple structure, rapid response and strong compatibility [4–7]. With the continuous improvement of control accuracy requirements, the existence of gap nonlinearity has gradually been paid attention to [8,9].

Backlash nonlinearity is inevitable in many practical systems and it may seriously affect the dynamic performance, control accuracy and stability of the controlled system [10–12]. The gear group is an important part of the EHB deceleration and torque increase mechanism. The position-control accuracy of EHB gear-drive servo systems is seriously affected due to clearance, parameter uncertainty and interference [13]. Because of the nonlinear problem of gear backlash, the parameters of the system will change [14]. Because of the unknown time-varying disturbance, the position control of the EHB gear-drive servo system will be more difficult. Therefore, it is necessary to find a control method suitable for the nonlinear backlash of the EHB gear transmission system.

Many experts and scholars have embarked on research on the problem of backlash in the process of gear transmission. In practical engineering, the stiffness and moment of inertia of the gear-drive servo system are often considered. Therefore, the dead-zone model is often used to describe the backlash nonlinear problem of the gear-drive servo system. Within this system, the existence of a backlash problem will lead to the change of different parameters. Additionally, due to its non-differentiable characteristics, it will have an impact on position control [15]. In order to overcome the non-differentiable characteristics of the backlash nonlinearity in the gear transmission system, a new differentiable dead-zone model was established in the literature [16]. State feedback is used to transform it into a linear system, thus avoiding the complex model-switching system and inverse model compensation. The output tracking problem of the gear-drive servo system is realized. In reference [17], the author proposes an inversion algorithm based on a differentiable dead-zone model. The non-differentiable characteristic of the dead-zone model is overcome; the backlash nonlinearity is eliminated and the high-precision output tracking is realized. Papers [18,19], respectively, put forward a differentiable dead-zone model for the gear-drive servo system to solve the output-control problem and verify the effectiveness of the controller through simulation. However, the above method is only applicable to the determination of dead-zone model parameters, which obviously makes it difficult to meet the above conditions in actual working conditions.

An EHB gear-drive servo system will inevitably change the system parameters due to various factors in the working process, which presents challenges in accurate output position tracking. Researchers at the University of Texas designed a dead-zone compensator based on a fuzzy logic controller for the dead-zone model with unknown system parameters. By designing the optimization algorithm of fuzzy logic parameters, small tracking error and bounded parameter estimation are guaranteed [20]. In reference [21], the author designed an adaptive controller by introducing a smooth inverse function of dead interval for the unknown dead interval nonlinearity of uncertain systems. The stability and transient performance of the system are guaranteed. In reference [22], the author uses fuzzy logic to identify unknown nonlinear functions. An adaptive fuzzy output feedback control method is proposed, which ensures the control precision and the stability of the closed-loop system. The unknown disturbance will inevitably exist in the actual system, which will affect the control precision. Some scholars compensate for the influence of unknown disturbance by disturbance observer or adversarial attack [23–25]. The authors of reference [26] further consider the influence of unknown perturbations while considering unknown parameters. By adding a robust term to the parameter-updating laws, the authors effectively compensate for the effect of unknown interference. In reference [27], the author proposed a linear nonsingular transformation of state variables and designed an adaptive controller for gear transmission systems. The tracking performance can be guaranteed in the case of parameter uncertainty and unknown disturbance. Therefore, according to the above analysis, the adaptive control algorithm is a good solution for the unknown parameters of the dead-zone model and the unknown disturbance [28,29].

Based on the above analysis, it is necessary to consider the influence of unknown system parameters and disturbance for the output position-tracking control of the gear-drive servo system. In the actual EHB condition, the influence of system stability and convergence rate should also be considered. In reference [30], the authors consider the problem that each subsystem has dead input and is not modeled. A decentralized finite-time controller based on recursion is designed, which makes all state variables return to zero in finite time. The controller designed by the author can ensure the stability of the system in fixed time, in which the stability time depends on the design parameters of the control system. In literature [31], the authors further consider the uncertainty and time delay of system parameters on the basis of the dead-zone problem. In this paper, an adaptive finite-time control scheme is proposed to solve the uncertainty and unknown nonlinearity of the system with neural network. In reference [32], the author proposed an adaptive finite-time control scheme. By designing an adaptive dummy variable, an adaptive law and a controller, the error is guaranteed to converge to a certain range in a certain time. Therefore, the combination of an adaptive control algorithm and finite-time control algorithm can effectively improve the control accuracy and convergence rate of a gear-drive servo system with unknown parameters, unknown disturbance and dead-zone problems.

Based on the above analysis, the dead zone will inevitably exist in the EHB gear-drive servo system, which will affect the precision of position control. System parameters may change due to the dead-zone problem. At the same time, the influence of the unknown disturbance and the convergence rate of the system should be considered. Therefore, on the basis of the above content in this paper, a finite-time adaptive control strategy for a symmetrical dead zone of EHB gear transmission is proposed. The contributions are as follows:

(1) The unknown disturbance and parameter uncertainty of the EHB gear-drive servo system are considered. In this paper, the parameter-updating law is designed for each parameter in view of the parameter uncertainty in the system. In order to compensate for the unknown disturbance, the boundary estimates are introduced into the parameter-updating law and the control law.

(2) In order to improve the convergence rate of the system, this paper combines the adaptive control theory with the finite-time control theory to introduce the power exponent β. By adjusting the value of β, the system error can be guaranteed to converge to a certain range in a limited time.

The rest of this article is arranged as follows. In Section 2, the position-control problem of an EHB gear-drive servo system with symmetric dead-zone nonlinearity is described. In Section 3, based on the above problems, a finite-time adaptive controller is designed, its stability is analyzed, and the reliability of the controller is proved. The simulation is verified in Section 4, and the HIL test is carried out in Section 5. Finally, the article is summarized in Section 6.

2. System Modeling

The EHB structure is shown in Figure 1. An EHB system is mainly composed of a brake pedal, a pedal stroke sensor, a transmission mechanism, a power source, an oil storage pot and a brake master cylinder. The gear-drive servo system is an important part of the transmission mechanism. The pinion of the gear-drive servo system is connected to the power source, and the big gear is connected with the motion-conversion mechanism, which plays the role of slowing down and increasing torque; it is an important part of EHB. The existence of a dead-zone problem will affect the position-control accuracy of the gear-drive servo system and then affect the pressure control of the EHB brake master cylinder, thus affecting the braking effect. Therefore, the existence of the dead-zone problem will pose a threat to vehicle safety and become a factor that cannot be ignored in the development process of intelligent vehicles.

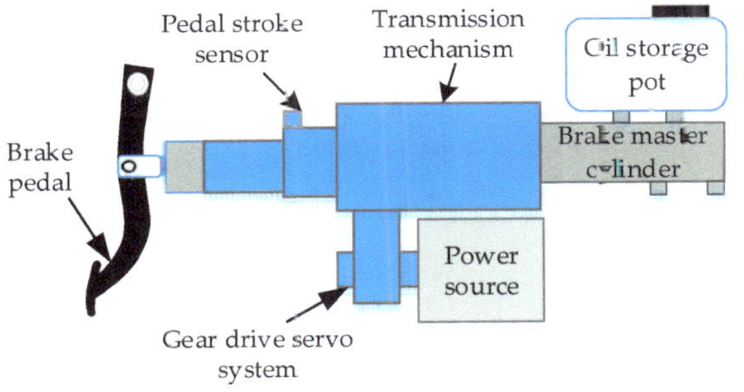

Figure 1. EHB structure diagram.

The symmetric dead zone of EHB gear transmission considered in this paper is shown in Figure 2.

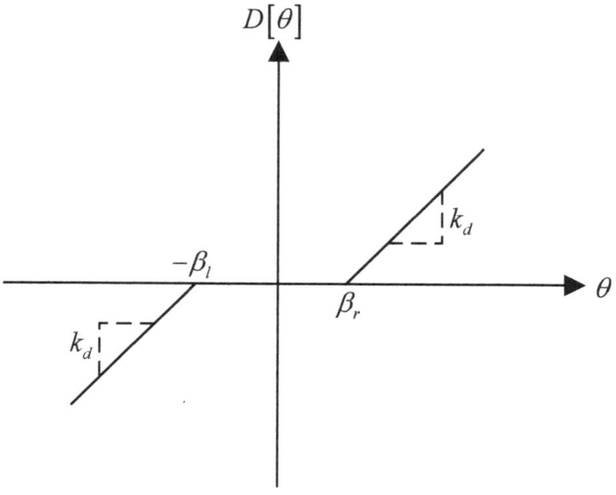

Figure 2. Symmetry dead zone.

The EHB gear-drive servo system is as follows [33]:

$$\begin{cases} J_l\ddot{\theta}_l + C_l\dot{\theta}_l = iD[\theta] \\ J_m\ddot{\theta}_m + C_m\dot{\theta}_m = u - D[\theta] \end{cases} \tag{1}$$

Among them, J_l, θ_l and C_l are the moment of inertia, position and viscous friction coefficient of the driven side gear, respectively. J_m, θ_m and C_m are the moment of inertia, position and viscous friction coefficient of the driving side gear, respectively. i is the transmission ratio, u is the input torque, θ is the relative displacement ($\theta = \theta_m - i\theta_l$), $D[\theta]$ is the transmission torque.

Transmission torque $D[\theta]$ is defined as:

$$D[\theta] = \begin{cases} k_d(\theta - \beta_r), & if \quad \theta \geq \beta_r \\ 0, & if \quad -\beta_l < \theta < \beta_r \\ k_d(\theta + \beta_l), & if \quad \theta \leq -\beta_l \end{cases} \tag{2}$$

Among them, $k_d > 0$ denotes the stiffness coefficient and $\beta_r > 0$, $\beta_l > 0$ denote the break points.

Remark 1. *The parameters J_m, θ_m, C_m, J_l, θ_l, C_l, k_d, β_r and β_l are unknown in the system. The control objective is to adjust the position of the driven side to the ideal position while ensuring the boundedness of all closed-loop signals.*

To deal with the effects caused by symmetry dead zones, rewrite $D[\theta]$ as:

$$D[\theta] = k_d\theta + \Delta d[\theta] \tag{3}$$

Among them,

$$\Delta d[\theta] = \begin{cases} -k_d\beta_r, & if \quad \theta \geq \beta_r \\ -k_d\theta, & if \quad -\beta_l < \theta < \beta_r \\ k_d\beta_l, & if \quad \theta \leq -\beta_l \end{cases} \tag{4}$$

It can be concluded that:

$$|\Delta d[\theta]| \leq k_d\overline{\beta} \tag{5}$$

Among them, $\bar{\beta} = \max\{\beta_r, \beta_l\}$.

Let $x_1 = \theta_l$, $x_2 = \dot{\theta}_l$, $x_3 = \theta_m$, $x_4 = \dot{\theta}_m$, $a_1 = ik_d/J_l$, $a_2 = i^2k_d/J_l$, $a_3 = C_l/J_l$, $b_1 = ik_d/J_m$, $b_2 = k_d/J_m$, $b_3 = C_m/J_m$, $b_4 = 1/J_m$, $d_1 = i\Delta d[\theta]/J_l$, $d_2 = -\Delta d[\theta]/J_m$. Finally, the following feedback system is established

$$\begin{cases} \dot{x}_1 = x_2 \\ \dot{x}_2 = a_1x_3 - a_2x_1 - a_3x_2 + d_1 \\ \dot{x}_3 = x_4 \\ \dot{x}_4 = b_1x_1 - b_2x_3 - b_3x_4 + d_2 + b_4u \end{cases} \tag{6}$$

Among them,

$$d_1 \le \frac{ik_d}{J_l}\bar{\beta} \le D_1 \tag{7}$$

$$d_2 \le \frac{k_d}{J_m}\bar{\beta} \le D_2 \tag{8}$$

where d_1 and d_2 are additional disturbances.

Remark 2. *The system will be subject to unknown interference in actual operation, such as vibration and temperature change. These unknown disturbances may cause changes in system parameters and affect preset dead-zone breakpoints. As shown in Equations (6)–(8), there are set bounded values for additional disturbances d_1 and d_2 to compensate for the effects of disturbances.*

3. Controller Designing and Stability Analysis

The control logic block diagram is shown in Figure 3.

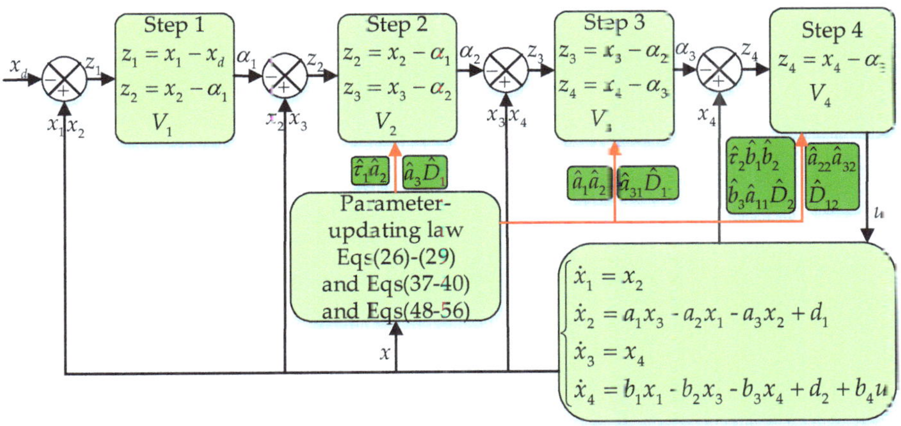

Figure 3. Control logic block diagram.

3.1. Design of Adaptive Controller

The multi-variable EHB gear-drive servo system is decomposed into several subsystems. The Lyapunov function of each step is constructed in the design process, and the parameter-updating law and control law are solved. Ensure that the position x_1 of the driven gear keeps track of the expected value x_d.

Step 1: First, construct the following error:

$$z_1 = x_1 - x_d \tag{9}$$

$$z_2 = x_2 - \alpha_1 \tag{10}$$

Among them, α_1 is the control law to be designed in this step.

Taking the derivative of z_1 gives us:

$$\dot{z}_1 = z_2 + \alpha_1 - \dot{x}_d \tag{11}$$

Design the Lyapunov function for this step.

$$V_1 = \frac{1}{2}z_1^2 \tag{12}$$

So,

$$\dot{V}_1 = z_1(z_2 + \alpha_1 - \dot{x}_d) \tag{13}$$

Design control law α_1 is:

$$\alpha_1 = -\lambda_1 z_1^{2\beta-1} - s_1 z_1 + \dot{x}_d \tag{14}$$

Among them, λ_1 and s_1 are normal numbers, and $0 < \beta < 1$.

$$\dot{V}_1 = -\lambda_1 z_1^{2\beta} - s_1 z_1^2 + z_1 z_2 \tag{15}$$

Step 2: Construct the error of this step.

$$z_3 = x_3 - \alpha_2 \tag{16}$$

From Equations (10), (14) and (16), we can get:

$$\dot{z}_2 = a_1 z_3 + a_1 \alpha_2 - a_2 x_1 - a_3 x_2 + d_1 - \frac{\partial \alpha_1}{\partial x_1}x_2 - \frac{\partial \alpha_1}{\partial x_d}\dot{x}_d - \frac{\partial \alpha_1}{\partial \dot{x}_d}\ddot{x}_d \tag{17}$$

Let

$$\tau_1 = \frac{1}{a_1} \tag{18}$$

$$\alpha_2 = \hat{\tau}_1 \overline{\alpha}_2 \tag{19}$$

So, design $\overline{\alpha}_2$ is:

$$\overline{\alpha}_2 = -z_1 - \lambda_2 z_2^{2\beta-1} - s_2 z_2 + \hat{a}_2 x_1 + \hat{a}_3 x_2 - \tanh\left(\frac{z_2}{\varepsilon_1}\right)\hat{D}_1 + \frac{\partial \alpha_1}{\partial x_1}x_2 + \frac{\partial \alpha_1}{\partial x_d}\dot{x}_d + \frac{\partial \alpha_1}{\partial \dot{x}_d}\ddot{x}_d \tag{20}$$

Among them, λ_2 and s_2 are normal numbers, $\hat{a}_2$ is the estimate of a_2, $\hat{a}_3$ is the estimate of a_3, $\hat{D}_1$ is the estimate of D_1, ε_1 is an arbitrary normal number.

The Lyapunov function for constructing this step is:

$$V_2 = V_1 + \frac{1}{2}z_2^2 + \frac{a_1}{2\gamma_1}\widetilde{\tau}_1^2 + \frac{1}{2\gamma_2}\widetilde{a}_2^2 + \frac{1}{2\gamma_3}\widetilde{a}_3^2 + \frac{1}{2\gamma_4}\widetilde{D}_1^2 \tag{21}$$

Among them: $\gamma_1 - \gamma_4$ is a normal number, $\widetilde{\tau}_1 = \tau_1 - \hat{\tau}_1$, $\widetilde{a}_2 = a_2 - \hat{a}_2$, $\widetilde{a}_3 = a_3 - \hat{a}_3$, $\widetilde{D}_1 = D_1 - \hat{D}_1$.

Taking the derivative of V_2 gives us:

$$\begin{aligned}\dot{V}_2 = -\lambda_1 z_1^{2\beta} - s_1 z_1^2 + z_1 z_2 + z_2(a_1 z_3 + \overline{\alpha}_2 - a_1 \widetilde{\tau}_1 \overline{\alpha}_2 - a_2 x_1 - a_3 x_2 + d_1 - \tfrac{\partial \alpha_1}{\partial x_1}x_2 - \tfrac{\partial \alpha_1}{\partial x_d}\dot{x}_d \\ - \tfrac{\partial \alpha_1}{\partial \dot{x}_d}\ddot{x}_d) - \tfrac{a_1}{\gamma_1}\widetilde{\tau}_1\dot{\hat{\tau}}_1 - \tfrac{1}{\gamma_2}\widetilde{a}_2\dot{\hat{a}}_2 - \tfrac{1}{\gamma_3}\widetilde{a}_3\dot{\hat{a}}_3 - \tfrac{1}{\gamma_4}\widetilde{D}_1\dot{\hat{D}}_1\end{aligned} \tag{22}$$

It can be obtained from Equation (20):

$$\begin{aligned}\dot{V}_2 &= -\lambda_1 z_1^{2\beta} - \lambda_2 z_2^{2\beta} - s_1 z_1^2 - s_2 z_2^2 + a_1 z_2 z_3 - z_2 a_1 \widetilde{\tau}_1 \overline{\alpha}_2 + z_2 \widetilde{a}_2 x_1 + z_2 \widetilde{a}_3 x_2 \\ &+ z_2 d_1 - z_2 \tanh(\tfrac{z_2}{\varepsilon_1}) \hat{D}_1 - \tfrac{a_1}{\gamma_1} \widetilde{\tau}_1 \dot{\hat{\tau}}_1 - \tfrac{1}{\gamma_2} \widetilde{a}_2 \dot{\hat{a}}_2 - \tfrac{1}{\gamma_3} \widetilde{a}_3 \dot{\hat{a}}_3 - \tfrac{1}{\gamma_4} \widetilde{D}_1 \dot{\hat{D}}_1 \\ &\leq -\lambda_1 z_1^{2\beta} - \lambda_2 z_2^{2\beta} - s_1 z_1^2 - s_2 z_2^2 + a_1 z_2 z_3 - z_2 a_1 \widetilde{\tau}_1 \overline{\alpha}_2 + z_2 \widetilde{a}_2 x_1 + z_2 \widetilde{a}_3 x_2 \\ &+ \left[|z_2| - z_2 \tanh(\tfrac{z_2}{\varepsilon_1}) \right] D_1 + z_2 \tanh(\tfrac{z_2}{\varepsilon_1}) \widetilde{D}_1 - \tfrac{a_1}{\gamma_1} \widetilde{\tau}_1 \dot{\hat{\tau}}_1 - \tfrac{1}{\gamma_2} \widetilde{a}_2 \dot{\hat{a}}_2 - \tfrac{1}{\gamma_3} \widetilde{a}_3 \dot{\hat{a}}_3 - \tfrac{1}{\gamma_4} \widetilde{D}_1 \dot{\hat{D}}_1\end{aligned} \tag{23}$$

The following inequality applies to any $\varphi > 0$ and $\varphi \in R$ [34–36].

$$0 \leq |\varphi| - \varphi \tanh(\frac{\varphi}{\varepsilon}) \leq 0.2785\varepsilon \tag{24}$$

where φ is a constant, and $\varphi = e^{-(\varphi+1)}$. And that gives us $\varphi = 0.2785$.

So, Formula (23) becomes:

$$\begin{aligned}\dot{V}_2 &\leq -\lambda_1 z_1^{2\beta} - \lambda_2 z_2^{2\beta} - s_1 z_1^2 - s_2 z_2^2 + a_1 z_2 z_3 + 0.2785\varepsilon_1 D_1 - z_2 a_1 \widetilde{\tau}_1 \overline{\alpha}_2 + z_2 \widetilde{a}_2 x_1 \\ &+ z_2 \widetilde{a}_3 x_2 + z_2 \tanh(\tfrac{z_2}{\varepsilon_1}) \widetilde{D}_1 - \tfrac{a_1}{\gamma_1} \widetilde{\tau}_1 \dot{\hat{\tau}}_1 - \tfrac{1}{\gamma_2} \widetilde{a}_2 \dot{\hat{a}}_2 - \tfrac{1}{\gamma_3} \widetilde{a}_3 \dot{\hat{a}}_3 - \tfrac{1}{\gamma_4} \widetilde{D}_1 \dot{\hat{D}}_1 \\ &= -\lambda_1 z_1^{2\beta} - \lambda_2 z_2^{2\beta} - s_1 z_1^2 - s_2 z_2^2 + a_1 z_2 z_3 + 0.2785\varepsilon_1 D_1 + \tfrac{a_1}{\gamma_1} \widetilde{\tau}_1 (-z_2 \gamma_1 \overline{\alpha}_2 - \dot{\hat{\tau}}_1) \\ &+ \tfrac{1}{\gamma_2} \widetilde{a}_2 (z_2 \gamma_2 x_1 - \dot{\hat{a}}_2) + \tfrac{1}{\gamma_3} \widetilde{a}_3 (z_2 \gamma_3 x_2 - \dot{\hat{a}}_2) + \tfrac{1}{\gamma_4} \widetilde{D}_1 [z_2 \gamma_4 \tanh(\tfrac{z_2}{\varepsilon_1}) - \dot{\hat{D}}_1]\end{aligned} \tag{25}$$

According to Equation (25), the updating laws of design parameters is shown as follows:

$$\dot{\hat{\tau}}_1 = -z_2 \gamma_1 \overline{\alpha}_2 + \gamma_1 \lambda_3 \widetilde{\tau}_1^{2\beta-1} + \gamma_1 s_3 \widetilde{\tau}_1 \tag{26}$$

$$\dot{\hat{a}}_2 = z_2 \gamma_2 x_1 + \gamma_2 \lambda_4 \widetilde{a}_2^{2\beta-1} + \gamma_2 s_4 \widetilde{a}_2 \tag{27}$$

$$\dot{\hat{a}}_3 = z_2 \gamma_3 x_2 + \gamma_3 \lambda_5 \widetilde{a}_3^{2\beta-1} + \gamma_3 s_5 \widetilde{a}_3 \tag{28}$$

$$\dot{\hat{D}}_1 = z_2 \gamma_4 \tanh(\frac{z_2}{\varepsilon_1}) + \gamma_4 \lambda_6 \widetilde{D}_1^{2\beta-1} - \gamma_4 s_6 \widetilde{D}_1 \tag{29}$$

Combining Formula (26)–(29), Formula (25) becomes:

$$\begin{aligned}\dot{V}_2 \leq -\lambda_1 z_1^{2\beta} - \lambda_2 z_2^{2\beta} - a_1 \lambda_3 \widetilde{\tau}_1^{2\beta} - \lambda_4 \widetilde{a}_2^{2\beta} - \lambda_5 \widetilde{a}_3^{2\beta} - \lambda_6 \widetilde{D}_1^{2\beta} - s_1 z_1^2 - s_2 z_2^2 - a_1 s_3 \widetilde{\tau}_1^2 - s_4 \widetilde{a}_2^2 \\ - s_5 \widetilde{a}_3^2 - s_6 \widetilde{D}_1^2 + a_1 z_2 z_3 + 0.2785\varepsilon_1 D_1\end{aligned} \tag{30}$$

Step 3: The error of constructing this step is:

$$z_4 = x_4 - \alpha_3 \tag{31}$$

From Formulas (6), (14), (16) and (18)–(20), we can obtain:

$$\dot{z}_3 = z_4 + \alpha_3 - a_1 \frac{\partial \alpha_2}{\partial x_2} x_3 + a_2 \frac{\partial \alpha_2}{\partial x_2} x_1 + a_3 \frac{\partial \alpha_2}{\partial x_2} x_2 + \frac{\partial \alpha_2}{\partial x_2} d_1 - H_1 \tag{32}$$

Among them,

$$H_1 = \frac{\partial \alpha_2}{\partial x_1} x_2 + \frac{\partial \alpha_2}{\partial \hat{\tau}_1} \dot{\hat{\tau}}_1 + \frac{\partial \alpha_2}{\partial \hat{D}_1} \dot{\hat{D}}_1 + \frac{\partial \alpha_2}{\partial x_d} \dot{x}_d + \frac{\partial \alpha_2}{\partial \dot{x}_d} \ddot{x}_d + \frac{\partial \alpha_2}{\partial \ddot{x}_d} \dddot{x}_d \tag{33}$$

Construct the Lyapunov function as follows:

$$V_3 = V_2 + \frac{1}{2} z_3^2 + \frac{1}{2\gamma_5} \widetilde{a}_1^2 + \frac{1}{2\gamma_6} \widetilde{a}_{21}^2 + \frac{1}{2\gamma_7} \widetilde{a}_{31}^2 + \frac{1}{2\gamma_8} \widetilde{D}_{11}^2 \tag{34}$$

Among them, $\gamma_5-\gamma_8$ are the normal number, $\widetilde{a}_1=a_1-\hat{a}_1$, $\widetilde{a}_{21}=a_2-\hat{a}_{21}$, $\widetilde{a}_{31}=a_3-\hat{a}_{31}$, $\widetilde{D}_{11}=D_1-\hat{D}_{11}$.

Design control law:

$$\begin{aligned}\alpha_3=-\hat{a}_1z_2-\lambda_7z_3^{2\beta-1}-s_7z_3+\hat{a}_1\tfrac{\partial\alpha_2}{\partial x_2}x_3-\hat{a}_{21}\tfrac{\partial\alpha_2}{\partial x_2}x_1-\hat{a}_{31}\tfrac{\partial\alpha_2}{\partial x_2}x_2\\-\tfrac{\partial\alpha_2}{\partial x_2}\tanh[\tfrac{(\partial\alpha_2/\partial x_2)z_3}{\varepsilon_2}]\hat{D}_{11}+H_1\end{aligned}\tag{35}$$

Among them, λ_7 and s_7 are normal numbers, $\hat{a}_1$ is the estimate of a_1, $\hat{a}_{21}$ is the estimate of a_2, $\hat{a}_{31}$ is the estimate of a_3, $\hat{D}_{11}$ is the estimate of D_1, ε_2 is an arbitrary normal number.

According to Formulas (31)–(35), the following can be obtained:

$$\begin{aligned}\dot{V}_3\ &\le-\lambda_1z_1^{2\beta}-\lambda_2z_2^{2\beta}-a_1\lambda_3\widetilde{\tau}_1^{2\beta}-\lambda_4\widetilde{a}_2^{2\beta}-\lambda_5\widetilde{a}_3^{2\beta}-\lambda_6\widetilde{D}_1^{2\beta}-\lambda_7z_3^{2\beta}-s_1z_1^2\\&-s_2z_2^2-a_1s_3\widetilde{\tau}_1^2-s_4\widetilde{a}_2^2-s_5\widetilde{a}_3^2-s_6\widetilde{D}_1^2-s_7z_3^2+z_3z_4+0.2785\varepsilon_1D_1+\widetilde{a}_1z_2z_3\\&-z_3\widetilde{a}_1\tfrac{\partial\alpha_2}{\partial x_2}x_3+z_3\widetilde{a}_{21}\tfrac{\partial\alpha_2}{\partial x_2}x_1+z_3\widetilde{a}_{31}\tfrac{\partial\alpha_2}{\partial x_2}x_2-\tfrac{\partial\alpha_2}{\partial x_2}z_3\tanh[\tfrac{(\partial\alpha_2/\partial x_2)z_3}{\varepsilon_2}]\hat{D}_{11}\\&+\tfrac{\partial\alpha_2}{\partial x_2}z_3D_1-\tfrac{1}{\gamma_5}\widetilde{a}_1\dot{\hat{a}}_1-\tfrac{1}{\gamma_6}\widetilde{a}_{21}\dot{\hat{a}}_{21}-\tfrac{1}{\gamma_7}\widetilde{a}_{31}\dot{\hat{a}}_{31}-\tfrac{1}{\gamma_8}\widetilde{D}_{11}\dot{\hat{D}}_{11}\\&\le-\lambda_1z_1^{2\beta}-\lambda_2z_2^{2\beta}-a_1\lambda_3\widetilde{\tau}_1^{2\beta}-\lambda_4\widetilde{a}_2^{2\beta}-\lambda_5\widetilde{a}_3^{2\beta}-\lambda_6\widetilde{D}_1^{2\beta}-\lambda_7z_3^{2\beta}-s_1z_1^2-s_2z_2^2\\&-a_1s_3\widetilde{\tau}_1^2-s_4\widetilde{a}_2^2-s_5\widetilde{a}_3^2-s_6\widetilde{D}_1^2-s_7z_3^2+z_3z_4+0.2785\varepsilon_1D_1+\widetilde{a}_1z_2z_3-z_3\widetilde{a}_1\tfrac{\partial\alpha_2}{\partial x_2}x_3\\&+z_3\widetilde{a}_{21}\tfrac{\partial\alpha_2}{\partial x_2}x_1+z_3\widetilde{a}_{31}\tfrac{\partial\alpha_2}{\partial x_2}x_2+\left\{\left|\tfrac{\partial\alpha_2}{\partial x_2}z_3\right|-\tfrac{\partial\alpha_2}{\partial x_2}z_3\tanh[\tfrac{(\partial\alpha_2/\partial x_2)z_3}{\varepsilon_2}]\right\}D_1\\&+\tfrac{\partial\alpha_2}{\partial x_2}z_3\tanh[\tfrac{(\partial\alpha_2/\partial x_2)z_3}{\varepsilon_2}]\widetilde{D}_{11}-\tfrac{1}{\gamma_5}\widetilde{a}_1\dot{\hat{a}}_1-\tfrac{1}{\gamma_6}\widetilde{a}_{21}\dot{\hat{a}}_{21}-\tfrac{1}{\gamma_7}\widetilde{a}_{31}\dot{\hat{a}}_{31}-\tfrac{1}{\gamma_8}\widetilde{D}_{11}\dot{\hat{D}}_{11}\\&\le-\lambda_1z_1^{2\beta}-\lambda_2z_2^{2\beta}-a_1\lambda_3\widetilde{\tau}_1^{2\beta}-\lambda_4\widetilde{a}_2^{2\beta}-\lambda_5\widetilde{a}_3^{2\beta}-\lambda_6\widetilde{D}_1^{2\beta}-\lambda_7z_3^{2\beta}\\&-s_1z_1^2-s_2z_2^2-a_1s_3\widetilde{\tau}_1^2-s_4\widetilde{a}_2^2-s_5\widetilde{a}_3^2-s_6\widetilde{D}_1^2-s_7z_3^2+z_3z_4+0.2785(\varepsilon_1+\varepsilon_2)D_1\\&+\tfrac{1}{\gamma_5}\widetilde{a}_1(\gamma_5z_2z_3-\gamma_5z_3\tfrac{\partial\alpha_2}{\partial x_2}x_2-\dot{\hat{a}}_1)+\tfrac{1}{\gamma_6}\widetilde{a}_{21}(\gamma_6z_3\tfrac{\partial\alpha_2}{\partial x_2}x_1-\dot{\hat{a}}_{21})\\&+\tfrac{1}{\gamma_7}\widetilde{a}_{31}(\gamma_7z_3\tfrac{\partial\alpha_2}{\partial x_2}x_2-\dot{\hat{a}}_{31})+\tfrac{1}{\gamma_8}\widetilde{D}_{11}\left\{\gamma_8\tfrac{\partial\alpha_2}{\partial x_2}z_3\tanh[\tfrac{(\partial\alpha_2/\partial x_2)z_3}{\varepsilon_2}]-\dot{\hat{D}}_{11}\right\}\end{aligned}\tag{36}$$

Updating the laws of design parameters is shown as follows:

$$\dot{\hat{a}}_1=\gamma_5z_2z_3-\gamma_5z_3\frac{\partial\alpha_2}{\partial x_2}x_2+\gamma_5\lambda_8\widetilde{a}_1^{2\beta-1}+\gamma_5s_8\widetilde{a}_1\tag{37}$$

$$\dot{\hat{a}}_{21}=\gamma_6z_3\frac{\partial\alpha_2}{\partial x_2}x_1+\gamma_6\lambda_9\widetilde{a}_{21}^{2\beta-1}+\gamma_6s_9\widetilde{a}_{21}\tag{38}$$

$$\dot{\hat{a}}_{31}=\gamma_7z_3\frac{\partial\alpha_2}{\partial x_2}x_2+\gamma_7\lambda_{10}\widetilde{a}_{31}^{2\beta-1}+\gamma_7s_{10}\widetilde{a}_{31}\tag{39}$$

$$\dot{\hat{D}}_{11}=\gamma_8\frac{\partial\alpha_2}{\partial x_2}z_3\tanh[\frac{(\partial\alpha_2/\partial x_2)z_3}{\varepsilon_2}]+\gamma_8\lambda_{11}\widetilde{D}_{11}^{2\beta-1}+\gamma_8s_{11}\widetilde{D}_{11}\tag{40}$$

Then, Formula (36) becomes:

$$\begin{aligned}\dot{V}_3\le\ &-\lambda_1z_1^{2\beta}-\lambda_2z_2^{2\beta}-a_1\lambda_3\widetilde{\tau}_1^{2\beta}-\lambda_4\widetilde{a}_2^{2\beta}-\lambda_5\widetilde{a}_3^{2\beta}-\lambda_6\widetilde{D}_1^{2\beta}-\lambda_7z_3^{2\beta}\\&-\lambda_8\widetilde{a}_1^{2\beta}-\lambda_9\widetilde{a}_{21}^{2\beta}-\lambda_{10}\widetilde{a}_{31}^{2\beta}-\lambda_{11}\widetilde{D}_{11}^{2\beta}-s_1z_1^2-s_2z_2^2-a_1s_3\widetilde{\tau}_1^2-s_4\widetilde{a}_2^2\\&-s_5\widetilde{a}_3^2-s_6\widetilde{D}_1^2-s_7z_3^2-s_8\widetilde{a}_1^2-s_9\widetilde{a}_{21}^2-s_{10}\widetilde{a}_{31}^2-s_{11}\widetilde{D}_{11}^2+z_3z_4+0.2785(\varepsilon_1+\varepsilon_2)D_1\end{aligned}\tag{41}$$

Step 4: Finally, deal with the last differential equation in Equation (6). According to Formulas (6) and (31), the following can be obtained.

$$\dot{z}_4 = b_1x_1 - b_2x_3 - b_3x_4 + d_2 + b_4u - a_1\frac{\partial\alpha_3}{\partial x_2}x_3 + a_2\frac{\partial\alpha_3}{\partial x_2}x_1 + a_3\frac{\partial\alpha_3}{\partial x_2}x_2 - \frac{\partial\alpha_3}{\partial x_2}d_1 - H_2 \tag{42}$$

Among them,

$$\begin{aligned} H_2 = {} & \tfrac{\partial\alpha_3}{\partial x_1}x_2 + \tfrac{\partial\alpha_3}{\partial x_3}x_4 + \tfrac{\partial\alpha_3}{\partial \hat{a}_1}\dot{\hat{a}}_1 + \tfrac{\partial\alpha_3}{\partial \hat{a}_{21}}\dot{\hat{a}}_{21} + \tfrac{\partial\alpha_3}{\partial \hat{a}_{31}}\dot{\hat{a}}_{31} + \tfrac{\partial\alpha_3}{\partial \hat{D}_{11}}\dot{\hat{D}}_{11} + \tfrac{\partial\alpha_3}{\partial \hat{\tau}_1}\dot{\hat{\tau}}_1 + \tfrac{\partial\alpha_3}{\partial \dot{\hat{D}}_1}\dot{\hat{D}}_1 \\ & + \tfrac{\partial\alpha_3}{\partial \dot{\hat{\tau}}}\ddot{\hat{\tau}} + \tfrac{\partial\alpha_3}{\partial \dot{\hat{D}}_1}\ddot{\hat{D}}_1 + \tfrac{\partial\alpha_3}{\partial x_d}\dot{x}_d + \tfrac{\partial\alpha_3}{\partial \dot{x}_d}\ddot{x}_d + \tfrac{\partial\alpha_3}{\partial \ddot{x}_d}\dddot{x}_d + \tfrac{\partial\alpha_3}{\partial \dddot{x}_d}\ddddot{x}_d \end{aligned} \tag{43}$$

Let

$$\tau_2 = \frac{1}{b_4} \tag{44}$$

$$u = \hat{\tau}_2\overline{u} \tag{45}$$

Among them, $\hat{\tau}_2$ is the estimate of τ_2.

The design of finite-time adaptive controller $\overline{u}$ is:

$$\begin{aligned} \overline{u} = {} & -z_3 - \lambda_{12}z_4^{2\beta-1} - s_{12}z_4 - \hat{b}_1x_1 + \hat{b}_2x_3 - \hat{b}_3x_4 - \tanh(\tfrac{z_4}{\varepsilon_3})\hat{D}_2 + \hat{a}_{11}\tfrac{\partial\alpha_3}{\partial x_2}x_3 - \hat{a}_{22}\tfrac{\partial\alpha_3}{\partial x_2}x_1 \\ & -\hat{a}_{32}\tfrac{\partial\alpha_3}{\partial x_2}x_2 - \tfrac{\partial\alpha_3}{\partial x_2}\tanh\tfrac{(\partial\alpha_3/\partial x_2)z_4}{\varepsilon_4}]\hat{D}_{12} + H_2 \end{aligned} \tag{46}$$

Among them, λ_{12} and s_{12} are normal numbers, $\hat{b}_1$ is the estimate of b_1, $\hat{b}_2$ is the estimate of b_2, $\hat{b}_3$ is the estimate of b_3, $\hat{D}_2$ is the estimate of D_2, $\hat{a}_{11}$ is the estimate of a_1, $\hat{a}_{22}$ is the estimate of a_2, $\hat{a}_{32}$ is the estimate of a_3, $\hat{D}_{12}$ is the estimate of D_1, ε_3 and ε_4 are any normal numbers.

The Lyapunov function for constructing this step is:

$$\begin{aligned} V_4 = {} & V_3 + \tfrac{1}{2}z_4^2 + \tfrac{b_4}{2\gamma_9}\widetilde{\tau}_2^2 + \tfrac{1}{2\gamma_{10}}\widetilde{b}_1^2 + \tfrac{1}{2\gamma_{11}}\widetilde{b}_2^2 + \tfrac{1}{2\gamma_{12}}\widetilde{b}_3^2 + \tfrac{1}{2\gamma_{13}}\widetilde{D}_2^2 + \tfrac{1}{2\gamma_{14}}\widetilde{a}_{11}^2 \\ & + \tfrac{1}{2\gamma_{15}}\widetilde{a}_{22}^2 + \tfrac{1}{2\gamma_{16}}\widetilde{a}_{32}^2 + \tfrac{1}{2\gamma_{17}}\widetilde{D}_{12}^2 \end{aligned} \tag{47}$$

Among them, $\gamma_9 - \gamma_{17}$ are normal numbers, $\widetilde{\tau}_2 = \tau_2 - \hat{\tau}_2$, $\widetilde{b}_1 = b_1 - \hat{b}_1$, $\widetilde{b}_2 = b_2 - \hat{b}_2$, $\widetilde{b}_3 = b_3 - \hat{b}_3$, $\widetilde{D}_2 = D_2 - \hat{D}_2$, $\widetilde{a}_{11} = a_1 - \hat{a}_{11}$, $\widetilde{a}_{22} = a_2 - \hat{a}_{22}$, $\widetilde{a}_{32} = a_3 - \hat{a}_{32}$, $\widetilde{D}_{12} = D_1 - \hat{D}_{12}$.

Updating the laws of design parameters is shown as follows:

$$\dot{\hat{\tau}}_2 = -\gamma_9z_4\overline{u} + \gamma_9\lambda_{13}\widetilde{\tau}_2^{2\beta-1} + \gamma_9s_{13}\widetilde{\tau}_2 \tag{48}$$

$$\dot{\hat{b}}_1 = \gamma_{10}z_4x_1 + \gamma_{10}\lambda_{14}\widetilde{b}_1^{2\beta-1} + \gamma_{10}s_{14}\widetilde{b}_1 \tag{49}$$

$$\dot{\hat{b}}_2 = -\gamma_{11}z_4x_3 + \gamma_{11}\lambda_{15}\widetilde{b}_2^{2\beta-1} + \gamma_{11}s_{15}\widetilde{b}_2 \tag{50}$$

$$\dot{\hat{b}}_3 = -\gamma_{12}z_4x_4 + \gamma_{12}\lambda_{15}\widetilde{b}_3^{2\beta-1} + \gamma_{12}s_{16}\widetilde{b}_3 \tag{51}$$

$$\dot{\hat{D}}_2 = \gamma_{13}z_4\tanh(\frac{z_4}{\varepsilon_3}) + \gamma_{13}\lambda_{17}\widetilde{D}_2^{2\beta-1} + \gamma_{13}s_{17}\widetilde{D}_2 \tag{52}$$

$$\dot{\hat{a}}_{11} = -\gamma_{14}z_4\frac{\partial\alpha_3}{\partial x_2}x_3 + \gamma_{14}\lambda_{18}\widetilde{a}_{11}^{2\beta-1} + \gamma_{14}s_{18}\widetilde{a}_{11} \tag{53}$$

$$\dot{\hat{a}}_{22} = \gamma_{15} z_4 \frac{\partial \alpha_3}{\partial x_2} x_1 + \gamma_{15}\lambda_{19}\widetilde{a}_{22}^{2\beta-1} + \gamma_{15}s_{19}\widetilde{a}_{22} \tag{54}$$

$$\dot{\hat{a}}_{32} = \gamma_{16} z_4 \frac{\partial \alpha_3}{\partial x_2} x_2 + \gamma_{16}\lambda_{20}\widetilde{a}_{32}^{2\beta-1} + \gamma_{16}s_{20}\widetilde{a}_{32} \tag{55}$$

$$\dot{\hat{D}}_{12} = \gamma_{17} z_4 \frac{\partial \alpha_3}{\partial x_2} \tanh[\frac{(\partial \alpha_3/\partial x_2) z_4}{\varepsilon_4}] + \gamma_{17}\lambda_{21}\widetilde{D}_{12} + \gamma_{17}s_{21}\widetilde{D}_{12} \tag{56}$$

Among them, $\lambda_{13} - \lambda_{21}$ and $s_{13} - s_{21}$ are normal numbers.

Similar to Step 3, we end up with:

$$\begin{aligned}\dot{V}_4 \quad & \leq -\lambda_1 z_1^{2\beta} - \lambda_2 z_2^{2\beta} - a_1\lambda_3\widetilde{\tau}_1^{2\beta} - \lambda_4\widetilde{a}_2^{2\beta} - \lambda_5\widetilde{a}_3^{2\beta} - \lambda_6\widetilde{D}_1^{2\beta} - \lambda_7 z_3^{2\beta} - \lambda_8\widetilde{a}_1^{2\beta} - \lambda_9\widetilde{a}_{21}^{2\beta} \\ & -\lambda_{10}\widetilde{a}_{31}^{2\beta} - \lambda_{11}\widetilde{D}_{11}^{2\beta} - \lambda_{12} z_4^{2\beta} - b_4\lambda_{13}\widetilde{\tau}_2^{2\beta} - \lambda_{14}\widetilde{b}_1^{2\beta} - \lambda_{15}\widetilde{b}_2^{2\beta} - \lambda_{16}\widetilde{b}_3^{2\beta} - \lambda_{17}\widetilde{D}_2^{2\beta} - \lambda_{18}\widetilde{a}_{11}^{2\beta} \\ & -\lambda_{19}\widetilde{a}_{22}^{2\beta} - \lambda_{20}\widetilde{a}_{32}^{2\beta} - \lambda_{21}\widetilde{D}_{12}^{2\beta} - s_1 z_2^2 - s_2 z_2^2 - a_1 s_3\widetilde{\tau}_1^2 - s_4\widetilde{a}_2^2 - s_5\widetilde{a}_3^2 - s_6\widetilde{D}_1^2 - s_7 z_3^2 \\ & -s_8\widetilde{a}_1^2 - s_9\widetilde{a}_{21}^2 - s_{10}\widetilde{a}_{31}^2 - s_{11}\widetilde{D}_{11}^2 - s_{12} z_4^2 - b_4 s_{13}\widetilde{\tau}_2^2 - s_{14}\widetilde{b}_1^2 - s_{15}\widetilde{b}_2^2 - s_{16}\widetilde{b}_3^2 - s_{17}\widetilde{D}_2^2 \\ & -s_{18}\widetilde{a}_{11}^2 - s_{19}\widetilde{a}_{22}^2 - s_{20}\widetilde{a}_{32}^2 - s_{21}\widetilde{D}_{12}^2 + 0.2785(\varepsilon_1 + \varepsilon_2 + \varepsilon_4)D_1 + 0.2785\varepsilon_3 D_2\end{aligned} \tag{57}$$

3.2. Stability Analysis

According to references [37,38], there are the following criteria:

Theorem 1. *If there exists a continuously differentiable function $L: p \to R$, such that it satisfies the following conditions:*

(1) Lis positive definite.

(2) The existence of positive real numbers $q > 0$ and $e \in (0, 1)$, and the existence of an open neighborhood $p_0 \subset p$ containing the origin, makes the following true:

$$\dot{L}(x) + qV^e(x) \leq 0, x \in p_0 \backslash \{0\} \tag{58}$$

If $p = p_0 = R^n$, then it is globally finitely time-stable.

From Equations (12), (21), (34) and (47), we can see:

$$\begin{aligned}V_4 \quad & = \tfrac{1}{2}z_1^2 + \tfrac{1}{2}z_2^2 + \tfrac{a_1}{2\gamma_1}\widetilde{\tau}_1^2 + \tfrac{1}{2\gamma_2}\widetilde{a}_2^2 + \tfrac{1}{2\gamma_3}\widetilde{a}_3^2 + \tfrac{1}{2\gamma_4}\widetilde{D}_1^2 + \tfrac{1}{2}z_3^2 + \tfrac{1}{2\gamma_5}\widetilde{a}_1^2 + \tfrac{1}{2\gamma_6}\widetilde{a}_{21}^2 \\ & + \tfrac{1}{2\gamma_7}\widetilde{a}_{31}^2 + \tfrac{1}{2\gamma_8}\widetilde{D}_{11}^2 + \tfrac{1}{2}z_4^2 + \tfrac{b_4}{2\gamma_9}\widetilde{\tau}_2^2 + \tfrac{1}{2\gamma_{10}}\widetilde{b}_1^2 + \tfrac{1}{2\gamma_{11}}\widetilde{b}_2^2 + \tfrac{1}{2\gamma_{12}}\widetilde{b}_3^2 + \tfrac{1}{2\gamma_{13}}\widetilde{D}_2^2 \\ & + \tfrac{1}{2\gamma_{14}}\widetilde{a}_{11}^2 + \tfrac{1}{2\gamma_{15}}\widetilde{a}_{22}^2 + \tfrac{1}{2\gamma_{16}}\widetilde{a}_{32}^2 + \tfrac{1}{2\gamma_{17}}\widetilde{D}_{12}^2\end{aligned} \tag{59}$$

According to Section 3.1, V_4 is a continuously differentiable function and a positive definite function.

Let

$$V_4 = V \tag{60}$$

It can be obtained from Equation (57):

$$\begin{aligned}\dot{V}_4 \quad & \leq -\lambda_1 z_1^{2\beta} - \lambda_2 z_2^{2\beta} - a_1\lambda_3\widetilde{\tau}_1^{2\beta} - \lambda_4\widetilde{a}_2^{2\beta} - \lambda_5\widetilde{a}_3^{2\beta} - \lambda_6\widetilde{D}_1^{2\beta} - \lambda_7 z_3^{2\beta} - \lambda_8\widetilde{a}_1^{2\beta} - \lambda_9\widetilde{a}_{21}^{2\beta} \\ & -\lambda_{10}\widetilde{a}_{31}^{2\beta} - \lambda_{11}\widetilde{D}_{11}^{2\beta} - \lambda_{12} z_4^{2\beta} - b_4\lambda_{13}\widetilde{\tau}_2^{2\beta} - \lambda_{14}\widetilde{b}_1^{2\beta} - \lambda_{15}\widetilde{b}_2^{2\beta} - \lambda_{16}\widetilde{b}_3^{2\beta} - \lambda_{17}\widetilde{D}_2^{2\beta} - \lambda_{18}\widetilde{a}_{11}^{2\beta} \\ & -\lambda_{19}\widetilde{a}_{22}^{2\beta} - \lambda_{20}\widetilde{a}_{32}^{2\beta} - \lambda_{21}\widetilde{D}_{12}^{2\beta} - s_1 z_2^2 - s_2 z_2^2 - a_1 s_3\widetilde{\tau}_1^2 - s_4\widetilde{a}_2^2 - s_5\widetilde{a}_3^2 - s_6\widetilde{D}_1^2 - s_7 z_3^2 - s_8\widetilde{a}_1^2 \\ & -s_9\widetilde{a}_{21}^2 - s_{10}\widetilde{a}_{31}^2 - s_{11}\widetilde{D}_{11}^2 - s_{12} z_4^2 - b_4 s_{13}\widetilde{\tau}_2^2 - s_{14}\widetilde{b}_1^2 - s_{15}\widetilde{b}_2^2 - s_{16}\widetilde{b}_3^2 - s_{17}\widetilde{D}_2^2 - s_{18}\widetilde{a}_{11}^2 \\ & -s_{19}\widetilde{a}_{22}^2 - s_{20}\widetilde{a}_{32}^2 - s_{21}\widetilde{D}_{12}^2 + 0.2785(\varepsilon_1 + \varepsilon_2 + \varepsilon_4)D_1 + 0.2785\varepsilon_3 D_2 \\ & \leq -aV - bV^\beta + c\end{aligned} \tag{61}$$

Among them,

$$a=\min\begin{bmatrix} s_1,s_2,s_3,s_4,s_5,s_6,s_7,s_8,s_9,s_{10},s_{11}, \\ s_{12},s_{13},s_{14},s_{15},s_{16},s_{17},s_{18},s_{19},s_{20},s_{21} \end{bmatrix} \quad (62)$$

$$b=\min\begin{bmatrix} \lambda_1 2^{\beta-1},\lambda_2 2^{\beta-1},\lambda_3 2^{\beta-1},\lambda_4 2^{\beta-1},\lambda_5 2^{\beta-1},\lambda_6 2^{\beta-1},\lambda_7 2^{\beta-1}, \\ \lambda_8 2^{\beta-1},\lambda_9 2^{\beta-1},\lambda_{10} 2^{\beta-1},\lambda_{11} 2^{\beta-1},\lambda_{12} 2^{\beta-1},\lambda_{13} 2^{\beta-1},\lambda_{14} 2^{\beta-1}, \\ \lambda_{15} 2^{\beta-1},\lambda_{16} 2^{\beta-1},\lambda_{17} 2^{\beta-1},\lambda_{18} 2^{\beta-1},\lambda_{19} 2^{\beta-1},\lambda_{20} 2^{\beta-1},\lambda_{21} 2^{\beta-1} \end{bmatrix} \quad (63)$$

$$c=0.2785(\varepsilon_1+\varepsilon_2+\varepsilon_4)D_1+0.2785\varepsilon_3 D_2 \quad (64)$$

Thus, $\tilde{\tau}_1$, $\tilde{a}_2$, $\tilde{a}_3$, $\tilde{D}_1$, $\tilde{a}_1$, $\tilde{a}_{21}$, $\tilde{a}_{31}$, $\tilde{D}_{11}$, $\tilde{\tau}_2$, $\tilde{b}_1$, $\tilde{b}_2$, $\tilde{b}_3$, $\tilde{D}_2$, $\tilde{a}_{11}$, $\tilde{a}_{22}$, $\tilde{a}_{32}$, $\tilde{D}_{12}$, z_1, z_2, z_3, and z_4 are all bounded, and the estimated parameters obtained by the parameter-updating laws are all bounded. Since z_1 is bounded in formula (9), and the target value x_d is also bounded, it follows that x_1 is also bounded. In formulas (10) and (11), since z_1 and z_2 are bounded, and the control law α_1 and the target value x_d are also bounded, x_2 is also bounded. In Formulas (16), (17), (19), and (20), since $\tilde{\tau}_1$, $\tilde{a}_2$, $\tilde{a}_3$, $\tilde{D}_1$, z_1, z_2, and z_3 are bounded, x_3 is also bounded. In Formulas (31), (32), and (35), since $\tilde{a}_1$, $\tilde{a}_{21}$, $\tilde{a}_{31}$, $\tilde{D}_{11}$, z_1, z_2, z_3, and z_4 are bounded, x_4 is also bounded. In Formulas (45) and (46), since $\tilde{\tau}_2$, $\tilde{b}_1$, $\tilde{b}_2$, $\tilde{b}_3$, $\tilde{D}_2$, $\tilde{a}_{11}$, $\tilde{a}_{22}$, $\tilde{a}_{32}$, and $\tilde{D}_{12}$ are all bounded, it can be judged that u is also bounded. Therefore, all closed-loop signals are bounded.

4. Simulation Results and Discussion

In order to verify the effectiveness of the finite-time adaptive controller designed in this paper, it is simulated and analyzed in this section. The object considered in this paper is EHB gear-drive servo system, and the simulation parameters set in establishing the simulation model are shown in Table 1.

Table 1. Simulation parameters of gear-drive servo system.

Symbol	Parameter	Value	Units
J_l	moment of inertia of the driven side	0.52	kg·m^2
J_m	driving side moment of inertia	0.21	kg·m^2
C_l	viscous friction on the driven side	0.124	Nm/rad
C_m	driving side viscous friction coefficient	0.11	Nm/rad
k_d	stiffness coefficient	0.33	Nm/rad
β_l	break point	0.062	rad
β_r	break point	0.051	rad

In the simulation experiment, in order to verify the position-tracking effect of the controller, we set up two sets of experiments based on MATLAB/Simulink for verification. In experiment 1, we set the expected value of x_d to be a half-sinusoidal signal, and the sinusoidal disturbance term was added to simulate the influence of unknown disturbance on the control effect. In experiment 2, the expected value x_d was set as the ramp signal, and the proportional disturbance term was added to simulate the influence of unknown disturbance on the control effect.

4.1. Experiment 1

The control objective of this paper is the position-tracking target value of the driven side gear of the gear transmission servo system, that is, the state quantity x_1 tracks the target value x_d, and the tracking effect is shown in Figure 4. The arrows show the location-tracking details at the peak. It can be seen from the figure that the tracking effect is good, the convergence rate is fast, and the maximum error is less than 0.1 rad. The effect of gap nonlinearity and unknown interference is compensated, which can meet the demand of position accuracy.

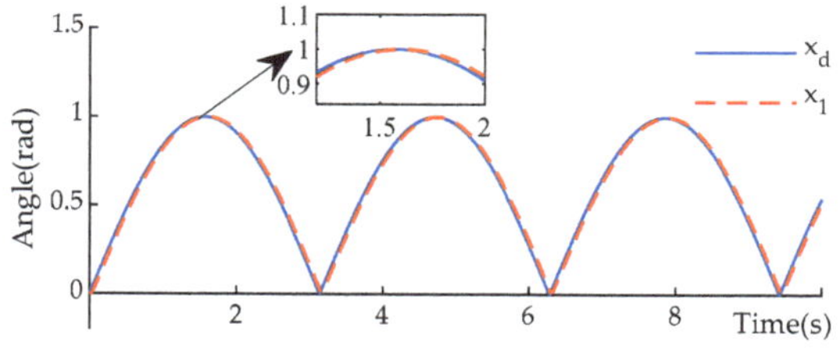

Figure 4. Position-tracking effect under half-sinusoidal input.

In order to observe the difference between the finite-time adaptive controller designed in this paper and the adaptive controller, the error z_1 under the control of each controller is compared. As shown in Figure 5. As can be seen from the figure, the finite-time adaptive controller has smaller overshoot, shorter convergence time and higher control accuracy than the adaptive controller. Compared with literature [27], the convergence rate of tracking error under the control strategy in this paper is faster.

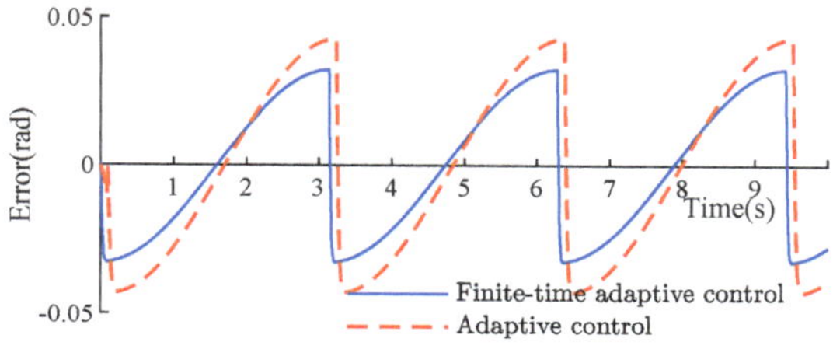

Figure 5. Error z_1 under half-sinusoidal input.

The acceleration of the driven side gear under the half-sinusoidal input condition is shown in Figure 6. From the figure we can see that the acceleration is in the ideal range.

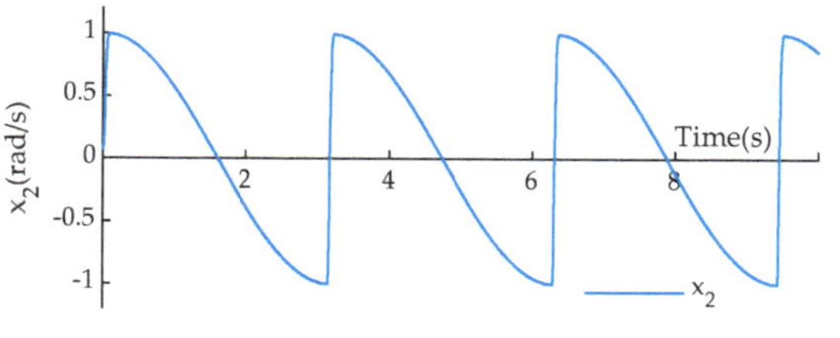

Figure 6. Acceleration of the driven side gear under half-sinusoidal input.

Figure 7 show the changes of estimated parameters of EHB gear-drive servo system. As can be seen from the figure, the estimated parameters of the system rapidly converge to a certain value and change in a small range with the change of the expected value of x_d. This means that the system parameters change while the system is running. At the same time, the validity of the parameter-update laws designed in this paper is proved.

In order to study the effect of parameter β on the convergence rate, we set the value of β to 0.6, 0.7 and 0.67, respectively. As shown in Figure 8. The arrows show the location-tracking details at the peak. As can be seen from the figure, the convergence rate is different when β takes different values. When the value of β is 0.67, the overshoot is the smallest and the convergence is the fastest.

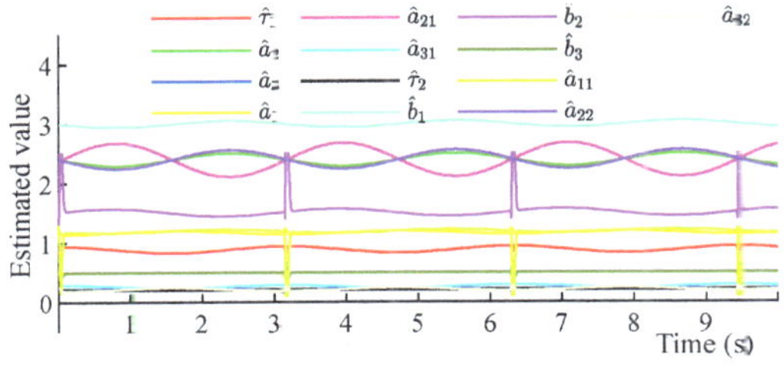

Figure 7. System parameter estimates under half-sinusoidal input.

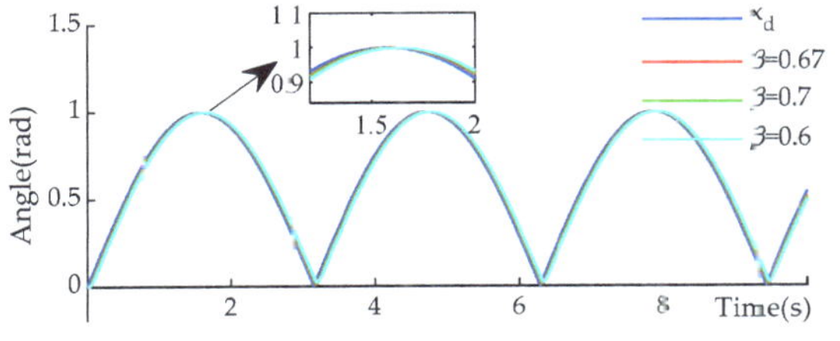

Figure 8. Influence of parameter β on tracking effect under half-sinusoidal input.

As can be seen from the above figures, the EHB gear transmission system parameters will change when working. The updated laws of the parameters designed in this paper can make the estimated parameters converge to a certain range in a certain time and adjust it adaptively according to the working conditions. The finite-time adaptive controller can effectively compensate the effects of unknown interference and dead-zone problems.

4.2. Experiment 2

In order to verify the control effect of the finite-time adaptive controller during fast braking, the expected value x_d is set as the ramp signal. The ramp signal is set to rise from 1s, with a slope of 5 and a final value of 1. As can be seen from Figure 9, the convergence rate of the driven side gear x_1 tracking expected value x_d is faster. We can see from the enlarged figure at the arrow that the maximum error is less than 0.1 rad, which can meet the requirements of control accuracy. The effect of dead zone and unknown interference is compensated effectively, and the control effect is ideal.

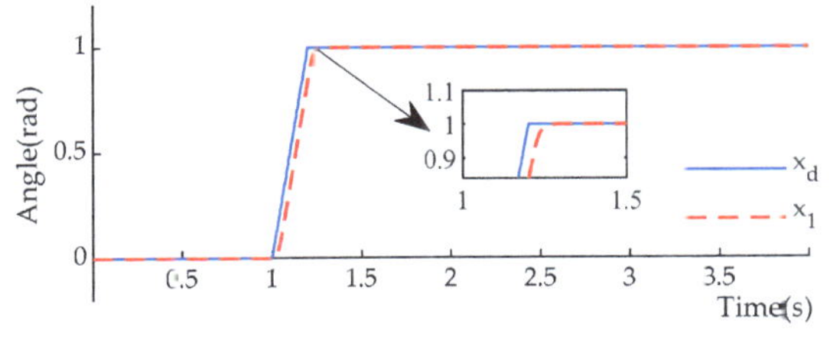

Figure 9. Position-tracking effect under ramp input.

In order to observe the superiority of the finite-time adaptive controller over the adaptive controller, the error z_1 of each controller is compared, as shown in Figure 10. It can be seen from the figure that the finite-time adaptive controller has less overshoot than the adaptive controller. The overshoot is reduced by about 0.05 rad, which is about 30 percent of the overshoot of the adaptive controller. The convergence rate of the finite-time adaptive controller is faster than that of the adaptive controller. It can be verified that the control strategy designed in this paper has better control effects and higher precision. Compared with the tracking error under adaptive control in the literature [33], the tracking error under the control strategy in this paper can be guaranteed to converge to a certain range within a certain period of time.

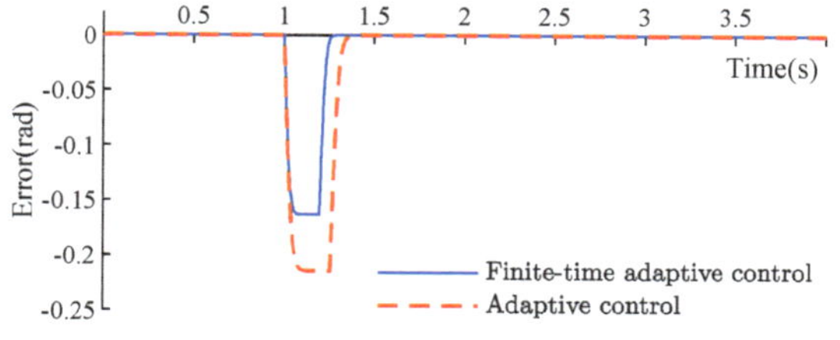

Figure 10. Error z_1 under ramp input.

The acceleration x_2 of the driven side gear is shown in Figure 11. It can be seen from the figure that the convergence rate is fast, and the control effect is excellent.

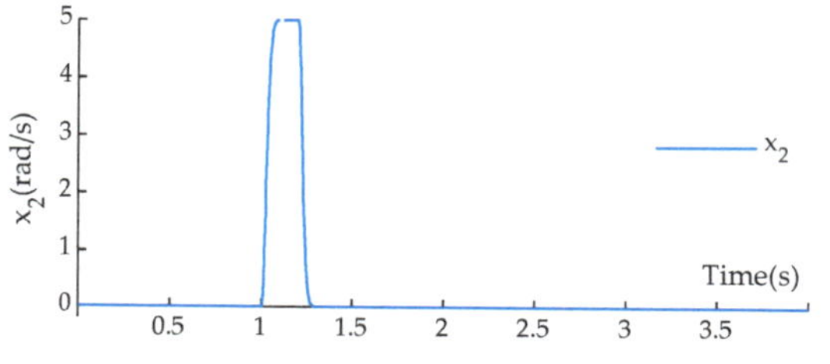

Figure 11. Acceleration of the driven side gear under ramp input.

Figure 12 shows the change in the estimated parameters of the EHB gear-drive servo system under the ramp input. As can be seen from the figure, the estimated parameters of the system will also change with the change in working conditions. A change in the expected value x_d at 1s results in a change in the parameter-update law. When the expected value x_d is stabilized, the new law of the parameters also converges to a certain range. This verifies the validity of the parameter-updating law designed in this paper.

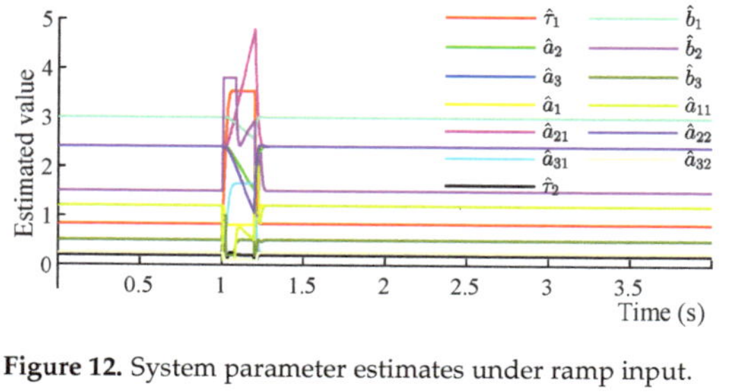

Figure 12. System parameter estimates under ramp input.

In order to study the effect of parameter β on the convergence rate under the ramp input condition, we set β values to 0.6, 0.7 and 0.67, respectively, as shown in Figure 13. It can be seen from the figure that parameter β has a significant effect on the convergence rate. As can be seen from the enlarged figure at the arrow, convergence is fastest when the value of β is 0.67. At the same time, it is proved that the finite-time adaptive controller designed in this paper is effective in improving the convergence rate.

The simulation results show that the control strategy designed in this paper can make the driven side gear of the EHB gear-drive servo system reach the desired position in a limited time and can meet the requirements of position accuracy.

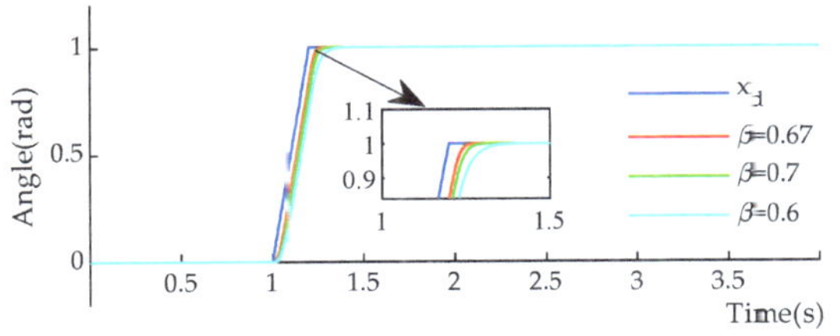

Figure 13. Influence of parameter β on tracking effect under ramp input.

5. Implementation and Results of the HIL Test

In Section 4, simulation tests are carried out to prove the effectiveness of the control strategy designed in this paper. However, the control effect in actual engineering needs further verification, so the HIL test is carried out in this section.

Figure 14 displays the HIL test platform, which is based on NI-PXI and dSPACE. Measuring the position of the gear-drive servo system presents a challenge. However, since the driving side gear is coaxial with the permanent magnet synchronous motor, we can derive the motor rotor's position information from the Hall sensor. This, in turn, provides us with the position information of the driving side gear. Utilizing the transmission relationship, we can calculate the position information of the driven side gear based on the displacement of the master-cylinder push rod. The displacement data of the master-cylinder push rod are obtained from the pulse width modulation signal generated by the digital sensor. By applying the above position relationships and conversions, we acquire the position information of the gear-drive servo system. This information serves to validate the effectiveness of the control strategy proposed in this paper.

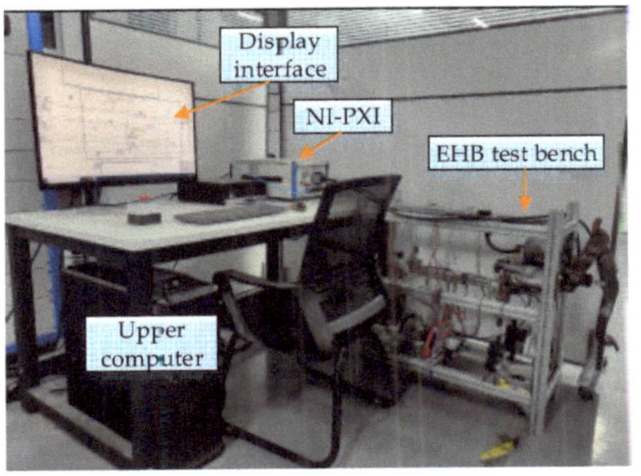

Figure 14. HIL test platform.

In HIL test 1, we set up a half-sine tracking experiment with an amplitude of 2 rad and a period of πs, as shown in Figure 15. Figure 15**a** shows the displacement y of the master-cylinder push rod, with a peak value of about 1.6 mm. The angle of the driven side gear can be obtained from the transmission relation $y = (s/2\pi)x_1$, where s is the lead of the ball screw mechanism, and $s = 5$. Figure 15b shows the rotation angle of the driving side gear, and it can be seen that its peak value is about 40 rad. It can be seen from Figure 15c that the angle of the driven side gear has a good tracking effect, and the maximum error is less than 0.1 rad. The effect caused by the dead zone is compensated effectively, the control effect is good, and the convergence rate is fast.

To further verify the effectiveness of the control strategy designed in this paper, we designed the amplitudes as 4 rad and 6 rad (see Figures 16 and 17). It can be seen from the two figures that the maximum error of the angle of the driven side gear is less than 0.2 rad when tracking the expected value, and the convergence rate is faster. Under the condition of unknown parameters, unknown disturbance and the dead zone, the finite-time adaptive controller designed in this paper can achieve a relatively stable control effect and can meet the position-tracking requirements of the EHB gear-drive servo system.

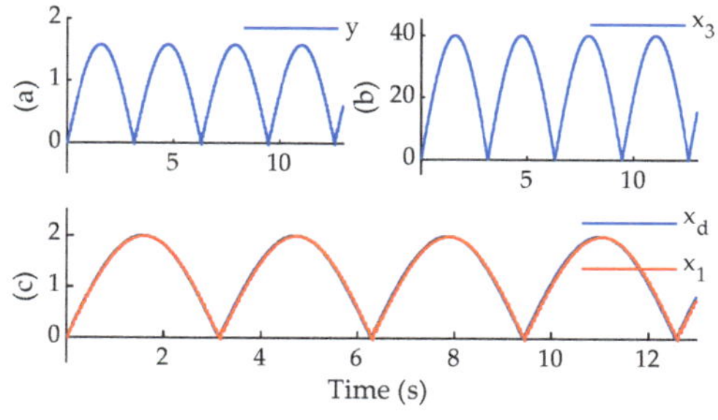

Figure 15. HIL test 1. (**a**) Master-cylinder push rod displacement (mm). (**b**) Drive side gear angle (rad). (**c**) Expected value and driven side gear angle (rad).

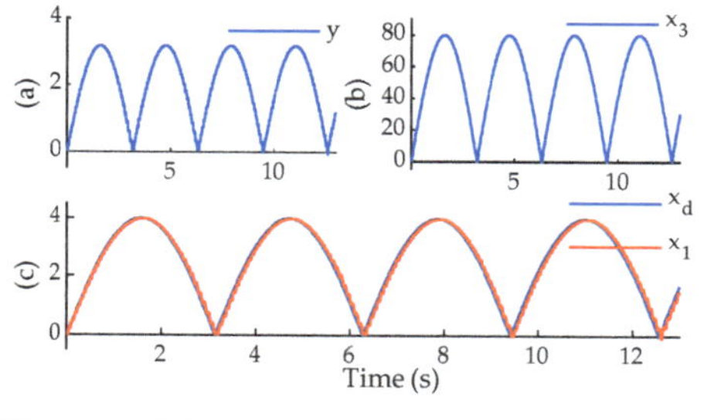

Figure 16. HIL test 2. (**a**) Master-cylinder push rod displacement (mm). (**b**) Drive side gear angle (rad). (**c**) Expected value and driven side gear angle (rad).

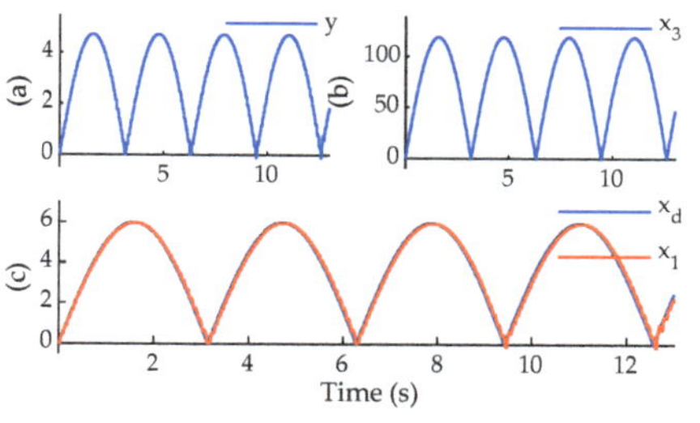

Figure 17. HIL test 3. (**a**) Master-cylinder push rod displacement (mm). (**b**) Drive side gear angle (rad). (**c**) Expected value and driven side gear angle (rad).

Through the HIL test, we have verified that the control strategy proposed in this paper can effectively compensate for the influence of unknown parameters, unknown disturbances, and dead zones. Based on the above analysis, it is evident that the control strategy presented in this paper is effective and capable of enhancing tracking accuracy and convergence rate.

6. Conclusions

In this paper, the backlash nonlinearity of EHB gear-drive servo system is analyzed, and the symmetric dead-zone model of an EHB gear drive is established. Secondly, by combining the finite-time control theory with the adaptive control theory, the parameter-update laws and control laws are designed for the uncertain parameters and dead-zone problems of the system, and boundary estimates are introduced to compensate for the unknown disturbance. Then the stability analysis is carried out to prove that all closed-loop signals are bounded. Finally, simulation and an HIL test are carried out, and the results show that the finite-time adaptive controller designed in this paper is effective.

In future work, the command filtering technology can be introduced to reduce the complexity of the operation and avoid the problem of data explosion. If the conditions permit, the actual vehicle test can be further conducted to verify the practicability of the control strategy.

Author Contributions: Conceptualization, J.W. and S.W.; writing—original draft preparation, S.W.; writing—review and editing, J.W., S.W. and Q.C.; methodology, J.W. and S.W.; data curation, S.W.; software, S.W. and F.M.; visualization, J.W.; formal analysis, Q.C. and F.M. All authors have read and agreed to the published version of the manuscript.

Funding: This research was funded by Shandong Provincial Higher School Youth Innovation Technology Project of China, grant number 2021KJ039.

Data Availability Statement: Not applicable.

Conflicts of Interest: The authors declare no conflict of interest.

References

1. Xiang, W.; Richardson, P.C.; Zhao, C.; Mohammad, S. Automobile brake-by-wire control system design and analysis. *IEEE Trans. Veh. Technol.* **2008**, *57*, 138–145. [CrossRef]
2. Milanés, V.; González, C.; Naranjo, J.E.; Onieva, E.; De Pedro, T. Electro-hydraulic braking system for autonomous vehicles. *Int. J. Automot. Technol.* **2010**, *11*, 89–95. [CrossRef]
3. Yeo, H.; Koo, C.; Jung, W.; Kim, D.; Cheon, J.S. *Development of Smart Booster Brake Systems for Regenerative Brake Cooperative Control*; 0148-7191; SAE Technical Paper; SAE International: Warrendale, PA, USA, 2011.
4. Feigel, H.-J. Integrated brake system without compromises in functionality. *ATZ Worldw.* **2012**, *114*, 46–50. [CrossRef]
5. Tan, Z.-H.; Chen, Z.-F.; Pei, X.-F.; Guo, X.-X.; Pei, S.-H. Development of integrated electro-hydraulic braking system and its ABS application. *Int. J. Precis. Eng. Manuf.* **2016**, *17*, 337–346. [CrossRef]
6. Yu, Z.; Xu, S.; Xiong, L.; Han, W. *An Integrated-Electro-Hydraulic Brake System for Active Safety*; 0148-7191; SAE Technical Paper; SAE International: Warrendale, PA, USA, 2016.
7. Zhao, J.; Hu, Z.; Zhu, B. *Pressure Control for Hydraulic Brake System Equipped with an Electro-Mechanical Brake Booster*; 0148-7191; SAE Technical Paper; SAE International: Warrendale, PA, USA, 2018.
8. Liu, S.; Zhang, L.; Niu, B.; Zhao, X.; Ahmad, A.M. Adaptive neural finite-time hierarchical sliding mode control of uncertain under-actuated switched nonlinear systems with backlash-like hysteresis. *Inf. Sci.* **2022**, *599*, 147–169. [CrossRef]
9. Cao, Z.; Niu, B.; Zong, G.; Xu, N. Small-gain technique-based adaptive output constrained control design of switched networked nonlinear systems via event-triggered communications. *Nonlinear Anal. Hybrid Syst.* **2023**, *47*, 101299. [CrossRef]
10. Theodorakopoulos, A.; Rovithakis, G.A. Guaranteeing preselected tracking quality for uncertain strict-feedback systems with deadzone input nonlinearity and disturbances via low-complexity control. *Automatica* **2015**, *54*, 135–145. [CrossRef]
11. Tarbouriech, S.; Prieur, C.; Queinnec, I. Stability analysis for linear systems with input backlash through sufficient LMI conditions. *Automatica* **2010**, *46*, 1911–1915. [CrossRef]
12. Li, M.; Chen, H.; Zhang, R. An input dead zones considered adaptive fuzzy control approach for double pendulum cranes with variable rope lengths. *IEEE/ASME Trans. Mechatron.* **2022**, *27*, 3385–3396. [CrossRef]
13. Margielewicz, J.; Gąska, D.; Litak, G. Modelling of the gear backlash. *Nonlinear Dyn.* **2019**, *97*, 355–368. [CrossRef]
14. Vörös, J. Modeling and identification of systems with backlash. *Automatica* **2010**, *46*, 369–374. [CrossRef]
15. Corradini, M.L.; Manni, A.; Parlangeli, G. Variable structure control of nonlinear uncertain sandwich systems with nonsmooth nonlinearities. In Proceedings of the 2007 46th IEEE Conference on Decision and Control, New Orleans, LA, USA, 12–14 December 2007; pp. 2023–2028.
16. Shi, Z. Global asymptotic tracking for gear transmission servo systems with differentiable backlash nonlinearity. In Proceedings of the 2014 Seventh International Symposium on Computational Intelligence and Design, Hangzhou, China, 13–14 December 2014; pp. 253–257.
17. Shi, Z.; Zuo, Z. Backstepping control for gear transmission servo systems with backlash nonlinearity. *IEEE Trans. Autom. Sci. Eng.* **2014**, *12*, 752–757. [CrossRef]
18. Zuo, Z.; Ju, X.; Ding, Z. Control of gear transmission servo systems with asymmetric deadzone nonlinearity. *IEEE Trans. Control Syst. Technol.* **2015**, *24*, 1472–1479. [CrossRef]
19. Shi, Z.; Zuo, Z.; Liu, H. Backstepping control for gear transmission servo systems with unknown partially nonsymmetric deadzone nonlinearity. *Proc. Inst. Mech. Eng. Part C J. Mech. Eng. Sci.* **2017**, *231*, 2580–2589. [CrossRef]

20. Lewis, F.L.; Tim, W.K.; Wang, L.-Z.; Li, Z. Deadzone compensation in motion control systems using adaptive fuzzy logic control. *IEEE Trans. Control Syst. Technol.* **1999**, *7*, 731–742. [CrossRef]
21. Zhou, J.; Wen, C.; Zhang, Y. Adaptive output control of nonlinear systems with uncertain dead-zone nonlinearity. *IEEE Trans. Autom. Control* **2006**, *51*, 504–511. [CrossRef]
22. Li, Y.; Tong, S. Adaptive fuzzy output-feedback control of pure-feedback uncertain nonlinear systems with unknown dead zone. *IEEE Trans. Fuzzy Syst.* **2013**, *22*, 1341–1347. [CrossRef]
23. He, X.; Lou, B.; Yang, H.; Lv, C. Robust decision making for autonomous vehicles at highway on-ramps: A constrained adversarial reinforcement learning approach. *IEEE Trans. Intell. Transp. Syst.* **2022**, *24*, 4103–4113. [CrossRef]
24. He, X.; Yang, H.; Hu, Z.; Lv, C. Robust lane change decision making for autonomous vehicles: An observation adversarial reinforcement learning approach. *IEEE Trans. Intell. Veh.* **2022**, *8*, 184–193. [CrossRef]
25. Javaid, U.; Dong, H.; Ijaz, S.; Alkarkhi, T.; Haque, M. High-performance adaptive attitude control of spacecraft with sliding mode disturbance observer. *IEEE Access* **2022**, *10*, 42004–42013. [CrossRef]
26. Xie, B.; Wang, W.; Zuo, Z. Adaptive output feedback control of uncertain gear transmission system with dead zone nonlinearity. In Proceedings of the 2018 13th IEEE Conference on Industrial Electronics and Applications (ICIEA), Wuhan, China, 31 May–2 June 2018; pp. 438–443.
27. Song, J.; Zuo, Z.; Ding, Z. Adaptive backstepping control of gear transmission systems with elastic deadzone. In Proceedings of the 2017 36th Chinese Control Conference (CCC), Dalian, China, 26–28 July 2017; pp. 878–883.
28. Wang, S.; Fan, Z.; Li, B.; Tang, X.; Chen, Z.; Zhou, Q.; Wu, J. Research on adaptive control of input time delay for vehicle electronic stability control system. In Proceedings of the 2022 6th CAA International Conference on Vehicular Control and Intelligence (CVCI), Nanjing, China, 28–30 October 2022; pp. 1–5.
29. Wu, L.-B.; Park, J.H.; Xie, X.-P.; Liu, Y.-J.; Yang, Z.-C. Event-triggered adaptive asymptotic tracking control of uncertain nonlinear systems with unknown dead-zone constraints. *Appl. Math. Comput.* **2020**, *386*, 125528. [CrossRef]
30. Hua, C.; Li, Y.; Guan, X. Finite/fixed-time stabilization for nonlinear interconnected systems with dead-zone input. *IEEE Trans. Autom. Control* **2016**, *62*, 2554–2560. [CrossRef]
31. Bao, J.; Wang, H.; Liu, P.X. Finite-time synchronization control for bilateral teleoperation systems with asymmetric time-varying delay and input dead zone. *IEEE/ASME Trans. Mechatron.* **2020**, *26*, 1570–1580. [CrossRef]
32. Yao, D.; Dou, C.; Zhao, N.; Zhang, T. Finite-time consensus control for a class of multi-agent systems with dead-zone input. *J. Frankl. Inst.* **2021**, *358*, 3512–3529. [CrossRef]
33. Wang, W.; Xie, B.; Zuo, Z.; Fan, H. Adaptive backstepping control of uncertain gear transmission servosystems with asymmetric dead-zone nonlinearity. *IEEE Trans. Ind. Electron.* **2018**, *66*, 3752–3762. [CrossRef]
34. Yin, S.; Shi, P.; Yang, H. Adaptive fuzzy control of strict-feedback nonlinear time-delay systems with unmodeled dynamics. *IEEE Trans. Cybern.* **2015**, *46*, 1926–1938. [CrossRef]
35. Wu, J.; Zhang, J.; Nie, B.; Liu, Y.; He, X. Adaptive control of PMSM servo system for steering-by-wire system with disturbances observation. *IEEE Trans. Transp. Electrif.* **2021**, *8*, 2015–2028. [CrossRef]
36. He, X.; Wu, J.; Huang, Z.; Hu, Z.; Wang, J.; Sangiovanni-Vincentelli, A.; Lv, C. Fear-neuro-inspired reinforcement learning for safe autonomous driving. *IEEE Trans. Pattern Anal. Mach. Intell.* **2023**, 1–13. [CrossRef]
37. Lu, W.; Liu, X.; Chen, T. A note on finite-time and fixed-time stability. *Neural Netw.* **2016**, *81*, 11–15. [CrossRef]
38. Li, X.; Yang, X.; Song, S. Lyapunov conditions for finite-time stability of time-varying time-delay systems. *Automatica* **2019**, *103*, 135–140. [CrossRef]

machines

MDPI

Article

Design and Implementation of a Recursive Feedforward-Based Virtual Reference Feedback Tuning (VRFT) Controller for Temperature Uniformity Control Applications

Juan Gabriel Araque [1], Luis Angel [1], Jairo Viola [2,*] and Yangquan Chen [2]

[1] Electronic Engineering, Pontifical Bolivarian University, Medellín 050031, Colombia; araquemora@gmail.com (J.G.A.); luis.angel@upb.edu.co (L.A.)
[2] School of Engineering, University of California, Merced, CA 95343, USA; ychen53@ucmerced.edu
* Correspondence: jviola@ucmerced.edu

Abstract: Data-driven controller synthesis methods use input/output information to find the coefficients of a proposed control architecture. Virtual Reference Feedback Tuning (VRFT) is one of the most popular frameworks due to its simplicity and one-shoot synthesis style based on open-loop system response for classic regulators such as PI or PID. This paper presents a recursive VRFT framework to extend VRFT into high-order controllers with more complex structures. The framework first defines a reference model and controller structure, then uses the open-loop data to compute the virtual reference and error signals, and, finally, uses these to find the controller parameters via an optimization algorithm. Likewise, the recursive VRFT controller performance is improved by adding a model-based feedforward loop to improve reference signal tracking. The recursive method is tested to design a temperature uniformity control system. The obtained results show that the recursive VRFT with a feedforward improves the system response while allowing more complex controller synthesis.

Keywords: recursive VRFT; model-reference feedforward; process control

Citation: Araque, J.G.; Angel, L.; Viola, J.; Chen, Y. Design and Implementation of a Recursive Feedforward-Based Virtual Reference Feedback Tuning (VRFT) Controller for Temperature Uniformity Control Applications. *Machines* **2023**, *11*, 975. https://doi.org/10.3390/machines11100975

Academic Editors: Jian Wu, Xiangkun He and Guangfei Xu

Received: 26 September 2023
Revised: 15 October 2023
Accepted: 17 October 2023
Published: 20 October 2023

1. Introduction

Model-based control methods cover an important part of control systems design. These are supported by mathematical process models for the design and synthesis of a closed-loop controller [1]. However, obtaining an acceptable process model can be challenging in processes with uncertainty and non-linear behavior, compromising the system's stability and performance [1–4]. This is especially relevant for processes that face variable environments that may require not only an acceptable model but also a constant parameter adaptation using optimization algorithms or adaptive control methodologies, such as [5–11]. Thus, the interest in data-driven control methods has been increasingly motivated by the increasing availability of process data given new technologies such as IoT, edge computing, or cheap sensing [12]. Some of these methods include Iterative Feedback Tuning [13], dynamic mode decomposition [14], Iterative Learning Control [15], extremum seeking [16], model-free adaptive control [17], unfalsified control [18], and simultaneous perturbation stochastic approximation [19].

One of the most studied data-driven control methods is the Virtual Reference Feedback Tuning (VRFT). It uses input–output data from a process to generate a virtual reference signal based on a desired reference model to find the optimal parameters that minimize the difference with the desired behavior [20,21].

Some applications of VRFT include motion controls, power electronics, or process control. For example, [22] employs a dual VRFT controller tuning method for velocity ripple compensation on a motor system. In [23], VRFT is employed to find the compensator for several DC–DC power converters with non-minimum phase (NMP) behavior as buck, boost, and buck–boost configurations with a reference model that compensates for the NMP zero. Likewise, [24] presents the use of VRFT on a cooperative control system with Neuroendocrine Ultrashort Feedback Controllers. Moreover, VRFT can be employed on multiple input multiple output (MIMO) configurations by defining a multivariable reference model [25] or combined with the internal model control (IMC) design method for dual temperature control [26].

In these applications, the synthesis process of VRFT follows the standard framework proposed by [20] following standard controller architectures such as PI and PID, and the model reference selection is based on prior knowledge of the system response. However, sometimes, the controller response has differences regarding its desired response. This can be explained by the fact that there are no criteria for reference model selection within the controller design method that ensure adequate closed-loop system performance.

This paper presents the design of a data-driven controller based on the Virtual Reference Feedback Tuning (VRFT) framework [3] to improve the setpoint tracking response of uniform temperature control systems.

Unlike other VRFT approaches, this work first generates the virtual error signal using the system's data-driven response with the reference model. Then, a compensator transfer function is selected (e.g., first-order, second-order, high-order), and parameters are determined through an optimization algorithm using the virtual error as input and control action as output signals. The controller synthesis process evaluates several reference model structures to select a suitable one that minimizes the system's energy consumption and steady-state error and allows a lower-order system implementation. Likewise, the proposed controller uses adaptive feedforward compensation based on the desired reference system model to improve the closed-loop system response based on the error between the reference model and system output. The proposed VRFT controller is tested for the temperature uniformity control of a Peltier thermoelectric system in simulation and experimentally using hardware in the loop configuration. The VRFT controller is evaluated against the classic PI controller tuned via pole IMC and Skogestad IMC (SIMC) [27].

The contributions of this paper are:

- Extending the VRFT framework for synthesizing different control structures through an optimization process using the virtual error signal obtained only from the system data.
- Present a method to assess reference models for VRFT controller design and different compensator architectures.
- Add a feedforward compensation to the system using the VRFT reference model to improve closed-loop response in setpoint tracking tasks.

The main benefit of the proposed algorithm is the recursive solution of the VRFT control problem, which can be extended into controllers with fixed complex structures, such as high-order compensators or non-linear controllers in SISO and MIMO configurations that include feedforward capabilities. Thus, the controller parameters are synthesized using an optimization algorithm using the virtual error that minimizes the control action and tracking error based on a desired reference model.

The manuscript is structured as follows. Section 2 introduces the foundations of VRFT controller design. Section 3 shows the recursive framework used to obtain the virtual error signal, the optimization process for VRFT synthesis, and the feedforward compensation scheme based on the reference model. Section 4 shows a case study corresponding to the temperature uniformity control system and its modeling using multiphysics simulation tools, the VRFT controller synthesis, feedforward design, and its validation in the simulation environment. Section 5 shows the experimental validation of the VRFT controller on the temperature uniformity control system. Finally, our conclusions and possibilities for future work are presented.

2. VRFT Control Framework

The VRFT framework is shown in Figure 1. Initially, an input–output $u(t), y(t)$ open-loop response dataset of the system P is obtained, and a reference model M is defined to establish the system's desired behavior. So, using the inverse reference model M^{-1}, the virtual reference $r_v(r)$ is created to compute the virtual error $e_v(t)$. Thus, after selecting a structure for the controller C_ρ, its coefficients are determined using the virtual error signal $e_v(t)$ to make the virtual control action $u_v(t)$ closer to the real system's control action $u(t)$ [28,29]. During the data acquisition process, it is important that the input signal $u(t)$ satisfies the persistent excitation conditions to ensure a representative output $y(t)$ in terms of system identification. Likewise, filtering $y(t)$ is crucial for calculation of the virtual error $e_v(t)$ and the controller's synthesis quality [21,28]. Thus, using the filtered version of the input–output signals $u_f(t) = u(t) * F(t)$ y $y_f(t) = y(t) * F(t)$, where $F(t)$ is the filter transfer function, the virtual reference signals r_{vf} and e_{vf} can be computed as (1)

$$\begin{aligned} r_{vf}(t) &= y_f(t) * M^{-1}, \\ e_{vf}(t) &= r_{vf}(t) - y_f(t). \end{aligned} \tag{1}$$

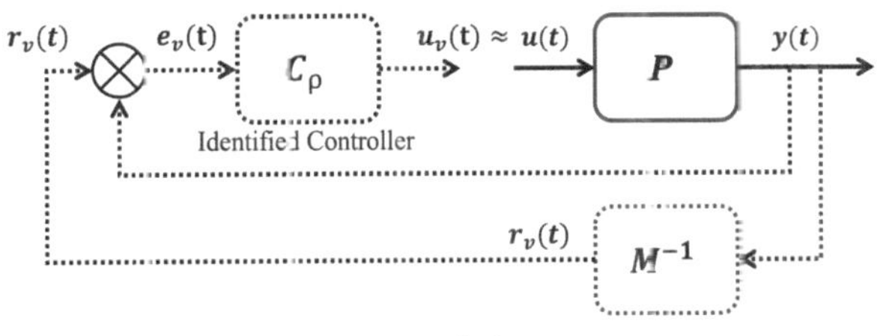

Figure 1. Classic VRFT framework [28].

So, using the filtered signals $[e_{vf}(t), u_f(t)]$, the parameters of the selected C_ρ can be obtained by minimizing (2), where N is the total amount of data samples and θ is the vector of controller parameters.

$$J_{VR} = \lim_{N \to \infty} \sum_{t=1}^{N} (u_f(t) - C_\rho(z;\theta) e_{vf}(t))^2. \tag{2}$$

Assuming that $C_\rho(z;\theta)$ is in the discrete domain z and is in the form $C_\rho(z;\theta) = \beta(z)\theta$, (2) can be rewritten as (3)

$$\begin{aligned} J_{VR} &= \lim_{N \to \infty} \textstyle\sum_{t=1}^{N} (u_f(t) - \varphi_f^T \theta))^2 \\ \varphi_f &= \beta(z) e_{vf}(t). \end{aligned} \tag{3}$$

Thus, (3) can be rewritten as a least squares problem, defined as (4), where $\widehat{\theta}_N$ is the estimated controller's parameters' vector.

$$\widehat{\theta}_N = \left(\sum_{t=1}^{N} \varphi_f(t) \varphi_f^T(t) \right)^{-1} \sum_{t=1}^{N} \varphi_f(t) u_f(t). \tag{4}$$

It is important to notice that for standard controller structures, such as PI, PID, or PD, the VRFT problem can be straightforwardly expressed using (3) and (4). However, for more complex architectures, such as high-order transfer function-based or non-linear controllers, expressing the VRFT problem is harder. Therefore, this paper proposes solving (3) by using standard optimization procedures, leveraging the virtual error and reference signals for controller synthesis.

3. Recursive VRFT Controller Synthesis

The procedure to perform a recursive VRFT controller synthesis is shown in Figure 2. The first step defines the reference model M and the controller structure C_ρ and collects the input–output data of the open-loop stable process.

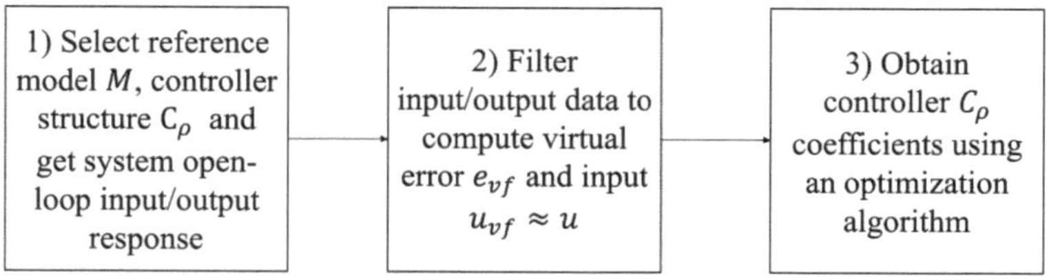

Figure 2. Recursive VRFT controller synthesis framework for a general controller architecture design based on an optimization algorithm to find controller parameters according to (4).

In the second step, the input–output data are filtered to compute the reference signal's E_v using M. This procedure is illustrated on Figure 3. According to [21,28], the virtual error E_v can be computed using $Y(s)$ and the inverse reference model M^{-1}, as shown in Figure 3a, which can be simplified using the equivalent reference system G^{-1} shown in Figure 3b. Thus, M and G can be related in the frequency domain as (5).

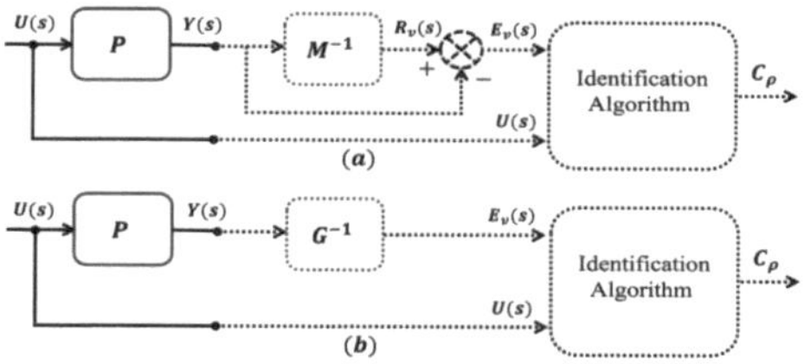

Figure 3. Recursive VRFT signal architecture. (**a**) shows the the virtual error $E_v(s)$ calculation using the closed-loop model $M(s)$, which feeds the identification algorithm to find the controller parameters. Likewise, (**b**) simplifies the computation of $E_v(s)$ by using the open-loop $G(s)$ reference model.

$$M(s) = \frac{G(s)}{1+G(s)}. \tag{5}$$

Likewise , (1) can be rewritten in the frequency domain as (6) and (7)

$$R_v(s) = Y(s)M^{-1}(s), \tag{6}$$

$$E_v(s) = R_v(s) - Y(s). \tag{7}$$

Replacing (6) with (7), the virtual error in terms of M is given by (8)

$$E_v(s) = Y(s)M^{-1}(s) - Y(s) = Y(s)[M^{-1}(s) - 1]. \tag{8}$$

Now, substituting (5) in (8), the virtual error E_v is expressed in terms of $G(s)$ by (9), as shown in Figure 3b.

$$E_v(s) = Y(s)\left(\frac{1+G(s)}{G(s)} - 1\right) = Y(s)G^{-1}(s). \tag{9}$$

Finally, in the third step, the coefficients of the controller C_ρ can be calculated using an optimization algorithm using e_{vf} and u_{vf} as input/output signals, respectively. In this manuscript, the optimization stage is performed using the Matlab system identification toolbox in order to synthesize complex high-order controllers rather than classic PI or PID regulators.

Feedforward VRFT-SISO Controller Recursive Synthesis

As can be observed in Figure 4a, if the controller is ideal, it means $C_\rho = C_{ideal} = G/P$, and the error between the desired response of the system $y_d(t)$ and the current one $y(t)$ is zero. However, the obtained controller C_ρ is not ideal in almost all the cases, making $e_a \neq 0$, showing a difference between the desired and actual system response. For this reason, the Feedforward VRFT-SISO controller shown in Figure 4b can be implemented to account for this mismatching. As can be observed, it uses the error signal $E_a(s) = Y_a(s) - Y(s)$ multiplied by $C_a(s)$ to produce an additional correction $U_a(s)$ to the control action $U(s)$ to minimize the difference between the desired and actual process response. To analyze the controller response, the system response is computed as (10):

$$Y(s) = [E(s)C_\rho(s) + E_a(s)C_a(s)]P(s) = [[R(s) - Y(s)]C_\rho(s)[Y_d(s) - Y(s)]C_a(s)]P(s), \quad (10)$$

replacing $Y_d(s) = R(s)M$ in (10),

$$Y(s)[1 + P(s)[C_\rho(s) + C_a(s)]] = R(s)[C_\rho(s) + M(s)C_a(s)]P(s). \quad (11)$$

So, the system's transfer function is given by (12)

$$\frac{Y(s)}{R(s)} = \frac{P(s)[C_\rho(s) + M(s)C_a(s)]}{1 + P(s)[C_\rho(s) + C_a(s)]}. \quad (12)$$

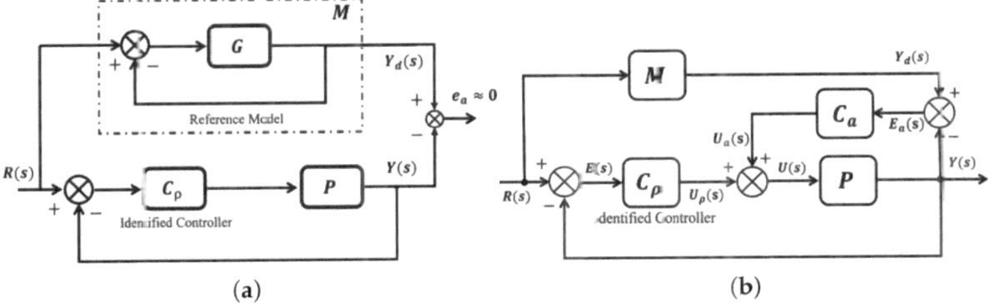

(a) (b)

Figure 4. VRFT feedforward controller proposal. (**a**) Model reference scheme of VRFT; (**b**) feedback feedforward VRFT control system.

Replacing $C_a(s) = k \cdot C_\rho(s)$ and (5) in (12), where k is a proportional gain, the closed-loop response is given as (13):

$$\frac{Y(s)}{R(s)} = \frac{P(s)C_\rho(s)[1 + (k+1)G(s)]}{[1 + G(s)][1 + P(s)C_\rho(s)(k+1)]}. \quad (13)$$

If the controller is ideal, $C_\rho(s) = C_{ideal}(s) = G(s)/P(s)$, and the feedforward will not affect the controller performance. However, if C_ρ(s) is not ideal, the feedforward improves the process response. The effect will be more noticiable for a value of k that makes $G(s)[k+1] \gg 1$, so (13) can be approximated as (14).

$$\frac{Y(s)}{R(s)} \approx \frac{P(s)C_\rho(s)[G(s)(k-1)]}{[1 + G(s)][P(s)C_\rho(s)(k+1)]} \approx \frac{G(s)}{1 + G(s)}. \quad (14)$$

4. A Case Study: Temperature Uniformity Control System

The temperature uniformity control system using real-time thermal infrared (TIR) vision feedback presented in Figure 5 is employed as a case study for VRFT controller design and implementation. The system is composed of a Peltier thermoelectric module (M1) acting as a heating/cooling element, a thermal infrared camera (M2) with TCP/IP communication protocol as a temperature feedback sensor running on a Raspberry Pi and communicated using TCP/IP communication protocol, a LattePanda Embedded board

(M3) running Windows 10 that executes Matlab in a hardware in the loop configuration (HIL). The power applied to the Peltier module is controlled using the power driver (M4) via PWM. The platform is equipped with a battery (M5) that provides the power for all of the components in the box with four hours of autonomy. A detailed description of the components of the temperature uniformity control system is presented in Table 1. Likewise, a guide for the system implementation can be found in [30,31].

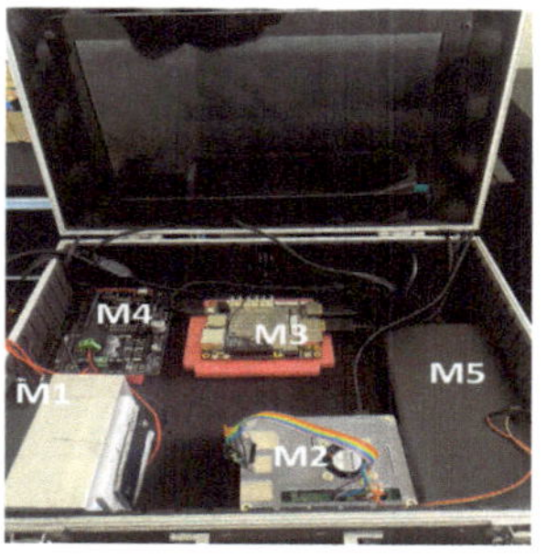

(a)

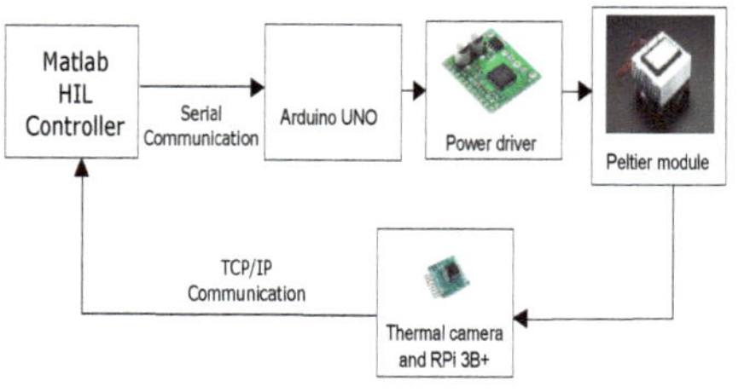

(b)

Figure 5. Case study: real-time vision feedback infrared temperature uniformity control system. (**a**) Physical asset; (**b**) HIL closed-loop controller implementation using Simulink/Arduino.

Table 1. Temperature Uniformity Control System Specifications.

Component	Features
FLIR lepton thread Infrared thermal camera	Wavelength: 8 to 14 µm Resolution: 80 × 60 pixels Accuracy: ±0.5 °C
TEC1-12706 Peltier Module	$Q_{max} = 50$ W $\Delta_{Tmax} = 75$ °C $I_{Max} = 6.4$ A $V_{max} = 16.4$ V
MC33926 DC Power Driver	Input: 0–5 V Output: 0–12 V Peak Current: 5 A
LattePanda board	5 inch Windows 10 64 bits PC Intel Atom μp 4 GB of RAM Built-in Arduino Leonardo board

The main components of the system are the Peltier thermoelectric module and the power driver, whose detailed representations are shown in Figure 6a. They consist of the cascade connection of the driver, the Peltier cell, and an aluminum plate divided in V_{ij} volumetric elements sorted in an 8×5 array, where the Peltier and plate are thermally coupled through the volume element V_{34}, providing forty measurement points equally separated by 0.03 m.

The system is simulated using Matlab/Simscape, and the corresponding simulation model is shown in Figure 6b containing three main elements: the power driver, the Peltier device, and the plate thermal circuit.

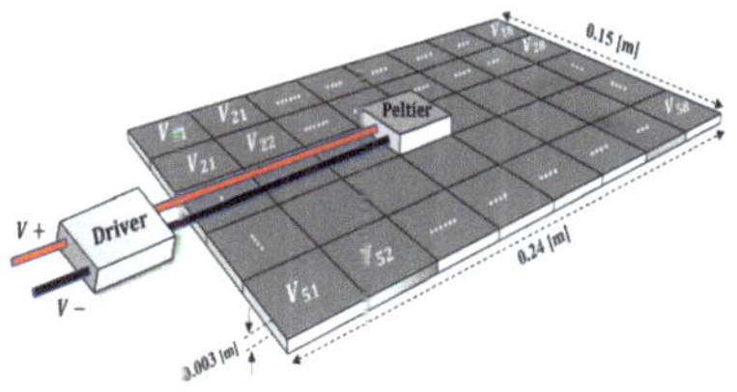

(a)

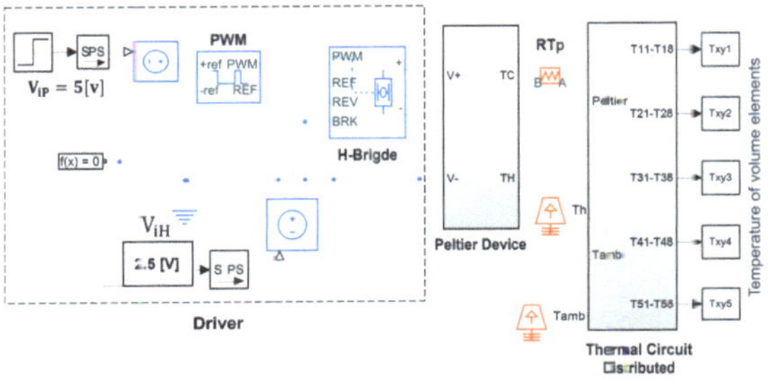

(b)

Figure 6. Temperature uniformity system process. (**a**) The 3D schematic including driver, Peltier module, and the volumetric modeled thermal plate; (**b**) Simulink/Simscape process model representation.

The power driver is configured so that the Peltier cell cools the plate when $V_{iH} < 2.5$ [V]. For $V_{iH} \geq 2.5$ [V], the Peltier acts as a heater. The electrical power input to the Peltier cell is regulated by Pulse Width Modulation (PWM), modulating the input voltage in the range $0\,[\text{V}] \leq V_{iP} \leq 5\,[\text{V}]$.

Likewise, the Peltier equivalent circuit is implemented in Simscape using discrete thermal and electrical components as shown in Figure 7. It shows the implemented thermoelectric circuit for the Peltier cell, where the gray blocks represent the Seebeck, Joule, and Peltier effects given by (15)–(17), respectively, where α is the Seebeck coefficient, T_c, q_c and T_h, q_h are temperatures and heat on the cold and hot sides of the Peltier, and I, R_E are the Peltier current and resistance. The manufacturer's datasheet specifications for the Peltier coefficients can be found in [32]. Likewise, a complete and detailed description of each parameter of the Peltier and thermal plate modeling can be found in [33]. For this paper, the values employed to simulate the Peltier are $\alpha = 0.0421$ [V/K], $R_E = 2.65$ [Ω].

$$FEM = \alpha(T_h - T_c), \tag{15}$$

$$q_j = I^2 R_E, \tag{16}$$

$$q_c = \alpha T_c I \quad q_h = \alpha T_h I. \tag{17}$$

Finally, the plate heat transfer is modeled by conduction and convection based on the finite element method in a volumetric grid as shown in Figure 8a, where each element is modeled as a thermal resistive–capacitive (RC) network, shown in Figure 8b, with a constant temperature per element volumetric element, as shown in [33]. The Matlab/Simscape implementation of the plate thermal circuit is shown in Figure 8b, where each volume element V_{nm} is represented with a thermal capacitance C_{Tnm} and resistance R_{Tnm}.

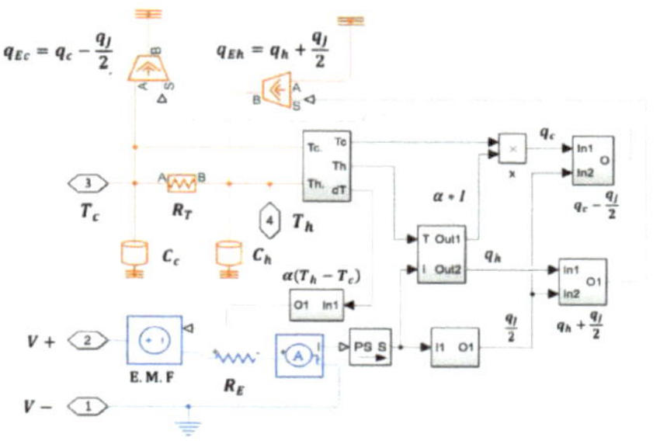

Figure 7. Peltier RC thermal circuit equivalent assembly in Simulink/Simscape including Peltier, Seebeck, and Joule effects.

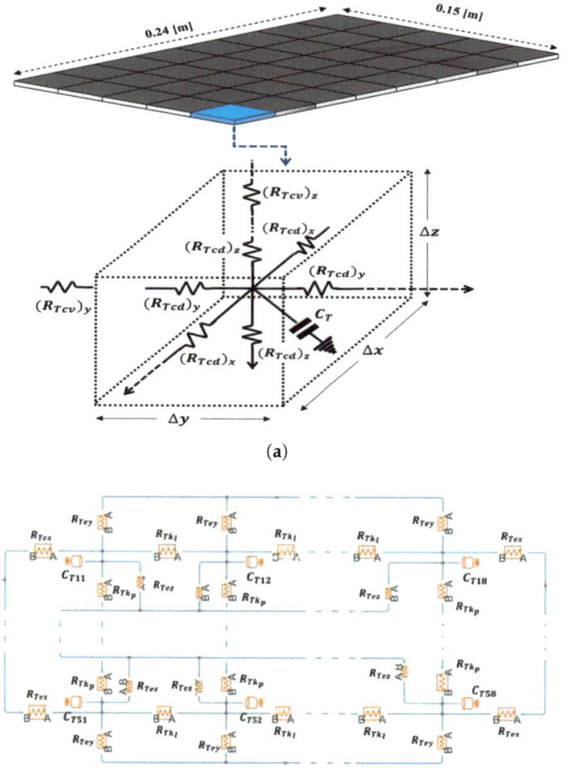

(**a**)

(**b**)

Figure 8. Thermal plate interaction modeling: (**a**) based on RC network for individual volumetric elements and (**b**) Simulink/Simscape model.

4.1. Feedforward VRFT Controller Design

The VRFT controller synthesis is performed using the procedure shown in Figure 2. Initially, the reference models are defined, which are given by (18):

$$M_1(s) = \frac{w_n}{s+w_n},\ M_2(s) = \frac{w_n^2}{s^2+2w_ns+w_n^2},\ M_3(s) = \frac{w_n^2(T_zs+1)}{s^2+2w_ns+w_n^2}. \tag{18}$$

By replacing (18) in (8) and (9), the open-loop transfer functions $G_{1,2,3}$ for the reference models $M_{1,2,3}$ are given by (19):

$$G_1(s) = \frac{w_n}{s},\ G_2(s) = \frac{w_n^2}{s(2w_n + s)},\ G_3(s) = \frac{w_n^2(T_z s + 1)}{s^2 + w_n(2 - w_n T_z)s}. \tag{19}$$

Once the models are defined, the open-loop response of the system is simulated by applying the PRBS signal shown in Figure 9a that sets a ±100% duty cycle in random periods to the Peltier system via the power driver to produce the output signal $y(t)$ shown in Figure 9b.

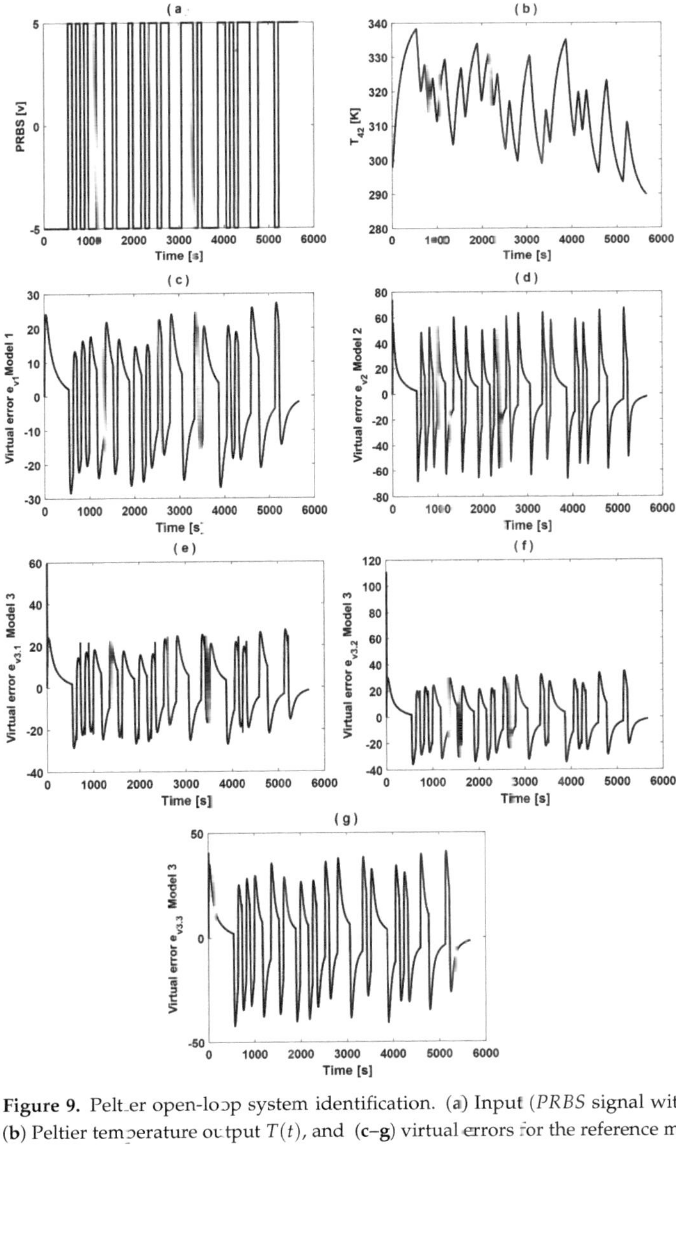

Figure 9. Peltier open-loop system identification. (**a**) Input (*PRBS* signal with 1/10 Hz periods), (**b**) Peltier temperature output $T(t)$, and (**c**–**g**) virtual errors for the reference models $G_{1-5}(s)$.

So, the virtual error signal e_{vr} is computed by evaluating the response of $G_{1,2,3}$ using $y(t)$ as input for $w_n = 11.66 \times 10^{-3}$ and $T_z = [\frac{1}{1.5w_n}, \frac{1}{2w_n}, \frac{1}{2.5w_n}]$, resulting in five virtual error signals $e_{vr_{1,\dots,5}}$, shown in Figure 9c–g. In this case, a first-order compensator is selected as controller $C_{\rho n}$'s structure, given by (20), where α, β are the controller coefficients. It is important to notice that for the proposed application, a lag-type compensator is selected ($\alpha_0 < \beta_0$) to achieve a smooth response.

$$C_{\rho n} = \frac{\alpha_1 s + \alpha_0}{\beta_1 s + \beta_0}. \tag{20}$$

Finally, based on the virtual error signals $e_{vr_{1-5}}$, the VRFT controller can be determined via optimization, which is performed using Matlab's system identification toolbox, using as the input signal the virtual error $e_{vr_{1,\dots,5}}$ and the input signal $u(t)$ as output following the logic shown in Figure 3. Thus, five controllers are obtained which are defined in (21) with an adjustment of 59.51%, 64.42%, 60.48%, 62.13%, and 59.95%, respectively.

$$\begin{aligned}
C_{\rho 1} &= \frac{-278.61 \cdot 10^{-3} s - 1.181 \cdot 10^{-3}}{s + 150.48 \cdot 10^{-6}}, \quad C_{\rho 2} &= \frac{-136.39 \cdot 10^{-3} s - 1.17 \cdot 10^{-3}}{s + 260.29 \cdot 10^{-6}}, \\
C_{\rho 3.1} &= \frac{-277.86 \cdot 10^{-3} s - 1.35 \cdot 10^{-3}}{s + 175.89 \cdot 10^{-6}}, \quad C_{\rho 3.2} &= \frac{-225.23 \cdot 10^{-3} s - 1.281 \cdot 10^{-3}}{s + 172.61 \cdot 10^{-6}}, \\
C_{\rho 3.3} &= \frac{-196.79 \cdot 10^{-3} s - 1.27 \cdot 10^{-3}}{s + 190.58 \cdot 10^{-6}}.
\end{aligned} \tag{21}$$

The controllers (21) are evaluated using the simulation model shown in Figure 10, which incorporates the feedforward compensation that uses the reference model and the controller transfer function with a tuning gain k as shown in Figure 4b. The closed-loop responses of the Peltier system and the control actions for the VRFT controllers can be observed in Figure 11. When $k = 0$, the feedforward is not active on the system, indicating an error regarding the desired trajectory y_d. Likewise, as k increases for $k \in [0; 2; 10]$, the error is minimized and becomes closer to the reference temperature.

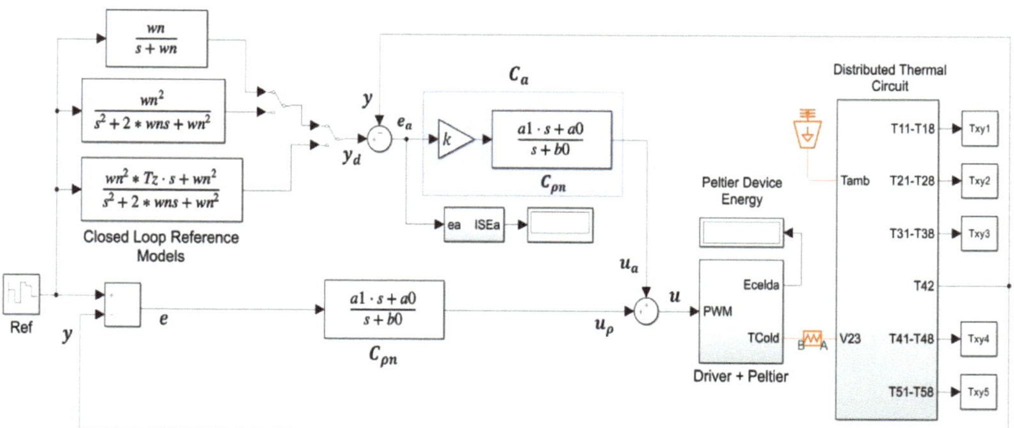

Figure 10. Feedforward VRFT controller implemented in Simulink composed by the reference models $M_{1,\dots,5}$, the feedforward error compensation C_a, and the temperature uniformity control system.

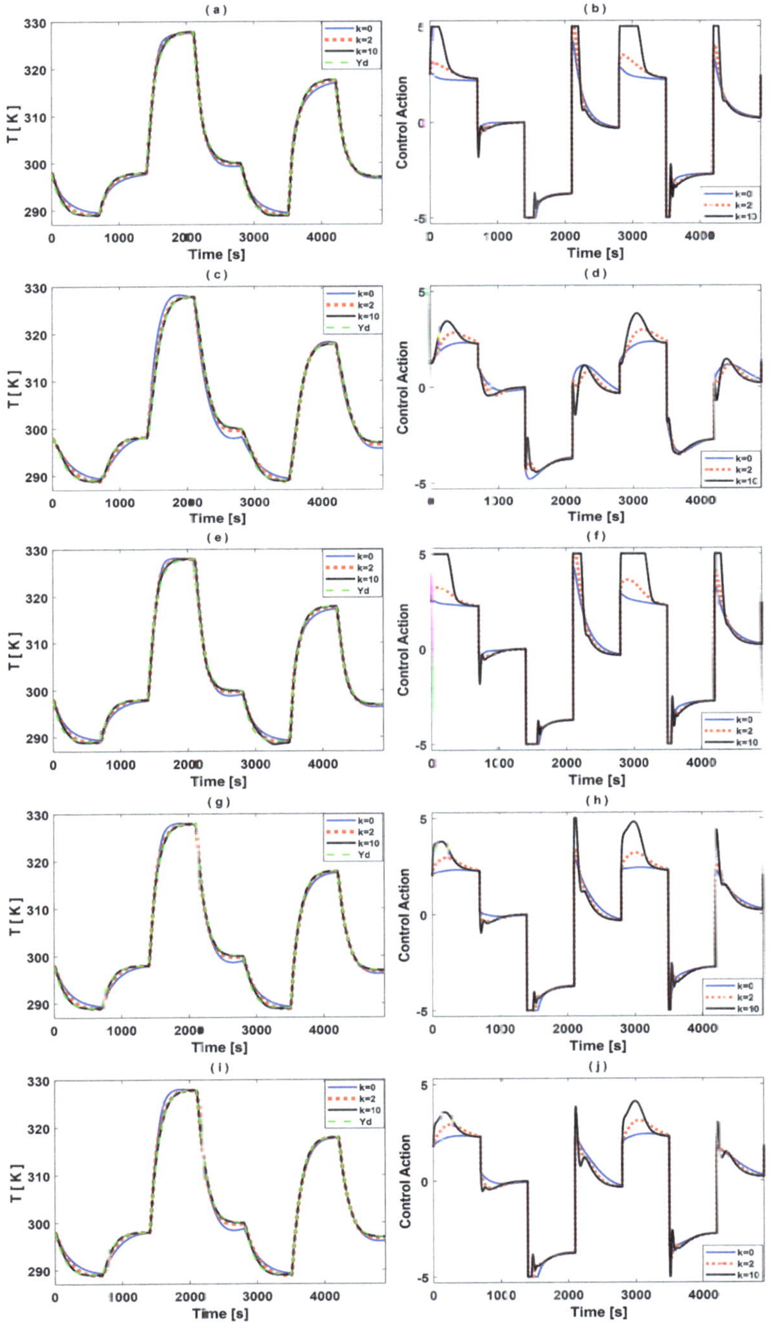

Figure 11. Temperature performance of controllers $C_{\rho_{1,\ldots,5}}$ (**a**–**e**) and their control actions (**f**–**j**) for different feedforward gains k. The feedforward compensates the tracking error but induces control action saturation as k increases.

The Integral Square Error ISE for the adaption error e_a and the Peltier cumulative energy u indices E_p given by (22) and (23) are used as key performance indicators (KPI) for the VRFT controller's quantitative analysis, where m is the number of samples. Likewise, the performance quantification against the feedforward gain is defined with the indices PD_{ISE} and PI_{EP}, defined in (24)–(25). These indices compute the percentage ratio between the ISE and E_P without feedforward $ISE(0)$, $E_P(0)$ and those related to the current feedforward employed $ISE(k)$ $E_P(k)$. The ISE and ITAE performance indices are standard KPIs to evaluate the transient response of the temperature in the Peltier cell given by the proposed VRFT compensators' C_ρ and the PI-IMC and PID-IMC controllers used as benchmark strategies. They also measure the tracking of the system to the desired response given by the reference model. Likewise, E_p is a performance index used to quantify the energy consumption of the Peltier Cell. Thus, the proposed indices can be used to provide a metric to determine a reasonable trade-off between the desired model's reference tracking and the energy consumption to find a suitable tuning rule for the feedforward gain k. Table 2 presents the performance indices for the synthesized VRFT controllers $C_{\rho_{1,\dots,5}}$. For example, in the controller $C_{\rho_{3.3}}$, for $k = 0$, the adaptation error ISE $ISE_a(0) = 6008.65$ is bigger for any k. For values of k other than zero, for example, with $k = 2$, the adjustment to the desired response is improved by 75%, but the energy consumption increases by 8.7%.

The effect of the feedforward gain k is further analyzed using the ISE and E_P indices as shown in Figure 12. As can be observed, the sweeping of k for $1 \leq k \leq 20$ for the controller $C_{\rho 3.3}$ shows that more energy is required as the adjustment error is minimized. Moreover, for the five controllers $C_{\rho_{1,\dots,5}}$, it is noticeable that controller $C_{\rho 1}$ provides the best fit to the reference model without a feedforward. On the other hand, the controller $C_{\rho 2}$ presents the greatest deviation compared to the desired response, although it should be noted that it has the lowest energy consumption. Likewise, it is noticeable that the most significant changes in the adjustment to the desired response as k increases occur in the interval $0 < k \leqslant 6$. However, the performance can be affected for considerably large values of k, which can generate problems with noise and saturation of the control action. Additionally, it can be observed that for $k = 4$, the controller $C_{\rho 3.3}$ achieves the best performance.

On the other hand, the VRTF controller's robustness is evaluated in the presence of external disturbance in the control action at $t = [1500, 2500, 3500, 4500]$ s, whose response is shown in Figure 13, and its quantitative evaluation is presented in Table 3. It can be observed that increasing k improves the tracking response and disturbance rejection. However, by increasing k, the control action becomes more aggressive, which, in turn, results in higher energy consumption.

$$ISE = \frac{1}{m}\sum_{j=0}^{m} e_a^2(j), \tag{22}$$

$$E_P = \frac{1}{m}\sum_{j=0}^{m} u^2(j), \tag{23}$$

$$PD_{ISE} = \left| \frac{ISE(0) - ISE(k)}{ISE(0)} \right| 100\%, \tag{24}$$

$$PI_{EP} = \left| \frac{E_p(0) - E_p(k)}{E_p(0)} \right| 100\%. \tag{25}$$

Table 2. VRFT controllers quantitative performance for different feedforward gain k on setpoint tracking tasks.

Controller	Gain k	$ISEa$	PD_{ISEa} [%]	E_P [J]	PI_{E_P} [%]
$C_{\rho 1}$	0	5227.24	0.0	46,981.76	0.0
	2	1997.81	61.8	54,220.91	15.4
	5	1391.21	73.4	61,055.75	30.0
	10	1328.20	74.6	70,605.25	50.3
$C_{\rho 2}$	0	12,341.97	0.0	41,406.34	0.0
	2	2745.85	77.8	43,332.89	4.7
	5	1028.79	91.7	45,161.89	9.1
	10	406.08	96.7	46,495.30	12.3
$C_{\rho 3.1}$	0	5440.11	0.0	49,806.09	0.0
	2	2250.38	58.6	58,327.48	17.1
	5	1733.08	68.1	68,945.06	38.4
	10	1724.12	68.3	78,772.28	58.2
$C_{\rho 3.2}$	0	5225.76	0.0	45,982.25	0.0
	2	1466.74	71.9	50,930.10	10.8
	5	666.45	87.2	54,343.65	18.2
	10	409.33	92.2	59,674.20	29.8
$C_{\rho 3.3}$	0	6008.65	0.0	44,689.47	0.0
	2	1503.10	75.0	48,560.12	8.7
	5	578.69	90.4	50,752.74	13.6
	10	264.06	95.6	52,740.13	18.0

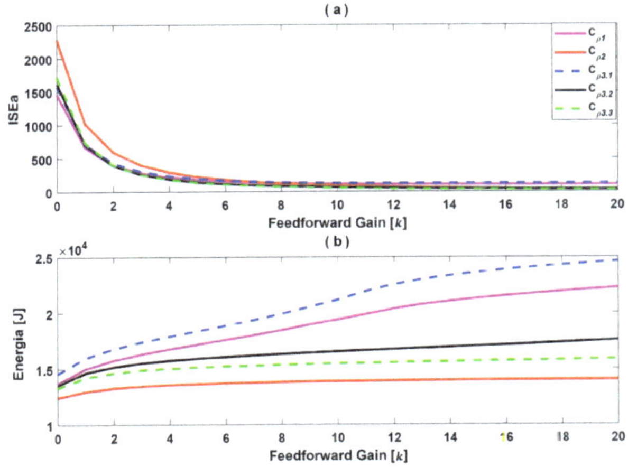

Figure 12. Sweeping of the adjustment error gain k vs. ISE and E_p for the synthesized controllers C_ρ. (**a**) explains the variation of ISE against k showing a error reduction for higher values of k. Likewise, (**b**) describes the variation of E_p against k showing an evident increment of energy consumption for higher k. Therefore, an acceptable performance is reached for $6 \leq k \leq 8$ with a good trade-off between error minimization and energy consumption.

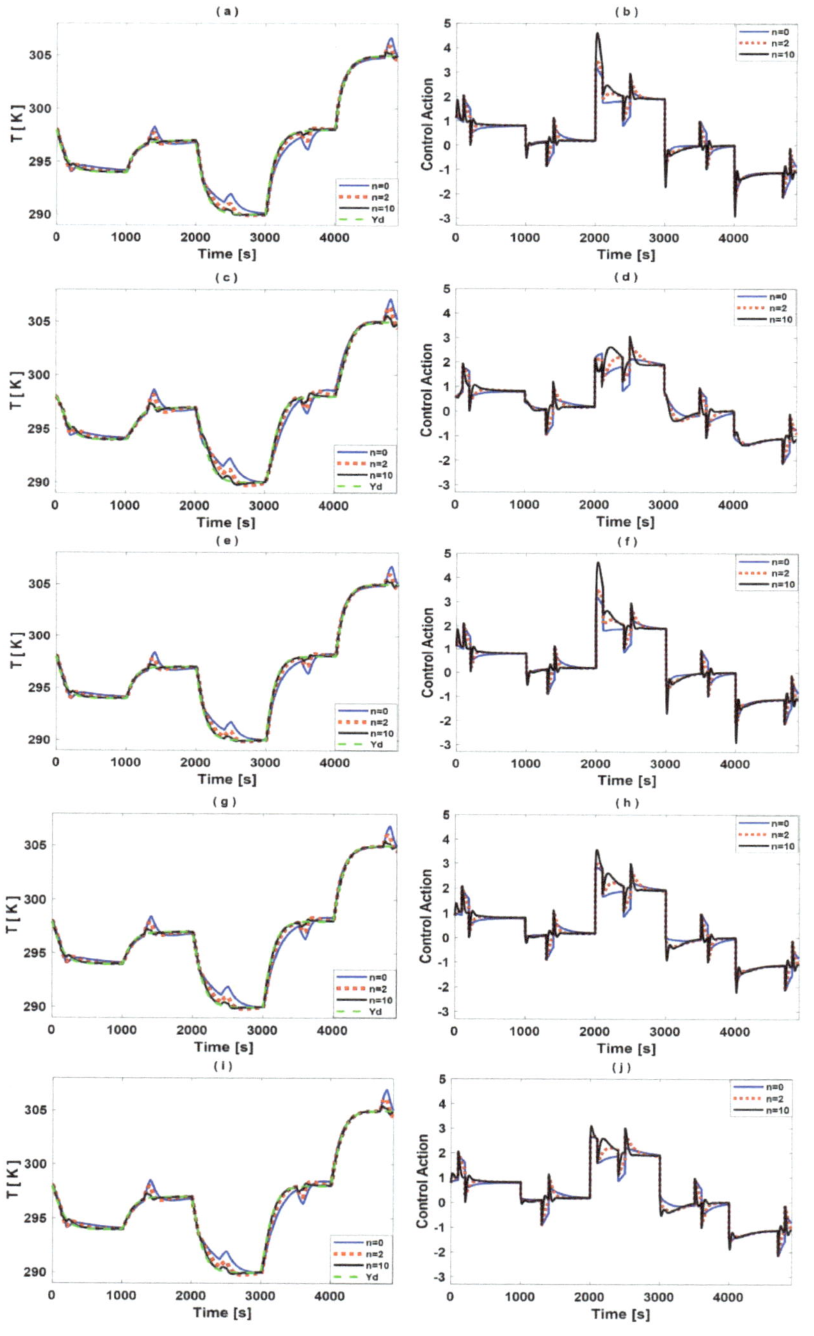

Figure 13. Temperature performance of controllers $C_{\rho_{1,\ldots,5}}$ (**a–e**) and their control actions (**f–j**) for different feedforward gains k in the presence of external disturbance in the control action at $t = [1500, 2500, 3500, 4500]$ s. The feedforward provides a disturbance rejection but increases the rejection energy required as k is bigger.

Table 3. VRFT controllers quantitative performance for different feedforward gain k on setpoint tracking tasks in the presence of thermal load disturbance.

Controller	Gain k	$ISEa$	PD_{ISEa} [%]	E_P [J]	PI_{E_P} [%]
$C_{\rho 1}$	0	1982.44	0.0	10,466.43	0.0
	2	496.87	74.9	11,650.06	11.3
	5	176.86	91.1	12,217.27	16.7
	10	68.73	96.5	12,947.27	23.7
$C_{\rho 2}$	0	2534.21	0	9702.82	0.0
	2	918.82	63.7	10,496.29	8.2
	5	420.37	83.4	10,647.11	9.7
	10	179.47	92.9	10,708.07	10.4
$C_{\rho 3.1}$	0	1849.71	0.0	10,893.32	0.0
	2	467.95	74.7	12,094.35	11.0
	5	169.15	90.9	12,697.67	16.6
	10	66.78	96.4	13,523.16	24.1
$C_{\rho 3.2}$	0	2069.79	0.0	10,468.95	0.0
	2	562.13	72.8	11,443.26	9.3
	5	208.87	89.9	11,744.04	12.2
	10	77.59	96.3	11,944.28	14.1
$C_{\rho 3.3}$	0	2153.58	0.0	10,299.65	0.0
	2	625.07	71.0	11,186.99	8.5
	5	243.00	88.7	11,399.55	10.7
	10	91.99	95.7	11,485.10	11.5

4.2. VRFT against PID Controllers Comparison

The obtained VRFT controllers are compared against three standard PID controllers: one tuned using the VRFT methodology proposed by [34] and PI and PID controllers tuned using the Internal Model Control (IMC) Method. For the PID-VRFT controller tuning, the system input/output response, shown in Figure 9a,b, is used with the reference model (26) and the PID controller structure (27). After the synthesis process, the controller parameters are $K_p = -18.08 \cdot 10^{-3}$, $K_d = -2.2 \cdot 10^{-3}$, and $K_i = -365.03 \cdot 10^{-6}$.

$$M(z^{-1}) = \frac{2.34 \cdot 10^{-3} + 67.27 \cdot 10^{-6} z^{-1} - 2.27 \cdot 10^{-3} z^{-2}}{1 - 1.977z^{-}1 + 0.9769z^{-2}}, \tag{26}$$

$$C_{VRFT-PID} = K_p - \frac{K_i}{s} + \frac{K_d s}{1 + T_s s}. \tag{27}$$

The design of the PI-IMC and PID-IMC controllers follows the method proposed in [35,36]. As shown in Figure 14a, the controller G_c is the inverse of the process model $\tilde{P}^{-1}$ followed for a nth-order filter to ensure a proper controller implementation, making λ the only tuning parameter, where $G(s)$ is given by (28) and $\tilde{P}$ is the approximated process model.

$$G_c(s) = \frac{\tilde{P}}{(\lambda s + 1)^n}. \tag{28}$$

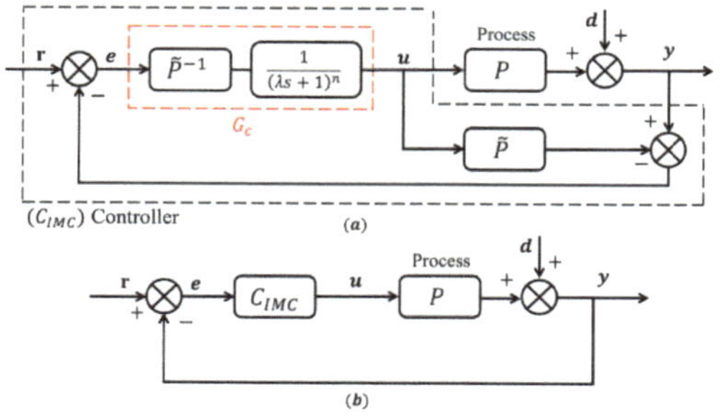

Figure 14. IMC controller framework design. (**a**) G_c is the inverse of the estimated process $\tilde{P}^{-1}$ plus a low-pass filter to obtain a proper and realizable controller. (**b**) shows the rearrangement of the IMC controller as a PI feedback regulator C_{IMC} given by (32).

So, (28) can be simplified, as shown in Figure 14b, where the resulting controller C_{IMC} can be represented as (29), according to [36].

$$C_{IMC}(s) = \frac{G_c}{(1-\tilde{P}G_c)} = \frac{\tilde{P}^{-1}}{(\lambda s+1)^n - 1}. \quad (29)$$

Thus, for the PI-IMC controller design, $\tilde{P_1}$ is selected as a first-order system calculated using the system response shown in Figure 9, resulting in (30),

$$\tilde{P_1} = \frac{a_0}{b_1 s + 1} = \frac{-5.58}{211.51s+1}, \quad (30)$$

which can be rewritten as (31)

$$C_{IMC_1}(s) = -\left(\frac{b_1}{a_0\lambda_1} + \frac{1}{a_0\lambda_1 s}\right). \quad (31)$$

and, considering the PI controller form in (32), its terms can be compared with (31) to obtain the PI controller's parameters as $K_{p1} = \frac{b_1}{a_0\lambda_1}$, $I_1 = \frac{1}{b_1}$, resulting in $k_{p1} = -261.21e^{-3}$, $k_{i1} = 4.727e^{-3}$ for $\lambda = 145$.

$$PI(s) = K_{p1}\left(1 + \frac{I_1}{s}\right). \quad (32)$$

For the PID-IMC, a second-order model of the process is obtained, which is given by (33)

$$\tilde{P_2} = \frac{c_0}{s^2 + d_1 s + 1 + d_0} = \frac{-2.929 \cdot 10^{-3}}{s^2 + 108.97 \cdot 10^{-3}s + 545.65 \cdot 10^{-6}}, \quad (33)$$

and replacing (33) in (29) with $n = 1$, the IMC controller can be expressed as (34)

$$C_{IMC_2}(s) = -\frac{s^2 + d_1 s + d_0}{(c_1 s + c_0)\lambda_2 s} = -\left(\frac{s^2 + d_1 s + d_0}{c_0\lambda_2 s}\right)\left(\frac{1}{\frac{c_1}{c_0} + 1}\right). \quad (34)$$

Thus, comparing the PID controller form given by (35) with (34), the PID controller's gains can be calculated as $K_{p2} = \frac{d_1}{2c_0\lambda_2}$, $T_{d2} = \frac{1}{d_1}$; $I_2 = \frac{d_0}{d_1}$; $T_{f2} = \frac{\lambda_2}{2}$ resulting in $K_{p2} = -206e^{-3}$, $I_2 = 9.176$, $T_{d2} = 5e^{-3}$, and $T_{f2} = 45$ for $\lambda_2 = 45$.

$$PID(s) = K_{p2}\left(1 + \frac{I_2}{s} + T_{d2}s\right)\left(\frac{1}{T_{f2}s + 1}\right). \quad (35)$$

The performances of the PI-IMC and PID-IMC controllers are contrasted against the VRFT controller $C_{\rho 3.3}$ in three operating conditions: nominal, random noise, and external disturbance (adding external thermal load) using the ISE (22), $ITAE$ (36), and E_p (23) performance indices. The heat load disturbance used for the testing, shown in Figure 15c, modifies the thermal properties of the V_{ij} volumetric element. The thermal disturbance added to each V_{ij} element is given by $C_T = mc_p$, where m is a thermal mass of 1 kg and c_p is the specific heat of $c_p = 3671$ J/kg/K. The volumetric elements affected are $V_{ij} = [V_{24}, V_{25}, V_{36}, V_{33}]$, and the disturbance is set at each element at the times $t = [1500, 2500, 3500, 4500]$ s, respectively.

The controller's response to different temperature setpoints without disturbance (nominal) is shown in Figure 15a and Table 4. It is observed how, with the feedforward, the VRFT control performs better setpoint tracking compared to the PID-VRFT, PI-IMC and PID-IMC controllers. Likewise, the VRFT control action reacts faster to changes in process operation with an energy consumption similar to the other control systems. In the presence of random noise, the controllers response is shown in Figure 15b and Table 5. In this case, the feedforward VRFT is more sensitive to random noise compared with the PI-IMC, and PID-IMC controllers. Finally, when thermal load disturbance at $t = [1500, 2000, 3000, 4000]$ s is added to the system, the controller's response is given by Figure 15c and Table 6. It is observed how the feedforward VRFT controller rejects the disturbances more efficiently than the PI-IMC and PID-IMC controllers but requires a greater energy consumption to perform the disturbance rejection.

$$ITAE = \frac{t_s}{m} \sum_{j=0}^{m} e_a^2(j). \quad (36)$$

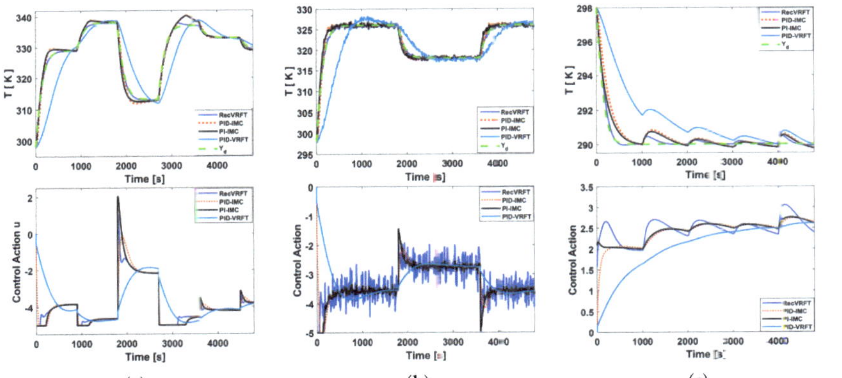

Figure 15. Recursive feedforward VRFT and PI/PID controllers' response: (**a**) temperature setpoint tracking; (**b**) random noise in the feedback loop; and (**c**) external disturbance produced by thermal load on the plate. The VRFT controller with a feedforward provides a more robust response for tracking and external disturbance.

Table 4. Process performance by changing the operating point.

Control System	ISE_a	$ITAE_a$	Ep
Recursive-VRFT	1009.65	3,543,789.93	120,879.63
IMC-PID	8796.40	10,322,383.31	122,416.94
IMC-PI	10,426.44	10,251,645.23	124,000.37
VRFT-PID	187,220.93	35,687,688.87	113,657.34

Table 5. Control system performance with feedback noise.

Control System	ISE_a	$ITAE_a$	E_P
Recursive-VRFT	718.74	3,275,236.08	83,427.47
IMC-PID	2644.18	4,284,296.65	83,738.90
IMC-PI	4079.30	4,801,885.60	84,462.43
VRFT-PID	91,064.21	14,019,821.75	77,352.45

Table 6. Process performance applying thermal load on the metallic plate.

Control System	ISE_a	$ITAE_a$	E_P
Recursive-VRFT	193.76	1,228,571.91	49,499.74
IMC-PID	1619.68	2,599,650.84	46,954.49
IMC-PI	736.25	2,306,393.53	47,797.99
VRFT-PID	16,352.03	7,674,062.47	36,716.94

5. Experimental Validation of Recursive VRFT

The experimental platform shown in Figure 5 is employed to perform experimental validation of the proposed recursive VRFT. Thus, the procedure described in Figure 2 is followed again to synthesize the controllers. First, second- and first-order reference models (37) are employed, which are based in G_1, G_2 defined in (19) with $w_n = 8 \times 10^{-3}$ for G_1 and $w_n = 11.66 \times 10^{-3}$ for G_2. Each model includes a low-pass filter $H(s) = \frac{1}{0.1371s+1}$ to compensate for the thermal camera feedback noise. In this case, the low-pass filter deals with the existing temperature randomness. However, it may not be accurate for more complex systems involving more variables with more robust stochastic behavior. Therefore, including a stochasticity awareness mechanism based on the randomness modeling that performs entropy minimization on the VRFT controller synthesis, such as [37,38], could lead to a more robust and stable compensator.

$$\begin{aligned} G_1(s) &= \frac{w_n^2}{s(2w_n+s)}H(s) \\ G_2(s) &= \frac{w_n}{s}H(s). \end{aligned} \tag{37}$$

Then, the open-loop response of the system is obtained by applying a PRBS signal, which is shown in Figure 16. So, the virtual error signals are computed using (37) along with the data in Figure 16. Thus, the VRFT controllers are synthesized using the compensator structures (38) and (39), along with the reference models (37), to obtain the first- and second-order controllers, respectively.

$$C_1(s) = \frac{A_1 s + A_0}{B_1 s + B_0} = \frac{2.75s + 0.1}{s + 640 \cdot 10^{-6}}, \tag{38}$$

$$C_2(s) = \frac{A_2 s^2 + A_1 s + A_0}{B_2 s^2 + B_1 s + B_0} = \frac{2.12s^2 + 0.11s + 543 \cdot 10^{-6}}{s^2 + 6.6 \cdot 10^{-3}s + 6.38 \cdot 10^{-12}}. \tag{39}$$

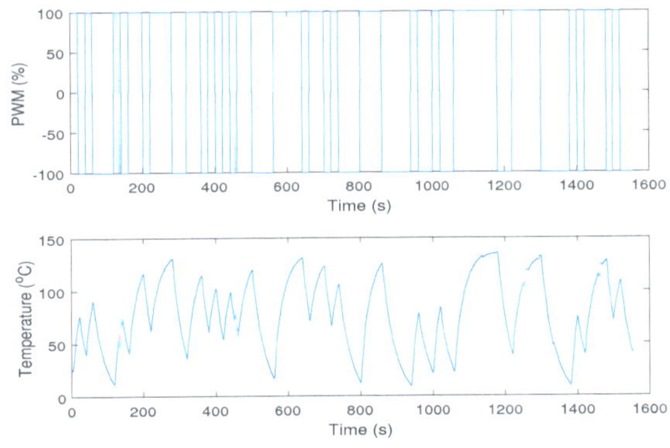

Figure 16. Temperature uniformity system open-loop response for a PRBS input signal.

Likewise, a PI controller is tuned using the Ziegler–Nichols method to compare the performance of the obtained VRFT controllers whose gains are $k_p = 25.05$ $k_i = 0.713$. The controller implementation is performed using a hardware in the loop configuration, shown in Figure 17, that can be downloaded from https://github.com/tartanus/Recursive-VRFT (accessed on 16 October 2023). As can be observed, the VRFT controllers are implemented in Simulink with real-time TCP/IP feedback for the infrared camera. Likewise, the PWM is applied using an Arduino board communicated via serial communication.

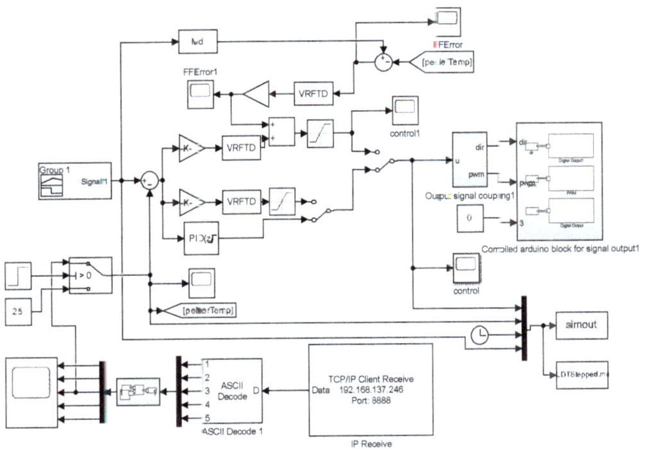

Figure 17. Matlab/Simulink feedforward VRFT HIL implementation. The implementation includes the TCP-IP feedback from thermal camera, VRFT controller, and feedback scheme with output interface using Arduino.

The response and control action for the VRFT controllers with a feedforward gain ($k = 3$), without a feedforward ($k = 0$), and the PI controller are shown in Figure 18a,b. As can be observed, the first-order VRFT controller with a feedforward has a better response, reaching a settling time of 100 s with a reasonable overshoot. Moreover, the second-order VRFT controller performance induces a slower response without feedforward compensation. Likewise, the PI controller has a faster settling time but a high overshoot and sensitivity in the presence of small feedback fluctuations. In this case, the presence of jitter during the time range [150–300] s is caused by the infrared thermal camera. It performs an automatic recalibration at 169 s to adjust the brightness and irradiance, which can be considered an external disturbance to the system. Thus, the high proportional and integral gains obtained for the ZN-PI controller compared with the gains on the C1 and C2 controllers synthesized using the recursive VRFT method make the system more sensitive to external disturbances and the presence of random noise, causing the aggressive response on the ZN-PI controller. In terms of control action, the first-order VRFT requires slightly more energy in the beginning compared with the second-order controller but provides a fast correction of the setpoint. Therefore, we can say that for the Peltier system, the VRFT controller with a feedforward can improve the system response as well as improve its robustness in the presence of small disturbances.

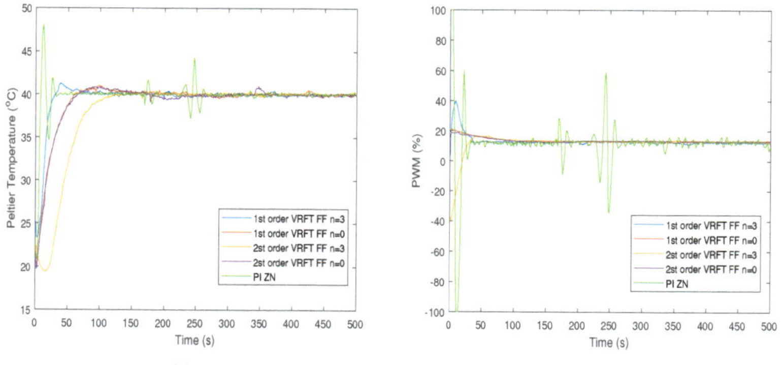

Figure 18. Temperature uniformity control response (**a**) for the VRFT and PI controllers computed control action (**b**) for a 40 °C setpoint. As can be observed, the time response (**a**) for the first-order VRFT controller with Feedforward gain $k = 3$ provides a faster response with less sensitivity to random noise compared with the other controllers. Also, the control action (**b**) of the VRFT first and second controller controllers provides an smooth and robust response compared with the PI-ZN in the presence of external disturbances as at $t = 250$ s caused by the thermal camera automatic recalibration.

6. Conclusions and Future Works

This paper presented a framework for synthesizing VRFT controllers with a feedforward using an optimization approach that allows the parameter calculation for different controller architectures based on a desired reference model and the system's open-loop response. It incorporates a feedforward strategy that minimizes the adjustment error between the desired reference model and system output. It employs a proportional gain k times the synthesized controller, providing an additional corrective control action that minimizes the adjustment error. The method was tested for the VRFT controller design of a temperature uniformity control system, using first and second-order reference models. The simulation and experimental results indicate that the VRFT controller with a feedforward improves the system's setpoint tracking response and external disturbance rejection, which increases as the feedforward gain k is increased. Due to the system's nature, the first-order compensator responds better than the high-order synthesized controllers. However, it is important to note that for more complex systems and controller architectures, the proposed VRFT synthesis can provide a reasonable and robust controller response based only on the available open-loop system's data. In future work, the proposed VRFT controller synthesis method can be modified for online controller tuning based on the open-loop or closed-loop system responses while ensuring its robustness and stability. Likewise, the evaluation of different reference models and controller architectures, such as non-linear or multivariable controls, is proposed.

Author Contributions: Conceptualization, J.G.A., L.A., J.V. and Y.C.; methodology, J.G.A. and L.A.; software, J.G.A. and L.A.; validation, J.G.A., L.A. and J.V.; formal analysis, L.A. and Y.C.; investigation, J.G.A. and L.A.; resources, J.G.A., L.A., J.V. and Y.C.; data curation, J.G.A., L.A. and J.V.; writing—original draft preparation, J.G.A., L.A. and J.V.; writing—review and editing, J.G.A., L.A., J.V. and Y.C.; visualization, J.G.A. and L.A.; supervision, L.A. and Y.C.; project administration, L.A.; funding acquisition, L.A. All authors have read and agreed to the published version of the manuscript.

Funding: This research received no external funding.

Institutional Review Board Statement: Not applicable.

Informed Consent Statement: Not applicable.

Data Availability Statement: Not applicable.

Conflicts of Interest: The authors declare no conflict of interest.

References

1. Hou, Z.S.; Wang, Z. From model-based control to data-driven control: Survey, classification and perspective. *Inf. Sci.* **2013**, *235*, 3–35. [CrossRef]
2. Guardabassi, G.O.; Savaresi, S.M. Virtual reference direct design method: An off-line approach to data-based control system design. *IEEE Trans. Autom. Control* **2000**, *45*, 954–959. [CrossRef]
3. David, J.; Fernández, R. Extensions and Applications of the Virtual Reference Feedback Tuning. Ph.D. Thesis, Universitat Autònoma de Barcelona, Barcelona, Spain, 2011.
4. Beninca, M.R. Virtual Reference Feedback Tuning of Controllers Parametrized Using Orthonormal Basis Functions. Ph.D. Thesis, Paraná Federal Univeristy, Parana, Brazil, 2015.
5. Imchen, S.; Das, D.K. Scheduling of distributed generators in an isolated microgrid using opposition based Kho-Kho optimization technique. *Expert Syst. Appl.* **2023**, *229*, 120452. [CrossRef]
6. Tian, J.; Zeng, Y.; Ji, L.; Zhu, H.; Guo, Z. Control Method of Cold and Hot Shock Test of Sensors in Medium. *Sensors* **2023**, *23*, 6536. . [CrossRef] [PubMed]
7. Wang, S.; Zhao, B.; Yi, S.; Zhou, Z.; Zhao, X. GAPSO-Optimized Fuzzy PID Controller for Electric-Driven Seeding. *Sensors* **2022**, *22*, 6678. . [CrossRef] [PubMed]
8. Pradhan, S.K.; Subudhi, B. Position control of a flexible manipulator using a new nonlinear self-tuning PID controller. *IEEE/CAA J. Autom. Sin.* **2020**, *7*, 136–149. [CrossRef]
9. Zhu, G.; Ma, Y.; Hu, S. Event-Triggered Adaptive PID Fault-Tolerant Control of Underactuated ASVs Under Saturation Constraint. *IEEE Trans. Syst. Man Cybern. Syst.* **2023**, *53*, 4922–4933. [CrossRef]
10. Chen, Q.; Wang, Y.; Song, Y. Tracking Control of Self-Restructuring Systems: A Low-Complexity Neuroadaptive PID Approach With Guaranteed Performance. *IEEE Trans. Cybern.* **2023**, *53*, 3176–3189. [CrossRef]
11. Sánchez-Palma, J.; Ordoñez-Ávila, J.L. A PID Control Algorithm With Adaptive Tuning Using Continuous Artificial Hydrocarbon Networks for a Two-Tank System. *IEEE Access* **2022**, *10*, 114694–114710. [CrossRef]
12. Shiro, M.; Matsui, Y. Data-Driven Control and Learning Systems. *IEEE Trans. Ind. Electron.* **2017**, *64*, 4070–4075. . [CrossRef]
13. Hjalmarsson, H.; Gevers, M.; Gunnarsson, S.; Lequin, O. Iterative feedback tuning: Theory and applications. *IEEE Control Syst. Mag.* **1998**, *18*, 26–41. [CrossRef]
14. Kutz, J.N.; Brunton, S.L.; Brunton, B.W.; Proctor, J.L. *Dynamic Mode Decomposition*; Society for Industrial and Applied Mathematics: Philadelphia, PA, USA, 2016. [CrossRef]
15. Chen, Y.; Wen, C. *Iterative Learning Control: Convergence, Robustness and Applications*, 1st ed.; Springer: London, UK, 1999; p. 204. [CrossRef]
16. Ariyur, K.; Krstic, M. *Real-Time Optimization by Extremum-Seeking Control*; Wiley-Interscience: Hoboken, NJ, USA, 2003.
17. Hou, Z.; Jin, S. *Model Free Adaptive Control: Theory and Applications*; CRC Press: Boca Raton, FL, USA, 2013.
18. Safonov, M.; Tsao, T.C. The unfalsified control concept and learning. *IEEE Trans. Autom. Control* **1997**, *42*, 843–847. [CrossRef]
19. Spall, J.C. *Introduction to Stochastic Search and Optimization*, 1st ed.; John Wiley & Sons, Inc.: Hoboken, NJ, USA, 2003.
20. Lecchini, A.; Campi, M.C.; Savaresi, S.M. Sensitivity shaping via virtual reference feedback tuning. In Proceedings of the Proceedings of the 40th IEEE Conference on Decision and Control (Cat. No.01CH37228), Orlando, FL, USA, 4–7 December 2001; Volume 1, pp. 750–755. [CrossRef]
21. Campi, M.C.; Lecchini, A.; Savaresi, S.M. Virtual reference feedback tuning: A direct method for the design of feedback controllers. *Automatica* **2002**, *38*, 1337–1346. [CrossRef]
22. Sato, T.; Sakai, Y.; Kawaguchi, N.; Arrieta, O. Dual-Rate Data-Driven Virtual Reference Feedback Tuning: Improvement in Fast-Tracking Performance and Ripple-Free Design. *IEEE Access* **2021**, *9*, 144426–144437. [CrossRef]
23. Remes, C.L.; Gomes, R.B.; Flores, J.V.; Líbano, F.B.; Campestrini, L. Virtual Reference Feedback Tuning Applied to DC–DC Converters. *IEEE Trans. Ind. Electron.* **2021**, *68*, 544–552. [CrossRef]

24. Ding, Y.; Xu, N.; Ren, L.; Hao, K. Data-Driven Neuroendocrine Ultrashort Feedback-Based Cooperative Control System. *IEEE Trans. Control. Syst. Technol.* **2015**, *23*, 1205–1212. [CrossRef]
25. Campestrini, L.; Eckhard, D.; Chía, L.A.; Boeira, E. Unbiased MIMO VRFT with application to process control. *J. Process Control* **2016**, *39*, 35–49. . [CrossRef]
26. Chiluka, S.K.; Ambati, S.R.; Sonawane, S.H.; Seepana, M.M.; Babu Gara, U.B. Robust IMC-PID Controller Design using VRFT: Theoretical and Experimental Investigation; IFAC-PapersOnLine . In Proceedings of the 7th International Conference on Advances in Control and Optimization of Dynamical Systems ACODS 2022, Silchar, India, 22–25 February 2022; Volume 55, pp. 241–246. [CrossRef]
27. Vilanova, R.; Visioli, A. *PID Control in the Third Millennium: Lessons Learned and New Approaches*; Springer: London, UK, 2012.
28. Formentin, S.; Campi, M.C.; Carè, A.; Savaresi, S.M. Deterministic Continuous-Time Virtual Reference Feedback Tuning (VRFT) with Application to PID Design. *Syst. Control Lett.* **2019**, *127*, 25–34. . [CrossRef]
29. Freddy F.; Valderrama Gutierrez, F.O.R. Novel Data-Driven Control Approaches. Ph.D. Thesis, Pontifical Javerian University, Bogota, Colombia, 2019.
30. Viola, J.; Radici, A.; Dehghan, S.; Chen, Y. Low-cost real-time vision platform for spatial temperature control research education developments. In Proceedings of the International Design Engineering Technical Conferences and Computers and Information in Engineering Conference, Anaheim, CA, USA, 18–21 August 2019; American Society of Mechanical Engineers: New York, NY, USA, 2019; Volume 59292, p. V009T12A030.
31. Viola, J.; Oziablo, P.; Chen, Y. A Portable and Affordable Networked Temperature Distribution Control Platform for Education and Research. *IFAC-PapersOnLine* **2020**, *53*, 17530–17535. [CrossRef]
32. Lineykin, S.; Ben-Yaakov, S. Modeling and analysis of thermoelectric modules. *IEEE Trans. Ind. Appl.* **2007**, *43*, 505–512. [CrossRef]
33. Araque–Mora, J.; Angel, L. Distributed Equivalent Circuit for Modeling Heat Transfer Process in a Thermoelectric System. In Proceedings of the 2021 IEEE 5th Colombian Conference on Automatic Control (CCAC), Ibague, Colombia, 19–22 October 2021; pp. 326–331. [CrossRef]
34. Carè, A.; Torricelli, F.; Campi, M.C.; Savaresi, S.M. A Toolbox for Virtual Reference Feedback Tuning (VRFT). In Proceedings of the 2019 18th European Control Conference (ECC), Naples, Italy, 25–28 June 2019; pp. 4252–4257. [CrossRef]
35. Angel, L.; Viola, J.; Paez, M. Speed control of a motor-generator system using internal model control techniques. In Proceedings of the 2017 IEEE 3rd Colombian Conference on Automatic Control (CCAC), Cartagena, Colombia, 18–20 October 2017; pp. 1–6. [CrossRef]
36. Nath, U.M.; Dey, C.; Mudi, R.K. Review on IMC-based PID Controller Design Approach with Experimental Validations. *IETE J. Res.* **2021**, *69*, 1640–1660. [CrossRef]
37. Mifeng Ren, Q.Z.; Zhang, J. An introductory survey of probability density function control. *Syst. Sci. Control. Eng.* **2019**, *7*, 158–170. [CrossRef]
38. Zhang, Q.; Zhou, Y. Recent Advances in Non-Gaussian Stochastic Systems Control Theory and Its Applications. *Int. J. Netw. Dyn. Intell.* **2022**, *1*, 111–119. [CrossRef]

machines

Article

High-Performance Lightweight Fall Detection with an Improved YOLOv5s Algorithm

Yuanpeng Wang [1,†], Zhaozhan Chi [2,†], Meng Liu [2], Guangxian Li [2,3,*] and Songlin Ding [3,*]

1 School of Computer, Electronics and Information, Guangxi University, Nanning 530004, China; ypwang@st.gxu.edu.cn
2 School of Mechanical Engineering, Guangxi University, Nanning 530004, China; 2001300543@st.gxu.edu.cn (Z.C.); 2111301036@st.gxu.edu.cn (M.L.)
3 School of Engineering, RMIT University, Melbourne 3000, Australia
* Correspondence: liguangxian@gxu.edu.cn (G.L.); songlin.ding@rmit.edu.au (S.D.)
† These authors contributed equally to this work.

Abstract: The aging population has drastically increased in the past two decades, stimulating the development of devices for healthcare and medical purposes. As one of the leading potential risks, the injuries caused by accidental falls at home are hazardous to the health (and even lifespan) of elderly people. In this paper, an improved YOLOv5s algorithm is proposed, aiming to improve the efficiency and accuracy of lightweight fall detection via the following modifications that elevate its accuracy and speed: first, a k-means++ clustering algorithm was applied to increase the accuracy of the anchor boxes; the backbone network was replaced with a lightweight ShuffleNetV2 network to embed simplified devices with limited computing ability; an SE attention mechanism module was added to the last layer of the backbone to improve the feature extraction capability; the GIOU loss function was replaced by a SIOU loss function to increase the accuracy of detection and the training speed. The results of testing show that the mAP of the improved algorithm was improved by 3.5%, the model size was reduced by 75%, and the time consumed for computation was reduced by 79.4% compared with the conventional YOLOv5s. The algorithm proposed in this paper has higher detection accuracy and detection speed. It is suitable for deployment in embedded devices with limited performance and with lower cost.

Keywords: fall detection; k-means++; ShufflenetV2; SE attention mechanism; SIOU loss function

Citation: Wang, Y.; Chi, Z.; Liu, M.; Li, G.; Ding, S. High-Performance Lightweight Fall Detection with an Improved YOLOv5s Algorithm. *Machines* **2023**, *11*, 818. https://doi.org/10.3390/machines11080818

Academic Editors: Jian Wu, Xiangkun He and Guangfei Xu

Received: 22 May 2023
Revised: 31 July 2023
Accepted: 1 August 2023
Published: 10 August 2023

1. Introduction

The global population of individuals aged 65 and above is increasing rapidly [1]. Unattended falls can be a life-threatening risk to these individuals if they are unable to call for help. According to WHO's global report on fall prevention among seniors, unintentional falls within the home environment rank as the second leading cause of accidental injuries and subsequent fatalities [2], and around half of this demographic are not able to stand up independently after they fall [3]. If the elderly lose consciousness due to the injury, they may miss the chance of timely treatment and face a higher risk of death [4]. However, if the elderly are equipped with fall detection devices, their physical conditions can be monitored and responded to in real time, and they can be rescued swiftly if their lives and health are threatened by a fatal injury caused by the accidental fall.

With the development of artificial intelligence, object detection algorithms are becoming more and more important to our lives and have been applied in many fields, such as security, healthcare, robotics [5], and autonomous driving [6–8]. Currently, there are two categories of fall detection methods: non-computer-vision-based detection methods, and computer-vision-based detection methods. The non-computer-vision-based methods detect fall movement via various sensors. For example, the elderly can wear wearable devices with built-in sensors (e.g., accelerometers) on the wrist, chest, or waist, and the fall movements can be detected by analyzing the human posture data (velocity, acceleration, etc.) obtained by these sensors. In a study of Mthie et al. [9], fall detection was implemented by analyzing the changes in acceleration signals of different movements, and falling and standing up movements were effectively distinguished. Lu et al. [10] analyzed the pressure signals to determine whether an accidental fall had occurred. However, these methods have the drawback of relying on the performance of the sensors, which increases the cost of the device. Furthermore, the installation/wearing of devices equipped with multiple sensors could reduce the comfort of the elderly in their daily lives. Fall detection via computer vision relies on figures and videos of the daily activities of elderly individuals recorded by cameras. When the person in the camera is detected to fall or to have fallen, the surveillance system can immediately send a distress message proactively so that they can be rescued in time. In the detection process, image processing, pattern recognition, and other related techniques are utilized to extract the information of human motion, and the human behavior detection model is constructed to identify the fall movement. Compared with the non-machine-vision methods, the machine vision methods have three main outstanding advantages: (1) they are non-intrusive, so the elderly are free from being inconvenienced by wearing the devices; (2) they are not affected by environmental noise, which prevents missed detection or misjudgment caused by the interference of wearable sensors; (3) they can monitor other abnormal emergencies at the same time.

With the advances in artificial intelligence, detection algorithms are becoming more and more important to our lives and have been applied in many fields, such as finance, medical treatment, robotics, and autonomous driving [6,11]. Currently, deep learning algorithms have become the mainstream in computer-vision-based fall detection and have been extensively investigated. Deep-learning-based fall detection algorithms can be categorized as either two-stage algorithms or one-stage algorithms. The two-stage detection algorithms, such as R-CNN [12], Fast R-CNN [13], and Faster R-CNN [14], initially generate a series of candidate frames as samples and then classify the samples via a convolutional neural network (CNN). Liu et al. [15] successfully used Faster R-CNN for the detection of the elderly falling off furniture. The algorithm detected and tracked human activity characteristics, measured the changes in these characteristics, and then determined whether the individual fell by analyzing the locations of the individual and the furniture in the Cartesian system. One obvious drawback of such algorithms is their slow detection speed. The one-stage algorithms predict the localization and classification of targets directly through the target detection network. The algorithms can implement the end-to-end prediction of target frames and target classes at once, making the detection faster and more efficient than that of the two-stage algorithms. The mainstream one-stage algorithms include SSD [16] and the YOLO [17–19] series algorithms. The YOLO series algorithms have the advantages of being fast and efficient, highly accurate, and easy to deploy and use. Among them, YOLOv5 is one of the leading target detection algorithms in the industry, with faster speed and higher accuracy than YOLOv3 and YOLOv4. Yin et al. [20] achieved real-time accurate fall detection using the improved YOLOv5s model, but the speed and volume of the algorithm were still deficient. This disadvantage led to an increase in cost in their study; that is, they needed to upload the video data to the cloud to judge whether a fall had taken place.

Although video monitoring systems are widely used in public places, this method still has limits in the detection of indoor falling. This is because the existing fall algorithms require a large net metering and bandwidth, whereas the slow speed of the home network limits the fall detection efficiency, making the algorithm difficult to deploy in embedded devices. In order to solve the problem of the efficiency and accuracy of detection, this paper proposes a lightweight fall detection algorithm based on YOLOv5s. The algorithm has the advantages of high detection accuracy, fast detection speed, less hardware requirement and computational burden, and is feasible for deployment in embedded devices. The structure of the paper is organized as follows: Section 2 illustrates the modelling of the lightweight YOLOv5s algorithm;. Section 3 describes the training and validation of the algorithm;

Section 4 is the results and relevant discussions on the mechanisms of high-accuracy and high-efficiency detection achieved via the lightweight algorithm.

2. Methodology

2.1. The series of YOLOv5 Algorithms

Typically, there are four categories of YOLOv5 algorithm: YOLOv5s, YOLOv5m, YOLOv5L and YOLOv5x; and these models differ in two parameters: the depth and the width of the network. The depth refers to the number of layers of the network, and the width refers to the number of channels output by the network layer.

Increasing the width and depth of the model can improve the performance of detection; however, this also increases the computation time, consumes more memory, and increases the inference time. Among the algorithms, YOLOv5s has the smallest depth and width, and has fewer channels per layer of the output mapping networks, which significantly reduces the number of parameters and, eventually, computational tasks. The typical network structure of the YOLOv5s algorithm consists of four different parts: input, backbone, neck, and prediction (head), as shown in Figure 1.

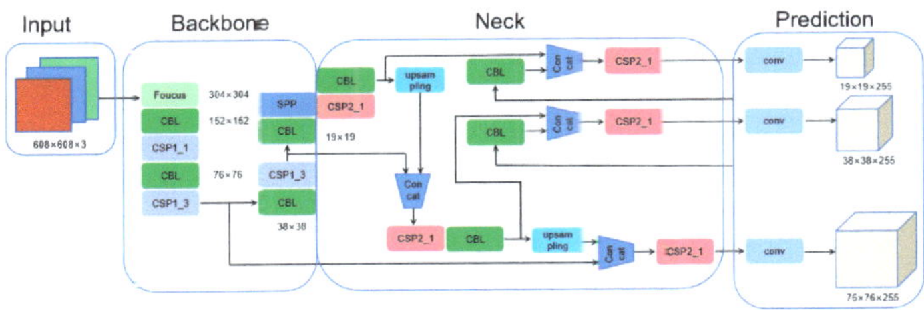

Figure 1. The network structure of YOLOv5s.

The inputs of the YOLOv5 are images, which are processed via Mosaic-4 for data enhancement, including cropping, stitching, and scaling operations of the figures. This increases the training efficiency by enriching the detection dataset and enlarging the small-scale targets.

The Backbone part uses CSPDarknet53 as the feature extraction network. The network consists of Conv structure and C3 structure, which are used to extract the feature maps of the input images and provide the basis for the subsequent processing of the model. Then, the extracted feature maps are directly input to the spatial pyramid pooling module (SPP) and are transformed into feature vectors with fixed sizes.

The neck network adopts a PANet+FPN [21] structure to achieve multi-scale feature fusion. FPN enhances the effect of target detection by fusing high underlying features, especially for small-sized targets; PANet [22] is the modified FPN with an additional bottom-up information flow path and the information delivery path is shortened, which enhances the accuracy of the underlying localization information to the entire feature extraction network. The structure of PANet is shown in Figure 2.

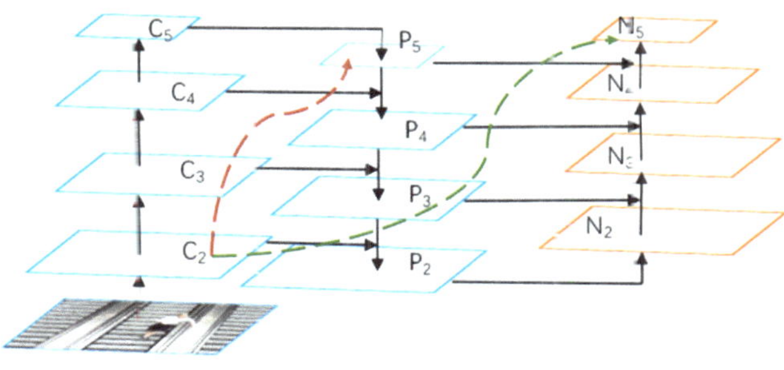

Figure 2. Network structure of PANet.

The head predicts the features of the target. Anchor boxes are applied on the targeted feature map to generate the final output vector with category probabilities and target boxes.

2.2. Improved YOLOv5s Network

Although YOLOv5s is relatively small in size among different versions of YOLOv5, it is still a challenge to deploy the YOLOv5s model directly into embedded devices due to the following potential issues: (1) real-time inference may be slow; (2) the model size may exceed the available memory of the embedded device, resulting in the model not being able to be loaded or run; (3) the device may overheat during inference, which affects the inference performance and lifespan of the device. Thus, further lightweight processing of the YOLOv5s algorithm is needed to reduce the computational task of the model.

To optimize the algorithm by avoiding the issues, the K-means++ algorithm is used on the fall dataset to optimize the scale of predefined anchors, which improves the matching degree between anchor points and real samples. The backbone is replaced by the lightweight ShuffleNetV2 network to simplify the fall detection model. Then, the SE attention module is embedded at the end of the backbone to make up for the loss of accuracy caused by model simplification. Finally, the SIOU loss function is introduced to improve the detection accuracy of the model and accelerate the convergence speed. The structure of the improved YOLOv5s network is shown in Figure 3.

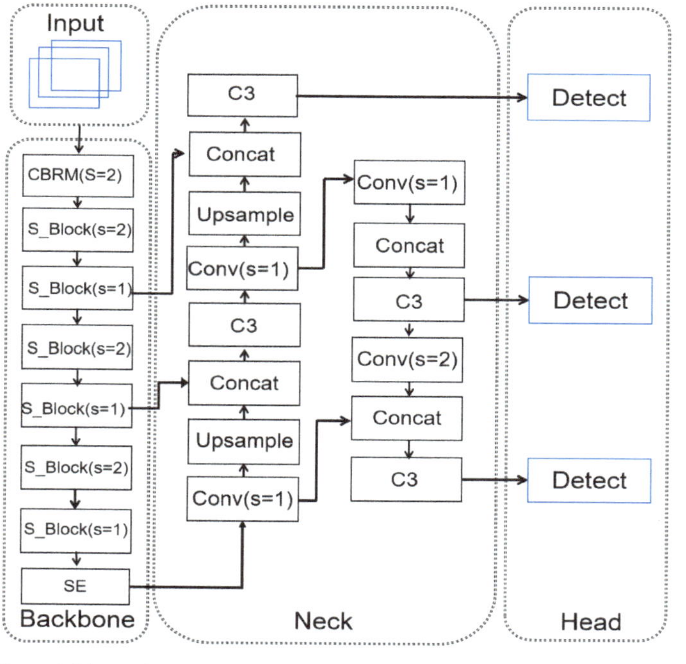

Figure 3. Network structure of improved YOLOv5s.

2.2.1. K-Means++ Algorithm

The anchor boxes in YOLOv5 are conventionally clustered using the K-means algorithm. However, this algorithm randomly assigns initial cluster centers, which could deviate from the optimal cluster centers, lead to local optimum solution, and impact the efficacy of the clustering. Therefore, the K-means++ clustering approach is applied to generate more proper anchor values, aiming to improve the training convergence without additional parameters and computation. The mechanism of K-means++ can be expressed as follows:

Firstly, a sample is randomly selected from the sample data set as the initial cluster center. The shortest distances between the samples and the selected clustering center are then calculated, and the samples are assigned to the category corresponding to its closest cluster center. The probability of each sample to be selected as the next cluster center is calculated according to Equation (1).

$$p = \frac{\mathrm{D(n)}^2}{\sum_{n \in N} \mathrm{D(n)}^2} \quad (1)$$

where D(n) represents the shortest distance from the sample to the cluster center. When a new sample is assigned into the cluster, the clustering center is recalculated based on the updated clustering samples. This process is repeated to select all the clusters to obtain the K clustering center set.

Table 1 shows the default and optimized sizes of the priori anchor boxes. The K-means++ algorithm selects the initial clustering centers using a smarter initialization, which avoids the instability of random initialization and avoids falling into local optimal solutions. Additionally, the K-means++ converges quickly after selecting the appropriate cluster centers, which plays a role in the training convergence of the fall detection dataset.

Table 1. Prior anchor frame size before and after improvement.

Feature Map Scale	Default Anchor Box Size	Optimized Anchor Box Size
Small scale (P3, 80 × 80)	(10,13) (16,30) (33,23)	(23,58) (44,125) (113,278)
Mesoscale (P4, 40 × 40)	(30,61) (62,45) (59,119)	(191,468) (238,257) (356,204)
Large scale (P5, 20 × 20)	(116,90) (156,198) (373 326)	(367,518) (479,259) (561,539)

In Table 1, 80 × 80 represents the size of the shallow feature map (P3), which contains more low-level information and is suitable for the detection of small-sized targets. In contrast, 20 × 20 represents the size of the deep feature map (P5), which contains more high-level information, such as contour, structure, and other information, and it is suitable for the detection of large-sized targets. The other 40 × 40 is the size of the mesoscale feature map (P4), which uses an anchor size between the two mentioned above and is used for detecting medium-sized targets. The second column indicates the preset anchor box size for the three scales, and the two numbers in parentheses indicate the width and height of the anchor box. The third column shows the optimized anchor box size.

2.2.2. Lightweight ShuffleNetV2 Backbone Network

The CSPDarknet53 feature extraction network is commonly used in YOLOv5s. Although the features extraction is efficient, it is difficult to deploy the algorithm into the embedded devices due to its heavy computational duties. In this study, the ShuffleNetv2 [23] network was used in the improved algorithm to replace the original backbone of YOLOv5s, which met the requirements of being lightweight and highly accurate. ShuffleNetV2 inherits the deep separable convolution [24] and channel shuffle from ShuffleNetV1 [25], and can split the channel. The two basic units of ShuffleNetV2 are shown in Figure 4. It can be seen that the channels of the units are divided into two branches because the channel splits before concatenation (Figure 4a), which can effectively reduce the redundant features and increase the network computational efficiency. Adding the channel shuffle module after shortcut connections avoids the problem that the output of a channel only comes from a small part of the original feature map, which realizes the exchange of feature information between different branches and improves the detection accuracy. The application of channel split and channel shuffle compresses the computation and memory usage of the model, significantly simplifying the model.

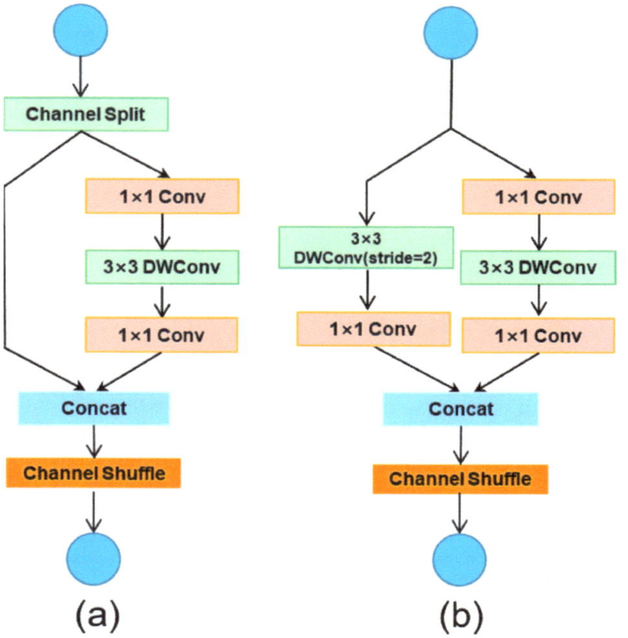

Figure 4. Basic unit of ShuffleNetv2. (**a**) S_Block1 (**b**) S_Block2.

For S_Block1, the left branch constantly maps through shortcut connections to increase the network structural depth, which reduces the fragmentation and accelerates the training speed. The right branch performs convolution through multiple layers to ensure the equal channel numbers of input and output. This minimizes the memory access to improve the modelling speed.

S_Block2 is a downsampling module where the two branches are introduced directly into the input without the splitting operation. It adjusts the number of channels using 1×1 convolution and downsamples of the depth convolution in steps of 2. The two branches are concatenated together to halve the size of the feature map and double the number of channels.

2.2.3. SE Attention Module

Inspired by the way that human eyes can naturally and efficiently find important areas in complex scenes, the "attention mechanism" has been introduced into the field of computer vision [26]. Due to its excellent performance, the attention mechanism is widely used to solve various tasks in computer vision, such as image recognition, object detection, semantic segmentation, etc. Currently, most studies focus on the extraction of spatial features but lack attention to different channels. To enhance the network's perception of character motion features in videos, the relationship among the channels should be considered.

The SE attention module [27] can automatically learn the importance of each channel through training, and assign different weights to the spatial and channel dimensions of the network. The more important the information about the channel, the larger the weighting factor. This module directs the network to focus on important features and ignore irrelevant features, which improves its ability to distinguish the features. The structure of the SE attention module is shown in Figure 5.

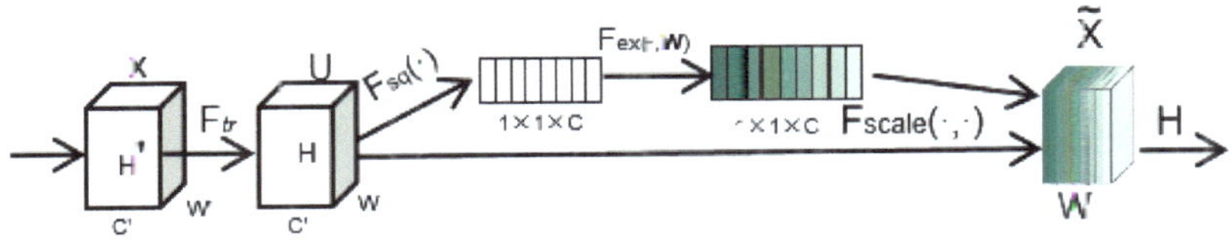

Figure 5. SE net structure.

The SE attention module consists of three main operations: Squeeze, Excitation, and Rescale. The Squeeze operation transforms the input feature map into a global description vector through global averaging pooling. The fall usually involves changes in human posture and the surrounding environment. With the Squeeze operation, the network is able to capture the global features from inputs, which provides a more comprehensive contextual understanding, and improves the accuracy on distinguishing fall movement from other motions. While keeping the number of channels C unchanged, the size of the input feature map is compressed from (*H*,*W*) to (1,1), and then global pooling is used to encode the overall spatial feature of one channel into a global feature. The value indicating the *c*th channel after the squeezing operation can be calculated via Equation (2).

$$z_c = F_{sq}(u_c) = \frac{1}{H \times W}\sum_{i=1}^{H}\sum_{j=1}^{W} u_c(i,j) \tag{2}$$

where z_c represents the one-dimensional vector of the *c*th feature. *H* represents the height of the feature map and *W* represents the width of the feature map. $u_c(i,j)$ denotes the *c*th global feature.

The Excitation operation emphasizes the attention of features related to the fall movement by learning the relationships among feature map channels. The features of the fall movement involve overall body movements and posture changes, which are often transferred via specific channels. Through the excitation operation, the network is able to self-adaptively learn the weights of each channel to highlight the focus on fall movement features, which enhances the network's response to the channels related to fall movement, improving the accuracy and sensitivity of detection. The weights can be calculated via Equation (3).

$$s = F_{ex}(z,W) = \sigma(g(z,W)) = \sigma(W_2\delta(W_2\delta(W_1 z))) \tag{3}$$

where σ refers to the Sigmoid function, δ refers to the ReLU function, and $g(z,W)$ refers to the bottleneck structure consists of two fully connected layers, $W_1 \in \mathbb{R}^{\frac{C}{R}\times C}$ and $W_2 \in \mathbb{R}^{C\times\frac{C}{R}}$.

The Rescale operation re-weights the feature map based on the learned excitation vectors. The detection of fall movement is usually disturbed by the complex background interference. By re-weighting the feature map, the rescale operation highlights important features and suppresses interference with less important features, which improves the localization and detection of fall movement and enhances the network's ability to perceive key actions of falling. The rescaling the weights of the channels can be calculated via Equation (4).

$$\widetilde{x_c} = F_{scale}(u_c, s_c) = s_c u_c \tag{4}$$

where $\widetilde{x_c}$ indicates the output result. $F_{scale}(u_c, s_c)$ denotes the product of the feature map u_c and channel weights s_c.

2.2.4. Improvement of the Loss Function

In the YOLOv5s network, the loss function consists of three components: the rectangular frame loss, the classification loss, and the confidence loss. The rectangular box loss uses GIOU loss [28], which adds the smallest rectangular box surrounded by the real and predicted boxes in the calculation of the loss based on IOU. This solves the problem of gradient vanish when there is no overlapping area between the two boxes. Assuming that A represents the ground truth, B represents the prediction frame, and C is the smallest bounding box that can cover them, the GIOU loss can be calculated by the following equations:

$$\mathrm{IOU} = \frac{|\mathrm{A} \cap \mathrm{B}|}{|\mathrm{A} \cup \mathrm{B}|} \tag{5}$$

$$\mathrm{GIOU} = \mathrm{IOU} - \frac{\mathrm{C} - (\mathrm{A} \cup \mathrm{B})}{\mathrm{C}} \tag{6}$$

$$\mathrm{L_{GIOU}} = 1 - \mathrm{GIOU} \tag{7}$$

However, GIOU could degenerate to IOU when the predicting box contains the ground truth. Additionally, slower convergence and less-accurate regression are the major drawback of the GIOU. Therefore, the improved algorithm uses SIOU loss [29] to replace the original GIOU loss function in this study. SIOU loss considers the vector angle between the desired regressions and the redefined penalty indicator. The SIOU loss function consists of four losses: angular loss, distance loss, shape loss, and IOU loss.

For the angle loss, SIOU loss adds the angle perception between the center of the real frame B_{gt} and the predicted frame B, which reduces the distance-related extra variables. The angle loss can be calculated with the following equations:

$$d = \sqrt{\left(b_x^{gt} - b_{c_x}\right)^2 + \left(b_{c_y}^{gt} - b_{c_y}\right)^2} \tag{8}$$

$$B_h = max\left(b_{c_y}^{gt}, b_{c_y}\right) - min\left(b_{c_y}^{gt}, b_{c_y}\right) \tag{9}$$

$$x = \frac{B_h}{d} = \sin(\alpha) \tag{10}$$

$$\Lambda = 1 - 2 \times \sin^2(\arcsin(x) - \frac{\pi}{4}) \tag{11}$$

where Λ is the final calculation result of the angular loss. x is the sine of the angle α between the center points of the real and predicted frames in the figure; d is the distance between them; and B_h is the relative height difference between the two points.

The distance loss of SIOU (Δ) is different from that of GIOU due to the new addition of the angle calculation, which is expressed as follows:

$$\rho_x = (\frac{b_{c_x}^{gt} - b_{c_x}}{c_w})^2, \rho_y = (\frac{b_{c_y}^{gt} - b_{c_y}}{c_h}), \gamma = 2 - \Lambda \tag{12}$$

$$\Delta = \sum\nolimits_{t=x,y} (1 - \mathrm{e}^{-\gamma\rho_t}) \tag{13}$$

where ρ_x and ρ_y are the squared ratios of the relative distances of the centroids of the real and predicted boxes in the X and Y directions to the width and height of the smallest outer rectangles. e is the Euler constant.

The formula for calculating shape loss is shown in Equations (14) and (15).

$$W_w = \frac{|w - w^{gt}|}{\max(w, w^{gt})}, W_h = \frac{|h - h^{gt}|}{\max(h, h^{gt})} \tag{14}$$

$$\Omega = \sum\nolimits_{t=w,h} \left(1 - e^{-wt}\right)^\theta \tag{15}$$

where (w, h) and (w^{gt}, h^{gt}) are the widths and heights of the prediction frame and the real frame, respectively; θ is the attention coefficient in the shape loss calculation formula, and the values were defined between 2 and 6 for different data sets.

The final SIOU loss can be calculated via Equation (16).

$$L_{SIOU} = 1 - IOU + \frac{\Delta + \Omega}{2} \tag{16}$$

The SIOU loss function effectively increases the convergence speed of the model and improves the performance of the fall detection model.

3. Training and Testing of the Algorithm

3.1. Training Environment

In this study, the training of the lightweight algorithm was implemented on a commercial desktop, and the specifications of the hardware are listed in Table 2. The major parameters of the improved YOLOv5s in the training process were set as follows: the epoch was 300, the batch size was 16, the learning rate was 0.01, the cosine annealing hyperparameter was 0.15, the stochastic gradient descent optimizer SGD (Stochastic Gradient Descent) was used, the learning rate momentum of the optimizer was 0.937, and the weight decay coefficient was 0.0003.

Table 2. Training environment configuration.

Configuration	Parameter
CPU	AMD Ryzen7 5800H
GPU	6GB NVIDIA RTX 3600 Laptop
Accelerated environment	CUDA 11.4 CUDNN 8.2.4
Development language	Python 3.8
Operating system	Windows 11

3.2. Testing Environment

In order to facilitate the deployment of deep learning models into the right edge development hardware, NVIDIA has launched a number of miniature AI development kits, such as Jetson AGX Orin, Jetson Orin NX, Jetson AGX Xavier series, Jetson Xavier NX series, Jetson TX2 series, Jetson Nano, etc. Jetson Nano is cost-effective, and the power consumption is lower among these development kits. It was used as the embedded device for deployment in this study. Table 3 shows the configuration details of the Jetson Nano used in the testing experiments.

Table 3. Testing environment configuration.

Configuration	Parameter
CPU	4-core ARM® Cortex®-A57 MPCore
GPU	NVIDIA Maxwell™ with 128 NVIDIA CUDA® core
Memory	4 GB 64 bit LPDDR4
CUDA	Pytorch
Programming Language	Python3.6

3.3. Dataset

The fall detection dataset, which was the set of images of falling-down figures, was used to simulate falls in different circumstances. To enhance the robustness and generalization ability of the training algorithm, and the fall detection performance under different conditions, we used a data augmentation method on the dataset. As shown in Figure 6, the data enhancement operations, such as darkening, brightening, flipping, Gaussian noise addition, salt and pepper noise addition, and equalization, were used to simulate various situations that may occur in the real world, including different indoor lighting levels and different viewing angles. After data augmentation, the final dataset contained 9,834 images, which were made up of 7,984 training images and 1,850 testing images (Table 4).

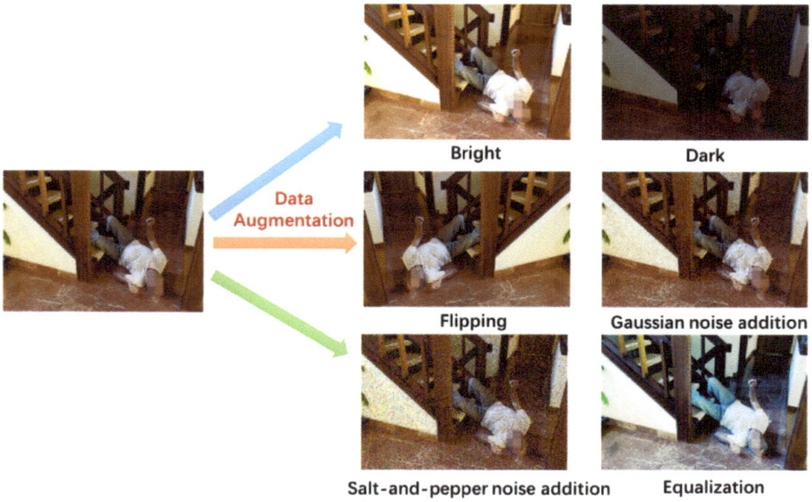

Figure 6. Data augmentation methods for self-made dataset.

Table 4. The training dataset: categories of movement and the amount of training/test data.

Sort	Training	Testing
Fall detected	3234	854
Walking	2401	506
Siting	2349	490
Total	7984	1850

The dataset contained three categories, "Fall detected", "Walking", and "Sitting". Figure 7 shows some images from this dataset. The sample distributions of the above three categories are shown in Table 4.

Figure 7. Some images from the dataset.

4. Results and Discussion

4.1. Evaluation of the Improved Yolov5s Algorithm

In this study, ablation experiments were conducted using the self-built fall detection dataset to test the validity of the algorithm, and the performance of the algorithm was evaluated via the following parameters: the number of parameters (Param), floating-point operations per second (FLOPs), and the size of the model weight files (Weights) were used to evaluate whether a model is lightweight. Specifically, the Param is the sum of the parameters in a model, which is an important indicator of memory usage and the initialization time of the program. The computational volume, also known as FLOPs, is the sum of the number of multiplications and additions when performing forward inference, which reflects the requirement of the hardware due to the computational units and the size of the model. The model weight file (Weights) is required for the final deployment. For devices with limited space resources, it is necessary to keep the model weight file as small as possible.

In addition, the detection effect was evaluated using Precision (*P*), Recall (*R*), mean Average Precision (*mAP*), and forward pass time (*FP*). Precision is the proportion of true positive examples in the prediction result; Recall is the proportion of all positive examples that are correctly predicted; *mAP* is the average precision value for each class; *FP* time is the time that is required for data to pass through the neural network, from input to output. The four indicators can be quantified via the following equations:

$$P = \frac{TP}{TP + FP} \tag{17}$$

$$R = \frac{TP}{TP + FN} \tag{18}$$

$$AP = \int_0^1 PdR \tag{19}$$

$$mAP = \frac{\sum_i^N AP_i}{N} \tag{20}$$

where *TP* is the number of targets detected correctly by the model, *FP* is the number of targets detected incorrectly by the model, and *FN* is the number of correct targets missed by the model. Therefore, the *FP* time is the sum of preprocessing time, inference time, and NMS time.

4.2. The results of Ablation Experiment

To validate the model, ablation experiments were conducted using the self-built fall detection dataset in this study. The specific results of the ablation experiments are shown in Table 5. The "√" indicates that the corresponding method was used to improve the model, and the "-" denotes that the corresponding method was not applied in the fall detection model.

Table 5. The results of the ablation experiments.

K-Means++	ShuffleNetV2	SE	SIOU	P/%	R	mAP/%	Param/10^6	FLOPs/10^9	Weight/MB
-	-	-	-	93.2	91.5	92	7.1	16.5	13.6
√	-	-	-	94.2	91.9	93.2	7.1	16.5	13.6
-	√	-	-	90.2	88.7	89.4	2.4	3.3	3.5
-	-	√	-	93.8	91.7	93.1	7.2	16.6	13.7
-	-	-	√	94.3	92.1	93.3	7.1	16.5	13.6
√	√	√	√	96.5	94.3	95.2	2.4	3.4	3.4

According to Table 5, compared with the conventional YOLOv5s model, the detection accuracy of the improved algorithm is improved with the application of the K-means++ clustering algorithm which generated anchor frames. By replacing the original CSPDarknet53 feature extraction network with the lightweight ShuffleNetV2 network, the model lost some accuracy and the mAP decreased by 2.8%, compared with the original YOLOv5s model. Table 5 also shows the results of different modified YOLOv5s models. It can be found that the value of Param of the model is reduced by 66.2% and the Weight is reduced by 74%. This significantly reduces the complexity and computational task of the algorithm and boosts the detection speed while maintaining the light weight. With the addition of the SE attention module, the mAP is improved by 1.2% with a minimum influence on the size and complexity of the model; this verifies the effectiveness of adding the SE attention module, which trades a small amount of computation for a large improvement in detection performance. By replacing the original GIOU loss function with the SIOU loss function, the change in mAP is insignificant (1.4%), thus verifying the effectiveness of the SIOU loss function. Furthermore, the improved model improves detection accuracy by 3.5%, reduces weight by 75%, reduces number of parameters by 66.2%, and reduces computation by 79.4%, compared with the YOLOv5s model. The detection speed was boosted, and the weight was significantly reduced, while ensuring accuracy, which reflects the feasibility of deploying the improved real-time fall detection algorithm in embedded devices with limited computational resources.

4.3. Comparison with Different Algorithms

To further evaluate the performance of the improved algorithm, a variety of current mainstream target detection algorithms, including Faster R-CNN, YOLOv3, YOLOv4, and YOLOv5, were used as comparisons in an in-house fall detection dataset, and the results of the various detection models are shown in Table 6.

Table 6. Comparison of the detection performance of several models on the test set.

Model	mAP/%	Param/10^6	FLOPs/10^9	Weight/MB
Faster R-CNN	79.6	28.4	304.1	108
YOLOv3	84.5	61.5	154.5	234
YOLOv4	90.4	64.3	132.6	244
YOLOv5	95.6	7.1	16.5	13.6
Improved-YOLOv5	96.5	2.4	3.4	3.4

Among the listed models, the faster R-CNN cannot meet the requirements of high accuracy and high speed of fall detection because the algorithm requires two stages to complete the inference task, making the calculation too complex to perform real-time reasoning tasks. Although YOLOv3 and YOLOv4 algorithms have higher detecting accuracy, they are not appropriate for the deployment in embedded devices with limited computing resources due to the larger size of their models, number of parameters, and higher computation burden. By comparing the performance of the YOLOv5 and the improved-YOLOv5m models, it can be found that the overall performance of the improved model, Improved-YOLOv5, is better; it outperforms the base model YOLOv5s, which is the base model of those listed in Table 6.

The experimental results shown in Figure 8 demonstrate the parametric performance curves of the model before and after improvement. The figure not only shows performance curves such as accuracy, recall rate, and mAP, but also presents the loss curve of model parameter optimization. The Box curve represents the boundary box loss, with a smaller value indicating the target with higher accuracy. The Objectness curve represents the inferred average loss of the target; the smaller the value, the more accurate the target detection. The classification curve is the inferred average loss for classification. The smaller the value, the more accurate the target classification. From the results before and after optimization, it can be seen that the optimized model has better overall loss and detection performance.

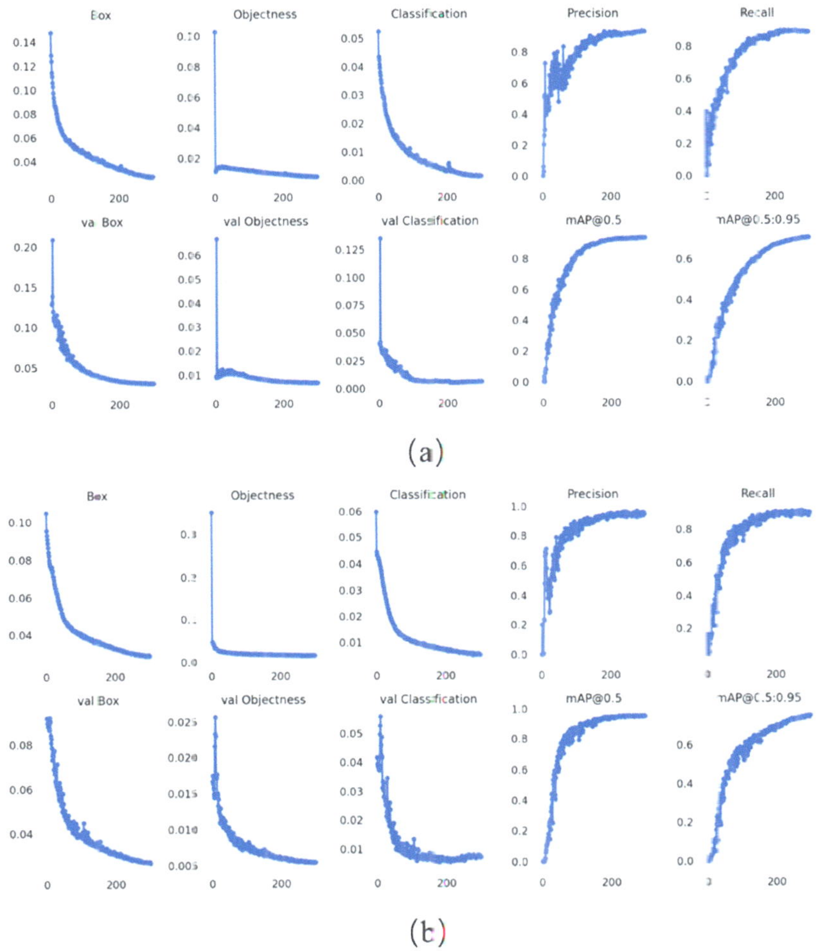

Figure 8. Comparison of training metrics. (**a**) The training metrics of YOLOv5s; (**b**) The training metrics of Improved-YOLOv5.

Figure 9 shows the visualization of the above model tested on randomly selected images in the test set. According to the results of the comparative experiments in Table 5 and the results of the visualization tests in Figure 9, the Improved-YOLOv5 model has higher detection accuracy than other mainstream detection models. Additionally, according to the three metrics of Param, FLOPs, and Weight, it can be seen that the Improved-YOLOv5 model has lower complexity which reduced the computational tasks. This means the improved-YOLOv5s algorithm meets the lightweight design requirements. The fall detection can be conducted both quickly and accurately, and it is suitable for deployment in embedded devices.

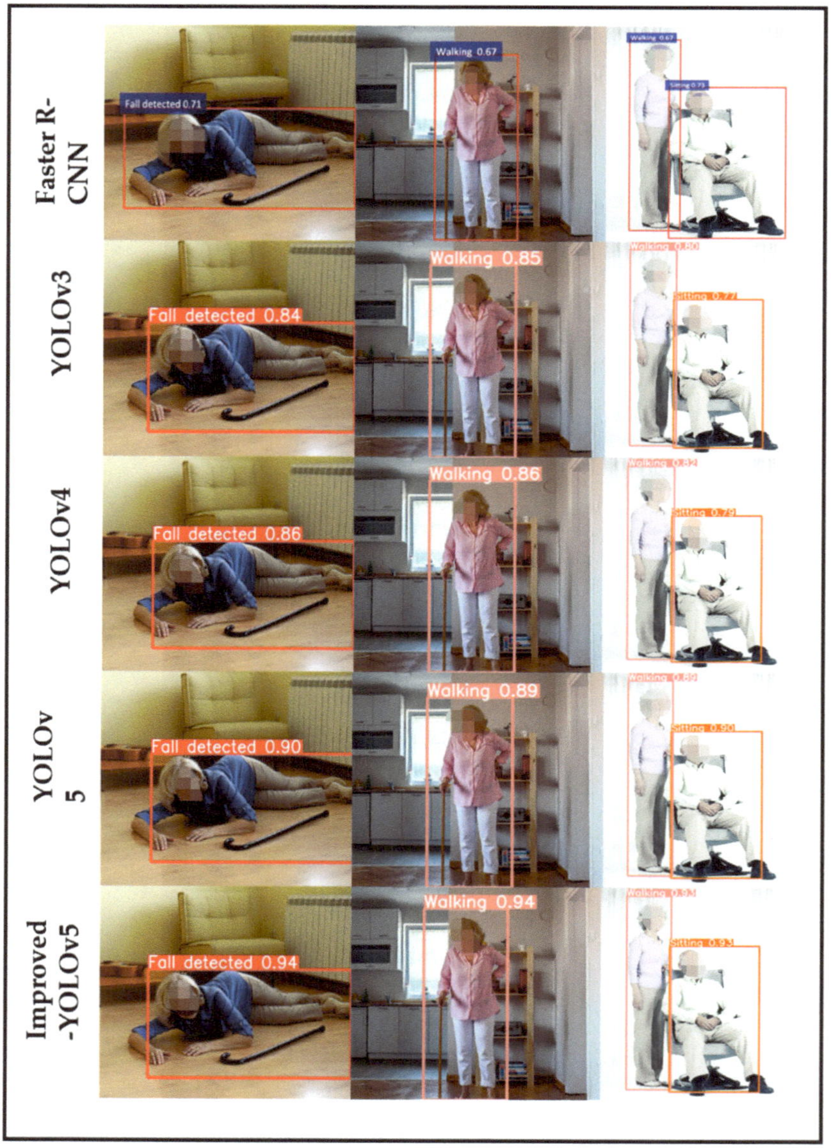

Figure 9. The detection performance of different models.

4.4. Embedded Device Deployment Experiment

The experiments in this section are designed to evaluate the detecting speed of the improved model on an embedded device in practice. Jetson Nano was used as the hardware for deploying the model (Figure 10). Both YOLOv5s model and the improved YOLOv5s model were deployed on the device, and their inference speed was evaluated using the *FP* time. The shorter the forward pass time, the faster the model's inference.

The main factors affecting FP time include: (1) the size of the input image. The larger the size of the input image, the more pixel points needed to be processed, and the more pre-propagation time is needed; (2) the complexity of the model and the number of parameters. Larger models usually require more computational resources and time; (3) hardware devices. The speed of forward pass is affected by hardware devices. The use of GPU can significantly improve the forward pass speed of the model. In general, the better the performance of the GPU, the shorter the time required for forward pass; (4) batch size, which is the number of images input at one time. Larger batch sizes usually result in higher parallelism and increase the speed of forward propagation. However, the batch size is also limited by the size of the available graphic memory.

Figure 10. Jetson Nano embedded device.

In the experiments, the YOLOv5s model and the Improved-YOLOv5 model were deployed into the same Jetson Nano, and the sizes of the testing images were 640 × 640, 512 × 512, and 320 × 320. The comparative results of the detecting speeds of the two models are shown in Table 7. It can be seen that, when the size of the input image is 640 × 640, the FP time of the improved model is reduced by 22.3% compared to the YOLOv5s model. When the size of the input image is 512 × 512, the FP time of the improved model is reduced by 14.9% compared to the YOLOv5s model. When the size of the input image is 320 × 320, the FP time of the improved model is reduced by 14.0% compared with the YOLOv5s model. With the reduction of the input image size, the FP time of both models decreases. It is also verified that the FP time is affected by the input image size. The experimental results show that the model Improved-YOLOv5 has faster detection speed than YOLOv5s. The improved model achieves a well-balanced detection speed and accuracy.

Table 7. Deployment experiment results.

Image Size	Model	Preprocessing/ms	Inference/ms	NMS/ms	FP Time/ms
640 × 640	YOLOv5s	1.6	192.3	12.1	206.0
	Improved-YOLOv5s	1.4	150.2	8.6	160.2
512 × 512	YOLOv5s	1.2	131.2	10.2	142.6
	Improved-YOLOv5s	1.1	110.8	9.3	121.2
320 × 320	YOLOv5s	0.7	65.9	7.5	74.1
	Improved-YOLOv5s	0.7	56.2	6.8	63.7

5. Conclusions

An effective lightweight fall detection algorithm based on YOLOv5s, which is feasible for the deployment in an embedded device, the Jetson Nano, is proposed in this study. The model used the K-means++ algorithm on a fall dataset, and optimized the scale of predefined anchors and improved the matching degree between anchor points and real samples. The backbone was replaced by the Lightweight ShuffleNetV2 network to simplify the fall detection model. The SE attention module was embedded in the end of the backbone to make up for the loss of accuracy caused by model simplification. The SIOU loss function was applied to improve the detection accuracy of the model and to accelerate the convergence speed. The experiment results of testing showed that the mAP of the improved algorithm was improved by 3.5%, the model size was reduced by 75%, and the time consumption of computation was reduced by 79.4% compared with the conventional YOLOv5s. The improved model has a higher accuracy and faster detection speed, and is appropriate for deployment in embedded devices such as Jetson Nano. It can detect the activity status of the elderly at home in real time and can quickly detect a fall, so that the fallen person can be helped as soon as possible.

Author Contributions: Conceptualization, Z.C. and Y.W.; Methodology, Z.C. and Y.W.; Validation, M.L. and G.L.; Investigation, Z.C. and Y.W.; Resources, Y.W. and Z.C.; Data curation, M.L.; Writing—original draft, Z.C. and Y.W.; Writing—review and editing, G.L. and S.D.; Visualization, M.L. and G.L.; Supervision, G.L. and S.D. All authors have read and agreed to the published version of the manuscript.

Funding: This research received no external funding.

Data Availability Statement: No new data were created or analyzed in this study. Data sharing is not applicable to this article.

Conflicts of Interest: The authors declare no conflict of interest.

Abbreviations

Symbol	Definition
$D(n)$	The shortest distance from the sample to the cluster center
z_c	Global eigenvalues of channel *c*
H	Height of the feature map
W	Width of the feature map
$u_c(i,j)$	Eigenvalues of channel *c* at point *(i,j)*
δ	ReLU function
σ	Sigmoid function
W_1	Weight matrix of the first fully connected layer
W_2	Weight matrix of the second fully connected layer
s	Weight vectors
$F_{sq}()$	Squeeze function
$F_{ex}()$	Excitation function
$F_{scale}()$	Recalibration function
B_{gt}	Center of the real frame
B	Predicted frame
B_h	Relative height difference
Λ	Angular loss
x	Sine of the angle α between the center points of the real and predicted frames
ρ_x	Squared ratios of the relative distances of the centroids of the real and predicted boxes on the X direction to the width and height of their smallest outer rectangles
ρ_y	Squared ratios of the relative distances of the centroids of the real and predicted boxes on the Y direction to the width and height of their smallest outer rectangles
(w,h)	Width and height of the prediction frame
(w^{gt},h^{gt})	Width and height of the real frame
Ω	Shape loss
θ	Attention coefficient in the shape loss

References

1. Raza, A.; Yousaf, M.H.; Velastin, S.A. Human Fall Detection using YOLO: A Real-Time and AI-on-the-Edge Perspective. In Proceedings of the 2022 12th International Conference on Pattern Recognition Systems (ICPRS), Saint-Etienne, France, 7–10 June 2022; pp. 1–6.
2. Kong, Y.; Huang, J.; Huang, S.; Wei, Z.; Wang, S. Learning Spatiotemporal Representations for Human Fall Detection in Surveillance Video. *J. Vis. Commun. Image Represent.* **2019**, *59*, 215–230. [CrossRef]
3. Roush, R.E.; Teasdale, T.A.; Murphy, J.N.; Kirk, M.S. Impact of a personal emergency response system on hospital utilization by community-residing elders. *South. Med. J.* **1995**, *88*, 917–922. [CrossRef] [PubMed]

4. West, J.; Hippisley-Cox, J.; Coupland, C.A.; Price, G.M.; Groom, L.M.; Kendrick, D.; Webber, E. Do rates of hospital admission for falls and hip fracture in elderly people vary. *Public Health* **2004**, *118*, 576–581. [CrossRef] [PubMed]
5. Hu, Z.; Zhang, Y.; Lv, C. Affine Layer-Enabled Transfer Learning for Eye Tracking with Facial Feature Detection in Human–Machine Interactions. *Machines* **2022**, *10*, 853. [CrossRef]
6. He, X.; Lou, B.; Yang, H.; Lv, C. Robust Decision Making for Autonomous Vehicles at Highway On-Ramps: A Constrained Adversarial Reinforcement Learning Approach. *IEEE Trans. Intell. Transp. Syst.* **2023**, *24* 4103–4113. [CrossRef]
7. He, X.; Yang, H.; Hu, Z.; Lv, C. Robust Lane Change Decision Making for Autonomous Vehicles: An Observation Adversarial Reinforcement Learning Approach. *IEEE Trans. Intell. Veh.* **2023**, *8*, 184–193. [CrossRef]
8. He, X.; Liu, Y.; Lv, C.; Ji, X.; Liu, Y. Emergency steering control of autonomous vehicle for collision avoidance and stabilisation. *Veh. Syst. Dyn.* **2019**, *57*, 1163–1187. [CrossRef]
9. Mathie, M.J.; Coster, A.C.F.; Lovell, N.H.; Celler, B.G. Accelerometry: Providing an integrated, practical method for long-term, ambulatory monitoring of human movement. *Physiol. Meas.* **2004**, *25*, R1. [CrossRef] [PubMed]
10. Lu, W.; Wang, C.; Stevens, M.C.; Redmond, S.J.; Lovell, N.H. Low-power operation of a barometric pressure sensor for use in an automatic fall detector. In Proceedings of the 2016 38th Annual International Conference of the IEEE Engineering in Medicine and Biology Society (EMBC), Orlando, FL, USA, 16–20 August 2016.
11. He, X.; Lv, C. Towards Energy-Efficient Autonomous Driving: A Multi-Objective Reinforcement Learning Approach. *IEEE/CAA J. Autom. Sin.* **2023**, *10*, 2–10. [CrossRef]
12. Girshick, R.; Donahue, J.; Darrell, T.; Malik, J. Rich feature hierarchies for accurate object detection and semantic segmentation. In Proceedings of the 27th IEEE Conference on Computer Vision and Pattern Recognition (CVPR), Columbus, OH, USA, 23–28 June 2014; pp. 580–587.
13. Girshick, R. Fast R-CNN. In Proceedings of the IEEE International Conference on Computer Vision, Santiago, Chile, 11–18 December 2015; pp. 1440–1448.
14. Ren, S.Q.; He, K.M.; Girshick, R.; Sun, J. Faster R-CNN: Towards Real-Time Object Detection with Region Proposal Networks. *IEEE Trans. Pattern Anal. Mach. Intell.* **2017**, *39*, 1137–1149. [CrossRef] [PubMed]
15. Min, W.; Cui, H.; Rao, H.; Li, Z.; Yao, L. Detection of Human Falls on Furniture Using Scene Analysis Based on Deep Learning and Activity Characteristics. *IEEE Access* **2018**, *6*, 9324–9335. [CrossRef]
16. Liu, W.; Anguelov, D.; Erhan, D.; Szegedy, C.; Reed, S.; Fu, C.Y.; Berg, A.C. SSD: Single Shot MultiBox Detector. In Proceedings of the 14th European Conference on Computer Vision (ECCV), Amsterdam, The Netherlands, 8–16 October 2016; pp. 21–37.
17. Redmon, J.; Divvala, S.; Girshick, R.; Farhadi, A. You Only Look Once: Unified, Real-Time Object Detection. In Proceedings of the 2016 IEEE Conference on Computer Vision and Pattern Recognition (CVPR), Seattle, WA, USA, 27–30 June 2016; pp. 779–788.
18. Redmon, J.; Farhadi, A. YOLO9000: Better, Faster, Stronger. In Proceedings of the 30th IEEE/CVF Conference on Computer Vision and Pattern Recognition (CVPR), Honolulu, HI, USA, 21–26 July 2017; pp. 6517–6525.
19. Redmon, J.; Farhadi, A. YOLOv3: An Incremental Improvement. *arXiv* **2018**, arXiv:1804.02767.
20. Yin, Y.; Lei, L.; Liang, M.; Li, X.; He, Y.; Qin, L. Research on Fall Detection Algorithm for the Elderly Living Alone Based on YOLO. In Proceedings of the 2021 IEEE International Conference on Emergency Science and Information Technology (ICESIT), Chongqing, China, 22–24 November 2021; pp. 403–408.
21. Rezatofighi, H.; Tsoi, N.; Gwak, J.; Sadeghian, A.; Reid, I.; Savarese, S. Generalized Intersection Over Union: A Metric and a Loss for Bounding Box Regression. In Proceedings of the IEEE/CVF Conference on Computer Vision and Pattern Recognition, Long Beach, CA, USA, 15 –20 June 2019; pp. 658–666.
22. Al-Smadi, Y.; Alauthman, M.; Al-Qerem, A.; Aldweesh, A.; Quaddoura, R.; Aburub, F.; Mansour, K.; Alhmiedat, T. Early Wildfire Smoke Detection Using Different YOLO Models. *Machines* **2023**, *11*, 246. [CrossRef]
23. Ma, N.N.; Zhang, X.Y.; Zheng, H.T.; Sun, J. ShuffleNet V2: Practical Guidelines for Efficient CNN Architecture Design. In Proceedings of the 15th European Conference on Computer Vision (ECCV), Munich, Germany, 8–14 September 2018; pp. 122–138.
24. Gao, H.; Zhang, Y.; Lv, W.; Yin, J.; Qasim, T.; Wang, D. A Deep Convolutional Generative Adversarial Networks-Based Method for Defect Detection in Small Sample Industrial Parts Images. *Appl. Sci.* **2022**, *12*, 6569. [CrossRef]
25. Zhang, X.; Zhou, X.Y.; Lin, M.X.; Sun, R. ShuffleNet: An Extremely Efficient Convolutional Neural Network for Mobile Devices. In Proceedings of the 31st IEEE/CVF Conference on Computer Vision and Pattern Recognition (CVPR), Salt Lake City, UT, USA, 18–23 June 2018; pp. 6848–6856.
26. Hu, Z.; Zhang, Y.; Li, Q.; Lv, C. A Novel Heterogeneous Network for Modeling Driver Attention With Multi-Level Visual Content. *IEEE Trans. Intell. Transp. Syst.* **2022**, *23*, 24343–24354. [CrossRef]
27. Hu, J.; Shen, L.; Sun, G. Squeeze-and-Excitation Networks. *IEEE Trans. Pattern Anal. Mach. Intell.* **2020**, *42*, 2011–2023. [CrossRef] [PubMed]
28. Belmont, J.W.; Gibbs, R.A. Genome-wide linkage disequilibrium and haplotype maps. *Am. J. Pharmacogenom. Genom.-Relat. Res. Drug Dev. Clin. Pract.* **2004**, *4*, 253–262. [CrossRef]
29. Gevorgyan, Z. SIoU Loss: More Powerful Learning for Bounding Box Regression. *arXiv* **2022**, arXiv:2205.12740.

machines

MDPI

Article

Design and Experiment of Apple Foam Net Sleeve Packaging Machine with Posture Adjustment Function

Shisheng He [1], Qun Sun [1,*], Quanjin Wang [1], Zhiqin Liang [1], Ying Zhao [1], Xiuhao Yu [1] and Haigang Xu [2]

1 School of Mechanical and Automotive Engineering, Liaocheng University, Liaocheng 252059, China; heshishengde@163.com (S.H.); w17515117895@163.com (Q.W.); sisterinmyheart@163.com (Z.L.); zhaoying@lcu.edu.cn (Y.Z.); m17864878187@163.com (X.Y.)

2 Shandong Shifeng (Group) Company Ltd., Liaocheng 252899, China; tuchuanwen2@163.com

* Correspondence: sunqun@lcu.edu.cn

Abstract: Background: In this study, an apple foam net packing machine is designed to improve the efficiency of apple foam net packing, reduce the intensity of manual work, and meet the market demand for diverse forms of apple packing. Methods: Based on the analysis of apple shape characteristics, a secondary posture adjustment method for apples with a vertical posture adjustment mechanism and a side posture adjustment mechanism is proposed. Based on the observation of human hand-set nets and the analysis of the mechanical characteristics of nets, a dual-system net opening method combining a net slackening system and a net opening system is proposed. Results: The experiment shows that: the apple foam net sleeve packing machine can reach 99% of the apple's posture adjustment rate, the net opening rate can reach 99%, and the packing rate can reach 98.10%. The packing speed is 10 s/pcs, and the packing efficiency is 355–365 pcs/h. The length of the fused mesh net sleeve is mainly distributed between 138 and 143 mm, which accounts for approximately 80% of the total number of samples. Conclusions: The apples packed by this machine are neatly packaged according to specifications. It can improve the efficiency of apple foam net sleeve packing and replace manual work.

Keywords: apple; posture adjustment; foam net sleeve; packaging; dual system open net

Citation: He, S.; Sun, Q.; Wang, Q.; Liang, Z.; Zhao, Y.; Yu, X.; Xu, H. Design and Experiment of Apple Foam Net Sleeve Packaging Machine with Posture Adjustment Function. *Machines* **2023**, *11*, 436. https://doi.org/10.3390/machines11040436

Academic Editor: Dan Zhang

Received: 14 February 2023
Revised: 18 March 2023
Accepted: 26 March 2023
Published: 29 March 2023

1. Introduction

Apples rank second globally in terms of fruit production. China is the world's largest producer and consumer of apples, accounting for more than 50% of the world's total apple acreage and production [1,2]. Apples are rich in dietary fiber, phenols, and organic acids [3,4]. It has many health benefits, such as lowering blood pressure, preventing obesity complications, and reducing cardiovascular diseases [5,6]. Global apple production continues to grow and consumption is stable, with fresh consumption of apples accounting for 70–75% of total production [7]. If fresh apples are not well packaged, fresh apples will develop wounds, peel cracks, and other mechanical damage and water loss phenomenon during logistics and handling [8,9]. This can cause apple rot, which affects the appearance of apples and reduces their freshness of apples, resulting in large losses in transportation, storage, and sales, and the loss of the most basic value of apples [10]. Before the apples are boxed, they are covered with a foam mesh sleeve, as shown in Figure 1a. The foam netting is made of polyethylene foam material. Polyethylene foam materials were developed in Europe as early as the 1970s [11,12] and have now become a new environment-friendly packaging material. It has good buffering, high thermal insulation, shock absorption, and certain freshness preservation functions, and is widely used in the field of packaging easily damaged goods, which can be a good way to reduce the problem of transport loss of apples [13–15]. At present, there are two ways of apple net sleeve packaging [16]. The first way is the packing method of apple vertical as shown in Figure 1b. That is, the net sleeve completely wrapped the flesh of the apple so that the stem and umbilical end of the apple leaked out slightly. Apples using vertical packaging can maximize the flesh of apples so that they are better protected. The second way is the packing method of the apple sideways, as shown in Figure 1c; that is, the net sleeve completely wraps the stem and navel end of the apple so that the flesh of the apple is exposed. The side of apples allows the apples to be partially exposed all around for easy labeling. After the apples are placed in the store or supermarket, it is easy to display information such as the color of the apples and whether they have defects, which is beautiful and easy for consumers to select. However, the process of apple sleeve net packaging is still dominated by the traditional manual method as shown in Figure 1d. This leads to a low efficiency in apple packaging and inconsistent packaging specifications.

Developed countries have researched fruit, vegetables, and other food processing and packaging machine for a longer period of time, combining advanced technology, personalized packaging forms, and a high degree of intelligence to get results [17–19]. Lino Antoni Giefer et al. [20] used convolutional neural networks for automatic localization and pose estimation of camera-captured apples, laying the foundation for the automation of fruit packaging machine systems. Anuja Bhargava et al. [21] developed an automatic grading system based on apple quality to identify defective apples. Nabil S. M. Elkaoud et al. [22] developed a fruit-size sorting machine that can perform sorting operations on fruits such as apples and oranges. Z. Zhang et al. [23] summarized the research progress of mechanical apple harvesting technology and described the current technical bottlenecks and future development trends. M Vallone et al. [24] developed a novel wireless device for real-time mechanical impact assessment in citrus packing lines that allows timely intervention to ensure fruit quality in post-harvest operations. However, because of the different standards and environmental requirements of apple packaging in various countries, no in-depth research has been conducted on apple foam net sleeve packaging machines in developed countries.

Xu Qinchao et al. [25] developed a plastic foam mesh sleeve packaging machine for spherical fruits, which can achieve continuous packaging function for spherical fruits such as apples and navel oranges by using a flexible expanding cone and inverted pricking chuck. However, the packaging machine packaging function is single, which can only achieve irregular packaging of apples, not according to market demand to achieve two different forms of apple vertical and sideways net set packaging. Julie Hu et al. [26] developed automation equipment for apple protect foam net packaging, which used a propping apparatus and electro-mechanical clamps to realize the opening operation of the net sleeve. Hongmei Xia et al. [27] designed an Ordinary Multilayer Fruit Paper Bag Supplying Device for Fruit Bagging, which can continuously take out multilayer fruit paper bags and finish opening them from the inside. With the synchronous driving trajectory and allowable high moving speed, the developed supplying device could achieve more than 90% opening success rate and less than a 2-second opening time. Liu Ruige et al. [28] designed an Automatic Fruit Packing Machine which is compact in size and low-cost to produce. It is powered by a removable battery, can work at any site, and is capable of packing fruit in a net sleeve. However, there is no description related to the principle of net opening, and no specific machine was built. Pan Fan et al. [29] demonstrated the best mode of apple picking, which provides a theoretical basis for further research on apple-picking robots. Wei Ji et al. [30] proposed a variable damping impedance control strategy for gripper grasping force tracking based on the establishment of a contact force model to achieve the compliant grasping of apple harvesting robots and reduce mechanical damage during the process of grasping.

At present, domestic research on foam net sleeve packing machines for apples is still focused on prototype design. For the few existing packaging machines on the market, the net sleeve damage rate is high, and the packaging continuity is low. Therefore, the two net-sleeve packaging methods for apples cannot be packaged, and the packaging specifications are not uniform. In order to solve the above problems, this study designs a multi-functional apple foam net sleeve packaging machine, which can realize two different net sleeve packaging methods for apples to meet the market demand. Meanwhile, it can solve the existing problems of inconsistent apple packaging specifications, the high manufacturing cost of the packaging machine, and the high damage rate of the net sleeve.

Figure 1. (**a**) Diagram of foam net sleeve; (**b**) Apple vertical release of the net sleeve packaging form; (**c**) Apple side release of the net sleeve packaging form; (**d**) Traditional manual completion of apple net sleeve packaging.

2. Foam Net Sleeve Packaging Machine for Apple Composition and Working Principle

Preliminary research (research place: Yantai City, Shandong Province, China) has shown that there are two ways of handling apples picked from fruit trees. The first kind of apple farmers will be picked and stored in cold storage waiting for apple buyers to buy apples, and apple buyers the acquisition of apples will give the apples a set of foam net covers, this way more for small fruit farmers. The second kind of apple will be picked by the farmers themselves to sell or fixed delivery to the fruit retail enterprises, farmers, or enterprises in the sale of apples will be packaged on the foam net cover, and more so for large and medium-sized fruit farmers and fixed orchards. The packaging machine used in this study was an apple foam net sleeve packaging segment.

2.1. Foam Net Sleeve Packaging Machine for Apple Composition

As shown in Figure 2, the overall structure of the apple foam net sleeve packing machine consists of five parts: feeding device, apple posture adjustment device, net device, boxing device, and aluminum profile frame. The apple posture adjustment device includes an apple vertical posture adjustment mechanism and an apple side posture adjustment mechanism. The vertical posture adjustment mechanism is located in the center of the right frame and is used to adjust the posture of the apple for vertical placement. The side posture adjustment mechanism is located at the top of the right frame to adjust the side posture of the apple in the vertical posture. The net device includes a net slackening mechanism, opening mechanism, fusing mechanism, and traction conveying mechanism for net sleeve operation on apples. The net slackening mechanism is located at the center of the frame for the initial slackening and lifting of the net sleeve. The opening mechanism is located at the uppermost part of the frame and is used to open the net sleeve finally. The fusing mechanism is located at the right front of the frame and is used to fuse the net sleeve. The traction mechanism is located at the bottom of the frame to store the net sleeve and to pull the net sleeve back to the initial state after one working cycle. The packaging machine adopts STM32 as the core control chip, using a combination of pneumatic and electric power supply and through the action of the pneumatic circuit to make the net slack fingers, net open fingers, and various pneumatic actuators to achieve the purpose of clamping, opening, extending and returning. The main technical parameters of the apple foam net sleeve packaging machine are shown in Table 1.

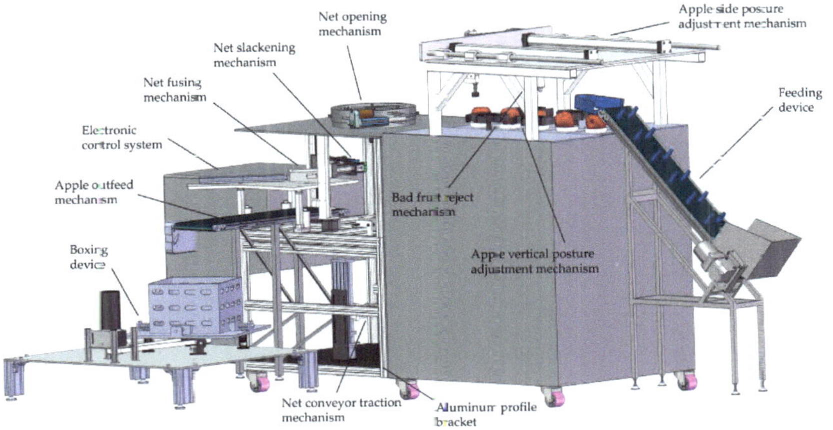

Figure 2. Three-dimensional model of apple foam net sleeve packing machine.

Table 1. Technical parameters of the proposed device.

Technical Parameters	Value
Overall size(L · W · H)/(mm · mm · mm)	3900 × 1550 × 1620
Overall machine mass/(kg)	132.4
Net slackening mechanism stepper motor torque/(N·M)	2.3
Side posture adjustment mechanism stepper motor torque/(N·M)	1.2
Net traction conveyor mechanism stepper motor torque/(N·M)	2.3
Feeding device conveyor belt line speed/(m/s)	0–0.17
Discharge mechanism conveyor belt line speed/(m/s)	0.17
Side posture adjustment mechanism rudder power/(kg·cm)	20.5
Apple posture adjustment disc rotation line speed	0.31
Electric cylinder line speed/(m/s)	0.1
Work efficiency/(pcs/h)	355–365

2.2. Foam Net Sleeve Packaging Machine for Apple Working Principle

Design the apple foam net sleeve packaging machine according to the design principle in Figure 3. In the initial state of the apple foam sleeve packaging machine, the net sleeve slackening finger is in the lower stop position, the traction finger is in the upper stop position, and the net opening mechanism is in the closed state. Before the machine starts to work, the non-cutting foam net sleeve is firstly wound by hand on the net sleeve storage tray of the traction conveying mechanism then the work object apple is put into the hopper of the feeding device and one end of the net sleeve is transported to the initial position through the guide roller and the traction roller. Turning on the machine through the control panel choose whether to use the apple vertical placement of the net sleeve packaging form or the side placement of the net sleeve packaging form. If you choose the net sleeve packaging form of apple vertical placement, the feeding device will transport the apples into the posture adjustment slot of the vertical placement posture adjustment device. The chain conveyor transports the apples in a forward motion. When the infrared tube sensor detects that the apples are transported to the last chain plate, a signal is sent to the microcontroller to stop the movement of the chain plate conveyor. After that, the vertical posture adjustment mechanism's rotating apple posture adjustment disc starts rotating movement. Then, the vertical posture adjustment mechanism's rotating apple posture adjustment disc starts rotating movement. Its wings toggle the apples in the posture adjustment slot so that the apples rotate in the slot. When the recessed area of the apple is matched with the posture adjustment ball, the apple vertical posture adjustment is completed. If the OpenCV vision module detects a bad apple during the rotation process, the bad fruit reject mechanism will reject the apple. The feeding device transports the next apple into the posture adjustment slot while the vertical posture adjustment mechanism is in operation, the net slackening mechanism of the net device is driven by a stepper motor, driven by a single line rail ball screw die set, and under the action of two pairs of slackening fingers to make the net sleeve initially loosen, then lifts the net sleeve upward 20 cm, so that the slack net sleeve covers the four net sleeve open fingers, the net slackening mechanism is reset, and the four spreader fingers of the net sleeve spreading mechanism hold the net sleeve open; the fusion push cylinder pushes the fruit carrier to move, and the electric wire in front of the fruit carrier fuses the net sleeve and then the traction conveying mechanism pulls the net sleeve back to the initial position. While the sleeve opening mechanism and the fusing mechanism operate, the vacuum suction cup of the side posture adjustment mechanism picks up the apples and delivers them to the top of the open net sleeve under the action of the push cylinder. If you choose the form of apple side net sleeve packaging, the vacuum suction cups suck up the apples and then the servo rotates 90° to complete the side adjustment, and the apples finish the net sleeve wrapping work in the air during the process of falling, and the apples with the net sleeve fall into the fruit box through the apple sending mechanism.

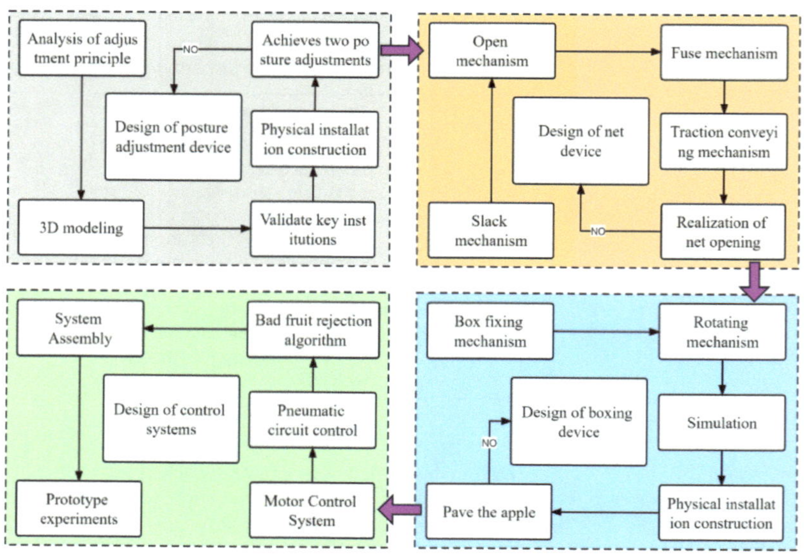

Figure 3. Machine design schematic.

3. Analysis of Apple Shape Characteristics and Net Sleeve Mechanical Properties

3.1. Apple Shape Feature Analysis

In order to determine the final area of the net opening, the type selection of the vacuum generator, and the shape of the adjustment ball in the posture adjustment slot, the shape characteristics of the apple need to be analyzed. The project group's paper [31] has given fruit shape parameters for apple fruit weight (maximum 350 g), lateral diameter (average diameter of 80.2 mm, mainly in the range of 76–88 mm), and longitudinal height (average height of 76.4 mm), therefore, in this paper, only the width and depth of the apple fruit navel are measured, and apple fruit shape parameters are as shown in Figure 4. From the market, 50 red Fuji apples were randomly selected to measure the depth of the depression and the width of the opening at the navel end of the apple using Vernier calipers, the measurement results are shown in Figure 5. According to Figure 5, the depth of the fruit navel was between 9.6 and 10.8 mm and the width was between 17 and 24 mm.

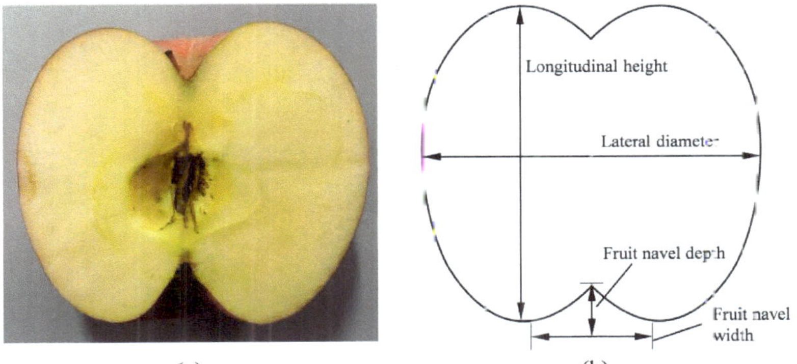

(a) (b)

Figure 4. (**a**) Real apples; (**b**) Fruit shape parameters.

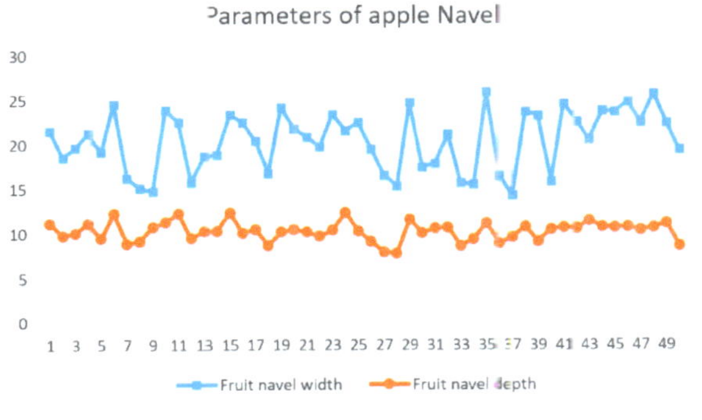

Figure 5. Apple fruit navel measurement results

3.2. Analysis of the Mechanical Properties of the Net Sleeve

The size of the opening of the net affects the wrapping property, protection, and aesthetics of the net cover for apples. As shown in Figure 6b, if the opening area of the net sleeve is too large, after the net sleeve is packed with apples, the net sleeve cannot shrink to the right position resulting in weak protection of the apples and poor aesthetics of the apples. For this purpose, the effect of the size of the opening of the net sleeve on its shrinkage characteristics has been experimented with. This experiment has the following four steps. In the first step, a roll of 20 m long and 7 cm wide non-cutting net sleeve was purchased randomly from the market, and several 14 cm long net sleeves were cut from this roll as the object to be tested. In the second step, since the average lateral diameter of an apple is 80.2 mm, we conduct a test experiment from the initial opening size of 80 mm. In the third step, a test experiment was conducted using the manner shown in Figure 6a, where the average lateral diameter apple was placed inside the net sleeve after the net sleeve was propped open, and the wrapping effect of the net sleeve was observed; In the fourth step, it was found that the net sleeve had the best wrapping effect on the apple when the opening size of the net sleeve was 130 mm as shown in Figure 6c, and the experimental results are shown in Table 2.

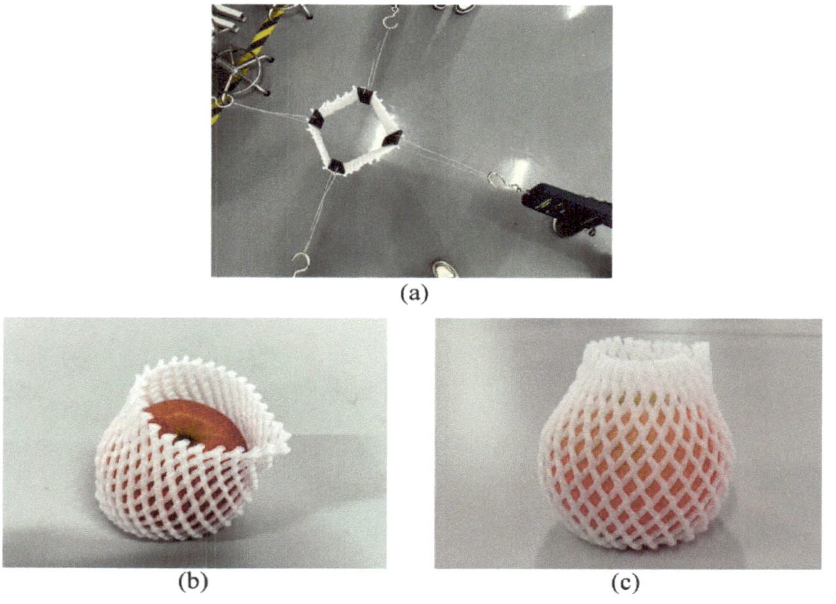

Figure 6. (**a**) Experiment on shrinkage characteristics of the net sleeve; (**b**) The net opening is too large and poorly wrapped; (**c**) Optimal wrapping of the net sleeve.

Table 2. Experimental results of the effect of the opening size of the foam mesh sleeve on its shrinkage characteristics.

Serial Number	Opening Size/(mm)	Required Tension(N)
1	80	1.82
2	85	2.03
3	90	2.24
4	95	2.38
5	100	2.42
6	105	2.53
7	110	2.74
8	115	2.93
9	120	3.15
10	125	3.28
11	130	3.42
12	135	3.55
13	140	3.67
14	145	3.78
15	150	3.96

4. Apple Posture Adjustment Device Adjustment Principal Analysis and Design

4.1. Analysis of the Adjustment Principle of the Vertical Posture Adjustment Mechanism

The apple starts to rotate and roll under the action of the posture adjustment disc, and the forces on the apple in the placement disc can be simplified, as shown in Figure 7. F_a and F_b are the friction forces of the posture adjustment disc on the apple; F_{N1} and F_{N2} are the support forces of the adjustment slot on the apple; O is the center of mass of the apple; M is the rotational force couple of the two posture adjustment discs on the apple; the points A and B are the contact points between the posture adjustment disc and the apple; the points C and D are the contact points between the adjustment slot and the apple. When the posture adjustment disc rotates under the drive of the sprocket, the apples are subjected to the frictional forces F_a and F_b of the posture adjustment discs on the apples, under the action of F_a and F_b, a rotational force couple M is generated at the apple center of mass O. Under the action of M, the apple rotates on the axis of the line where the center of mass O is located. During the rotation, the apple is in contact with the wall of the posture adjustment disc and a force F is given to the apple. The apple moves in the direction of F while rotating around the axis, when the concave part of the fruit navel is combined with the posture adjustment ball, the apple rotates on the axis of its own fruit axis without moving, and the posture adjustment of vertical placement is completed. The posture adjustment of the apple is performed entirely by the mechanical system, without the need to control the position of the apple with the visual system.

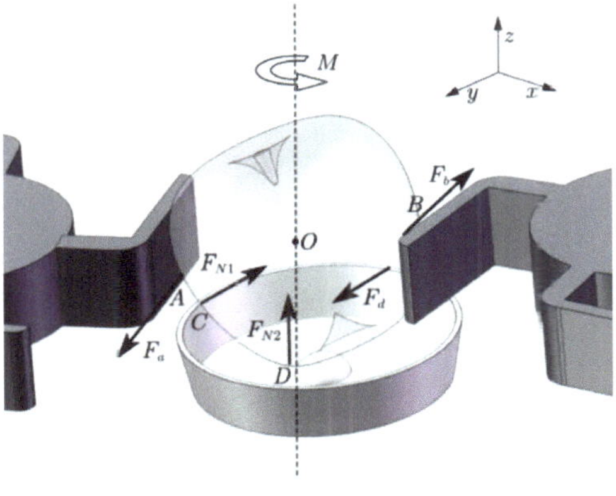

Figure 7. Apple vertical posture adjustment mechanics sketch.

In order to illustrate that the apple rotates mainly around the Z-axis, the principle of D'Alembert's principle is applied to analyze the principle of apple vertical adjustment. The equilibrium equation of the space force system is established:

$$\begin{cases} -F_{f1} + F_{f2} + F_a - F_b + F_{IY} = 0 \\ F_{N1}\sin\beta - F_{N2}\sin\lambda + F_{IX} = 0 \\ F_{N1}\cos\beta + F_{N2}\cos\lambda - G + F_{IZ} = 0 \\ M_{IY} = 0 \\ F_{f1}d_1\cos\beta - F_{f2}d_2\cos\lambda + M_{IX} = 0 \\ F_{f1}d_1\sin\beta + F_{f2}d_2\sin\lambda - F_a d_3 - F_b d_4 + M_{IZ} = 0 \end{cases} \quad (1)$$

Among them:

$$\begin{cases} F_{f1} = \mu_1 F_{N1} \\ F_{f2} = \mu_2 F_{N2} \\ G = mg \end{cases} \quad (2)$$

The solution gives:

$$\begin{cases} F_{IY} = \mu_1 F_{N1} - \mu_2 F_{N2} - F_a + F_b \\ F_{IX} = -F_{N1}\sin\beta + F_{N2}\sin\lambda \\ F_{IZ} = -F_{N1}\cos\beta - F_{N2}\cos\lambda + G \\ M_{IY} = 0 \\ M_{IX} = -\mu_1 F_{N1} d_1 \cos\beta + \mu_2 F_{N2} d_2 \cos\lambda \\ M_{IZ} = F_a d_3 + F_b d_4 - \mu_1 F_{N1} d_1 \sin\beta - \mu_2 F_{N2} d_2 \sin\lambda \end{cases} \quad (3)$$

where F_{IY}, F_{IX}, F_{IZ} represents the inertial forces added to the Y, X, and Z axes of the apple; M_{IY}, M_{IX}, M_{IZ} represents the moment of inertia added to the Y, X, and Z axes of the apple; F_{N1}, F_{N2} represents the support force of the apple posture adjustment slot on the apple; F_{f1}, F_{f2} represents the frictional force of the apple posture adjustment slot on the apple; β represents the angle between F_{N1} and Z; λ represents the angle between F_{N2} and Z; m represents the mass of the apple; d_1, d_2, d_3, and d_4 represent the distance from the contact point of C, D, A, and B to point O.

According to the actual design conditions, it is known that $F_a + F_b \gg F_{f1} + F_{f2}$, so from the solved data, we know that $M_{IY} < M_{IX} < M_{IZ}$. The apple rotates in the posture adjustment slot mainly around the Z-axis. When the recessed portion of the apple navel is combined with the stance adjustment ball, the vertical posture adjustment is completed.

4.2. Design of Vertical Posture Adjustment Mechanism

The specific structure of the vertical posture adjustment mechanism is shown in Figure 8, it consists of an apple posture adjustment disc, posture adjustment slot, chain plate, adjustment disc drive mechanism, and chain plate conveying mechanism. From the experimental analysis of apple shape characteristics, it was found that the lateral diameter of the apple was mostly in the range of 76–88 mm, the depth of the apple stalk was mostly 7–7.5 mm, and the width was mostly 20–25 mm. To ensure that the apples do not fall out of the posture adjustment disc during rotation, the height of the posture adjustment slot should be 1/3 to 1/2 of the lateral diameter of the apples, therefore, the height of the posture adjustment slot is designed as 38 mm, the height of the posture adjustment ball is designed as 10.2 mm, and the width is 20.4 mm. To reduce apple rotation damage, the wings and posture adjustment slots of the posture adjustment disc were 3D printed with flexible materials, and the apple posture adjustment slot and posture adjustment disk paste a layer of EPE pearl cotton used to protect the apple, which also increases the friction of the apple, reducing the speed of the apple rotation.

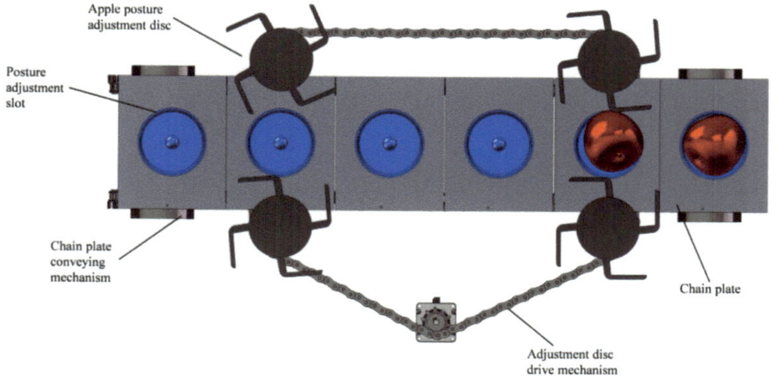

Figure 8. Apple vertical posture adjustment schematic.

4.3. Design of Side Posture Adjustment Mechanism

The role of the side posture adjustment mechanism is to adjust the side posture of the apples that have finished the vertical posture adjustment and to transport the apples that have finished the posture adjustment to the top of the net device. Its structure is shown in Figure 9, it consists of a push cylinder, guide shaft, linear slider, side drive module, suction cup drive cylinder, vacuum suction cup, etc.

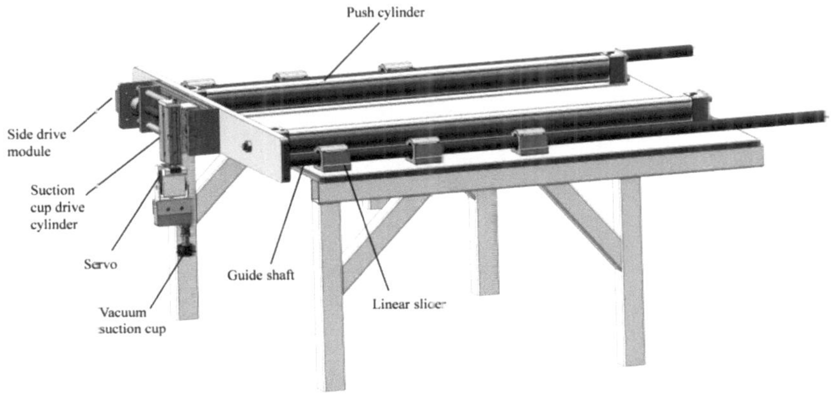

Figure 9. Apple side posture adjustment schematic.

The parameters of the vacuum generator need to be determined for the vacuum generator to have a suction force large enough to suck up the apples. The response time of the vacuum generator is 0.4 s in order to increase the packing speed. The maximum weight of the apple is 350 g.

The gravity of the apple as:

$$G_{\max} = M_{\max} g \tag{4}$$

The vacuum level P of the required vacuum generator is as:

$$p = \frac{4Wt}{\pi D^2 n} \tag{5}$$

where $G_{\max}$ represents the gravity of the load; t represents the safety factor; D represents suction cup diameter; n represents the number of suction cups.

The load gravity W is equal to the maximum weight of the apple $G_{\max}$. According to the working conditions, the pre-selection of a vacuum suction cup with a diameter of 40 mm and a safety factor of t = 8, then the data is substituted into Equation (5) to calculate the vacuum degree of 21.83 kPa.

Select the SMC ZH series vacuum generator whose maximum vacuum level is $P_v = 88$ kPa, because $P/P_v = 0.2509$, and the time required is $T = 0.32T_1$, which can be found in the vacuum generator manual. The average suction flow rate Q is:

$$Q = 0.32\frac{60v}{T} \tag{6}$$

The maximum suction flow rate Q is as follows

$$Q_e = C_q Q \tag{7}$$

where v represents the total adsorption volume; T represents the required vacuum response time; C_q represents the resistance coefficient.

The total adsorption volume is 0.1 L. Due to the small flow resistance of the vacuum line, C_q is taken as 2 and then the data is substituted into the Formula (7) to calculate $Q_e = 9.6$ L/min.

According to $Q_e = 9.6$ L/min, the manual shows that the vacuum generator ZH07BS with a nozzle diameter of 0.7 should be used, whose maximum inhalation flow is 12 L/min and average inhalation flow is 6 L/min. The actual suction response time is calculated by Equation (6) as 0.32 s, which is less than 0.4 s and meets the response requirement.

4.4. Finite Element Analysis of Guide Shaft of Side Posture Adjustment Mechanism

The guide shaft in the side posture adjustment mechanism is a cantilever beam when it is extended, and the force is mainly concentrated at the front end of the shaft. Finite element simulation analysis of the guide shaft is required to ensure that the guide shaft does not bend, deform, or break. Since the guide shaft needs to carry large forces, alloy steel with good hardness and toughness is chosen as the material whose physical properties are shown in Table 3. The total mass of the front part of the guide shaft, m = 3.3025 kg, was measured using an AnHeng ALH electronic scale (accuracy 0.01 g).

Table 3. Physical properties of alloy steel materials.

Properties	Value	Unit
Modulus of elasticity	2.1×10^{11}	N/m^2
Poisson's ratio	0.28	
Shear modulus	7.9×10^{10}	N/m^2
Mass density	7700	kg/m^3
Tension strength	723,825,600	N/m^2
Yield strength	620,422,000	N/m^2
Coefficient of thermal expansion	1.3×10^{-5}	/K
Thermal conductivity	50	W/(m·k)
Specific heat	460	J/(kg·k)

The total weight of the front-end parts of guide shaft G is as follows:

$$G = mg \tag{8}$$

Because the side-release posture adjustment mechanism consists of two guide shafts and one cylinder shaft, the force F applied to a single guide shaft is expressed as follows:

$$F = \frac{1}{3}G \tag{9}$$

where m represents the total mass of the front part of the guide shaft; g represents the acceleration of gravity, which in this case takes the value of 9.8 m/s^2.

The force on the guide shaft calculated by substituting the data into Equation (9) is 10.7881 N. Finite element simulation of the guide shaft by SolidWorks, one end of the guide shaft is fixed, and the other end is loaded with a force of 10.7881 N. The simulation results obtained are shown in Figure 10.

According to Figure 10, it can be seen that:

$$F_3 = 6.8 \times 10^6 \frac{\text{N}}{\text{m}^2}, F_4 = 6.204 \times 10^8 \frac{\text{N}}{\text{m}^2}, F_3 \ll F_4$$

where F_3 represents the maximum deformation stress of alloy steel material; F_4 stands for alloy steel material yield force.

The maximum strain of the guiding shaft is 2.384 × 10^{-5} within the allowable strain range, and the maximum deformation stress is within the yield stress range, which can bear the gravity of the front part and complete the guiding work smoothly. It meets the requirements of use.

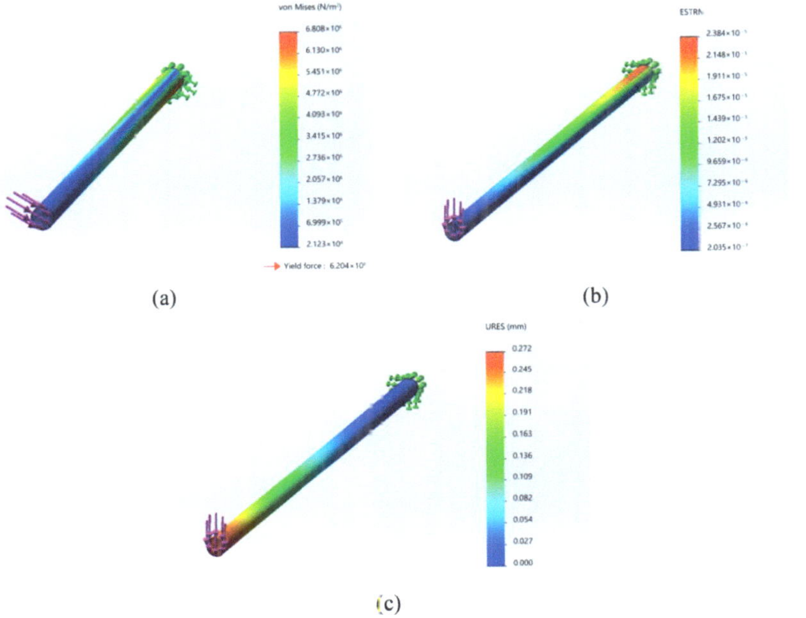

Figure 10. Results of finite element analysis: (**a**) Stress analysis; (**b**) Strain analysis; (**c**) Displacement analysis.

5. Design of Net Device

5.1. Design of Net Slack Mechanism

The purpose of the net slack mechanism is to open a small square opening in the original state of the web, providing the prerequisite for the final opening of the net, the specific structure is shown in Figure 11. The net slack mechanism mainly consists of the first pair of stepper motor filament modules, the second pair of stepper motor filament modules, the first pair of slack push cylinders, the second pair of slack push cylinders, the first pair of slack fingers and the second pair of slack fingers. As shown in Figure 12, the working process of the mesh sleeve relaxation mechanism has three steps as follows:

(1) The first pair of net slackening push cylinders drive the first pair of net slackening fingers to move, when the front of the net slackening finger is 5 mm from the edge of the net, the net slackening finger is closed to clamp the net.

(2) As shown in Figure 12a, the second pair of net slackening push cylinders push the second pair of net slackening fingers into motion, when the second pair of sleeve slack fingers moves to the end, the paired net slack fingers clamp the net by squeezing.

(3) As shown in Figure 12b, the two pairs of mesh sleeve slackening push cylinders return simultaneously to slacken the mesh sleeve.

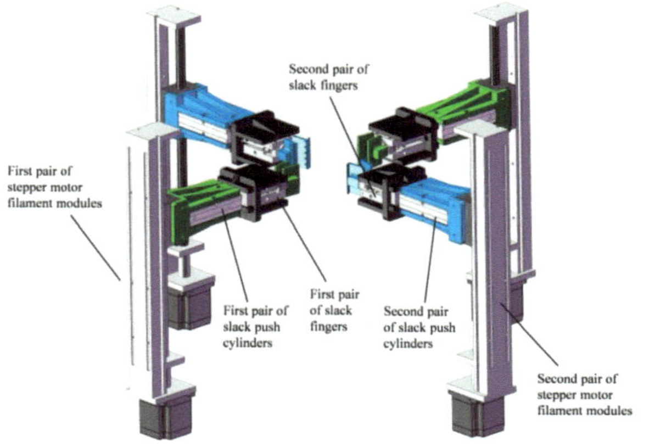

Figure 11. Schematic diagram of net slackening mechanism.

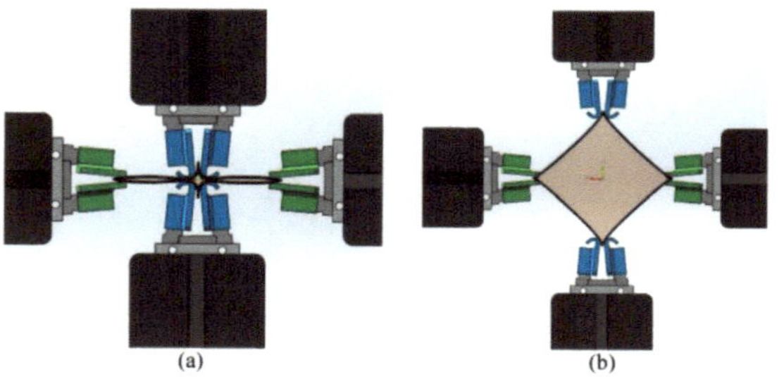

Figure 12. Net slackening mechanism working process: (**a**) Net sleeve finger clamping the net sleeve; (**b**) Net sleeve finger slack net sleeve.

In order to illustrate the principle of squeezing and clamping the net sleeve, the side of the net sleeve that is squeezed by the second pair of slack fingers of the net sleeve is selected as the object of study for force analysis of the net sleeve. The mechanical diagram is shown in Figure 13, at points A and B, two slack fingers squeeze each other to apply two pressures F_{n1} and F_{n2} to the net. When the two fingers are closed, a pair of frictional forces $F_{\mu1}$ and $F_{\mu2}$ are applied to the net under the action of pressure. As the AB length of the net is squeezed tight by the slack finger when the finger is closed the AB section of the net sleeve will do contraction movement to the center 0 under the action of friction, and finally, the net sleeve is squeezed and clamped within the finger.

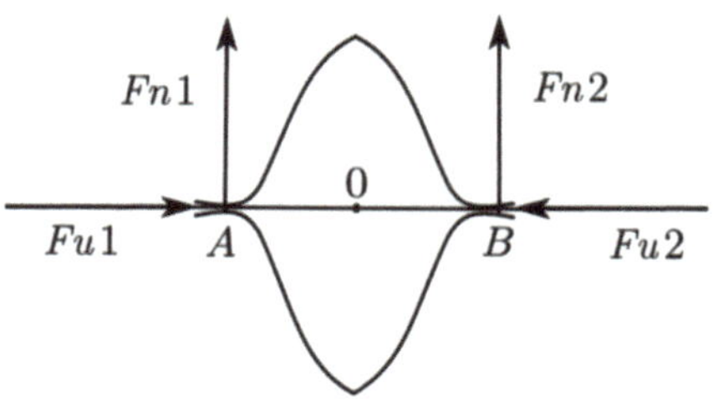

Figure 13. Mechanical sketch of net sleeve extrusion clamping.

5.2. Design of Net Opening Mechanism

The role of the net opening mechanism is to finally open the net, and its specific structure is shown in Figure 14a. It consists of an Electric cylinder, upper restraint wheel, lower restraint wheel, and open finger. Through the expansion and contraction of the electric actuator to achieve the opening and closing of the opening finger to open the purpose of the net sleeve located below the opening finger, the shape of the net sleeve is a rectangle after opening. As shown in Figure 14b in order to limit the movement of the mechanical fingers so that they can only move in the radial direction of the opening mechanism, the lower restraint wheel uses four restraint slots to limit the direction of finger movement. After fixing the lower restraint device, the upper restraint device can be rotated clockwise and counterclockwise under the action of the electric cylinder. For the upper restraint device to drive the four spreader fingers to open and close, the slide curve of the upper restraint device needs to be parameterized.

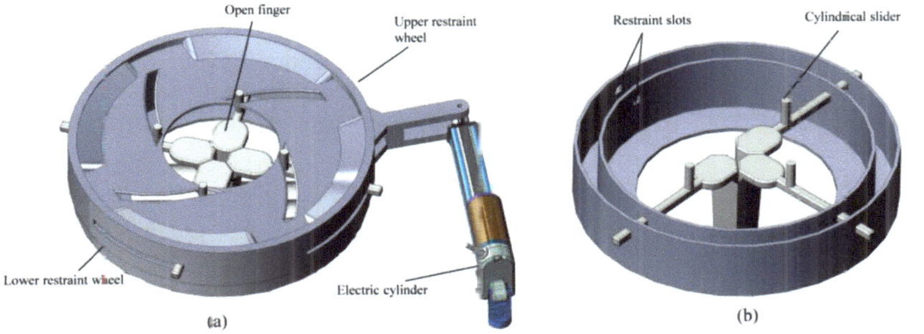

Figure 14. (**a**) Schematic diagram of the net opening mechanism; (**b**) Schematic diagram of lower restraint wheel.

As the cylindrical slider moves away from or near the center of the restraining wheel, it is the groove edge of the upper restraining wheel that provides power to the slider. As shown in Figure 15a, the force analysis of the propped finger is carried out with the position of the circle center of the cylindrical slider as the coordinate origin.

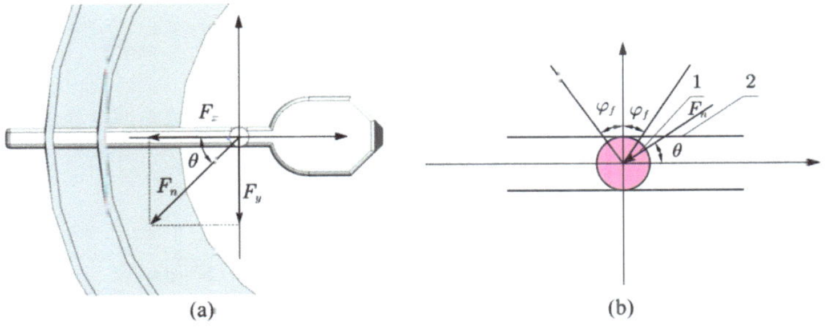

Figure 15. (**a**) Finger mechanics analysis; (**b**) Self-locking phenomenon.

Within the figure, F_n represents the combined force of the slide on the finger; F_x represents the component force of F_n in the direction of finger movement; F_y stands for frictional resistance; θ represents the angle between the combined force and the direction of motion.

The side walls of the slots and the sides of the cylindrical sliders belonging to the opening fingers are in contact with each other to prevent relative sliding. As shown in Figure 15b, due to the self-locking phenomenon, if the line of action of the combined force F_n of all the main forces acting on the opening finger is within the friction angle and is directed toward the support surface, no matter how strong the force is, the opening finger must remain stationary. If the line of action of the combined force F_n of all the main forces is outside the friction angle, no matter how small this corner is, the opening finger must slide. The friction angle was calculated as follows:

$$\varphi_f = \arctan f_s \tag{10}$$

$$\begin{aligned} \tan\varphi_f &= \frac{F_{\max}}{F_N} \\ &= \frac{f_S F_N}{F_N} \\ &= f_s \end{aligned} \tag{11}$$

where $F_{\max}$ represents the maximum frictional resistance; f_s stands for coefficient of sliding friction.

Therefore, it is only necessary to ensure the mechanical fingers slide smoothly along the slots by ensuring that $90° - \theta \geq \varphi_f$. Based on the self-locking principle, this paper provides the following two design solutions for the slide curve. The first one is the equiangular spiral line type chute curve; the second one is the circumferential type chute curve. The first of these is given only for theoretical verification, and the second scheme is used for actual manufacturing.

5.2.1. Design of Equiangular Spiral Type Chute Curve

Suppose there is a chute trajectory curve as shown in Figure 16a, in which O point is the center position of the opening mechanism, and the angle a remains constant during the movement of the open finger, then two straight lines tangent to the slot curve can be obtained through points A and A', and the angle between these two lines and OA and OB will always be$_{\alpha}$. Specify that k_1 represents a line tangent to the chute curve; k_2 represents the slope of the line segment formed by the point O joining the tangent point. A further simplified model is shown in Figure 16b.

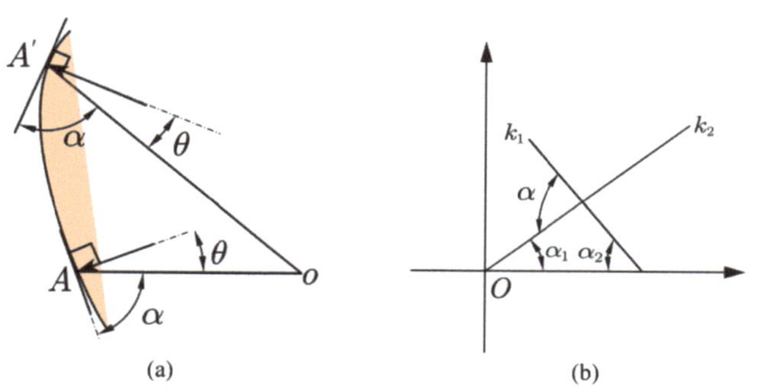

Figure 16. (**a**) Hypothetical chute curve; (**b**) Simplified model in polar coordinates.

According to the external angle property of the triangle and the formula of two angles and tangent, Equation (12) can be obtained.

$$\begin{cases} \tan\alpha = \tan(\alpha_1+\alpha_2) = \frac{\tan\alpha_1 + \tan\alpha_2}{1-\tan\alpha_1\tan\alpha_2} \\ \tan\alpha_1 = k_2, \tan\alpha_2 = -k_1 \\ \tan\alpha = \frac{k_2-k_1}{1+k_2k_1} = C \end{cases} \tag{12}$$

Establish the polar equation:

$$\rho = \rho(\alpha) \tag{13}$$

$$\begin{cases} x = \rho(\alpha) \cos \alpha \\ y = \rho(\alpha) \sin \alpha \end{cases} \tag{14}$$

$$\begin{cases} k_1 = \frac{dy}{dx} = \frac{\rho \cos \alpha + \rho' \sin \alpha}{\rho' \cos \alpha - \rho \sin \alpha} \\ k_2 = \frac{y}{x} = \tan \alpha \end{cases} \tag{15}$$

Substituting k_1 and k_2 into $\frac{k_2 - k_1}{1 + k_2 k_1} = C$ to organize them gives:

$$-\frac{\rho}{\rho'} = c \tag{16}$$

It can also be written as $\frac{\rho'}{\rho} = c_2$.
Simplify to obtain:

$$\begin{cases} (\ln \rho)' = c_2 \\ \ln \rho = c_2 \alpha + c_3 \\ \rho = e^{c_2 \alpha + c_3} = e^{c_3} e^{c_2 \alpha} \\ a = e^{c_3} \\ b = c_2 \end{cases}$$

From the above calculation, the equation of the curve can be obtained as:

$$\rho = a e^{b\alpha} (a \geq 0) \tag{17}$$

This curve is an equiangular spiral, and the curve parameters can be determined according to the actual working conditions during the application. It is only necessary to ensure that $90° - \theta \geq \varphi_f$, this curve can make the radial motion of the propped finger in the slot.

5.2.2. Design of Circumferential Type Chute Curve

As shown in Figure 17a, suppose the trajectory curve to be sought is part of a circle, the trajectory circle intersects the outer circle with point B, point B is always on the Y-axis, and the center A of the trajectory circle is always on the X-axis.

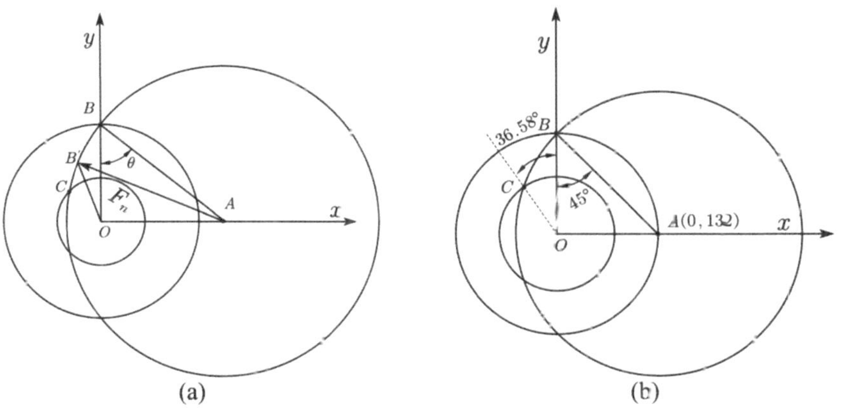

Figure 17. (**a**) Circumferential chute curve trajectory; (**b**) Circumferential chute curve trajectory after determining parameters.

In order to minimize frictional resistance, the upper constraining circle and the opening finger are made using smooth photosensitive resin material using 3D printing technology. The friction coefficient f_s of photosensitive resin material is equal to 0.21 [32]. The magnitude of the friction angle is calculated by Equation (10) as 11.86°.

When the upper constrained circle is rotated from point C to point B, Equation (18) is obtained from the sine theorem.

$$\frac{AO}{\sin \angle AB'O} = \frac{AB'}{\sin \angle B'OA} \tag{18}$$

During the movement of the upper constrained circle from point C to point B, $sin\angle B'OA$ gradually increases, and since AB' is a constant value, the value on the right side of equation (18) decreases. Since the value of AO remains unchanged, the value of $\angle AB'O$ gradually increases, and $\angle AB'O$ has a maximum value when the upper constraint circle moves to point B. Based on the above analysis, just make sure that $90° - \theta \geq \varphi_f$ at point B, that is, when AD is less than 78.14°, the finger can move in the chute.

Based on the test experiment of the effect of the size of the opening of the foam mesh sleeve on its shrinkage characteristics, in order to make the mesh sleeve can achieve the best effect of wrapping the apples. The distance between the two mechanical fingers facing each other when the mechanical fingers are open is designed to be 130 mm. Since the gap between the two mechanical fingers facing each other when the mechanical fingers are closed is 16 mm, the distance L_1 for the mechanical fingers to move is as follows: $L_1 = \frac{130-16}{2} = 57$ mm. The diameter of the smaller concentric circle D_1 is specified as 150 mm according to the actual demand, and the values of the diameters D_2 of the larger concentric circles are as follows: $D_2 = 150 \text{ mm} + 57 \text{ mm} \times 2 = 264$ mm. After determining the size of the two concentric circles and the size of circle A, $\overset{\frown}{BC}$ is the required track.

Specify $\angle OBA = 45°$, and according to the sine theorem, the value of AB is calculated as follows:

$$AB = \frac{BO}{\sin 45^{\circ}} \sin 90^{\circ} \tag{19}$$

Substituting BO = 132 into Equation (19), the radius of circle A is calculated to be 186.68. The coordinate of A is (0.132).

The equation of circle A is as follows:

$$(x-132)^2 + y^2 = 186.68^2$$

The equation of the smaller concentric circle and circle A are combined to obtain the following equation:

$$\begin{cases} (x-132)^2 + y^2 = 186.68^2 \\ x^2 + y^2 = 75^2 \end{cases}$$

Solving the system of Equation (19) yields the following calculation:

$$\begin{cases} x = -44.69 \\ y = 60.23 \end{cases}$$

$$\angle BOC = \left| \tan^{-1} \frac{x}{y} \right| \tag{20}$$

Substitute x = −44.69, y = 60.23 into Equation (20) to calculate $\angle BOC = 36.58°$, the angle occupied by $\overset{\frown}{BC}$ is 36.58°. Therefore, the upper restraint wheel only needs to turn 36.58° to complete the opening and closing action of the mechanical finger. The circumferential chute curve after the design parameters are determined is shown in Figure 17b.

5.2.3. ADAMS Simulation Analysis of the Circumferential Opening Mechanism

Dynamic analysis was performed using the ADAMS View software to verify the rationality of the chute curve design of the circumferential opening mechanism. Build a model of the opening mechanism in SolidWorks according to the circumferential chute curve. As shown in Figure 18, the model was imported into the ADAMS View software, and a coordinate system centered on the circular position of the opening mechanism was established. The simulation results are shown in Figure 18. First, the material parameters of the parts are edited, the upper constraint wheel is set as a rotating sub-connection and the lower constraint wheel as a fixed one, then the binding force is added to each part in turn, and finally, the simulation analysis is performed.

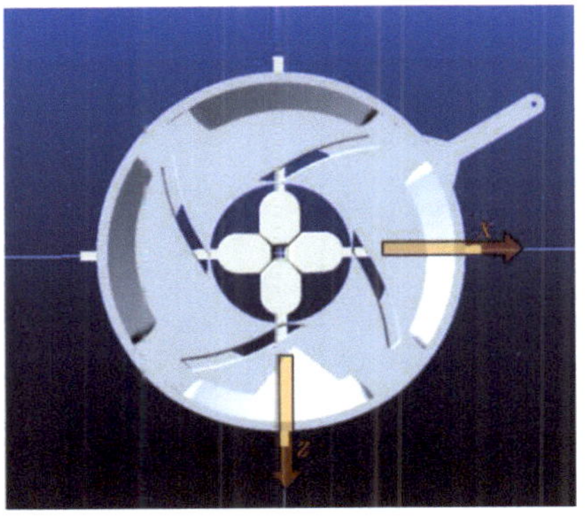

Figure 18. ADAMS View diagram of the coordinate system of the opening mechanism.

According to Figure 19a, the initial positions of the upper and lower opening fingers on the Z-axis are −43.125 and 43.125, respectively, and the end positions are −100.1702 and 100.1702, respectively. The distance traveled in the Z direction by both braced fingers is L_2 = 100.1702 − 43.124 = 57.0452 mm, within the allowable error range of the theoretical design parameter of 57 mm. According to Figure 19b, the initial coordinates of the upper constrained wheel center of mass are A (16.5638, −9.5717), the coordinates of the end position are B (7.5643, −17.5095), and the rotation angle is as follows: $\alpha = \tan^{-1}\left(\frac{-17.5095}{7.5643}\right) - \tan^{-1}\left(\frac{-9.5717}{16.5638}\right) = 36.61^{\circ}$, within the allowable error range of the theoretical design parameter of 36.58°mm.

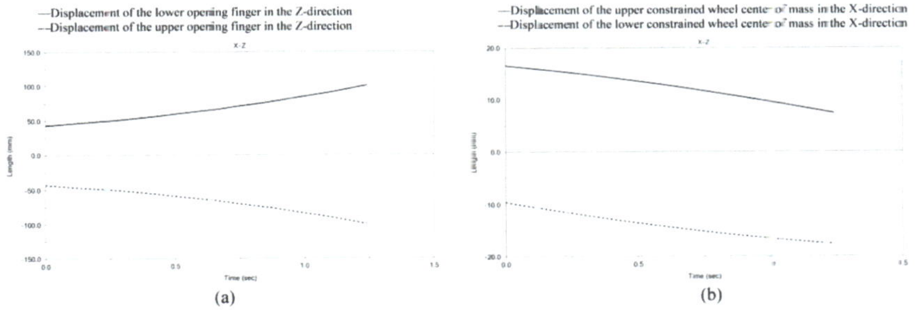

(a) (b)

Figure 19. ADAMS View simulation results: (**a**) Open finger displacement in the Z direction; (**b**) Displacement of the upper constraint wheel in the X and Z directions.

The above analysis results show that the theoretical design and simulation verification results of the chute curve of the circumferential opening mechanism are the same within the tolerance range. The circumferential opening mechanism made in this paper can brace the mesh sleeve while ensuring the best wrapping effect of the net sleeve.

5.3. Design of Net Traction Conveying Mechanism

The role of the net traction conveyor is to convey the net sleeve and pull the fused net sleeve back to the initial position. The specific structure is shown in Figure 20, which consists of a traction drive ball screw, traction finger, traction roller, guide roller, and net sleeve storage tray. To prevent the net sleeve from swinging during lifting or pulling down, cylindrical projections and grooves are designed to match each other on the cylindrical surface of the traction rollers. Based on this design, the net sleeve can be confined inside so that the net sleeve can only move in the direction of lifting or pulling down to prevent the net sleeve from swinging from side to side.

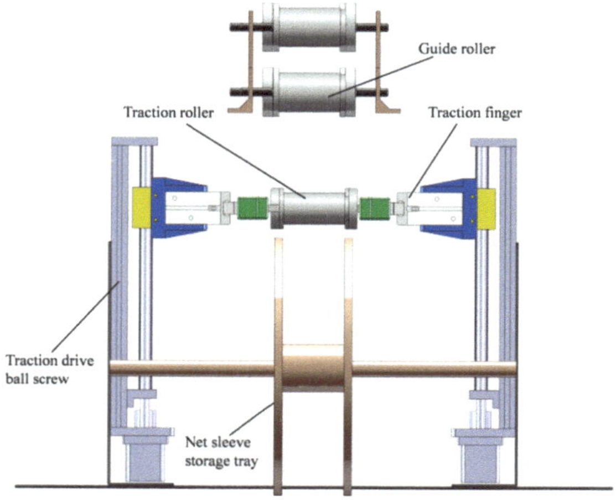

Figure 20. Schematic diagram of traction conveying mechanism.

6. Design of Control System

6.1. The Overall Design Idea of the Control System

This machine adopts an STM32F1 series microcontroller as the core chip of the apple foam net sleeve packaging machine control system. Control system hardware includes: Servo for controlling the rotation of the vacuum suction cup (Model: TD-8120MG, blocking torque 20 kg/cm^3); Dual optical axis ball screw stepper motor module to control apple side release adjustment and traction finger up/down movement (Model: GGP8080, effective stroke 200 mm); 57 stepper geared motor controlling the rotation of the apple posture adjustment disc (Model: 23HDB76-30PG, torque 2.3 Nm); Single line rail ball screw module for controlling up and down movement of slack fingers (Model: SGX, effective stroke 250 mm); Electric cylinder to control the opening and closing of the opening mechanism (Speed 100 mm/s, effective stroke 150 mm); DC geared motor for controlling the discharge conveyor (Model: 5D120-GU-2, power 120 W); Solenoid valves for controlling each pneumatic actuator(Model: two-position five-way 4V210-08); Relay to control solenoid valve on/off (Model: JQC-3FF-S-Z); DC24V switching power supply; Infrared pair of tube sensors (Model: Opposing photoelectric switch NPN type, opposing distance: 1000 mm); Stepper motor driver (Model: PFDE-DM542, built-in three generations of 32-bit DSP processor).

This machine uses a combination of electric and pneumatic components to actuate, with a combination of interrupt and timer control to link each control hardware of the packaging machine. Control the work of each electric element and pneumatic element with the feedback signal from the infrared pair of tube sensors as a sign. Precise control of stepper motor ball screw module slider position, stepper motor speed, servo rotation angle, relay and solenoid valve on and off, pneumatic finger opening and closing, cylinder extension and return. Whether the apples need to be conveyed in the packaging machine, whether the apple posture adjustment plate needs to adjust the posture of the apples, when the propping mechanism opens, and when the traction conveying mechanism pulls the net sleeve, all need to be judged according to the position of the apples. The machine uses an infrared tube sensor to detect the position of the apple and feedback to the microcontroller to control each mechanism to execute the corresponding action. The control timing of the machine is shown in Figure 21.

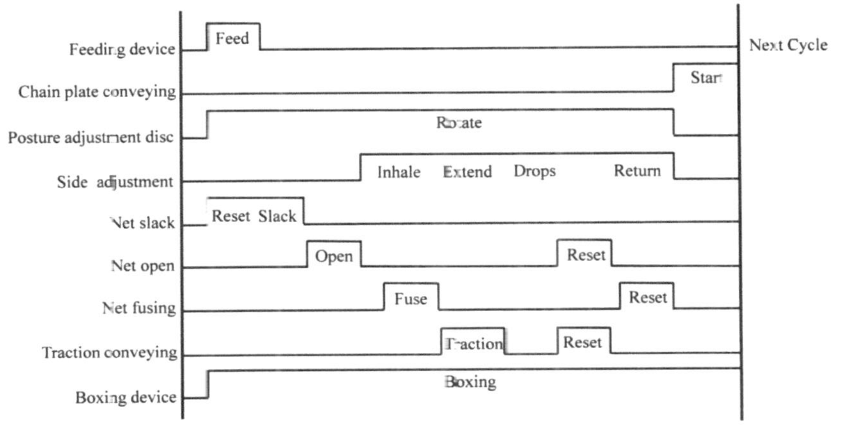

Figure 21. Control Timing Chart.

6.2. Design of Motor Drive System

The packaging machine uses the STM32F103VET6 main control chip with stepper motor driver output PWM pulse width adjustable square wave signal to accurately control the speed of the stepper motor. The stepper motor driver has three signal terminals: pulse signal (PUL), direction signal (DIR), and enable signal (ENA), respectively. The driver and the controller have two wiring methods: a common cathode connection and a common anode connection, the common anode connection is used in this paper. Since the model 57 stepper motor is used, the power supply voltage of the stepper motor driver is DC24V. The relationship between timer frequency and speed needs to be determined for accurate control of stepper motor speed, the speed n of the stepper motor can be obtained from Equation (21).

$$n = \frac{72 \times 10^6 S}{360X(TIM3\_ARR + 1)(TIM3\_PSC + 1)} \tag{21}$$

where S represents the step angle of the stepper motor; X represents the subdivision value of the drive; $TIM3\_ARR$ represents the value of the Auto Reload register; $TIM3\_PSC$ represents the value of the Prescaler.

The machine uses a stepper motor with a steep angle of $S = 1.8°$ and a drive subdivision value of $X = 16$, requiring a speed of $n = 12.5$ r/s, substitute the above data into Equation (21) to obtain $TIM3\_ARR = 299$ and $TIM3\_PSC = 599$.

The control of the servo uses the PWM1 output comparison. In order to make the microcontroller feedback a signal that can make the servo rotate 90°, it is necessary to calculate the relationship between the servo rotation angle and the pulse width (duty cycle). The machine uses a 180° servo and uses a pulse width of 20 ms to modulate PWM, with a pulse width of 0.5–2.5 ms partly used to control the servo angle. The angle duty cycle of the servo is between 0.025 and 0.125, and the servo rotation angle is 90° when the pulse width is 1.5 ms. Set the value of Auto Reload Register ($TIM4\_ARR$) to 999 and the value of Prescaler ($TIM4\_PSC$) to 1439 The value of duty cycle P can be calculated from Equation (22).

$$P = \frac{CRR}{ARR + 1} \tag{22}$$

$$Out\_angle = \frac{\left(0.5 + \frac{angle}{180} \times (2.5 - 0.5)\right)}{20.0} \times 1000 \quad (23)$$

The initial *CRR* value is 25, and the duty cycle *P* is 0.025 by substituting the *ARR* of 999 into Equation (22), indicating that the servo rotation angle is zero currently. When an angle is 90° use Equation (23) to calculate *Out_angle* is 75, that is, the value loaded to the comparison register *CRR* is 75. Substituting the value of *CRR* into Equation (22), the calculated *P* is 0.075, indicating that the rudder rotation angle is 90° currently. Using the above design, it is possible to achieve boundary processing for duty cycles that exceed the range, accurately output the required pulse width, and precisely control the servo rotation angle. The machine control architecture is shown in Figure 22.

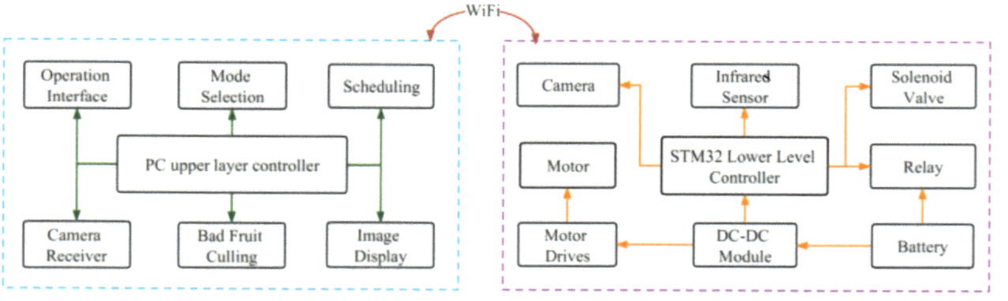

Figure 22. Control architecture for machines.

7. Packaging Machine Prototype Experiment

As shown in Figure 23, according to the above mechanism design ideas and principles using SolidWorks to complete the virtual prototype construction work and processed the prototype of the apple foam mesh sleeve packaging machine. In order to test the apple posture adjustment rate and netting success rate of the prototype, 100 red Fuji apples were selected from the market for the prototype experiment. The average lateral diameter of the sample apples was 80 mm, and the average longitudinal height was 70 mm. Since this machine has both an apple posture adjustment function and a net packing function, the product of the posture adjustment success rate and the net success rate is used as the packing success rate of this machine. The working screen of the nesting device is shown in Figure 24. The experimental results are shown in Table 4.

Figure 23. Prototype apple foam net sleeve packaging machine.

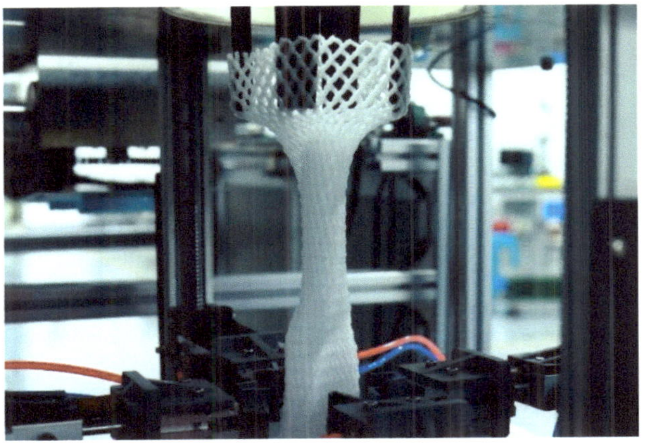

Figure 24. Working screen of the nesting device.

Table 4. Apple foam net packaging machine experimental results.

Total Sample /pcs	Length of Packing Net/mm	Quantity/ pcs	Posture Adjustment Rate/%	Net Rate/%	Packing Rate/%	Packaging Speed /(s/pcs)	Work Efficiency /(pcs/h)
100	135–137 mm	8	99%	99%	98.10%	10	355–365
	138–140 mm	41					
	141–143 mm	39					
	144–146 mm	12					

The experimental results show that the apple foam net sleeve packaging robot can complete the basic functions of apple feeding, posture adjustment, net sleeve conveying and traction, net sleeve slackening, net opening, and so on. As can be seen from Table 4, the packing rate is 98.10%, the cut mesh sleeve is mainly within 138–143 mm, the single fruit packing speed is 10 s/each, and the working efficiency is 355–365 pcs/h. The packaging machine-working objects are not only limited to apples. As shown in Figure 25, the posture adjustment device can be used for fruits with similar shape characteristics to apples and with depressions at the navel and stalk ends, such as crown pears and autumn moon pears, the netting device can be used for fruits with a diameter of less than 130 mm, such as oranges, pears, oranges, kiwis and other spherical fruits.

Figure 25. Packaging machine to work on other fruits: (**a**) Posture adjustment device to work on crown pear; (**b**) Posture adjustment device to work on Akizuki pear.

8. Discussion

In this study, we propose a secondary posture adjustment method based on the analysis of the shape characteristics of apples with a vertical posture adjustment mechanism and a side posture adjustment mechanism. Based on the observation of human hand-set nets and the analysis of the mechanical characteristics of nets, a dual-system net opening method combining a net slackening system and a net opening system was proposed for fruit top-down mesh sleeve bottom-up. The research in this study is as follows:

(1) Experimental basis: Experiments on the analysis of apple shape characteristic parameters and the effect of foam mesh sleeve opening size on its shrinkage characteristics were conducted to provide an experimental basis for the selection of packaging machine design parameters.
(2) Principal analysis: Analysis of the principle of vertical release attitude adjustment based on D'Alembert's principle and analysis of the principle of the chute curve of the opening mechanism based on the self-locking principle.
(3) Design simulation: SolidWorks-based 3D virtual prototype construction, mechanical analysis of the apple foam sleeve wrapping machine, and ADAMS View-based dynamics analysis of the sleeve opening mechanism chute curve.
(4) Prototype experiments: From Table 4 in Section 7, it can be seen that the packing rate is 98.10%, the single-fruit packing speed is 10 s/pcs, and the working efficiency is 355–365 pcs/h. The operation of the packaging machine is not only limited to apples but also to pears, peaches, oranges, and other spherical fruits.

8.1. The Advantages of This Study

At present, the apple foam net sleeve packaging machine net sleeve method mostly uses the fruit from the bottom-up net sleeve from the top-down cylindrical sleeve with mechanical clamping jaws of the multi-actuator net sleeve method and does not have the apple posture adjustment function. Xu et al. developed a packaging machine with a single fruit packing speed of 25 s/pcs, and Hu et al. developed a packaging machine with a single fruit packing speed of 4.6 s/pcs [25–28]. This nesting method is effective for low-speed nesting packages and when the foam nesting material is suitable. However, the clamping force of the mechanical jaws, frictional force on the outer surface of the cylindrical sleeve, and complex links of this method are ignored. For example, improper control of the clamping force of the mechanical gripping jaws causes the net sleeve to be easily pulled off when sliding on the cylindrical sleeve; excessive friction on the outer surface of the cylindrical sleeve causes the net sleeve to slide on the outer surface of the cylindrical sleeve at a slower rate or fail to slide; this method requires a specific sleeve link to wrap the open net sleeve over the apple after opening the net sleeve. These factors can reduce the utilization of the net sleeve and the packaging efficiency. Moreover, most of the existing apple foam net sleeve packaging machines do not have an apple posture adjustment function and cannot meet the market demand for apples to achieve two types of net sleeve packaging, namely, vertical net sleeve packaging and side net sleeve packaging, which makes it difficult for traditional packaging machines to adapt to the changes in the apple packaging market.

The advantages of this study are as follows: First, based on the observation of human manual set nets and net sleeve opening size on the shrinkage characteristics of the experiment, a fruit top-down net sleeve bottom-up net slackening system and a net opening system combined with a dual-system net opening method to improve the net sleeve packaging efficiency and reduce the rate of net sleeve damage. In this method, the impact force generated by the potential energy of the apples when they fall is cleverly used to allow the apples to be covered by foam netting in the process of falling, reducing the need for specific sleeve links; the opening mechanism designed by this research only uses one electric actuator to open the net sleeve, which reduces the manufacturing cost and improves the packaging efficiency compared with the traditional multiple cylinders moving simultaneously to open the net sleeve; compared with the traditional mechanical jaws and cylindrical sleeve opening method, this method eliminates the traditional cylindrical sleeve and mechanical jaws, and adopts a new net sleeve opening method with a slackening mechanism and an opening mechanism, which reduces the damage rate of the net sleeve. As shown in Table 5 the packing time of a single fruit is 4 s/pcs when the net device works alone, which significantly improves the packing speed compared to the conventional 25 s/pcs and 4.6 s/pcs. Second, based on the analysis of the shape characteristics of apples, a secondary posture adjustment method was proposed to realize two different net sleeve packaging methods for apples, namely, vertical net sleeve packaging and side net sleeve packaging, by combining a vertical posture adjustment mechanism and a side posture adjustment mechanism. As shown in Table 6 when the posture adjustment device works alone, the adjustment time for a single fruit is 9.5 s/pcs.

Table 5. Net device experimental results.

Total Sample /pcs	Length of Packing Net/mm	Quantity/ pcs	Packaging Speed /(s/pcs)	Net Breakage Rate/%	Work Efficiency /(pcs/h)
100	135–137 mm 138–140 mm 141–143 mm 144–146 mm	6 41 41 12	4	0%	895–905

Table 6. Apple posture adjustment device experimental results.

Total Sample /pcs	Posture Adjustment Speed /(s/pcs)	Work Efficiency /(pcs/h)
100	9.5	374–384

8.2. Application Prospects and Economic Benefits

At present, with the improvement of living standards, fresh fruit consumption has become the mainstream form of fruit consumption. With the development of the e-commerce and logistics industry, to reduce fruit damage during transportation, the demand for apple net sleeve packaging is increasing, but the cost of high-efficiency low manual packaging is difficult to adapt to the current environment. This study designs and builds an apple foam net sleeve packaging machine with a posture adjustment function applied to the apple packaging field, which can improve the efficiency of apple foam net sleeve packaging, free hands to reduce manual work intensity and meet the market demand for diversified apple net sleeve packaging. The net and attitude adjustment devices of this packaging machine can be used as separate parts, which makes it easy to realize docking with the apple packaging line and has great prospects for future applications.

This study solves the existing net sleeve packaging machine in net sleeve packaging with low packaging efficiency, net sleeve damage, packaging function single problem, traditional manual work labor intensity, and low-efficiency problems. The processing and manufacturing costs of the existing apple net sleeve packaging machine are high (up to 100,000 yuan per unit), but the apple foam net sleeve packaging machine built in this study uses electricity as the main energy source of approximately 2.1 kw·h, and the processing cost is low (40,000 yuan per unit). The apple packaging method provided in this study makes apple packaging more intelligent and efficient, in line with the needs of modern society, and has good economic benefits.

9. Conclusions

(1) Traditional manual packaging methods have the following problems: manual packaging labor is costly, inefficient, and cannot be operated continuously with limited daily working hours. The existing apple net packaging machine has the following problems:

High processing and manufacturing costs (up to 100,000 yuan per unit); Large size is not adapted to the national conditions of Chinese apple retail farming; The single form of packaging cannot meet the market demand for packaging diversification; The form of net support mostly adopts the form of a sleeve with a manipulator which is easy to pull off the net sleeve causing a low utilization rate of the net sleeve. The apple foam mesh sleeve packaging machine designed in this paper has the following advantages: The machine is small in size, can operate continuously, and has a low processing cost (40,000 yuan per unit); Two different types of packaging can be realized for apples: vertical packaging and side packaging, so the machine packaging form diversification to meet the market demand; using the combination of the net slackening mechanism and the net opening mechanism to improve the utilization rate of the net sleeve. Therefore, the packaging machine designed in this paper can be a good solution to the above problems.

(2) In this paper, experiments were conducted on the effect of the opening size of the net sleeve on its shrinkage characteristics. The experimental results showed that the best wrapping effect on apples was achieved when the net opening size was 130 mm. The design and analysis of the adjustment principle of the vertical posture adjustment mechanism, the selection of the vacuum generator of the side posture adjustment mechanism, the squeezing and clamping principle of the net slackening mechanism, and the chute curve of the net opening mechanism of the apple foam net packaging robot were carried out. The analysis results showed that the upper constraint wheel slideway curve can be equiangular helix type or circumferential type, this paper adopts circumferential type for physical design. The calculation results showed that the finger can be moved in the slide groove when the control ∠AOB was less than 78.14°. According to the actual working condition, ∠AOB = 45° was selected and the rotation angle of the upper constraint wheel was calculated to be 36.58°.

(3) SolidWorks software for static analysis of the guiding axis of the apple side release posture adjustment mechanism. According to the simulation results, we can see that $F_3 = 6.8 \times 10^6 \frac{\text{N}}{\text{m}^2}$, $F_4 = 6.204 \times 10^8 \frac{\text{N}}{\text{m}^2}$, $F_3 \ll F_4$. The maximum strain of the guiding shaft was 2.384 within the allowable strain range, and the maximum deformation stress was within the yield stress range, which can withstand the gravity of the front part and complete the guiding work smoothly to meet the requirements of use. ADAMS View software was used to analyze the dynamics of the circumferential bracing mechanism to verify the rationality of the slide curve design. According to the simulation results, it can be seen that the travel distance of the opening finger in the Z direction is 57.0452 mm, which is within the error tolerance of the theoretical design parameter of 57 mm, the rotation angle α = 36.61° is within the error tolerance of 36.58° of the theoretical design.

(4) Based on the 3D model and the results of theoretical analysis, a physical prototype of the apple foam net packaging machine was built, and apple packaging experiments were conducted. Prototype experimental results showed that the machine can realize the continuous operation of feeding, posture adjustment, and nesting and boxing of red Fuji apples. The success rate of machine posture adjustment was 99%, the success rate of the net sleeve was 99%, the packing rate was 98.10%, the fused net sleeve was mainly within 138–143 mm, the packing speed of single fruit was 10 s/pcs, the working efficiency was 355–365 pcs/h and the working object was not only limited to apple, but also to pear, peach, orange, and other spherical fruits.

In the future, the project team will further optimize the structure of the posture adjustment device to improve the machine's inclusiveness for different varieties of apples; analyze the damage mechanism of apples to reduce the damage problem of apples in the packaging process; and conduct kinematic analysis on the principle of apple posture adjustment to further understand the adjustment principle. In addition, the structure and software system will be further optimized to expand the machine into net sleeve packaging for more fruit varieties.

Author Contributions: Conceptualization, S.H. and Q.S.; methodology, S.H.; investigation, Q.W. and Z.L.; writing—original draft preparation, S.H.; writing—review and editing, S.H., Q.S. and Y.Z.; software, S.H., Q.W. and Z.L.; validation, S.H., Q.S. and X.Y.; formal analysis, S.H. and H.X.; project administration, S.H., Q.S. and Y.Z.; supervision, S.H. and Q.S.; funding acquisition, S.H. and Q.S. All authors have read and agreed to the published version of the manuscript.

Funding: This research was funded by Discipline with Strong Characteristics of Liaocheng University—Intelligent Science and Technology under Grant 319462208, and Liaocheng University Undergraduate Innovation and Entrepreneurship Training Program under Grant CXCY2022363.

Data Availability Statement: Not applicable.

Conflicts of Interest: The authors declare no conflict of interest.

References

1. Huo, X.; Liu, T.; Liu, J.; Wei, Y.; Yao, X.; Ma, X.; Lu, P. China Apple Industry Development Report in 2020. *China Fruit Veg.* **2022**, *42*, 1–6. [CrossRef]
2. Vasylieva, N.; James, H. Production and trade patterns in the world apple market. *Innov. Mark.* **2021**, *17*, 16–25. [CrossRef]
3. Bondonno, C.P.; Bondonno, N.P.; Shinde, S.; Shafaei, A.; Boyce, M.C.; Swinny, E.; Jacob, S.R.; Lacey, K.; Woodman, R.J.; Croft, K.D.; et al. Phenolic composition of 91 Australian apple varieties: Towards understanding their health attributes. *Food Funct.* **2020**, *11*, 7115–7125. [CrossRef]
4. Skinner, R.C.; Gigliotti, J.C.; Ku, K.M.; Tou, J.C. A comprehensive analysis of the composition, health benefits, and safety of apple pomace. *Nutr. Rev.* **2018**, *76*, 893–909. [CrossRef]
5. Kim, S.J.; Anh, N.H.; Jung, C.W.; Long, N.P.; Park, S.; Cho, Y.H.; Yoon, Y.C.; Lee, E.G.; Kim, M.; Son, E.Y.; et al. Metabolic and Cardiovascular Benefits of Apple and Apple-Derived Products: A Systematic Review and Meta-Analysis of Randomized Controlled Trials. *Front. Nutr.* **2022**, *9*, 766155. [CrossRef]
6. Liddle, D.M.; Lin, X.; Cox, L.C.; Ward, E.M.; Ansari, R.; Wright, A.J.; Robinson, L.E. Daily apple consumption reduces plasma and peripheral blood mononuclear cell-secreted inflammatory biomarkers in adults with overweight and obesity: A 6-week randomized, controlled, parallel-arm trial. *Am. J. Clin. Nutr.* **2021**, *114*, 752–763. [CrossRef]
7. Lyu, F.; Luiz, S.F.; Azeredo, D.R.P.; Cruz, A.G.; Ajlouni, S.; Ranadheera, C.S. Apple Pomace as a Functional and Healthy Ingredient in Food Products: A Review. *Processes* **2020**, *8*, 319. [CrossRef]
8. Basumatary, I.B.; Mukherjee, A.; Katiyar, V.; Kumar, S. Biopolymer-based nanocomposite films and coatings: Recent advances in shelf-life improvement of fruits and vegetables. *Crit. Rev. Food Sci. Nutr.* **2022**, *62*, 1912–1935. [CrossRef]
9. Hussein, Z.; Fawole, O.A.; Opara, U.L. Harvest and Postharvest Factors Affecting Bruise Damage of Fresh Fruits. *Hortic. Plant J.* **2020**, *6*, 1–13. [CrossRef]
10. Zhou, T.T.; Sun, X.L.; Sun, Z.Z.; Peng, H.H.; Sun, T.; Hu, D. Current Status and Future Perspective of Spectroscopy and Imaging Technique Applications in Bruise Detection of Fruits and Vegetables: A Review. *Spectrosc. Spectr. Anal.* **2022**, *42*, 2657–2665. [CrossRef]
11. Fried, M.; Zuern, L.; Stahnecker, E. Manufacture of Crosslinked Foamable Moldings from Olefin Polymers. U.S. Patent 4,166,890, 4 September 1979.
12. Kuhnigk, J.; Standau, T.; Dörr, D.; Brütting, C.; Altstädt, V.; Ruckdäschel, H. Progress in the development of bead foams—A review. *J. Cell. Plast.* **2022**, *58*, 707–735. [CrossRef]
13. Jiang, J.; Feng, W.; Zhao, D.; Zhai, W. Poly(ether imide)/Epoxy Foam Composites with a Microcellular Structure and Ultralow Density: Bead Foam Fabrication, Compression Molding, Mechanical Properties, Thermal Stability, and Flame-Retardant Properties. *ACS Omega* **2020**, *5*, 25784–25797. [CrossRef] [PubMed]
14. Lin, M.; Chen, J.; Chen, F.; Zhu, C.; Wu, D.; Wang, J.; Chen, K. Effects of cushioning materials and temperature on quality damage of ripe peaches according to the vibration test. *Food Packag. Shelf Life* **2020**, *25*, 100518. [CrossRef]
15. Raps, D.; Hossieny, N.; Park, C.B.; Altstädt, V. Past and present developments in polymer bead foams and bead foaming technology. *Polymer* **2015**, *56*, 5–19. [CrossRef]
16. Meng, X.; Liu, F.; Guo, C.; Ge, H.; Wu, H.; Wang, L.; Cai, F.; Xie, G. A Kind of Automatic Apple Sleeve Net Case Packing Machine. CN108327967B, 25 September 2018.
17. Occhiuzzi, C.; D'Uva, N.; Nappi, S.; Amendola, S.; Giallucca, C.; Chiabrando, V.; Garavaglia, L.; Giacalone, G.; Marrocco, G. Radio-Frequency-Identification-Based Intelligent Packaging: Electromagnetic Classification of Tropical Fruit Ripening. *IEEE Antennas Propag. Mag.* **2020**, *62*, 64–75. [CrossRef]
18. Rodriguez-Parada, L.; Mayuet, P.F.; Gamez, A.J. Custom Design of Packaging through Advanced Technologies: A Case Study Applied to Apples. *Materials* **2019**, *12*, 467. [CrossRef]
19. Suprem, A.; Mahalik, N.; Kim, K. A review on application of technology systems, standards and interfaces for agriculture and food sector. *Comput. Stand. Interfaces* **2013**, *35*, 355–364. [CrossRef]
20. Giefer, L.A.; Arango Castellanos, J.D.; Babr, M.M.; Freitag, M. Deep Learning-Based Pose Estimation of Apples for Inspection in Logistic Centers Using Single-Perspective Imaging. *Processes* **2019**, *7*, 424. [CrossRef]

21. Bhargava, A.; Bansal, A. Machine learning based quality evaluation of mono-colored apples. *Multimed. Tools Appl.* **2020**, *79*, 22989–23006. [CrossRef]
22. Elkaoud, N.S.M.; Mahmoud, R.K. Design and implementation of sequential fruit size sorting machine. *Rev. Bras. Eng. Agric. E Ambient.* **2022**, *26*, 722–728. [CrossRef]
23. Zhang, Z.; Igathinathane, C.; Li, J.; Cen, H.; Lu, Y.; Flores, P. Technology progress in mechanical harvest of fresh market apples. *Comput. Electron. Agric.* **2020**, *175*, 105606. [CrossRef]
24. Vallone, M.; Alleri, M.; Bono, F.; Catania, P. A New Wireless Device for Real-Time Mechanical Impact Evaluation in a Citrus Packing Line. *Trans. ASABE* **2020**, *63*, 1–9. [CrossRef]
25. XU, Q.; Chen, H.; Pan, H.; Li, S. Design and test of foam net packing machine for spherical fruit. *Trans. Chin. Soc. Agric. Eng.* **2019**, *35*, 56–61.
26. Hu, J.; Dai, Y.; WU, H.; Yang, Y. Design of automation equipment for apple protect foam net packaging. *Manuf. Autom.* **2021**, *43*, 102–106.
27. Xia, H.; Zhen, W.; Chen, D.; Zeng, W. An Ordinary Multilayer Fruit Paper Bag Supplying Device for Fruit Bagging. *HortScience* **2019**, *54*, 1644–1649. [CrossRef]
28. Liu, R.; Song, F.; Liu, X. The Design of the Automatic Fruit Packing Machine. *Mech. Electr. Eng. Technol.* **2017**, *46*, 79–81+127.
29. Fan, P.; Yan, B.; Wang, M.; Lei, X.; Liu, Z.; Yang, F. Three-finger grasp planning and experimental analysis of picking patterns for robotic apple harvesting. *Comput. Electron. Agric.* **2021**, *188*, 106353. [CrossRef]
30. Ji, W.; Tang, C.; Xu, B.; He, G. Contact force modeling and variable damping impedance control of apple harvesting robot. *Comput. Electron. Agric.* **2022**, *198*, 107026. [CrossRef]
31. Wu, C.; Wang, Y.; Sun, Q.; Zhao, Y.; Zhang, L. Design and Test of Auxiliary Harvesting Device of Apple. *Recent Pat. Eng.* **2021**, *15*, 107–116. [CrossRef]
32. Li, H.; Gao, P.; Cheng, W.; Yin, H.; Wu, H.; Wei, X. Tribological Properties of Wedge-Shaped Crown Texture on Surface of 3D Printed Photosensitive Resin. *Eng. Plast. Appl.* **2022**, *50*, 112–118+123.

machines

Article

Path Tracking Control of Commercial Vehicle Considering Roll Stability Based on Fuzzy Linear Quadratic Theory

Zhixian Fan [1,2,*], **Yang Yan** [2,3], **Xiangyu Wang** [2] **and Haizhu Xu** [2]

1 School of Vehicle and Mobility, Tsinghua University, Beijing 100084, China
2 Zhongtong Bus Co., Ltd., Liaocheng 252000, China
3 School of Mechanical and Automotive Engineering, Liaocheng University, Liaocheng 252000, China
* Correspondence: fzx19@mails.tsinghua.edu.cn

Abstract: Commercial vehicles generally drive at a higher speed on structured expressways, and their higher center of mass leads to a lower rollover threshold and a greater rollover risk while steering. Therefore, the design of a lateral trajectory-tracking control strategy for commercial vehicles should not only consider the accuracy of trajectory tracking but also consider roll stability. Based on this control objective, a fuzzy linear quadratic controller was designed in this study to ensure rolling stability in the path-tracking control process and improve the adaptability of the strategy to the driving scenario. Firstly, a steering and braking cooperative control model based on the four-degree-of-freedom model and the multi-point preview model was established. Then, a path tracking controller considering roll stability was designed based on the linear quadratic theory. On this basis, a fuzzy linear quadratic controller was designed to realize the online optimization of cost function weights. Finally, the effectiveness of the control strategy was verified using co-simulation and hardware-in-loop experiments. The results show that the designed controller can effectively adjust the weight of path-tracking and stability according to the vehicle's state. This effectively improves the vehicle's control distribution problem.

Keywords: path tracking control; roll stability control; fuzzy linear quadratic controller

Citation: Fan, Z.; Yan, Y.; Wang, X.; Xu, H. Path Tracking Control of Commercial Vehicle Considering Roll Stability Based on Fuzzy Linear Quadratic Theory. *Machines* **2023**, *11*, 382. https://doi.org/10.3390/machines11030382

Academic Editor: Nasser L. Azad

Received: 24 January 2023
Revised: 6 March 2023
Accepted: 9 March 2023
Published: 13 March 2023

1. Introduction

With the development of intelligent transportation and smart cities, the research and development of intelligent vehicle-driving technology are increasing urgently [1–3]. Commercial vehicles play an important role in the transportation industry. With the continuous increase in highway-cargo transportation, people's requirements for the comprehensive performance of commercial vehicles are also increasing day by day. Compared with the application environment of passenger cars [4], the autonomous driving scenario of commercial vehicles is relatively simple and fixed, which provides a good foundation for the promotion and application of intelligent driving technology. At the same time, the improvement in commercial vehicles' intelligence levels will also have an obvious effect on promoting a reduction in greenhouse gas emissions, driving comfort and other aspects. However, due to their special working attributes, commercial vehicles cause much more harm than passenger vehicles once traffic accidents occur [5]. Therefore, for reasons including comprehensive safety, saving energy, costs and others, the demand for the intelligent driving of commercial vehicles is more urgent [6].

In recent years, with developments in sensor and electronic control technology, trajectory-tracking technology has become a research hotspot in academia and industry. Many advanced control theories have been applied to improve the tracking accuracy of autonomous vehicles. A lateral path-tracking control strategy, based on command filtering adaptive torque control, has been proposed [7]. A hardware-in-loop experiment proves that path-tracking accuracy can quickly reach a high level even when encountering interference. A lateral path-tracking strategy which combines robustness theory and adaptive neural network control theory, has been proposed to ensure strong robustness in path tracking [8]. An interactive control strategy based on random games has been proposed to improve the path-tracking accuracy of the system by considering the uncertain input of the driver [9]. A path-tracking control strategy using model predictive control has been proposed to effectively improve tracking accuracy by considering the dynamic performance of the steering actuator [10]. An adaptive sliding mode control strategy has also been proposed [11]. Simulation results show that the algorithm has good tracking performance in excellent road conditions and a more than 4s path-tracking performance in bad road conditions. A robust path-tracking control strategy based on Lyapunov has been proposed to effectively improve the tracking accuracy of the strategy for low-adhesion road surfaces through the combination of a sliding mode controller and a backstepping controller [12]. A path-tracking control strategy combining adaptive and robust theory has been proposed to improve vehicle-tracking accuracy in extreme working conditions by considering uncertain parameters and unknown external disturbances [13]. A fuzzy neural network robust path-tracking controller has been proposed, which would not only improve the input buffeting of the controller but also effectively improve the vehicle's path-tracking performance [14]. All of the above control strategies can guarantee that the vehicle has better path-tracking performance. However, better tracking accuracy may cause the vehicle to lose stability in extreme obstacle-avoidance conditions. Especially for commercial vehicles, a higher center-of-mass position and a larger load may lead to greater roll risk while steering. Therefore, stability especially roll stability, must be considered in the design of a lateral trajectory-tracking control strategy for commercial vehicles.

Commercial vehicles are prone to rollover in driving conditions such as sharp steering, resulting in serious traffic accidents. Therefore, by deepening their research, scholars began to pay attention to the stability of vehicles in path tracking. By considering the sideslip in the feedforward loop, a feedback–feedforward robust steering controller has been designed to improve the control problem between path tracking accuracy and stability [15]. A path-tracking and stability hierarchical control method, which combines model predictive control and fuzzy PID, has been proposed to ensure control stability during path tracking [16]. A lateral path-tracking control method that combines the PID and adaptive model predictive control has been proposed to effectively improve path-tracking accuracy and vehicle stability by integrating automatic driving and in-wheel motor control [17]. The intelligent driving system and the stability control system were modeled as two agents in the dynamic game process. Based on the Stackelberg game theory, a coordination controller for path tracking and vehicle stability was designed [18,19]. Based on linear matrix inequalities, a robust path-tracking controller considering stability was designed [20]. The co-simulation results show that the robust path-tracking controller provides sufficient driving stability while ensuring tracking accuracy under unknown external disturbance conditions. By considering the driver's decision-time delay, an interactive controller for the vehicle's roll stability and path tracking with the driver in the loop was designed to reduce the driver's operating burden, as well as to improve the vehicle's roll stability and path-tracking accuracy [21]. To improve the stability of vehicle trajectory tracking, a multi-loop controller using the reverse-step method was designed. All of the above controller's considered dynamic stability during path tracking. This ensured high tracking accuracy and excellent body stability [22].

However, the cost function of the above controller needs to be designed in advance [23]. The weight allocation of the cost function involves tedious manual parameter adjustments. The highly complex and dynamic driving scenarios lead to the poor adaptability of the cost function with fixed weight coefficients. Based on the above analysis, the aim of this paper is divided into two points: (1) At the same time, the tracking accuracy and roll stability of the vehicle should be taken into account to improve the driving safety of the vehicle in the tracking process; (2) realizing the online adjustment of the controller, according to the driving scenario, to improve the adaptability of the controller to the complex and changeable driving scenario. Based on the above control objectives, a fuzzy linear quadratic (LQ) controller was designed. Firstly, a path-tracking controller considering vehicle roll stability was designed based on the LQ theory. Then, a fuzzy LQ controller with a self-adjusting weight coefficient was designed to analyze the influence of tracking errors and the roll angle on the vehicle's performance. The results of the simulation and hardware-in-loop experiments show that the fuzzy LQ controller could greatly improve the path-tracking accuracy on the premise of ensuring roll stability. The main contributions of this study are summarized as follows:

(1) To solve the problem of driving safety under extreme working conditions, a path-tracking controller considering roll stability was designed based on the LQ theory.

(2) The weight of the classical LQ controller's cost function is fixed, and the function's adaptability to the driving scenario is poor. Therefore, a fuzzy LQ controller with a self-adjusting weight coefficient was designed. The dynamic performance of the system can be improved effectively by optimizing the weight coefficient of the cost function online.

This paper is arranged as follows: The methodology adopted in this paper is described in Section 2. The steering and braking cooperative control model of intelligent commercial vehicles are designed in Section 3. The path-tracking cooperative controller that considers the vehicle's stability is proposed in Section 4. The Fuzzy control of a self-adjusting weight coefficient is deduced in Section 5. The co-simulation and hardware-in-loop experiment are carried out in Section 6.

2. Methodology

In this paper, a path-tracking controller that considers roll stability was designed based on the LQ theory. On this basis, the fuzzy control theory was used to realize the online optimization of the cost function weight to improve the adaptability of the strategy to the driving scene. Firstly, the LQ theory and fuzzy control theory are summarized.

2.1. Linear Quadratic Theory

The LQ theory is generally used to study the optimal control problem, which consists of the motion equation, state constraint, objective set, admissible control set and the cost function of the controlled system. By finding the optimal control strategy in the admissible control set, the state of the controlled system changes along with the desired state, and the value of the cost function is minimized. This theory has been widely applied in the field of robot control. This is because an LQ controller has the characteristics of reflecting the control requirements of practical engineering problems and takes into account the system performance and input consumption.

Firstly, the state-space equation for a linear system is given:

$$\begin{aligned} \dot{\mathbf{x}}_t &= \mathbf{A}_t\mathbf{x}_t - \mathbf{B}_t\mathbf{u}_t \\ \mathbf{y}_t &= \mathbf{C}_t\mathbf{x}_t \end{aligned} \tag{1}$$

where $\mathbf{x}_t$ is the state variable, $\mathbf{u}_t$ is the control input of the system, $\mathbf{y}_t$ is the output variable, and $\mathbf{A}_t$, $\mathbf{B}_t$ and $\mathbf{C}_t$ are the coefficient matrices of the corresponding dimension.

Secondly, the quadratic cost function needs to be set, as follows:

$$\mathbf{J} = \frac{1}{2}\mathbf{Z}_t^{\mathrm{T}}\boldsymbol{\rho}_t\mathbf{Z}_t + \frac{1}{2}\int_{t_0}^{t_f}[\mathbf{Z}_t^{\mathrm{T}}\boldsymbol{\rho}\mathbf{Z}_t + \mathbf{u}_t^{\mathrm{T}}\boldsymbol{\Gamma}\mathbf{u}_t]dt \tag{2}$$

where $\boldsymbol{\rho}_t$ is the semi-positive definite symmetric constant weight matrix, $\boldsymbol{\rho}$ is the semi-positive definite symmetric time-varying weight matrix, $\boldsymbol{\Gamma}$ is the positive definite symmetric time-varying weight matrix, t_f is the terminal time, $\frac{1}{2}\mathbf{Z}_t^{\mathrm{T}}\boldsymbol{\rho}_t\mathbf{Z}_t$ is the terminal cost, $\frac{1}{2}\int_{t_0}^{t_f}\mathbf{Z}_t^{\mathrm{T}}\boldsymbol{\rho}\mathbf{Z}_t dt$ is the process cost, and $\frac{1}{2}\int_{t_0}^{t_f}\mathbf{u}_t^{\mathrm{T}}\boldsymbol{\Gamma}\mathbf{u}_t dt$ is the control input cost.

According to the formula, the cost function also considers the tracking degree of the terminal state, the response speed of the system, and the cost. However, it is often difficult to combine these goals. For example, eliminating a terminal error and improving the system-response speed increases energy consumption. Therefore, in order to ensure good system performance, it is necessary to design a reasonable weight matrix. For example, if the specific gravity of an element in the medium $\mathbf{\Gamma}$ is increased, the amplitude and energy consumption of the corresponding control quantity will be inhibited. If a certain state quantity needs to reach the desired point quickly, the proportion of an element in the matrix $\boldsymbol{\rho}$ needs to be increased. It should be noted that although inequality constraints are not designed on the control input $\mathbf{u}_t$, the control input can still be limited within a reasonable range due to the existence of weights $\mathbf{\Gamma}$ and $\boldsymbol{\rho}$.

2.2. Fuzzy Control Theory

Fuzzy control is based on the fuzzy set theory, takes fuzzy language and fuzzy logic reasoning as tools, and designs intelligent controls according to the experience data from technicians in various fields. In 1973, the concept of fuzzy language rules was first put forward, and the emergence of the fuzzy control theory put forward effective solutions to practical problems faced in various fields. The characteristic of fuzzy control is that it can describe the relationship between the control variables, which cannot be described by the model formula with the method based on the experience to achieve effective control.

The design of a fuzzy controller mainly consists of four basic links: fuzzification, knowledge base, fuzzy reasoning and fuzzy resolving process, as shown in Figure 1.

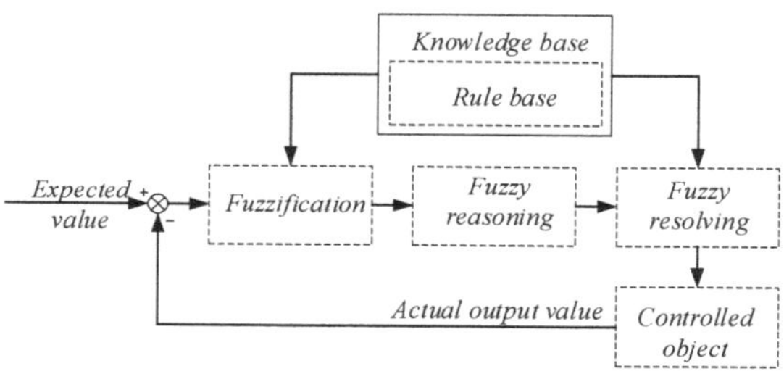

Figure 1. The basic structure diagram of a fuzzy controller.

In the figure, the main function of fuzzification is to fuzzify the input's precise quantity, convert it into a fuzzy quantity, and use the corresponding fuzzy set to describe it. The knowledge base is an important part of a fuzzy controller, which is composed of a database and a rule base. The rule base contains a set of control rules, which are derived from a large pool of expert knowledge and experience. The fuzzy-reasoning link is realized based on the control rules, the relationships are determined by knowledge and experience, and it has the ability to simulate human fuzzy reasoning. The fuzzy quantity generated by the logic judgment is converted into the precise quantity in the actual control application by the fuzzy solution.

3. Steering and Braking Cooperative Control Model

To avoid the delay effect of the complex system in the calculation, this study assumed a linear relationship between the tire lateral force and the tire slip angle. The four-degree-of-freedom(4-DOF) vehicle model was finally selected as the research object. As shown in Figure 2, this model is an ideal linear model that can effectively describe the vehicle's dynamic characteristics of the lateral, yaw and roll directions. The 4-DOF commercial-vehicle dynamics model assumes that the longitudinal speed is constant, and the tire sideslip characteristics are not considered.

$$\begin{cases} mv_x\dot{\beta} + m_sh\ddot{\phi} = (\frac{C_rl_r - C_fl_f}{v_x} - mv_x)\dot{\psi} - (C_f + C_r)\beta + C_f\delta_f \\ I_z\ddot{\psi} - I_{xz}\ddot{\phi} = (C_rl_r - C_fl_f)\beta - \frac{C_rl_r^2 + C_fl_f^2}{v_x}\dot{\psi} + C_fl_f\delta_f \\ m_shv_x\dot{\beta} + (I_x + m_sh^2)\ddot{\phi} = (m_sgh - k_s)\phi - m_sv_xh\dot{\psi} - b_s\dot{\phi} + k_s\phi_u + b_s\dot{\phi}_u + I_{xz}\ddot{\psi} \\ k_s\phi + b_s\dot{\phi} - m_u(h_u - r)v_x\dot{\beta} - b_s\dot{\phi}_u = [m_uv_x(h_u - r) + r\frac{l_rC_r}{v_x}]\dot{\psi} - rC_rv_x\beta + [k_u + k_s - m_ug(h_u - r)]\phi_u \end{cases} \quad (3)$$

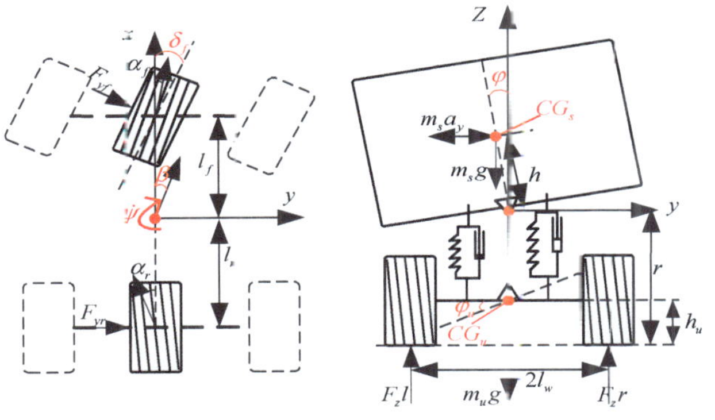

Figure 2. The four-degree-of-freedom commercial vehicle model.

Referring to [15], the lateral displacement and yaw angle can be written as shown in Equation (4).

$$\begin{aligned} \dot{y} &= v_x\psi + v_x\beta \\ \dot{\psi} &= \dot{\psi} \end{aligned} \tag{4}$$

By combining Equations (3) and (4), the steering control model with a front-wheel angle as the input was obtained.

$$\dot{\mathbf{x}} = \mathbf{A}_c\mathbf{x} + \mathbf{B}_{1c}\delta_f \tag{5}$$

where δ_f is the front-wheel angle and $\mathbf{x} = \begin{bmatrix} \beta & \dot{\psi} & \phi & \dot{\phi} & \phi_u & y & \psi \end{bmatrix}^{\mathrm{T}}$ is the system state variable. The coefficient matrix is as follows:

$$\mathbf{A}_c = \mathbf{E}_0^{-1}\mathbf{A}_0,\ \mathbf{B}_{1c} = \mathbf{E}_0^{-1}\mathbf{B}_{1,0},$$

$$\mathbf{E}_0 = \begin{bmatrix} e_{11} & 0 & 0 & e_{14} & 0 & 0 & 0 \\ 0 & e_{22} & 0 & e_{24} & 0 & 0 & 0 \\ e_{31} & e_{32} & 0 & e_{34} & e_{35} & 0 & 0 \\ e_{41} & 0 & 0 & 0 & e_{45} & 0 & 0 \\ 0 & 0 & e_{53} & 0 & 0 & 0 & 0 \\ 0 & 0 & 0 & 0 & 0 & e_{66} & 0 \\ 0 & 0 & 0 & 0 & 0 & 0 & e_{77} \end{bmatrix}, \mathbf{A}_0 = \begin{bmatrix} a_{11} & a_{12} & 0 & 0 & 0 & 0 & 0 \\ a_{21} & a_{22} & 0 & 0 & 0 & 0 & 0 \\ 0 & a_{32} & a_{33} & a_{34} & a_{35} & 0 & 0 \\ a_{41} & a_{42} & a_{43} & a_{44} & a_{45} & 0 & 0 \\ 0 & 0 & 0 & a_{54} & 0 & 0 & 0 \\ a_{61} & 0 & 0 & 0 & 0 & 0 & a_{67} \\ 0 & a_{72} & 0 & 0 & 0 & 0 & 0 \end{bmatrix}$$

$$\mathbf{B}_{1,0} = \begin{bmatrix} b_{11} & b_{21} & 0 & 0 & 0 & 0 & 0 \end{bmatrix}^T,$$

$$\begin{aligned}
&e_{11} = mv_x,\ e_{14} = m_s h,\ e_{22} = I_z,\ e_{24} = -I_{xz},\ e_{31} = m_s h v_x,\ e_{32} = -I_{xz}, \\
&e_{34} = I_x + m_s h^2,\ e_{35} = -b_s,\ e_{41} = -m_u(h_u - r)v_x,\ e_{45} = -b_s,\ e_{53} = 1, \\
&e_{66} = 1,\ e_{77} = 1,\ a_{11} = -(C_f + C_r),\ a_{12} = \frac{C_r l_r - C_f l_f}{v_x} - mv_x, \\
&a_{21} = C_r l_r - C_f l_f,\ a_{22} = -\frac{C_r l_r^2 + C_f l_f^2}{v_x},\ a_{32} = -m_s v_x h, \\
&a_{33} = m_s g h - k_s,\ a_{34} = -b_s,\ a_{35} = k_s,\ a_{41} = -rC_r v_x, \\
&a_{42} = m_u v_x(h_u - r) + r\frac{l_r C_r}{v_x},\ a_{43} = -k_s,\ a_{44} = -b_s, \\
&a_{45} = k_u + k_s - m_u g(h_u - r),\ a_{54} = 1,\ a_{61} = v_x,\ a_{67} = v_x, \\
&a_{72} = 1,\ b_{11} = C_f,\ b_{21} = C_f l_f,
\end{aligned}$$

For Equation (5), the steering and braking cooperative control model can be obtained by applying active yaw moment ΔM,

$$\begin{aligned}\dot{\mathbf{x}} &= \mathbf{A}_c x + \mathbf{B}_{1c}\delta_f + \mathbf{B}_{2c}\Delta M \\ &= \mathbf{A}_c x + \mathbf{B}_c \mathbf{u}\end{aligned} \tag{6}$$

where $\mathbf{u} = \begin{bmatrix}\delta_f & \Delta M\end{bmatrix}^{\mathrm{T}}$, $\mathbf{B}_{2c} = \mathbf{E}_0^{-1}\mathbf{B}_{2,0}$, $\mathbf{B}_{1,0} = \begin{bmatrix}0 & 1 & 0 & 0 & 0 & 0 & 0\end{bmatrix}^T$.

The discrete cooperative control model of steering and braking was obtained through the "C2D" command of MATLAB:

$$\mathbf{x}(k+1) = \mathbf{A}\mathbf{x}(k) + \mathbf{B}\mathbf{u}(k) \tag{7}$$

Referring to the multi-point preview theory [24], this study obtained the road update model. The dynamic process of road preview is expressed by Equation (8):

$$\mathbf{R}(k+1) = \mathbf{T}\mathbf{R}(k) + \mathbf{\Delta}^+\mathbf{r}^+ \ , \ \mathbf{r}^+ = \begin{bmatrix}y_d(k) & \psi_d(k)\end{bmatrix}^T \tag{8}$$

where $\mathbf{R}(k+1)$ is the matrix of road information at step $k+1$, $\mathbf{T}$ is the shift register matrix, $\mathbf{r}^+$ is the road update matrix, and $\mathbf{\Delta}^+$ is the coefficient of the road update matrix, as follows:

$$\mathbf{R}(k) = \begin{bmatrix}\mathbf{r}^{\mathrm{T}}(k) & \mathbf{r}^{\mathrm{T}}(k+1) & \cdots & \mathbf{r}^{\mathrm{T}}(k+Np-1)\end{bmatrix}^{\mathrm{T}},$$

$$\mathbf{T} = \begin{bmatrix}\mathbf{0} & \mathbf{1} \\ \mathbf{0}_{2\times 2} & \mathbf{0}\end{bmatrix}_{2Np\times 2Np}, \mathbf{\Delta}^+ = \begin{bmatrix}0 & 0 & \ldots & 0 & 1 \\ 0 & 0 & \ldots & 0 & 1\end{bmatrix}^{\mathrm{T}}.$$

By combining Equations (7) and (8), the augmented model containing the road information was obtained, as shown in Equation (9).

$$\mathbf{Z}(k+1) = \mathbf{A}_z\mathbf{Z}(k) + \mathbf{B}_z\mathbf{u}(k) + \mathbf{\Delta}_d{}^+\mathbf{r}^+ \tag{9}$$

where $\mathbf{Z}(k) = \begin{bmatrix}\mathbf{x}^{\mathrm{T}}(k+1) & \mathbf{R}^{\mathrm{T}}(k+1)\end{bmatrix}$, $\mathbf{A}_z = \begin{bmatrix}\mathbf{A} & \mathbf{0} \\ \mathbf{0} & \mathbf{T}\end{bmatrix}$,

$$\mathbf{B}_z = \begin{bmatrix}\mathbf{B}^{\mathrm{T}} & \mathbf{0}\end{bmatrix}, \ \mathbf{\Delta}_d{}^+ = \begin{bmatrix}\mathbf{0} & (\mathbf{\Delta}^+)^{\mathrm{T}}\end{bmatrix}^{\mathrm{T}}.$$

To simplify the derivation, the road update item was removed to obtain the steering and braking cooperative control model, including the road information.

$$\mathbf{Z}(k+1) = \mathbf{A}_z\mathbf{Z}(k) + \mathbf{B}_z\mathbf{u}(k) \tag{10}$$

A detailed description of each symbol in the equation is shown in Table 1.

Table 1. The basic parameters of the vehicle.

Symbol	Description	Value [Unit]
l_f, l_r	Distance from the center of mass	1.95 m; 1.54 m
l_w	Half of the wheelbase	1.3 m
m	ehicle mass	10,690 kg
m_s, m_u	Sprung and unsprung mass	9360 kg; 1330 kg
b_s	Suspension roll damping factor	8.26×10^4 Nms/rad

Table 1. *Cont.*

Symbol	Description	Value [Unit]
C_f, C_r	Equivalent cornering stiffness of front and rear axle	3.80×10^5 N/rad; 6.84×10^5 N/rad
k_u	Roll stiffness of unsprung mass	5.39×10^5 Nm/rad
v_x	Longitudinal velocity	70 km/h
g	Acceleration of gravity	9.8 m/s^2
k_s	Roll stiffness of suspension	1.06×10^6 Nm/rad
I_x	Roll inertia of sprung mass	7.70×10^3 kg m^2
I_z	Yaw inertia of sprung mass	3.01×10^4 kg m^2
I_{xz}	Yaw-roll inertia product of the sprung mass	0 kg m^2
r	Height of roll axis from the ground	0.63 m
h_u	Height of CG of unsprung mass from ground	0.51 m
ψ	Yaw angle	-
β	Sideslip angle	-
ϕ	Roll angle	-
δ_f	Front wheel angle input	-
ΔM	Active anti-roll moment	-

4. Cooperative Controller

The path tracking part is designed by weighing the tracking error and yaw angle deviation of the preview point, and the vehicle stability part is designed by weighting the roll angle deviation. Therefore, the path-tracking cooperative control cost function **J** considering vehicle stability is shown in Equation (11):

$$\begin{aligned}\mathbf{J}(N_p, k) = &\tfrac{1}{2}\mathbf{Z}^{\mathrm{T}}(k+N_p)\boldsymbol{\rho}_0\mathbf{Z}(k+N_p)\\ &+\tfrac{1}{2}\sum_{l=0}^{N_p-1}\left[\mathbf{Z}^{\mathrm{T}}(k+l)\boldsymbol{\rho}\mathbf{Z}(k+l)-\mathbf{u}^{\mathrm{T}}(k+j)\boldsymbol{\Gamma}\mathbf{u}(k+j)\right]\end{aligned} \tag{11}$$

where $\boldsymbol{\Gamma}$ is the weight matrix of the control input, Γ_y is the weight coefficient of the angle input, Γ_ϕ is the weight coefficient of the active yaw moment input, **E** is the weight matrix of the cost function, and $q_y \sim q_\phi$ are the weigh coefficients of lateral displacement, yaw angle and roll angle, respectively.

$$\boldsymbol{\rho} = \mathbf{M}^T\mathbf{E}\mathbf{M},\ \ \mathbf{E} = \begin{bmatrix} q_y & & \\ & q_\psi & \\ & & q_\phi \end{bmatrix}_{3*3},\ \boldsymbol{\Gamma} = \begin{bmatrix} \Gamma_y & 0 \\ 0 & \Gamma_\phi \end{bmatrix},$$

$$\mathbf{M} = \begin{bmatrix} 0 & 0 & 0 & 0 & 0 & 1 & 0 & -1 & 0 & \cdots & 0 \\ 0 & 0 & 0 & 0 & 0 & 0 & 1 & 0 & -1 & \cdots & 0 \\ 0 & 0 & 1 & 0 & 0 & 0 & 0 & 0 & 0 & \cdots & 0 \end{bmatrix}.$$

Referring to [19], $\boldsymbol{\rho}_0$ is set to $\boldsymbol{\rho}$ for convenient derivation and calculation. Based on the LQ theory, the optimal control input was obtained by iterating the Riccati difference equation with the initial condition $\boldsymbol{\rho}_0$.

$$\boldsymbol{\rho}_{j+1} = \boldsymbol{\rho}\left[\mathbf{A}_z{}^{\mathrm{T}}\boldsymbol{\rho}_j - \mathbf{A}_z{}^{\mathrm{T}}\boldsymbol{\rho}_j\mathbf{B}_z\left(\mathbf{B}_z{}^{\mathrm{T}}\boldsymbol{\rho}_j\mathbf{B}_z + \boldsymbol{\Gamma}\right)^{-1}\mathbf{B}_z{}^{\mathrm{T}}\boldsymbol{\rho}_j\right]\mathbf{A}_z \tag{12}$$

Therefore, the optimal control sequence is:

$$\begin{aligned}\mathbf{u}(k+N_p-j)_{opt} &= \left[-\left(\boldsymbol{\Gamma} + \mathbf{B}_z{}^{\mathrm{T}}\boldsymbol{\rho}_{j-1}\mathbf{B}_z\right)^{-1}\cdot \mathbf{B}_z{}^{T}\boldsymbol{\rho}_{j-1}\mathbf{A}_z\mathbf{Z}(k+N_p-j)\right] \\ &= \mathbf{K}(j-1)\mathbf{Z}(k+N_p-j)\end{aligned} \tag{13}$$

Through deduction, the optimal input $\mathbf{u}(k)_{opt}$ was obtained:

$$\mathbf{u}(k)_{opt} = \left[\delta_f{}^*(k) \quad \Delta M^*(k)\right]^{\mathrm{T}} = \mathbf{K}(N_p-1)\mathbf{Z}(k) \tag{14}$$

5. Fuzzy Control of Self-Adjusting Weight Coefficient

According to [12], the weight coefficients $\mathbf{E}$ and $\boldsymbol{\Gamma}$ in the cost function seriously affect vehicle path-tracking performance. $\mathbf{E}$ is set as a constant for design convenience. The vehicle path-tracking effect and stability were optimized by adjusting the size of Γ_y and Γ_ϕ online. In this study, the fuzzy logic toolbox in MATLAB was used to optimize the weight coefficient of the control input. According to practical experience, lateral displacement deviation and roll angle are important parameters that affect tracking accuracy and stability. Therefore, they were used as the control inputs for the fuzzy controller. ς_y is defined as the regulatory factor of Γ_y, and ς_ϕ is defined as the regulatory factor of Γ_ϕ. The two regulatory factors are the outputs of the fuzzy controller.

The fuzzy rule in this study was to reduce the path-tracking deviation as much as possible while ensuring better vehicle roll stability. When the vehicle is in a dangerous condition, to ensure driving safety, Γ_ϕ should be appropriately reduced. When vehicle roll stability is good, attention should be paid to tracking accuracy and Γ_y should be appropriately reduced.

Firstly, the lateral displacement deviation and the roll angle are normalized:

$$\bar{e} = \frac{e_{\max} - e}{e_{\max} - e_{\min}} \tag{15}$$

$$\bar{\phi} = \frac{\phi_{\max} - \phi}{\phi_{\max} - \phi_{\min}} \tag{16}$$

where e is the lateral displacement deviation and $\alpha_{\max}$ and $\alpha_{\min}$ represent the peak value of α.

The adjustment equation of the regulatory factor can be defined as shown in Equations (17) and (18):

$$\Gamma_y = \Gamma_{y0} * 4^{\varsigma_y} \tag{17}$$

$$\Gamma_\phi = \Gamma_{\phi 0} * 6^{\varsigma_\phi} \tag{18}$$

where Γ_{y0} and $\Gamma_{\phi 0}$ are the initial values of the weight coefficients.

Through the normalization process above, the control input's domain can be set to [0, 1]. The domain of the two output variables can be set to [−2, 2]. The fuzzy set of input and output variables t is represented as {NB NM NS NO PS PM PB}. The triangle function is selected for the membership function of input and output variables. The fuzzy control rules of ς_y and ς_ϕ are shown in Tables 2 and 3. The membership functions are shown in Figures 3 and 4.

Table 2. Fuzzy control rules of ς_y.

$\bar{e}$	$\bar{\phi}$						
	NB	**NM**	**NS**	**NO**	**PS**	**PM**	**PB**
NB	NO	PS	PS	PM	PM	PB	PB
NM	NO	NO	PS	PS	PM	PM	PM
NS	NS	NO	NO	PS	PS	PM	PM
NO	NM	NS	NS	NO	NO	PS	PS
PS	NM	NM	NS	NS	NO	NO	PS
PM	NB	NM	NM	NS	NS	NO	NO
PB	NB	NB	NM	NM	NS	NS	NO

Table 3. Fuzzy control rules of ς_ϕ.

$\bar{e}$	$\bar{\phi}$						
	NB	**NM**	**NS**	**NO**	**PS**	**PM**	**PB**
NB	NO	NO	NS	NM	NM	NB	NB
NM	PS	NO	NO	NS	NM	NM	NB
NS	PS	PS	NO	NS	NS	NM	NM
NO	PM	PS	PS	NO	NS	NS	NM
PS	PM	PM	PS	NO	NO	NS	NS
PM	PB	PM	PM	PS	NO	NO	NS
PB	PB	PM	PM	PS	PS	NO	NO

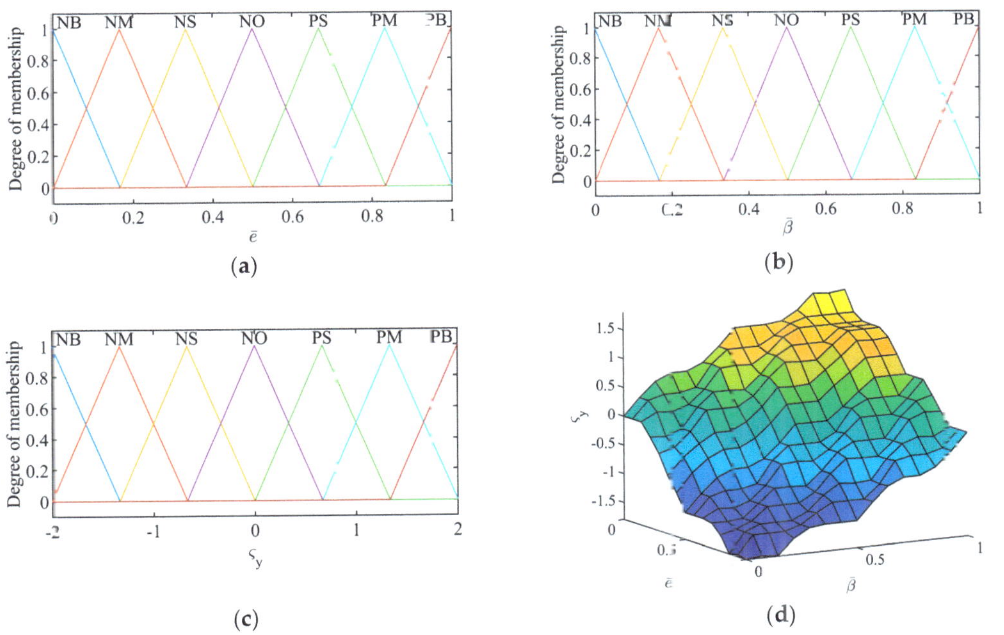

Figure 3. Fuzzy controller of Γ_y. (**a**) Membership function of $\bar{e}$; (**b**) Membership function of $\bar{\phi}$; (**c**) Membership function of ς_y; (**d**) Three-dimensional map.

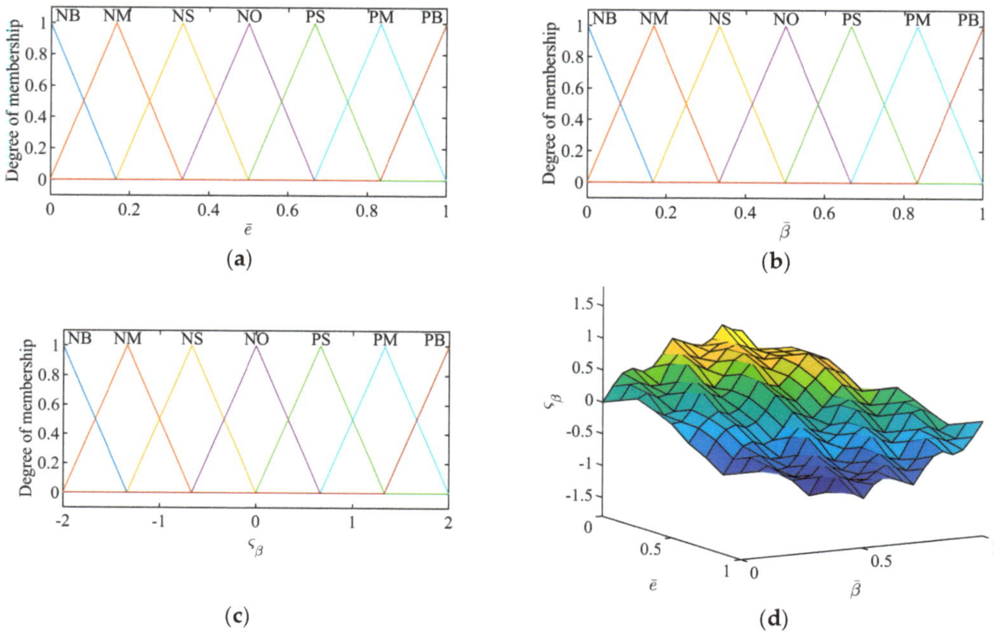

Figure 4. Fuzzy controller of Γ_ϕ. (**a**) Membership function of $\bar{e}$; (**b**) Membership function of $\overline{\phi}$; (**c**) Membership function of ς_ϕ; (**d**) Three-dimensional map.

6. Co-Simulation and Hardware-in-Loop Experiments

To demonstrate the superiority of the fuzzy LQ controller, a simulation system was designed, as shown in Figure 5. Then, the commercial vehicle hardware-in-loop experiment bench was built, as shown in Figure 6. The hardware-in-loop experiment bench uses a real commercial vehicle steering and brake-executive assembly. Firstly, the fuzzy LQ controller is compiled into the NI PXI real-time system for calculation, and the expected angle and braking pressure are obtained. Then, the expected decision is sent to the platform for tracking by the CAN signal, and the tracking value is fed back. Finally, the fuzzy LQ controller calculates the tracking value again with the highly simulated vehicle model in TruckSim to complete the closed loop. For details about the experiment bench, please refer to [19].

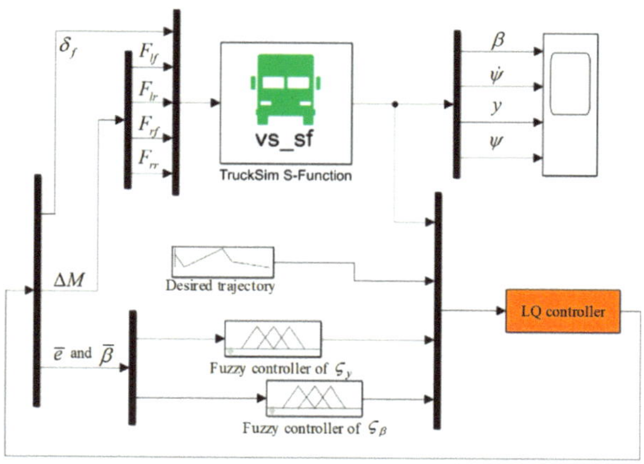

Figure 5. Simulation system based on fuzzy LQ theory.

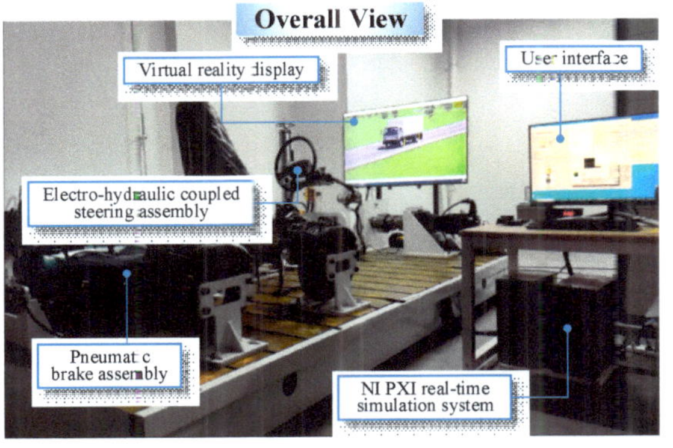

Figure 6. Commercial vehicle hardware-in-loop experiment bench.

The system control flow of the hardware-in-loop experiment is shown in Figure 7. The road information used in the experiment is shown in Figure 8. The single-line-change path is the test condition for Scenario 1, and the double-line change path is the test condition for Scenario 2.

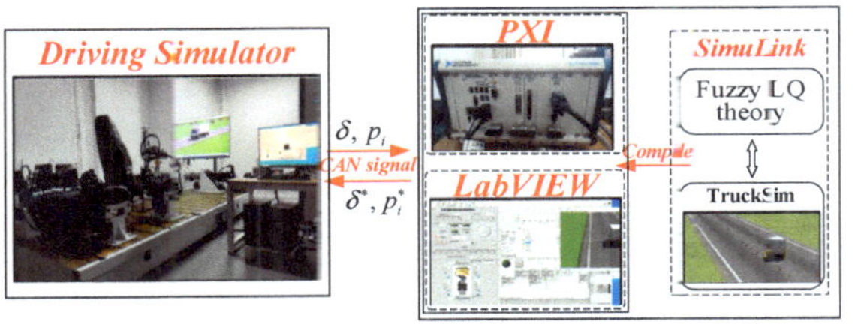

Figure 7. System control flow of hardware-in-loop experiment.

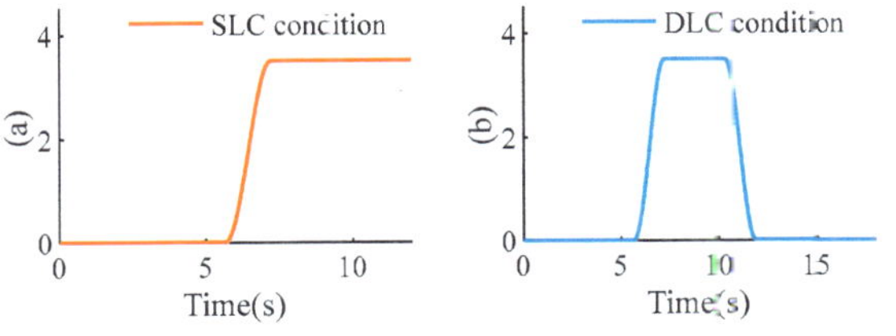

Figure 8. Desired path. (**a**) Single-line-change condition; (**b**) Double-line-change condition.

Scenario 1: The Single line-change condition at 80 km/h. Where DT represents the desired trajectory, LQ-S represents the simulation result of the linear quadratic controller, the fuzzy LQ-S represents the simulation result of the fuzzy linear quadratic controller, LQ-HIL represents the experimental result of the linear quadratic controller, and fuzzy LQ-HIL represents the experimental result of the fuzzy linear quadratic controller. The initial weight of the cost function is set as:

$$\mathbf{E} = \begin{bmatrix} 1 & & \\ & 1 & \\ & & 1.5 \end{bmatrix}, \Xi = \begin{bmatrix} 1 & \\ & 1.5 \end{bmatrix}.$$

As shown in Figure 9a,b, both tracking controllers can ensure a better tracking effect. Specifically, when the vehicle runs between 5.5 and 6.5 s, the fuzzy LQ controller adjusts Γ_y down and Γ_ϕ up to meet the requirements of path-tracking accuracy. By increasing the control weight of the front wheel angle, a more aggressive steering decision can be made, and a better path-tracking effect can be achieved.

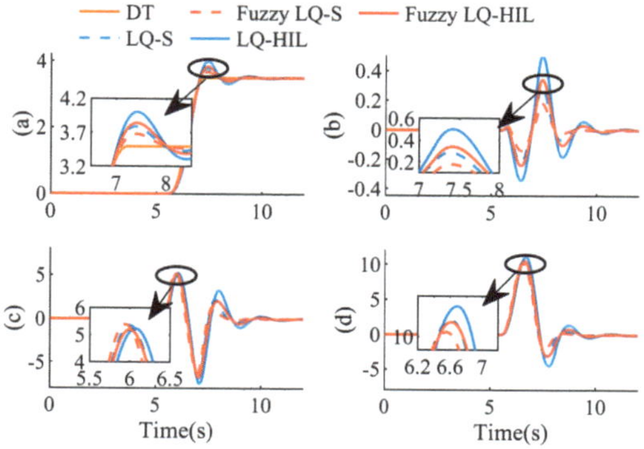

Figure 9. The control effect of single line-change condition. (**a**) Lateral displacement (m); (**b**) Lateral error (m); (**c**) Front-wheel angle (deg); (**d**) Yaw angle (deg).

As shown in Figure 10a, both controllers ensure that the roll angle is no greater than 3 deg. Between 7 and 8 s, the vehicle stability is slightly worse. The fuzzy LQ controller adjusts Γ_ϕ to become small. Driving safety can be improved by increasing the proportion of the stability control input. It can also be seen from Figure 10c that the fuzzy LQ control strategy has better stability than the LQ control strategy.

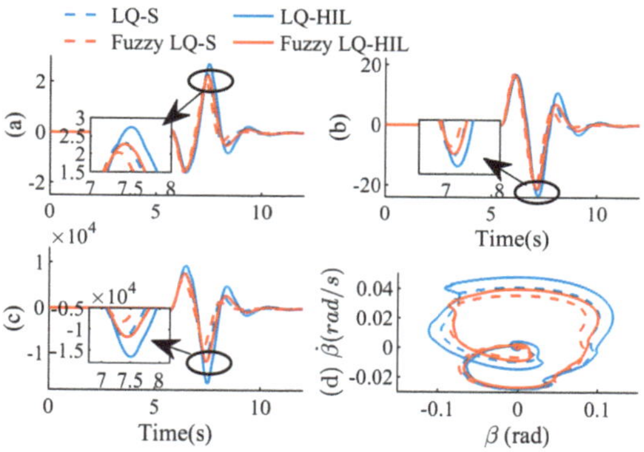

Figure 10. The control effect of single line—change condition. (**a**) Roll angle (deg); (**b**) Yaw rate (deg/s); (**c**) Active yaw moment (Nm); (**d**) $\dot{\beta} - \beta$ phase diagram.

In summary, the fuzzy LQ controller can adjust the control of the vehicle using path tracking and anti-roll adaptively, according to the driving scenario, to obtain a better control effect. When the vehicle-tracking accuracy is poor, the steering control weight is increased. When poor vehicle stability is detected, the steering weight is weakened, and the braking weight is increased to ensure driving safety.

Scenario 2: Double line-change condition at 80 km/h. The initial weights of the cost function are the same as in Scenario 1.

As can be seen from Figure 11a,b, the effect of the LQ controller decreased as the curvature of the tracking path increased while the control weight remained unchanged. The fuzzy LQ controller still guaranteed a good track-tracking effect. It proves that the fuzzy LQ controller has good environmental adaptability. As can be seen from Figure 11d, the yaw angle of the fuzzy LQ controller is still globally smaller than that of the LQ controller.

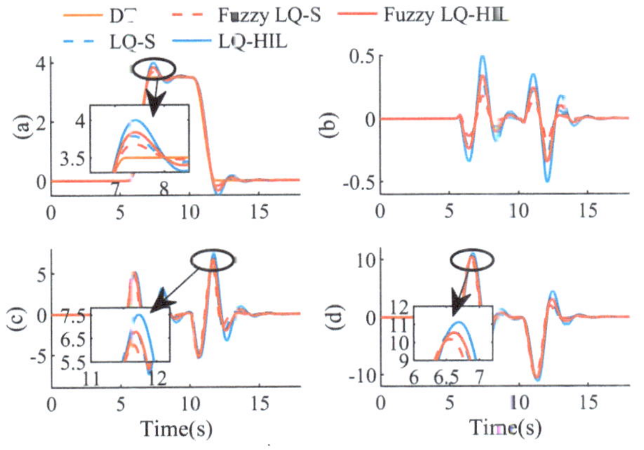

Figure 11. The control effect of double-line—change condition. (**a**) Lateral displacement (m); (**b**) Lateral error (m); (**c**) Front wheel angle (deg); (**d**) Yaw angle (deg).

By comparing Figures 9a and 12a, it can be seen that when the driving scenario deteriorated, the fuzzy LQ control adjusted the control value according to the driving scenario, which still ensures good roll stability. As can be seen from Figure 12c, the fuzzy LQ controller tends to calculate small brake intervention strength when the vehicle has good stability.

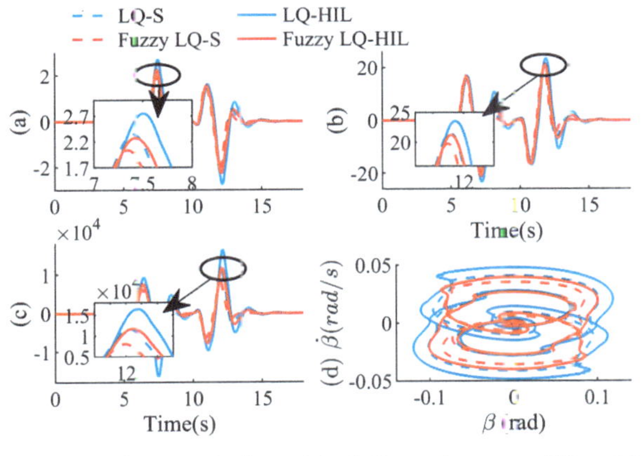

Figure 12. The control effect of single-line—change condition. (a) Roll angle (deg); (**b**) Yaw rate (deg/s); (**c**) Active yaw moment (Nm); (**d**) $\dot{\beta} - \beta$ phase diagram.

In conclusion, both controllers can ensure better path-tracking effects and vehicle roll stability. The fuzzy LQ controller can effectively improve the adaptability of the control strategy to the scene by optimizing the weight of the cost function online. The fuzzy LQ controller can ensure a better path-tracking effect and improve vehicle stability in the ultimate working condition. Its performance is better than the traditional LQ controller.

7. Conclusions

A fuzzy LQ controller was proposed in this paper, which could take into account the roll stability of the vehicle in the course of path-tracking and improve the adaptability of the controller to the driving scenario. Firstly, the LQ controller was designed in this study to solve the control object conflict problem between the path-tracking accuracy and the vehicle's roll stability. However, the classical LQ controller needed to design the cost function in advance. As a result, the vehicle in the actual simulation had poor adaptability to complex and changeable driving conditions. Therefore, a fuzzy LQ controller was designed based on the experience of real drivers. The weight of the cost function was optimized online using the road-tracking deviation and roll angle. Finally, the performance of the fuzzy LQ controller was verified using co-simulation and hardware-in-loop experiments.

The results show that, compared with the classical LQ controller, the fuzzy LQ controller can optimize the weight of the cost function online and improve the roll stability of the vehicle while ensuring better path-tracking accuracy. The following conclusions can be summarized from this study:

(1) To solve the problem of driving safely under extreme working conditions, a path-tracking controller that considers roll stability was designed based on the LQ theory.
(2) The weight of the classical LQ controller's cost function was fixed, and the function's adaptability to the driving scenario was poor. Therefore, a fuzzy LQ controller with a self-adjusting weight coefficient was designed. The dynamic performance of the system can be improved effectively by optimizing the weight coefficient of the cost function online.

In the future, we will try to include uncertain input elements of the driver and try to solve problems such as changes in the system-model parameters. To better verify the practicability of the control scheme, the fuzzy LQ controller designed in this paper will carry out real vehicle tests on various driving conditions.

Author Contributions: Supervision, Z.F. and X.W.; conceptualization, Z.F.; methodology, Z.F. and Y.Y.; software, H.X.; validation, X.W. and H.X.; formal analysis, Z.F., Y.Y. and H.X.; investigation, Z.F.; resources, Z.F., X.W. and H.X.; data curation, Z.F., Y.Y. and H.X.; writing—original draft preparation, Z.F. and Y.Y.; writing—review and editing, Z.F. and Y.Y. All authors have read and agreed to the published version of the manuscript.

Funding: This research received no external funding.

Data Availability Statement: Data are available on request from the authors. The data that support the findings of this study are available from the corresponding author, [Z.F.], upon reasonable request.

Conflicts of Interest: The authors declare no conflict of interest.

References

1. Wu, J.; Kong, Q.; Yang, K.; Liu, Y.; Cao, D.; Li, Z. Research on the Steering Torque Control for Intelligent Vehicles Co-Driving with the Penalty Factor of Human–Machine Intervention. *IEEE Trans. Syst. Man Cybern. Syst.* **2023**, *53*, 59–70. [CrossRef]
2. Wu, J.; Zhang, J.; Nie, B.; Liu, Y.; He, X. Adaptive Control of PMSM Servo System for Steering-by-Wire System with Disturbances Observation. *IEEE Trans. Transp. Electrif.* **2022**, *8*, 2015–2028. [CrossRef]
3. Nguyen, A.T.; Rath, J.; Guerra, T.M.; Palhares, R.; Zhang, H. Robust Set-Invariance Based Fuzzy Output Tracking Control for Vehicle Autonomous Driving Under Uncertain Lateral Forces and Steering Constraints. *IEEE Trans. Intell. Transp. Syst.* **2021**, *22*, 5849–5860. [CrossRef]
4. Benloucif, A.; Nguyen, A.T.; Sentouh, C.; Popieul, J.C. Cooperative Trajectory Planning for Haptic Shared Control between Driver and Automation in Highway Driving. *IEEE Trans. Ind. Electron.* **2019**, *66*, 9846–9857. [CrossRef]
5. He, X.; Yang, H.; Hu, Z.; Lv, C. Robust Lane Change Decision Making for Autonomous Vehicles: An Observation Adversarial Reinforcement Learning Approach. *IEEE Trans. Intell. Veh.* **2023**, *8*, 184–193. [CrossRef]
6. Zhang, L.; Yan, Y.; Zhu, Q.; Zhao, G.; Feng, D.; Wu, J. A Pneumatic Control Method for Commercial Vehicle Electronic Brake System Based on EPV Module. *Actuators* **2022**, *11*, 316. [CrossRef]
7. Zhang, J.; Wu, J.; Liu, J.; Zhou, Q.; Xia, J.; Sun, W.; He, X. Command-filter-adaptive-based lateral motion control for autonomous vehicle. *Control Eng. Pract.* **2022**, *121*, 105044. [CrossRef]
8. Ji, X.; He, X.; Lv, C.; Liu, Y.; Wu, J. Adaptive-neural-network-based robust lateral motion control for autonomous vehicle at driving limits. *Control Eng. Pract.* **2018**, *76*, 41–53. [CrossRef]
9. Ji, X.; Yang, K.; Na, X.; Lv, C.; Liu, Y. Shared steering torque control for lane change assistance: A stochastic game-theoretic approach. *IEEE Trans. Ind. Electron.* **2018**, *66*, 3093–3105. [CrossRef]
10. Kim, E.; Kim, J.; Sunwoo, M. Model predictive control strategy for smooth path tracking of autonomous vehicles with steering actuator dynamics. *Int. J. Automot. Technol.* **2014**, *15*, 1155–1164. [CrossRef]
11. Norouzi, A.; Kazemi, R.; Azadi, S. Vehicle lateral control in the presence of uncertainty for lane change maneuver using adaptive sliding mode control with fuzzy boundary layer. *Proc. Inst. Mech. Eng. Part I J. Syst. Control Eng.* **2018**, *232*, 12–28. [CrossRef]

12. Norouzi, A.; Masoumi, M.; Barari, A.; Farrokhpour Sani, S. Lateral control of an autonomous vehicle using integrated backstepping and sliding mode controller. *Proc. Inst. Mech. Eng. Part K J. Multi-Body Dyn.* **2019**, *233*, 141–151. [CrossRef]
13. Pang, H.; Yao, R.; Wang, P.; Xu, Z. Adaptive backstepping robust tracking control for stabilizing lateral dynamics of electric vehicles with uncertain parameters and external disturbances. *Control Eng. Pract.* **2021**, *110*, 104781. [CrossRef]
14. Taghavifar, H.; Rakheja, S. Path-tracking of autonomous vehicles using a novel adaptive robust exponential-like-sliding-mode fuzzy type-2 neural network controller. *Mech. Syst. Signal Process.* **2019**, *130*, 41–55. [CrossRef]
15. Kapania, N.R.; Gerdes, J.C. Design of a feedback-feedforward steering controller for accurate path tracking and stability at the limits of handling. *Veh. Syst. Dyn.* **2015**, *53*, 1687–1704. [CrossRef]
16. Lin, F.; Wang, S.; Zhao, Y.; Cai, Y. Research on autonomous vehicle path tracking control considering roll stability. *Proc. Inst. Mech. Eng. Part D J. Automob. Eng.* **2021**, *235*, 199–210. [CrossRef]
17. Ahn, T.; Lee, Y.; Park, K. Design of integrated autonomous driving control system that incorporates chassis controllers for improving path tracking performance and vehicle stability. *Electronics* **2021**, *10*, 144. [CrossRef]
18. Ji, X.; Liu, Y.; He, X.; Yang, K.; Na, X.; Lv, C.; Liu, Y. Interactive control paradigm-based robust lateral stability controller design for autonomous automobile path tracking with uncertain disturbance: A dynamic game approach. *IEEE Trans. Veh. Technol.* **2018**, *67*, 6906–6920. [CrossRef]
19. Yan, Y.; Wu, J.; Liu, X.; Zhao, Y.; Wang, S. A lateral trajectory tracking control method for intelligent commercial vehicles considering active anti-roll decision based on Stackelberg equilibrium. *IET Intell. Transp. Syst.* **2022**, *16*, 1193–1208. [CrossRef]
20. He, X.; Liu, Y.; Yang, K.; Wu, J.; Ji, X. Robust coordination control of AFS and ARS for autonomous vehicle path tracking and stability. In Proceedings of the 2018 IEEE International Conference on Mechatronics and Automation (ICMA), Changchun, China, 5–8 August 2018; pp. 924–929.
21. Xu, T.; Wang, X. Roll stability and path tracking control strategy considering driver in the loop. *IEEE Access* **2021**, *9*, 46210–46222. [CrossRef]
22. Shafiei, M.; Emami, M. Design of a robust path tracking controller for an unmanned bicycle with guaranteed stability of roll dynamics. *Syst. Sci. Control Eng.* **2019**, *7*, 12–19. [CrossRef]
23. Wu, J.; Wu, D.; Yan, Y.; Zhang, N.; Bao, C.; Wang, F. Steering and braking game control architecture based minimax robust stability control for emergency avoidance of autonomous vehicles. *Sci. China Technol. Sci.* **2022**, *65*, 943–955. [CrossRef]
24. Cole, D.J.; Pick, A.J.; Odhams, A.M.C. Predictive and linear quadratic methods for potential application to modelling driver steering control. *Veh. Syst. Dyn.* **2006**, *44*, 259–284. [CrossRef]

machines

MDPI

Article

Narrow Tilting Vehicle Drifting Robust Control

Dongxin Xu [1,*], Yueqiang Han [1], Xianghui Han [1], Ya Wang [2] and Guoye Wang [1,*]

[1] College of Engineering, China Agricultural University, Beijing 100083, China
[2] Beijing Zuoqi Technology Ltd., Beijing 101300, China
* Correspondence: xudongxin1996@163.com (D.X.); wgy1615@126.com (G.W.); Tel.: +86-010-6273-6856 (D.X.); +86-182-1093-7906 (G.W.)

Abstract: The narrow tilting vehicle receives extensive public attention because of traffic congestion and environmental pollution, and the active rolling motion control is a traffic safety precaution that reduces the rollover risk caused by the structure size of the narrow vehicle. The drifting motion control reflects the relatively updated attentive research of the regular-size vehicle, which can take full advantage of the vehicle's dynamic performance and improve driving safety, especially when tires reach their limits. The narrow tilting vehicle drifting control is worthy of research to improve the driving safety of the narrow tilting vehicle, especially when tires reach the limit. The nonlinear narrow tilting vehicle dynamic model is established with the UniTire model to describe the vehicle motion characteristics and is simplified to reduce the computation of the drifting controller design. The narrow tilting vehicle drifting controller is designed based on the robust theory with uncertain external disturbances. The controller has a wide application, validity, and robustness and whose performance is verified by realizing different drifting motions with different initial driving motions. The narrow tilting vehicle drifting robust control has some practical and theoretical significance for more research.

Keywords: narrow tilting vehicle; drifting motion; robust controller

Citation: Xu, D.; Han, Y.; Han, X.; Wang, Y.; Wang, G. Narrow Tilting Vehicle Drifting Robust Control. *Machines* **2023**, *11*, 90. https://doi.org/10.3390/machines11010090

Academic Editor: Jian Wu

Received: 12 December 2022
Revised: 8 January 2023
Accepted: 9 January 2023
Published: 10 January 2023

1. Introduction

With increasing population and rapid economic development, there are continuing high numbers of vehicle population as [1] and traffic accidents as [2,3]. The mini-size vehicle has been paid close attention to universally, and the narrow tilting vehicle entered the public consciousness. The narrow tilting vehicle is designed to solve traffic congestion and environmental pollution problems caused by the vehicle's rapid growth. Furthermore, the active rolling motion of the narrow tilting vehicle is the steady state realized by the active rolling controller based on the velocity and the steering angle to avoid the rollover.

Traffic safety is in constant focus [4], and most relative research includes the autonomous steering of the regular-size vehicle [5] and the active rolling of the narrow vehicle. The autonomous steering of regular-size vehicle research is almost wholly working under the linear vehicle and tire condition, such as the path following in [6] and the human-machine cooperation driving in [7]. Most traffic safety research on narrow vehicles has looked at how to reduce the rollover risk. The active rolling of the narrow vehicle can improve turning driving safety and avoid the rollover risk caused by the narrow size. The narrow vehicle is divided into direct tilt control and steering tilt control [8] based on active rolling methods. It is also divided into the four-wheel [9], the three-wheel [10], and the reversed three-wheel [11] based on the structures. The direct tilt control of the narrow tilting vehicle is directly applied to the vehicle body, but it has poor adaptability under different working conditions. The narrow tilting vehicle based on the steering tilting control realizes the active rolling based on the steering angle, but it needs the driving technique and is not well suited at the low longitudinal velocity. The narrow tilting vehicle studied in this paper is a four-wheel narrow tilting vehicle with direct tilt control based on the steering angle [12] to improve driving safety at a multi-level velocity in different steering angles. The active rolling target angle is the steady state calculated based on the centrifugal force balance analysis in [13], which is realized by relying on the active rolling mechanic [12,14]. The narrow tilting vehicle studied in this paper is verified whereby the active rolling angle can realize the steady-state target and lower the rollover risk significantly by the active rolling mechanic and controller [12–14]. Most other traffic safety research on the narrow tilting vehicle focuses on vehicle performance under normal working conditions, as seen in one study [15].

The drifting motion control is a unique control strategy available to enhance traffic safety, which is little used in narrow drifting motions. The drifting movement of the racing car can be controlled by the drifting controller, which realizes the target drifting motion analyzed based on the three-of-freedom vehicle dynamics model when tires reach the maximum in [16]. The regular-size vehicle drifting motion parameters, including velocity, sideslip angle, yaw rate, and steering angle can be calculated and obtained by analyzing the steady-state vehicle dynamics expressions in the adhesion limit [17], and the drifting controllers of a regular-size vehicle are designed based on the linear quadratic regulator theory [17–19], the model predictive control theory [20,21], and the adaptive control theory [22]. There is reinforcement learning [18], and neural networks [23] are applied to the vehicle drifting control based on existing vehicle drifting motion parameters. Most drifting research without the learning realize the target drifting motion, which is the core work, and ignores some important details which affect the practical implementation, especially disturbances. Most drifting research based on the learning realize the passed path under drifting motions, and the adaptability in different working conditions are not fully verified, and most vehicle dynamics models of drifting movements are very similar to the classic one [24] that suggests vehicle dynamics with no-active rolling. According to the narrow tilting vehicle characteristics, it is necessary to design a drifting controller with extensive suitability with external disturbances to improve driving safety when tires reach the maximum. The robust theory performs better in solving problems caused by disturbances and can accommodate engineering requirements in practice. The linear matrix inequality (LMI) is widely applied in solving the problems of controller and system designs, which has become popular in the controller solution, especially in the robust controller.

From the above analysis, this paper researches how to control the narrow tilting vehicle drifting motions. According to the narrow tilting vehicle characteristics, the non-linear narrow tilting vehicle dynamics model is established to describe the vehicle motion characteristics and is appropriately simplified to reduce the computing effort in the drifting controller design. The narrow tilting vehicle drifting controller is designed based on the robust theory with uncertain extern disturbances considering the practical application environment, which is relatively independent and does not affect the active rolling controller working. The performance of the drifting controller is tested to realize different target drifting motions based on different initial motions, which can express the reliability, availability, and universality of the designed controller.

2. Narrow Tilting Vehicle Dynamics System for Drifting Controller Design

The narrow tilting vehicle shown in Figure 1a, whose structural diagram is shown in Figure 1b in [25], avoids rollover danger by the active roll control, which is different from most vehicles. To ensure all wheels are on the ground and avoid rollover, the rolling controller of the narrow tilting vehicle creates a good balance between the gravity component caused by rolling and centrifugal force caused by turning [12–14].

The vehicle dynamics system is requested to describe the narrow tilting vehicle drifting motion characteristics and is applied to the controller design to realize target motions. The rolling motion of the narrow tilting vehicle is different from most vehicles; as a result, the major influence factors of the vertical load transfer are distinctive. The drifting motion is a unique motion state for the vehicle, and tire forces reach limits in the drifting motion. Therefore, the tire model also has an essential effect on the controller design.

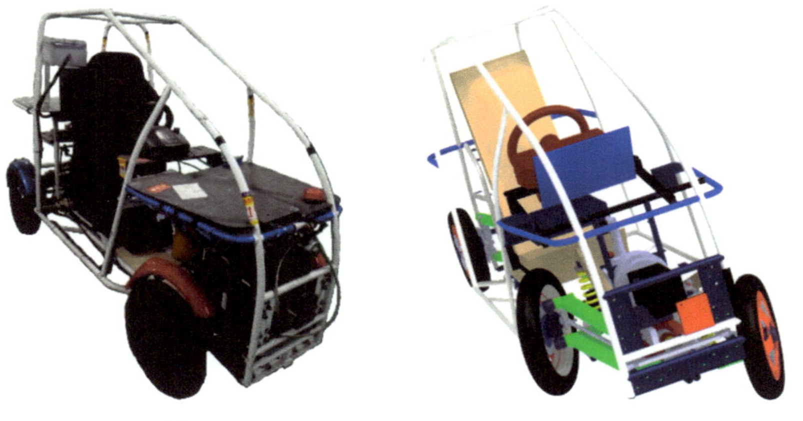

(**a**) (**b**)

Figure 1. The narrow tilting vehicle: (**a**) the physical; (**b**) the structural [25].

2.1. Narrow Tilting Vehicle Dynamics Equations

The active rolling movement meets the need for the active rolling angle for the centrifugal force equilibrium based on narrow tilting vehicle dynamics characteristics. The vehicle dynamics model, including the active rolling movement, is shown in Figure 2 and is described as Equations (1)–(4), where Equation (5) describes the normal rolling motion without the active rolling mechanic.

$$\begin{aligned}\dot{v} &= \left(F_{xfl}+F_{xfr}\right)\left(\frac{\cos\delta_f\cos\beta}{m}+\frac{\sin\delta_f\sin\beta}{m}\right)+\left(F_{yfl}+F_{yfr}\right)\left(-\frac{\sin\delta_f\cos\beta}{m}+\frac{\cos\delta_f\sin\beta}{m}\right)\\ &+\left(F_{xrl}+F_{xrr}\right)\frac{\cos\beta}{m}+\left(F_{yrl}+F_{yrr}\right)\frac{\sin\beta}{m}-\frac{C_dA_fv^2\cos^3\beta}{21.15m},\end{aligned} \tag{1}$$

$$\begin{aligned}\dot{\beta} = &-\gamma+\left(F_{xfl}+F_{xfr}\right)\left(\frac{\sin\delta_f\cos\beta}{mv}-\frac{\cos\delta_f\sin\beta}{mv}\right)+\left(F_{yfl}+F_{yfr}\right)\left(\frac{\cos\delta_f\cos\beta}{mv}+\frac{\sin\delta_f\sin\beta}{mv}\right)\\ &-\left(F_{xrl}+F_{xrr}\right)\frac{\sin\beta}{mv}+\left(F_{yrl}+F_{yrr}\right)\frac{\cos\beta}{mv}+\frac{C_dA_fv\cos^2\beta\sin\beta}{21.15m},\end{aligned} \tag{2}$$

$$\begin{aligned}\dot{\gamma} &= \tfrac{l_f}{I_z}\left(F_{xfl}+F_{xfr}\right)\sin\delta_f+\tfrac{l_f}{I_z}\left(F_{yfl}+F_{yfr}\right)\cos\delta_f-\tfrac{l_r}{I_z}\left(F_{yrl}+F_{yrr}\right)-\tfrac{w_f}{2I_z}\left(F_{xfl}-F_{xfr}\right)\cos\delta_f\\ &+\tfrac{w_f}{2I_z}\left(F_{yfl}-F_{yfr}\right)\sin\delta_f-\tfrac{w_r}{2I_z}\left(F_{xrl}-F_{xrr}\right),\end{aligned} \tag{3}$$

$$\psi = -\tan^{-1}\frac{v^2}{gR_r}, \tag{4}$$

$$\ddot{\psi} = \frac{I_{xz}}{I_x}\dot{\gamma}+\frac{m_bh_b}{I_x}\left(\dot{v}_y+v_x\gamma\right)+\frac{m_bgh_b}{I_x}\sin\psi-\frac{K_\psi\psi+C_\psi\dot{\psi}}{I_x} \tag{5}$$

where m is the vehicle mass, m_b is the vehicle body mass, h_b is the height of gravity center from the roll axis, l_f and l_r are the distances from gravity center to front and rear axle, w_f and w_r are the distances between left and right wheels of the front and rear axle, $R_r = \left(l_f+l_r\right)/\delta_f$ is the ideal turning radius calculated based on the steering angle, v and $\dot{v}$ are the vehicle velocity and accelerated velocity, β and $\dot{\beta}$ are the vehicle sideslip angle and rate, γ and $\dot{\gamma}$ are the vehicle yaw rate and angular acceleration, δ_f is the front-wheel steering angle, ψ is the vehicle steady roll angle and is implemented by the active roll controller, I_z is the moment of inertia with respect to yaw axis, C_d is the aerodynamics drag coefficient, A_f is the vehicle frontal area, and F_{xij} and F_{yij} $(i = f,r;\ j = l,r)$ are the longitudinal and lateral forces at each tire, respectively, the index $i = f,r$ is used to express the front and rear axles, and the index $j = l,r$ is used to express the left and right wheels.

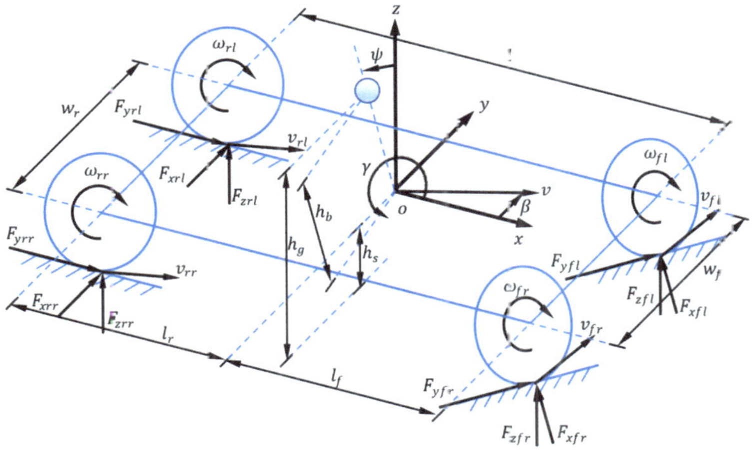

Figure 2. The vehicle dynamics model.

2.2. Vertical Load Transfer

The rolling movement of the narrow tilting vehicle is distinctive. The roll angle has a significant influence on the vertical load transfer. Considering the load transfer caused by longitudinal and lateral forces and the active roll control shown in Figure 3, the vertical force at each tire can be described as Equations (6)–(9).

$$F_{zfl} = \frac{mgl_r}{2\left(l_f + l_r\right)} - \frac{F_{sx}h_g}{2\left(l_f + l_r\right)} - \frac{l_r\left(F_{sy}h_g \cos\psi + m_b g h_b \sin\psi\right)}{\left(l_f + l_r\right)w_f}, \tag{6}$$

$$F_{zfr} = \frac{mgl_r}{2\left(l_f + l_r\right)} - \frac{F_{sx}h_g}{2\left(l_f + l_r\right)} + \frac{l_r\left(F_{sy}h_g \cos\psi + m_b g h_b \sin\psi\right)}{\left(l_f + l_r\right)w_f}, \tag{7}$$

$$F_{zrl} = \frac{mgl_f}{2\left(l_f + l_r\right)} + \frac{F_{sx}h_g}{2\left(l_f + l_r\right)} - \frac{l_f\left(F_{sy}h_g \cos\psi + m_b g h_b \sin\psi\right)}{\left(l_f + l_r\right)w_r}, \tag{8}$$

$$F_{zrr} = \frac{mgl_f}{2\left(l_f + l_r\right)} + \frac{F_{sx}h_g}{2\left(l_f + l_r\right)} + \frac{l_f\left(F_{sy}h_g \cos\psi + m_b g h_b \sin\psi\right)}{\left(l_f + l_r\right)w_r}. \tag{9}$$

where h_g is the height of gravity center, $F_{sx} = (F_{xfl} + F_{xfr})\cos\delta_f - (F_{yfl} + F_{yfr})\sin\delta_{fl} + (F_{xrl} + F_{xrr}) - C_d A_f v^2 \cos^3\beta/(21.15m)$, and $F_{sy} = (F_{xfl} + F_{xfr})\sin\delta_{fl} + (F_{yfl} + F_{yfr})\cos\delta_{fl} + (F_{yrl} + F_{yrr})$.

2.3. Tire Models and Equations

The vehicle drifting motion is nonlinear, and tire forces reach the maximum while the steering angle is non-vanishing in the drifting motion. Therefore, the nonlinear tire model is essential in this system to describe the force condition at each tire. The UniTire tire model can express tire forces more accurately not only in single working conditions but also in complex working conditions [26], which means it is ideally suited to the part of the vehicle drifting motion. If the UniTire tire model is applied directly to the narrow tilting vehicle drifting controller, there will be a large amount of computation in the controller design. The linearized tire model based on UniTire is proposed in the controller design to solve this problem and express tire forces accurately. Thus, UniTire describes tire forces during the narrow tilting vehicle motion, and the linearized tire model is applied to the narrow tilting drifting controller design.

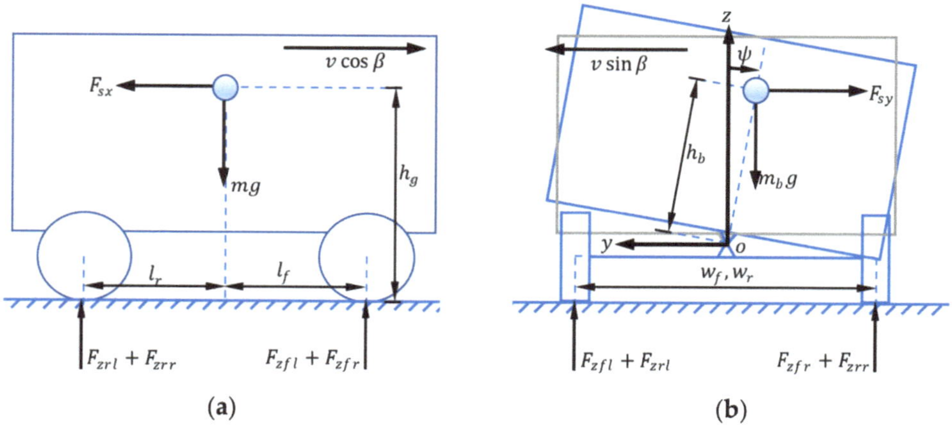

Figure 3. The vertical load transfer: (**a**) the longitudinal movement caused; (**b**) the lateral and active roll movements caused.

2.3.1. UniTire

UniTire can express longitudinal and lateral tire forces and overturning, rolling resistance, and aligning tire moments based on the tire coordinate system and is applicated to the vehicle dynamics simulation and the vehicle motion controller design [27]. In this paper, the tire model is applied to express longitudinal and lateral tire forces, which are the main characteristics of different working conditions. Thus, the simplified tire coordinate system is shown in Figure 4 with the consideration of longitudinal and lateral tire motions and the neglect of tire moments, and the equations of the tire model are described as Equation (10).

$$\begin{cases} \overline{F}_{ij} = 1 - e^{-\phi_{ij} - E\phi_{ij}{}^2 - (E^2 + \frac{1}{12})\phi_{ij}{}^3} \\ F_{xij} = \overline{F}_{ij}\frac{\phi_{xij}}{\phi_{ij}}\mu_{xij}F_{zij} \\ F_{yij} = \overline{F}_{ij}\frac{\phi_{yij}}{\phi_{ij}}\mu_{yij}F_{zij} \end{cases}, \tag{10}$$

where $\overline{F}_{ij}$ $(i = f, r;\ j = l, r)$ is the normalized force, E is the curvature factor of the combined slip resultant force and is the variable related to the tire slip ratios in [28], μ_{xij} and μ_{yij} $(i = f, r; j = l, r)$ are the longitudinal and lateral friction coefficient, $\phi_{xij} = K_{xij}S_{xij}/\left(\mu_{xij}F_{zij}\right)$, $\phi_{yij} = K_{yij}S_{yij}/\left(\mu_{yij}F_{zij}\right)$, and $\phi_{ijj} = \sqrt{\phi_{xij}{}^2 + \phi_{yij}{}^2}$ $(i = f, r; j = l, r)$ are normalized longitudinal, lateral and combined slip ratios, $S_{xij} = \left(\omega_{ij}r_{eij} - v_{ij}\cos\alpha_{ij}\right)/\left(\omega_{ij}r_{eij}\right)$ and $S_{yij} = -v_{ij}\sin\alpha_{ij}/\left(\omega_{ij}r_{eij}\right)$ $(i = f, r; j = l, r)$ are the longitudinal and lateral slip ratios in UniTire, K_{xij} and K_{yij} $(i = f, r; j = l, r)$ are the longitudinal slip and cornering stiffness of the tire based on S_{xij} and S_{yij}, v_{ij} $(i = f, r; j = l, r)$ is the wheel velocity, ω_{ij} $(i = f, r; j = l, r)$ is the wheel rotation angular velocity, r_{eij} $(i = f, r; j = l, r)$ is the wheel effective rolling radius, α_{ij} $(i = f, r; j = l, r)$ is the wheel slip angle, and the wheel slip angle of each wheel respectively satisfy $\alpha_{fl} = \tan^{-1}((-a_{fl}\sin\delta_f + (v\sin\beta + l_f\gamma)\cos\delta_f)(a_{fl}\cos\delta_f + (v\sin\beta + l_f\gamma)\sin\delta_f))$, $\alpha_{fr} = \tan^{-1}((-a_{fr}\sin\delta_f + (v\sin\beta + l_f\gamma)\cos\delta_f)/(a_{fr}\cos\delta_f + (v\sin\beta + l_f\gamma)\sin\delta_f))$, $\alpha_{rl} = \tan^{-1}((v\sin\beta - l_r\gamma)/(v\cos\beta - w_r\gamma/2))$, and $\alpha_{rr} = \tan^{-1}((v\sin\beta - l_r\gamma)/(v\cos\beta + w_r\gamma/2))$ in which $a_l = v\cos\beta - w_f\gamma/2$ and $a_r = v\cos\beta + w_f\gamma/2$.

2.3.2. The Linearized Tire Equations

UniTire is a nonlinear tire model, and its tire force equations are complicated for the controller design. The linearized tire equations are derived based on the representation of the nonlinear tire model. The TYDEX longitudinal slip ratio is the common expression, and the nonlinear tire model representation is shown in Figure 5.

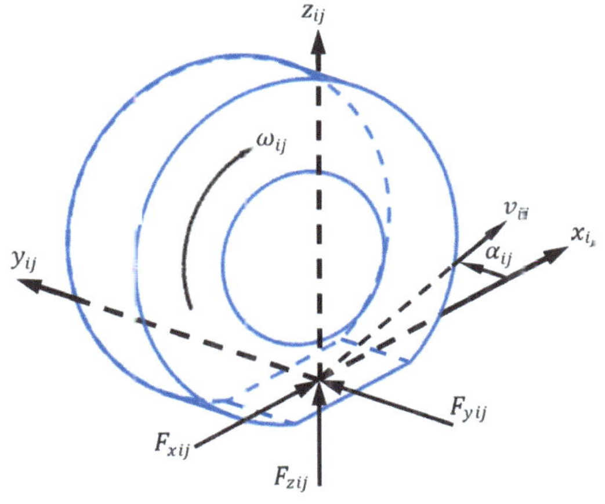

Figure 4. Simplified tire coordinate system.

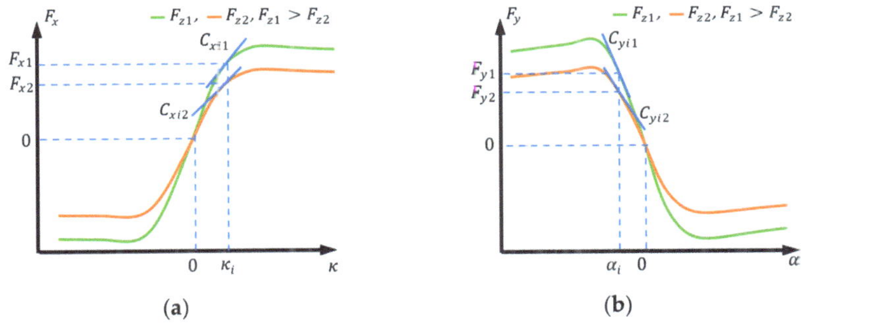

Figure 5. The representation of UniTire: (**a**) the longitudinal tire force in which κ_i is the TYDEX longitudinal slip ratio and C_{xi1} and C_{xi2} are slip stiffnesses under different vertical forces F_{z1} and F_{z2}; (**b**) the lateral tire force in which C_{yi1} and C_{yi2} cornering stiffnesses under different vertical forces F_{z1} and F_{z2}.

According to the UniTire representation, the nonlinear tire model can be approximately linearized and divided into many groups. The slip angles of the left and right wheels in the same axle are less differentiating based on the expression of slip angles, which has a slight influence on tire forces, and it suggests that the value of the vertical force makes a difference to the tire force value, the slip stiffness, and the cornering stiffness in Figure 5. The linearized tire equations are described as Equation (11) with the consideration of the simplified calculation of the controller system and the independent control of the front and rear axles.

$$\begin{cases} F_{xij} = \hat{C}_{xij}(\kappa_i - \hat{\kappa}_i) + \hat{F}_{xij} \\ F_{yij} = \hat{C}_{yij}(\alpha_i - \hat{\alpha}_i) + \hat{F}_{yij} \end{cases}, \tag{11}$$

where $\kappa_i = -(v_i \cos \alpha_i - \omega_i r_{ei})/(v_i \cos \alpha_i)$ $(i = f, r)$ is the TYDEX longitudinal slip ratio, $\hat{\kappa}_i$, $\hat{\alpha}_i$, $\hat{C}_{xij}$, $\hat{C}_{yij}$, $\hat{F}_{xij}$, and $\hat{F}_{yij}$ are the known slip ratio, slip angle, slip stiffness, cornering stiffness, longitudinal force, and lateral force of the approximation point, respectively, $\alpha_f = \tan^{-1}((-v \cos \beta \sin \delta_f + (v \sin \beta + l_f \gamma) \cos \delta_f)/(v \cos \beta \cos \delta_f + (v \sin \beta + l_f \gamma) \sin \delta_f))$ is the slip angle of the front axle, and $\alpha_r = \tan^{-1}((v \sin \beta - l_r \gamma)/(v \cos \beta))$ is the slip angle of the rear axle

3. Narrow Tilting Vehicle Drifting Controller Design

A block scheme of the narrow tilting vehicle drifting control strategy is shown as Figure 6. Considering uncertain extern disturbances, $\boldsymbol{d_e}$, the narrow tilting vehicle drifting controller is designed based on robust theory to realize the target drifting motion, including the target velocity v_{target}, the target sideslip angle β_{target}, the target yaw rate γ_{target}, the target steering angle δ_{f_target}, the target longitudinal slip ratio of the front axle κ_{f_target}, and the target longitudinal slip ratio of the rear axle, κ_{r_target}.

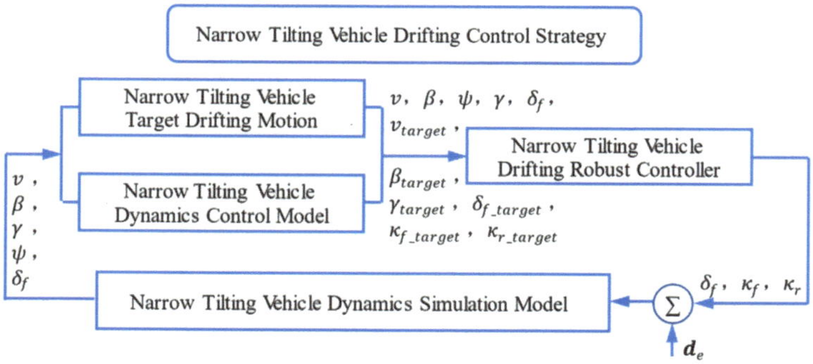

Figure 6. Narrow tilting vehicle drifting control strategy.

3.1. Controller System State-Space Expression

According to the above, the narrow tilting vehicle can be described as Equation (12).

$$\begin{bmatrix} \dot{v} \\ \dot{\beta} \\ \dot{\gamma} \\ \psi \end{bmatrix} = \begin{bmatrix} \frac{\partial \dot{v}}{\partial v} & \frac{\partial \dot{v}}{\partial \beta} & \frac{\partial \dot{v}}{\partial \gamma} & 0 \\ \frac{\partial \dot{\beta}}{\partial v} & \frac{\partial \dot{\beta}}{\partial \beta} & \frac{\partial \dot{\beta}}{\partial \gamma} & 0 \\ \frac{\partial \dot{\gamma}}{\partial v} & \frac{\partial \dot{\gamma}}{\partial \beta} & \frac{\partial \dot{\gamma}}{\partial \gamma} & 0 \\ \frac{\partial \psi}{\partial v} & 0 & 0 & 0 \end{bmatrix} \begin{bmatrix} v \\ \beta \\ \gamma \\ \Psi \end{bmatrix} + \begin{bmatrix} \frac{\partial \dot{v}}{\partial \delta_f} & \frac{\partial \dot{v}}{\partial \kappa_f} & \frac{\partial \dot{v}}{\partial \kappa_r} \\ \frac{\partial \dot{\beta}}{\partial \delta_f} & \frac{\partial \dot{\beta}}{\partial \kappa_f} & \frac{\partial \dot{\beta}}{\partial \kappa_r} \\ \frac{\partial \dot{\gamma}}{\partial \delta_f} & \frac{\partial \dot{\gamma}}{\partial \kappa_f} & \frac{\partial \dot{\gamma}}{\partial \kappa_r} \\ \frac{\partial \psi}{\partial \delta_f} & 0 & 0 \end{bmatrix} \begin{bmatrix} \delta_f \\ \kappa_f \\ \kappa_r \end{bmatrix}, \quad (12)$$

where Ψ is the integral value of the rolling angle ψ, and

$$\frac{\partial \dot{v}}{\partial v} = -\frac{2C_d A_f v \cos^3\beta}{21.15m} + \frac{\hat{C}_{yfl} + \hat{C}_{yfr}}{m}\left(-\sin\delta_f \cos\beta + \cos\delta_f \sin\beta\right) \times \frac{\partial \alpha_f}{\partial v} + \frac{\left(\hat{C}_{yrl} - \hat{C}_{yrr}\right)\sin\beta}{m} \times \frac{\partial \alpha_r}{\partial v},$$

$$\begin{aligned} \frac{\partial \dot{v}}{\partial \beta} &= \frac{3C_d A_f v^2 \cos^2\beta \sin\beta}{21.15m} + \left(\frac{F_{xfl}+F_{xfr}}{m} - \frac{\hat{C}_{yfl}+\hat{C}_{yfr}}{m} \times \frac{\partial \alpha_f}{\partial \beta}\right)\left(-\cos\delta_f \sin\beta + \sin\delta_f \cos\beta\right) \\ &+ \frac{F_{yfl}+F_{yfr}}{m}\left(\sin\delta_f \sin\beta + \cos\delta_f \cos\beta\right) - \left(\frac{F_{xrl}+F_{xrr}}{m} - \frac{\hat{C}_{yrl}+\hat{C}_{yrr}}{m} \times \frac{\partial \alpha_r}{\partial \beta}\right)\sin\beta + \frac{\left(F_{yrl}+F_{yrr}\right)\cos\beta}{m}, \end{aligned}$$

$$\begin{aligned} \frac{\partial \dot{v}}{\partial \beta} &= \frac{3C_d A_f v^2 \cos^2\beta \sin\beta}{21.15m} + \left(\frac{F_{xfl}+F_{xfr}}{m} - \frac{\hat{C}_{yfl}+\hat{C}_{yfr}}{m} \times \frac{\partial \alpha_f}{\partial \beta}\right)\left(-\cos\delta_f \sin\beta + \sin\delta_f \cos\beta\right) \\ &+ \frac{F_{yfl}+F_{yfr}}{m}\left(\sin\delta_f \sin\beta + \cos\delta_f \cos\beta\right) - \left(\frac{F_{xrl}+F_{xrr}}{m} - \frac{\hat{C}_{yrl}+\hat{C}_{yrr}}{m} \times \frac{\partial \alpha_r}{\partial \beta}\right)\sin\beta + \frac{\left(F_{yrl}+F_{yrr}\right)\cos\beta}{m}, \end{aligned}$$

$$\frac{\partial \dot{v}}{\partial \gamma} = \frac{\hat{C}_{yfl} + \hat{C}_{yfr}}{m}\left(-\sin\delta_f \cos\beta + \cos\delta_f \sin\beta\right) \times \frac{\partial \alpha_f}{\partial \gamma} + \frac{\left(\hat{C}_{yrl} + \hat{C}_{yrr}\right)\sin\beta}{m} \times \frac{\partial \alpha_r}{\partial \gamma},$$

$$\frac{\partial \dot{v}}{\partial \delta_f} = \left(\frac{F_{xfl} + F_{xfr}}{m} + \frac{\hat{C}_{yfl} + \hat{C}_{yfr}}{m} \times \frac{\partial \alpha_f}{\partial \delta_f}\right)\left(-\sin\delta_f \cos\beta + \cos\delta_f \sin\beta\right) - \frac{F_{yfl} + F_{yfr}}{m}\left(\cos\delta_f \cos\beta + \sin\delta_f \sin\beta\right),$$

$$\frac{\partial \dot{v}}{\partial \kappa_f} = \frac{\left(\hat{C}_{xfl} + \hat{C}_{xfr}\right)\left(\cos\delta_f \cos\beta + \sin\delta_f \sin\beta\right)}{m},$$

$$\begin{aligned}\frac{\partial \dot{\beta}}{\partial v} &= \frac{C_d A_f \cos^2\beta \sin\beta}{21.15m} + \frac{F_{xfl}+F_{xfr}}{mv^2}\left(-\sin\delta_f \cos\beta + \cos\delta_f \sin\beta\right) + \frac{F_{yfl}+F_{yfr}}{mv^2}\left(-\cos\delta_f \cos\beta - \sin\delta_f \sin\beta\right) \\ &+ \frac{\hat{C}_{yfl}+\hat{C}_{yfr}}{mv}\left(\cos\delta_f \cos\beta + \sin\delta_f \sin\beta\right) \times \frac{\partial \alpha_f}{\partial v} + \frac{\left(\hat{C}_{yrl}+\hat{C}_{yrr}\right)\cos\beta}{mv} \times \frac{\partial \alpha_r}{\partial v} + \frac{\left(F_{xrl}+F_{xrr}\right)\sin\beta}{mv^2} - \frac{\left(F_{yrl}+F_{yrr}\right)\cos\beta}{mv^2},\end{aligned}$$

$$\begin{aligned}\frac{\partial \dot{\beta}}{\partial \beta} &= \frac{C_d A_f v \cos^3\beta}{21.15m} - \frac{2C_d A_f v \sin^2\beta \cos\beta}{21.15m} - \left(\frac{F_{xfl}+F_{xfr}}{mv} - \frac{\hat{C}_{yfl}+\hat{C}_{yfr}}{mv} \times \frac{\partial \alpha_f}{\partial \beta}\right)\left(\sin\delta_f \sin\beta + \cos\delta_f \cos\beta\right) \\ &+ \frac{F_{yfl}+F_{yfr}}{mv}\left(-\cos\delta_f \sin\beta - \sin\delta_f \cos\beta\right) - \left(\frac{F_{xrl}+F_{xrr}}{mv} - \frac{\hat{C}_{yrl}+\hat{C}_{yrr}}{mv} \times \frac{\partial \alpha_r}{\partial \beta}\right)\cos\beta - \frac{\left(F_{yrl}+F_{yrr}\right)\sin\beta}{mv},\end{aligned}$$

$$\frac{\partial \dot{\beta}}{\partial \gamma} = -1 + \frac{\hat{C}_{yfl}+\hat{C}_{yfr}}{mv}\left(\cos\delta_f \cos\beta + \sin\delta_f \sin\beta\right) \times \frac{\partial \alpha_f}{\partial \gamma} + \frac{\left(\hat{C}_{yrl}-\hat{C}_{yrr}\right)\cos\beta}{mv} \times \frac{\partial \alpha_r}{\partial \gamma},$$

$$\frac{\partial \dot{\beta}}{\partial \delta_f} = \left(\frac{F_{xf}}{mv} + \frac{\hat{C}_{yfl}+\hat{C}_{yfr}}{mv} \times \frac{\partial \alpha_f}{\partial \delta_f}\right)\left(\cos\delta_f \cos\beta + \sin\delta_f \sin\beta\right) + \frac{F_{yf}}{mv}\left(-\sin\delta_f \cos\beta + \cos\delta_f \sin\beta\right),$$

$$\frac{\partial \dot{\beta}}{\partial \kappa_f} = \frac{\hat{C}_{xfl}+\hat{C}_{xfr}}{mv}\left(\sin\delta_f \cos\beta - \cos\delta_f \sin\beta\right), \quad \frac{\partial \dot{\beta}}{\partial \kappa_r} = -\frac{\left(\hat{C}_{xrl}+\hat{C}_{xrr}\right)\sin\beta}{mv},$$

$$\frac{\partial \dot{\gamma}}{\partial v} = \frac{l_f\left(\hat{C}_{yfl}+\hat{C}_{yfr}\right)}{I_z}\cos\delta_f \times \frac{\partial \alpha_f}{\partial v} - \frac{l_r\left(\hat{C}_{yrl}+\hat{C}_{yrr}\right)}{I_z} \times \frac{\partial \alpha_r}{\partial v} + \frac{w_f\left(\hat{C}_{yfl}-\hat{C}_{yfr}\right)}{2I_z}\sin\delta_f \times \frac{\partial \alpha_f}{\partial v},$$

$$\frac{\partial \dot{\gamma}}{\partial \beta} = \frac{l_f\left(\hat{C}_{yfl}+\hat{C}_{yfr}\right)}{I_z}\cos\delta_f \times \frac{\partial \alpha_f}{\partial \beta} - \frac{l_r\left(\hat{C}_{yrl}+\hat{C}_{yrr}\right)}{I_z} \times \frac{\partial \alpha_r}{\partial \beta} + \frac{w_f\left(\hat{C}_{yfl}-\hat{C}_{yfr}\right)}{2I_z}\sin\delta_f \times \frac{\partial \alpha_f}{\partial \beta},$$

$$\frac{\partial \dot{\gamma}}{\partial \gamma} = \frac{l_f\left(\hat{C}_{yfl}+\hat{C}_{yfr}\right)}{I_z}\cos\delta_f \times \frac{\partial \alpha_f}{\partial \gamma} - \frac{l_r\left(\hat{C}_{yrl}+\hat{C}_{yrr}\right)}{I_z} \times \frac{\partial \alpha_r}{\partial \gamma} + \frac{w_f\left(\hat{C}_{yfl}-\hat{C}_{yfr}\right)}{2I_z}\sin\delta_f \times \frac{\partial \alpha_f}{\partial \gamma},$$

$$\begin{aligned}\frac{\partial \dot{\gamma}}{\partial \delta_f} &= \frac{l_f \cos\delta_f}{I_z}\left(F_{xf} + \left(\hat{C}_{yfl}+\hat{C}_{yfr}\right) \times \frac{\partial \alpha_f}{\partial \delta_f}\right) + \frac{w_f \cos\delta_f}{2I_z}\left(\left(\hat{C}_{yfl}-\hat{C}_{yfr}\right)\left(\alpha_f - \hat{\alpha}_f\right) + \left(\hat{F}_{yfl}-\hat{F}_{yfr}\right)\right) \\ &+ \frac{w_f \sin\delta_f}{2I_z}\left(\left(\hat{C}_{xfl}-\hat{C}_{xfr}\right)\left(\kappa_f - \hat{\kappa}_f\right) + \left(\hat{F}_{xfl}-\hat{F}_{xfr}\right) + \left(\hat{C}_{yfl}-\hat{C}_{yfr}\right) \times \frac{\partial \alpha_f}{\partial \delta_f}\right) - \frac{F_{yf} l_f}{I_z}\sin\delta_f,\end{aligned}$$

$$\frac{\partial \dot{\gamma}}{\partial \kappa_f} = \frac{l_f\left(\hat{C}_{xfl}+\hat{C}_{xfr}\right)}{I_z}\sin\delta_f - \frac{w_f\left(\hat{C}_{xfl}-\hat{C}_{xfr}\right)}{2I_z}\cos\delta_f, \quad \frac{\partial \dot{\gamma}}{\partial \kappa_r} = -\frac{w_r\left(\hat{C}_{xrl}-\hat{C}_{xrr}\right)}{2I_z},$$

$$\frac{\partial \psi}{\partial v} = -\frac{2v\delta_f g\left(l_f+l_r\right)}{g^2\left(l_f+l_r\right)^2 + v^4{\delta_f}^2}, \quad \frac{\partial \psi}{\partial \delta_f} = \frac{v^2 g\left(l_f+l_r\right)}{g^2\left(l_f+l_r\right)^2 + v^4{\delta_f}^2}$$

in which

$$\frac{\partial \alpha_f}{\partial v} = -\frac{\gamma l_f \cos\beta}{v^2 + 2v\gamma l_f \sin\beta + \gamma^2 {l_f}^2}, \quad \frac{\partial \alpha_f}{\partial \beta} = \frac{v^2 + v\gamma l_f \sin\beta}{v^2 + 2v\gamma l_f \sin\beta + \gamma^2 {l_f}^2},$$

$$\frac{\partial \alpha_f}{\partial \gamma} = \frac{v l_f \cos\beta}{v^2 + 2v\gamma l_f \sin\beta + \gamma^2 {l_f}^2}, \quad \frac{\partial \alpha_f}{\partial \delta_f} = -1,$$

$$\frac{\partial \alpha_r}{\partial v} = \frac{\gamma l_r \cos\beta}{v^2 - 2v\gamma l_r \sin\beta + \gamma^2 {l_r}^2},$$

$$\frac{\partial \alpha_r}{\partial \beta} = \frac{v^2 - \gamma l_r \cos\beta}{v^2 - 2v\gamma l_r \sin\beta + \gamma^2 {l_r}^2}, \quad \frac{\partial \alpha_r}{\partial \gamma} = \frac{-l_r}{v^2 - 2v\gamma l_r \sin\beta + \gamma^2 {l_r}^2}.$$

The vehicle drifting motion is under the unique motion when forces between tires and the ground reach the maximum. According to the drifting motion characteristics, the status parameter values of the narrow tilting vehicle drifting motion can be calculated based on Equations (1)–(4) and play an important role as the target in the narrow tilting vehicle drifting realization. In order to simplify the controller design, considering the target drifting motion, the state-space expression of the narrow tilting dynamics model Equation (12) is derived as follows.

$$\dot{\boldsymbol{x}} = \boldsymbol{A}\boldsymbol{x} + \boldsymbol{B}_u\boldsymbol{u}, \tag{13}$$

where $\boldsymbol{x} = \begin{bmatrix} v - v_{target} & \beta - \beta_{target} & \gamma - \gamma_{target} \end{bmatrix}^{\mathrm{T}}$ is the state vector of system, $\boldsymbol{u} = \begin{bmatrix} \delta_f - \delta_{f\_target} & \kappa_f - \kappa_{f\_target} & \kappa_r - \kappa_{r\_target} \end{bmatrix}^{\mathrm{T}}$ is the control input of system, and coefficient matrices $\boldsymbol{A}$ and $\boldsymbol{B}_u$ are derived based on Equation (12) with the current and target motions.

3.2. Robust Controller

Combing Equation (13), the narrow tilting vehicle drifting control system with uncertain extern disturbances can be described as Equation (14).

$$\begin{cases} \dot{\boldsymbol{x}} = \boldsymbol{A}\boldsymbol{x} + \boldsymbol{B}_u\boldsymbol{u} + \boldsymbol{B}_d\boldsymbol{d}_e \\ \boldsymbol{y} = \boldsymbol{x} \\ \boldsymbol{z} = \left[\boldsymbol{C}_Q, \boldsymbol{0}_{3\times3}\right]^{\mathrm{T}}\boldsymbol{x} + \left[\boldsymbol{0}_{3\times3}, \boldsymbol{C}_R\right]^{\mathrm{T}}\boldsymbol{u} \end{cases}, \tag{14}$$

where $\boldsymbol{y}$ is the measurement output of system, $\boldsymbol{z}$ is the controlled output pf system, $\boldsymbol{d}_e = \begin{bmatrix} d_\delta & d_f & d_r \end{bmatrix}^{\mathrm{T}}$ and d_δ, d_f, and d_r are uncertain extern disturbances concerned with inputs, the coefficient matrix $\boldsymbol{B}_d$ expresses the correlation between the differential state vector and the disturbance vector, coefficient matrices $\boldsymbol{C}_Q$ and $\boldsymbol{C}_R$ are symmetric matrices and $\boldsymbol{C}_Q > 0$, and the symbol $\boldsymbol{0}_{3\times3}$ expresses the null matrix with three rows and three columns.

The robust state-feedback control law is designed as:

$$\boldsymbol{u} = \boldsymbol{K}\boldsymbol{x}, \tag{15}$$

where $\boldsymbol{K}$ is the control gain matrix.

The robust cost function is designed as:

$$J = \int_0^t \boldsymbol{z}^{\mathrm{T}}\boldsymbol{z}\mathrm{d}t = \int_0^t \left(\boldsymbol{x}^{\mathrm{T}}\boldsymbol{Q}\boldsymbol{x} + \boldsymbol{u}^{\mathrm{T}}\boldsymbol{R}\boldsymbol{u}\right)\mathrm{d}t, \tag{16}$$

where weight matrices $\boldsymbol{Q}$ and $\boldsymbol{R}$, respectively, equal $\boldsymbol{C}_Q{}^{\mathrm{T}}\boldsymbol{C}_Q$ and $\boldsymbol{C}_R{}^{\mathrm{T}}\boldsymbol{C}_R$.

To ensure the antijamming capability of the narrow tilting vehicle drifting control system, it need be suppressed that the uncertain extern disturbances impact on the controller performance. Thus, the cost function in Equation (16) needs to satisfy the following:

$$\int_0^t \left(\boldsymbol{x}^{\mathrm{T}}\boldsymbol{Q}\boldsymbol{x} + \boldsymbol{u}^{\mathrm{T}}\boldsymbol{R}\boldsymbol{u}\right)\mathrm{d}t < \rho^2 \int_0^t \boldsymbol{d}_e{}^{\mathrm{T}}\boldsymbol{d}_e\mathrm{d}t, \tag{17}$$

where ρ is the scalar to express the anti-disturbance capability and $\rho > 0$.

To guarantee the convergence of the control system towards the balance state, the Lyapunov stability is satisfied and the Lyapunov function and the derived function are selected as follows:

$$L = \boldsymbol{x}^{\mathrm{T}}\boldsymbol{P}\boldsymbol{x}, \tag{18}$$

$$\dot{L} = \dot{\boldsymbol{x}}^{\mathrm{T}}\boldsymbol{P}\boldsymbol{x} + \boldsymbol{x}^{\mathrm{T}}\boldsymbol{P}\dot{\boldsymbol{x}}, \tag{19}$$

where the matrix $\boldsymbol{P}$ is the positive definite symmetric matrix.

The Lyapunov derived function Equation (19) can be derived as:

$$\dot{L} = \lambda^{\mathrm{T}} \begin{bmatrix} P_s{}^{\mathrm{T}} + P_s & PB_d \\ B_d{}^{\mathrm{T}} P & 0_{3\times3} \end{bmatrix} \lambda, \tag{20}$$

where the vector λ is the set of the state vector and the uncertain extern disturbance vector and $\lambda = \left[x^{\mathrm{T}}, d_e{}^{\mathrm{T}}\right]^{\mathrm{T}}$, the matrix P_s represents the operation $P(A + B_u K)$, and the matrix $0_{3\times3}$ expresses the null matrix with three rows and three columns.

The derived function of the cost function Equation (16) can be described as:

$$x^{\mathrm{T}} Q x + u^{\mathrm{T}} R u - \rho^2 d_e{}^{\mathrm{T}} d_e = \lambda^{\mathrm{T}} \begin{bmatrix} Q + K^{\mathrm{T}} R K & 0_{3\times3} \\ 0_{3\times3} & -\rho^2 E_{3\times3} \end{bmatrix} \lambda \tag{21}$$

where the symbol $E_{3\times3}$ expresses the unit matrix with three rows and three columns.

When combining Equations (20) and (21), the following equation can be obtained:

$$\dot{L} + x^{\mathrm{T}} Q x + u^{\mathrm{T}} R u - \rho^2 d_e{}^{\mathrm{T}} d_e = \lambda^{\mathrm{T}} \begin{bmatrix} P_s{}^{\mathrm{T}} + P_s - Q + K^{\mathrm{T}} R K & PB_d \\ B_d{}^{\mathrm{T}} P & -\rho^2 E_{3\times3} \end{bmatrix} \lambda, \tag{22}$$

The result of the control system stability judgment can be obtained based on Equation (22), and the control system will be stable with uncertain external disturbances only if the symmetric matrix in Equation (22) satisfies the following:

$$\begin{bmatrix} P_s{}^{\mathrm{T}} + P_s + Q + K^{\mathrm{T}} R K & PB_d \\ B_d{}^{\mathrm{T}} P & -\rho^2 E_{3\times3} \end{bmatrix} < 0, \tag{23}$$

Then, Equation (22) can be derived as:

$$\dot{L} + x^{\mathrm{T}} Q x + u^{\mathrm{T}} R u - \rho^2 d_e{}^{\mathrm{T}} d_e \leq 0, \tag{24}$$

Integrating Equation (24), the cost function satisfies the following:

$$L \leq \rho^2 \int_0^t d_e{}^{\mathrm{T}} d_e \mathrm{d}t - \int_0^t \left(x^{\mathrm{T}} Q x + u^{\mathrm{T}} R u\right) \mathrm{d}t + L(0) \leq \rho^2 \int_0^t d_e{}^{\mathrm{T}} d_e \mathrm{d}t + L(0) \tag{25}$$

Thus, the following can be derived:

$$P_{min} ||x||^2 \leq x^{\mathrm{T}} P x \leq \rho^2 \int_0^t d_e{}^{\mathrm{T}} d_e \mathrm{d}t + L(0),$$

$$||x||^2 \leq \frac{\rho^2 \int_0^t d_e{}^{\mathrm{T}} d_e \mathrm{d}t + L(0)}{P_{min}}, \tag{26}$$

According to the Schur complement theorem in [29], the following can be derived and Equation (23) can be transformed as Equation (27).

$$\begin{bmatrix} P_s{}^{\mathrm{T}} + P_s + Q & PE_d & K^{\mathrm{T}} C_R \\ B_d{}^{\mathrm{T}} P & -\rho^2 E_{3\times3} & 0_{3\times3} \\ C_R K & 0_{3\times3} & -E_{3\times3} \end{bmatrix} < 0,$$

$$\begin{bmatrix} P_s{}^{\mathrm{T}} + P_s & PB_d & K^{\mathrm{T}} C_R & C_Q \\ B_d{}^{\mathrm{T}} P & -\rho^2 E_{3\times3} & 0_{3\times3} & 0_{3\times3} \\ C_R K & 0_{3\times3} & -E_{3\times3} & 0_{3\times3} \\ C_Q & 0_{3\times3} & 0_{3\times3} & -E_{3\times3} \end{bmatrix} < 0, \tag{27}$$

Because of the equations $P_s = P(A + B_u K)$ and ${P_s}^{\mathrm{T}} = (A + B_u K)^{\mathrm{T}} P$, Equation (27) becomes the following:

$$\begin{bmatrix} P(A+B_uK)+(A+B_uK)^{\mathrm{T}}P & PB_d & K^{\mathrm{T}}C_R & C_Q \\ {B_d}^{\mathrm{T}}P & -\rho^2 E_{3\times3} & 0_{3\times3} & 0_{3\times3} \\ C_R K & 0_{3\times3} & -E_{3\times3} & 0_{3\times3} \\ C_Q & 0_{5\times3} & 0_{3\times3} & -E_{3\times3} \end{bmatrix} < 0, \tag{28}$$

To solve the two unknown variable matrices P and K simply, both sides of Equation (28) are multiplied by the matrix $S = diag\{P^{-1} \quad E_{3\times3} \quad E_{3\times3} \quad E_{3\times3}\}$, and the solution can be calculated by the following LMIs.

$$P > 0,$$
$$\begin{bmatrix} AP^{-1}+B_uKP^{-1}+P^{-1}A^{\mathrm{T}}+P^{-1}K^{\mathrm{T}}{B_u}^{\mathrm{T}} & B_d & P^{-1}K^{\mathrm{T}}C_R & P^{-1}C_Q \\ {B_d}^{\mathrm{T}} & -\rho^2 E_{3\times3} & 0_{3\times3} & 0_{3\times3} \\ C_R K P^{-1} & 0_{3\times3} & -E_{3\times3} & 0_{3\times3} \\ C_Q P^{-1} & 0_{3\times3} & 0_{3\times3} & -E_{3\times3} \end{bmatrix} < 0, \tag{29}$$

The inverse operation and the multiplied operation related to the unknow matrices are expressed as the following for the simplified calculation of Equation (29).

$$R_1 = P^{-1}, \tag{30}$$

$$R_2 = KP^{-1}, \tag{31}$$

Thus, Equation (29) can be derived as Equation (32), and the solution of the equation can be transformed to the solution of Equation (28).

$$R_1 > 0,$$
$$\begin{bmatrix} AR_1+B_uR_2+R_1A^{\mathrm{T}}+{R_2}^{\mathrm{T}}{B_u}^{\mathrm{T}} & B_d & {R_2}^{\mathrm{T}}C_R & R_1C_Q \\ {B_d}^{\mathrm{T}} & -\rho^2 E_{3\times3} & 0_{3\times3} & 0_{3\times3} \\ C_R R_2 & 0_{3\times3} & -E_{3\times3} & 0_{3\times3} \\ C_Q R_1 & 0_{3\times3} & 0_{3\times3} & -E_{3\times3} \end{bmatrix} < 0, \tag{32}$$

According the solution, the control gain matrix K can be calculated and the actual control parameters can be obtained as Equation (33).

$$\left[\delta_f, \kappa_f, \kappa_r\right]^{\mathrm{T}} = Kx + \left[\delta_{f\_target}, \kappa_{f\_target}, \kappa_{r\_target}\right]^{\mathrm{T}}, \tag{33}$$

4. Narrow Tilting Vehicle Drifting Control Simulation Results

The narrow tilting vehicle main parameters are shown in Table 1 and the maximum steering angle is 0.52 rad in practice. The satisfying performance of the robust Controller is verified with uncertain disturbances in MATLAB/Simulink. The simulation system is established based on Equations (1)–(4) and Equations (6)–(10), and the uncertain external disturbances are described as random numbers and the coefficient matrix B_d equals the coefficient matrix B_u.

Table 1. The main parameters of the narrow tilting vehicle.

m	m_b	I_z	l_f	l_r	w_r	w_f	h_g
237.1 kg	195.4 kg	243.2 kg·m^2	0.73 m	0.80 m	0.68 m	0.70 m	0.42 m

The narrow tilting vehicle drifting controller is calculated and obtained with the steady-state target drifting motion in Table 2, which can be obtained by the analysis of the narrow tilting vehicle drifting motion state parameters in the condition that tire forces reach the maximum. Furthermore, the narrow tilting vehicle drifting controller starts to work under different initial motion conditions in Table 3 to verify that the designed controller has good performance in different conditions.

Table 2. Target drifting motion groups.

Target Group	v_{target}	β_{target}	γ_{target}	δ_{f_target}	κ_{f_target}	κ_{r_target}	μ
(1)	3.49 m/s	−0.32 rad	1.95 rad/s	0.1 rad	0	0.64	0.75
(2)	10 m/s	−0.73 rad	0.71 rad/s	−0.35 rad	0	1	0.75

Table 3. Initial motion groups.

Initial Group	v	β	γ	δ_f	κ_f	κ_r	μ
(1)	5 m/s	0 rad	0 rad/s	0 rad	0	0	0.75
(2)	8 m/s	0 rad	0 rad/s	0 rad	0	0	0.75
(3)	10 m/s	0 rad	0 rad/s	0 rad	0	0	0.75
(4)	12.5 m/s	0 rad	0 rad/s	0 rad	0	0	0.75
(5) *	6 m/s	−0.72 rad	1.16 rad/s	−0.34 rad	0	1	0.75

* A drifting motion.

Considering the steering angle and longitudinal slip ratio ranges in practice, their amplitudes and gains are limited in simulation. The designed controller is used to realize target drifting motions in different initial motion conditions. The simulation results are shown in Figures 7–11, including velocities, sideslip angles, yaw rates, steering angles, longitudinal slip ratios, and position modifications, which suggests the designed controller has a good performance. In Figure 7, the target drifting motion is a typical drifting working condition with a large yaw rate, and the initial motion is a uniform linear motion, and the simulation result suggests the designed controller can realize the drifting motion with a small turning radius at a low velocity and a large yaw rate. Figure 7f suggests that there are smaller vertical load transforms of the active rolling than the normal, which means the narrow tilting vehicle is safer than the normal narrow vehicle in the drifting motion. In Figures 8–10, the target drifting motion is another typical drifting working condition with the maximum longitudinal slip ratio of the rear axle wheels, and the initial motions are uniform linear motions with different velocities, and the simulation results suggest that the designed controller can realize the same target drifting motion with different initial velocities. In Figure 11, the initial motion is a steady-state drifting motion different from others, and the simulation result suggests the designed controller can switch implementations for different drifting motions. Because the controller dynamics model is simplified based on the simulation model, there are some distances from the final state values and the target state values in Figures 7–11, but most relative errors inspected are less than 5%, and the maximum relative error is less than 10%. In addition, because of the significant differences between the target and initial motions and the amplitude and gain limitations of the control inputs, the designed controller needs some implementation time and some slight adjustments for the final state values to accomplish the target motions.

The narrow tilting vehicle with the designed robust drifting controller can realize different target drifting motions in different initial states. The designed controller can realize the narrow tilting vehicle drifting motions with uncertain external disturbances and with the consideration of not only the target drifting state parameters, but also the target drifting control parameters. The controller realizes a typical drifting motion with a large yaw rate, and the final trajectory is a circle with a small turning radius at a low velocity. Additionally, the controller also can realize another typical drifting motion with a large longitudinal slip ratio of the rear axle wheels, and the initial states include different uniform linear motions and a drifting motion. The simulation results suggest that the designed controller makes it easy for the narrow tilting vehicle to realize different drifting motions with uncertain external disturbances.

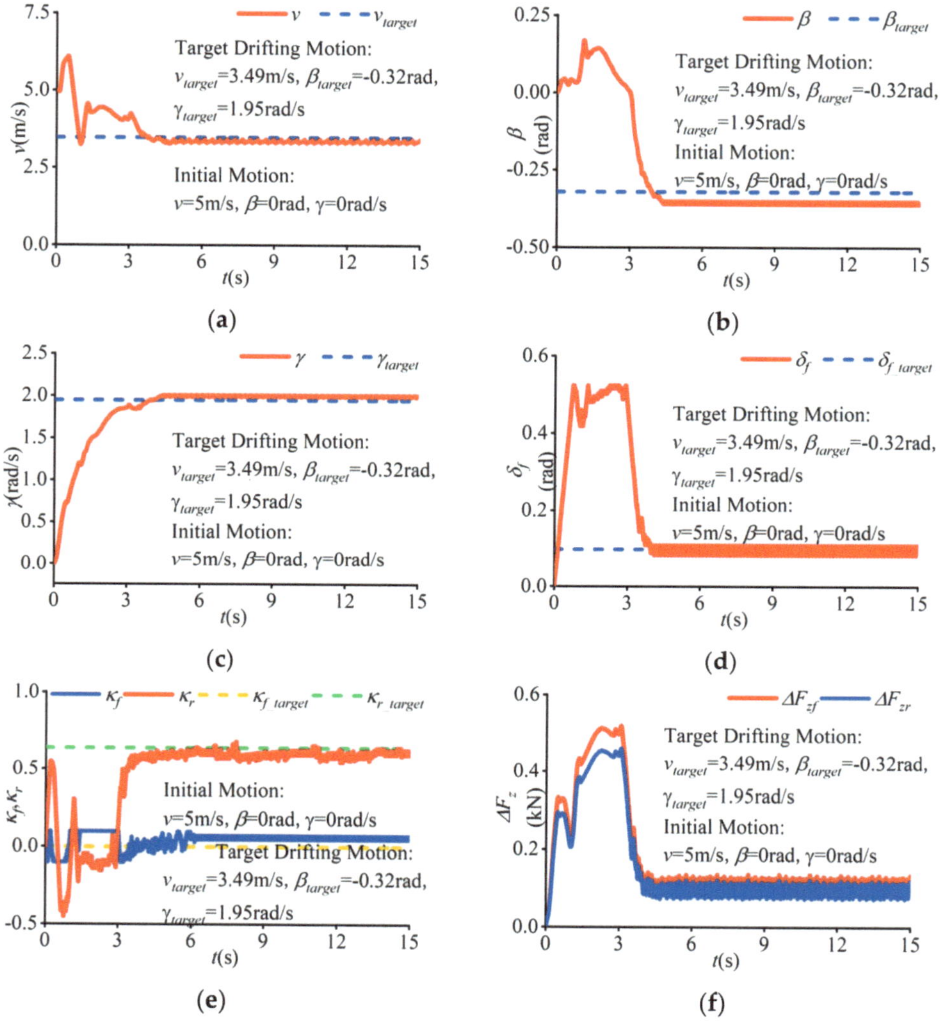

Figure 7. The simulation result with the initial motion as Group 1 in Table 3 to realize the target drifting motion as Group 1 in Table 2: (**a**) the variation curve of the velocity; (**b**) the variation cure of the sideslip angle; (**c**) the variation curve of the yaw rate; (**d**) the variation cure of the steering angle; (**e**) the variation curves of the TYDEX longitudinal slip ratios; (**f**) the differences between the steady-state vertical load transforms of the active rolling and the normal rolling based on Equations (4)–(9), where $\Delta F_{zf} = \left(\left| F_{zfl} - F_{zfr} \right| \Big|_{\text{normal rolling}} - \left| F_{zfl} - F_{zfr} \right| \Big|_{\text{active rolling}} \right) / 2$ and $\Delta F_{zf} = \left(|F_{zrl} - F_{zrr}| \big|_{\text{normal rolling}} - |F_{zrl} - F_{zrr}| \big|_{\text{active rolling}} \right) / 2$.

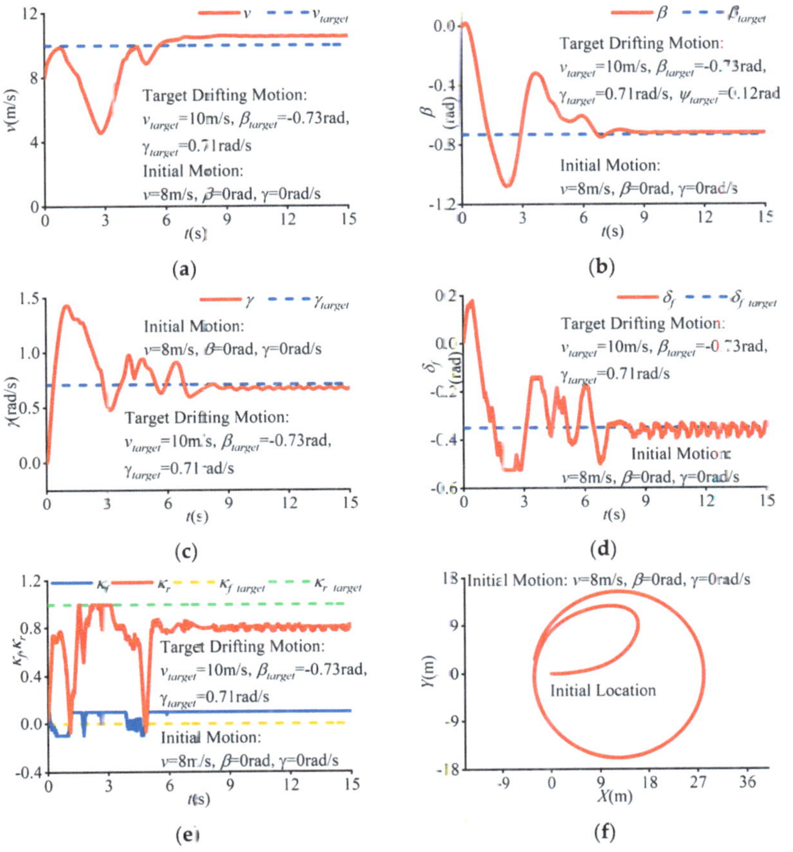

Figure 8. The simulation result with the initial motion as Group 2 in Table 3 to realize the target drifting motion as Group 2 in Table 2: (**a**) the variation curve of the velocity; (**b**) the variation cure of the sideslip angle; (**c**) the variation curve of the yaw rate; (**d**) the variation cure of the steering angle; (**e**) the variation curves of the TYDEX longitudinal slip ratios; (**f**) the position modification of the narrow tilting vehicle in the ground coordinates.

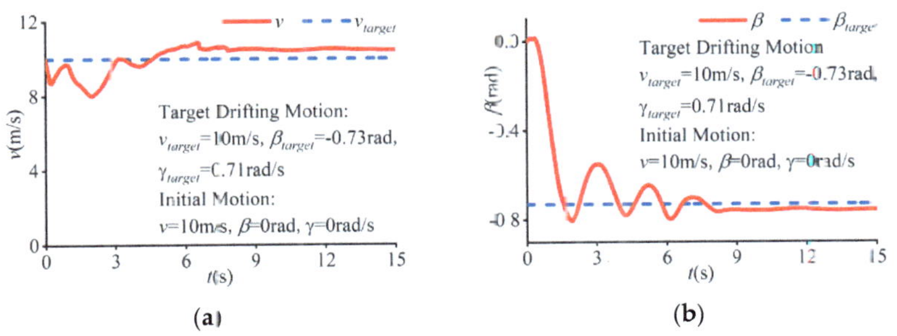

Figure 9. *Cont.*

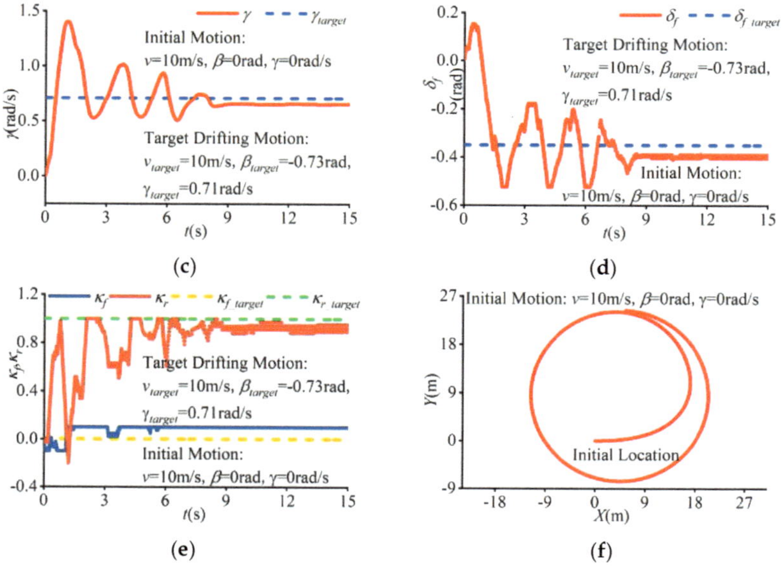

Figure 9. The simulation result with the initial motion as Group 3 in Table 3 to realize the target drifting motion as Group 2 in Table 2: (**a**) the variation curve of the velocity; (**b**) the variation cure of the sideslip angle; (**c**) the variation curve of the yaw rate; (**d**) the variation cure of the steering angle; (**e**) the variation curves of the TYDEX longitudinal slip ratios; (**f**) the position modification of the narrow tilting vehicle in the ground coordinates.

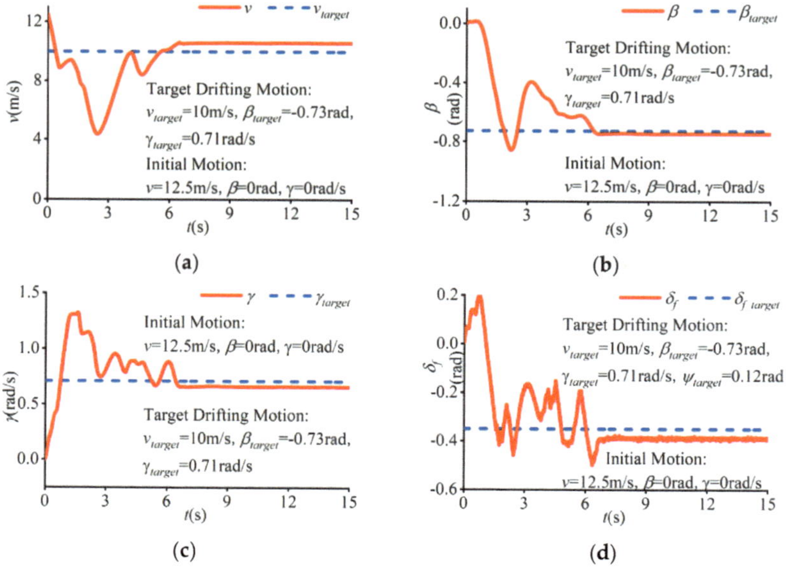

Figure 10. *Cont.*

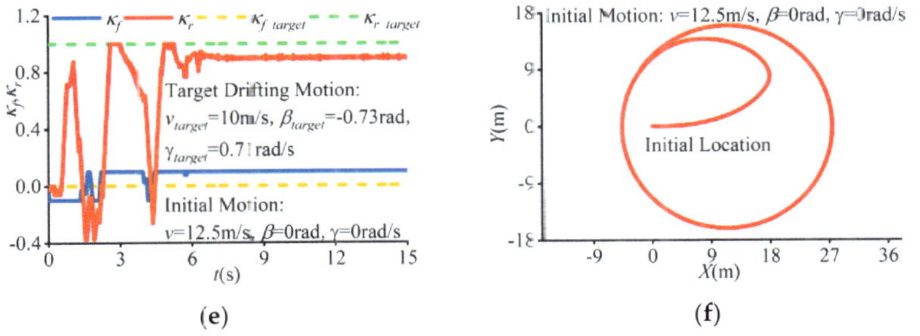

Figure 10. The simulation result with the initial motion as Group 4 in Table 3 to realize the target drifting motion as Group 2 in Table 2: (**a**) the variation curve of the velocity; (**b**) the variation cure of the sideslip angle; (**c**) the variation curve of the yaw rate; (**d**) the variation cure of the steering angle; (**e**) the variation curves of the TYDEX longitudinal slip ratios; (**f**) the position modification of the narrow tilting vehicle in the ground coordinates.

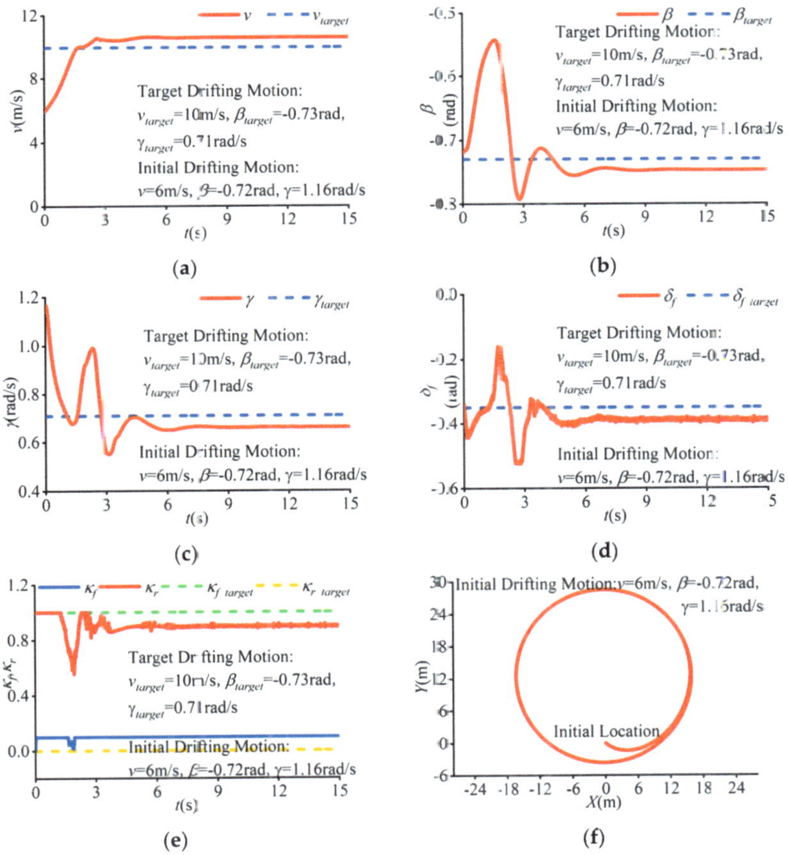

Figure 11. The simulation result with the initial drifting motion as Group 5 in Table 3 to realize the target drifting motion as Group 2 in Table 2: (**a**) the variation curve of the velocity; (**b**) the variation cure of the sideslip angle; (**c**) the variation curve of the yaw rate; (**d**) the variation cure of the steering angle; (**e**) the variation curves of the TYDEX longitudinal slip ratios; (**f**) the position modification of the narrow tilting vehicle in the ground coordinates.

5. Conclusions

This paper researches the drifting motion independent control of the narrow tilting vehicle based on a robust theory with uncertain external disturbances. The drifting control of the narrow titling vehicle is significant to improve traffic safety based on a narrow tilting vehicle and drifting motion characteristics. According to the active rolling chrematistics, the narrow tilting dynamic model is established to describe the characteristics, and simplified expressions are used in the drifting controller design. With uncertain external disturbances, the drifting controller of the narrow tilting vehicle is designed based on robust theory and calculated by the LMIs to realize narrow tilting vehicle drifting motions and tire limit operations. The controller is tested to realize different drifting motions with different initial motions, which suggests that the controller has a good performance of drifting motion realizations with a wide application. The narrow tilting vehicle drifting control can make the utmost of the vehicle's dynamic performance and improve driving safety, especially when tires reach the maximum, which is worth further research.

Author Contributions: Conceptualization, G.W.; methodology, D.X.; software, D.X.; validation, D.X.; formal analysis, G.W., D.X. and Y.W.; investigation, D.X.; resources, Y.W.; data curation, X.H.; writing—original draft preparation, D.X.; writing—review and editing, G.W., Y.H. and X.H.; visualization, Y.H.; supervision, G.W.; project administration, G.W.; funding acquisition, NSFC. All authors have read and agreed to the published version of the manuscript.

Funding: This research was funded by the National Natural Science Foundation of China, grant number 51775548.

Data Availability Statement: Not applicable.

Acknowledgments: The authors would like to thank to all the students taking part in the experiment from the College of Engineering, China Agricultural University. D.X. and G.W. conceived the motion mechanism analysis and the control algorithm, performed the simulations, and finished the manuscript. Y.H. and X.H. revised the paper. Y.W provided the research subject. The authors gratefully acknowledge the financial support from the National Natural Science Foundation of China (No. 51775548).

Conflicts of Interest: The authors declare no conflict of interest.

References

1. China Association of Automobile Manufacturers. Available online: http://en.caam.org.cn/Index/show/catid/57/id/1864.html (accessed on 15 November 2022).
2. World Health Organization. Available online: https://www.who.int/news-room/fact-sheets/detail/the-top-10-causes-of-death (accessed on 9 December 2020).
3. World Health Organization. Available online: https://www.who.int/publications/i/item/9789241565684 (accessed on 17 June 2018).
4. World Health Organization. Available online: https://www.who.int/zh/news/item/30-06-2022-new-political-declaration-to-halve-road-traffic-deaths-and-injuries-by-2030-is-a-milestone-achievement (accessed on 30 June 2022).
5. Wu, J.; Tian, Y.; Walker, P.; Li, Y. Attenuation Reference Model Based Adaptive Speed Control Tactic for Automatic Steering System. *Mech. Syst. Signal Process.* **2021**, *156*, 107631. [CrossRef]
6. Zhang, Z.; Li, Y.; Yu, Y.; Zhang, Z.; Zheng, L. Study on Local Path Planning and Tracking Algorithm of Intelligent Vehicle in Complex Dynamics Environment. *China J. Highw. Transp.* **2022**, *35*, 372–386. [CrossRef]
7. Wu, J.; Zhang, J.; Tian, Y.; Li, L. A Novel Adaptive Steering Torque Control Approach for Human-Machine-Cooperation Autonomous Vehicles. *IEEE Trans. Transp. Electrif.* **2021**, *7*, 2516–2525. [CrossRef]
8. Claveau, F.; Chevrel, P.; Mourad, L. Non-Linear Control of a Narrow Tilting Vehicle. In Proceedings of the 2014 IEEE International Conference on Systems, Man, and Cybernetics, San Diego, CA, USA, 5–8 October 2014. [CrossRef]
9. Mourad, F.; Claveau, F.; Chevrel, P. Design of a Two DOF Gain Scheduled Frequency Shaped LQ Controller for Narrow Tilting Vehicles. In Proceedings of the 2012 American Control Conference, Montreal, Canada, 27–29 June 2012. [CrossRef]
10. Barker, M.; Drew, B.; Darling, J.; Edge, K.A.; Owen, G.W. Steady-State Steering of a Tilting Three-Wheeled Vehicle. *Veh. Syst. Dyn.* **2010**, *48*, 815–830. [CrossRef]
11. Rajamani, R.; Gohl, L.; Alexander, L.; Starr, P. Dynamics of Narrow Tilting Vehicles. *Math. Comput. Model. Dyn. Syst.* **2003**, *9*, 209–231. [CrossRef]
12. Zhang, J. Research on Active Tilt Control Technology for Vehicle Steering. Master's Thesis, China Agricultural University, Beijing, China, 2022.

13. Liu, P.; Ke, C.; Gao, R.; Li, H.; Wei, W.; Wang, Y. Design and Test of Active Roll Vehicle. *Automot. Eng.* **2020**, *42*, 1552–1557, 1584. [CrossRef]
14. Zhang, J.; Li, H.; Gao, R.; Wang, Y.; Wei, W.; Wang, B. Research and Test on the Stability of Active Rollover Three-wheeled Vehicle. In Proceedings of the 2021 China SAE Congress and Exhibition (SAECCE), Shanghai, China, 19–21 October 2021.
15. Gao, R.; Li, H.; Zhang, J.; Wang, Y.; Wei, W.; Wang, B. Research on Steering Comfort of Active Tilting Vehicles. In Proceedings of the 2021 China SAE Congress and Exhibition (SAECCE), Shanghai, China, 19–21 October 2021.
16. Bianco, A.D.; Roberto, L.; Marco, M. Minimum Time Optimal Control Simulation of a GP2 Race Car. *Inst. Mech. Eng. J. Automob. Eng.* **2018**, *232*, 1180–2232. [CrossRef]
17. Velenis, E.; Frazzoli, E.; Tsiotras, P. On Steady-State Cornering Equilibria for Wheeled Vehicles with Drift. In Proceedings of the Joint 48th IEEE Conference on Decision and Control and 28th Chinese Control Conference, Shanghai, China, 16–18 December 2009.
18. Zhang, F. Trajectory Planning and Motion Control for Extreme Maneuvers of Autonomous Vehicles. Ph.D. Thesis, Tsinghua University, Beijing, China, 2018.
19. Li, B. Study on Rear Distributed Drive Vehicles Control under Extreme Drifting Maneuvers. Master's Thesis, Jilin University, Changchun, China, 2022.
20. Jelavic, E.; Gonzales, J.; Borrelli, F. Autonomous Drift Parking using a Switched Control Strategy with Onboard Sensors. *Sci. Direct* **2017**, *50*, 2714–3719. [CrossRef]
21. Qiao, W. Research on the Control Method of Vehicle Drifting Motion under Extreme Conditions. Master's Thesis, Haerbin Institute of Technology, Haerbin, China, 2022.
22. Wang, H.; Zuo, J.; Liu, S.; Zheng, W.; Wang, L. Model-free adaptive control of steady-state drift of unmanned vehicles. *Control Theory Appl.* **2021**, *38*, 23–32. [CrossRef]
23. Spielberg, N.A.; Brown, M.; Kapania, N.R.; Kegelman, J.C.; Gerdes, J.C. Neural network vehicle models for high-performance automated driving. *Sci. Robot.* **2019**, *4*, eaaw1975. [CrossRef] [PubMed]
24. Pacejka, H.B. *Tyre and Vehicle Model*, 2nd ed.; Elsevier: Oxford, UK, 2006; pp. 16–60.
25. Zuoqiweilai. Available online: http://www.allinride.com/ (accessed on 5 December 2022).
26. Guo, K.; Lu, D.; Chen, S.; Lin, W.C.; Lu, X. The UniTire model: A nonlinear and non-steady-state tyre model for vehicle dynamics simulation. *Veh. Syst. Dyn.* **2005**, *43*, 341–358. [CrossRef]
27. Guo, K. UniTire: Unified Tire Model. *J. Mech. Eng.* **2016**, *52*, 90–98. [CrossRef]
28. Xu, N. Study on the Steady State Tire Model under Combined Conditions. Ph.D. Thesis, Jilin University, Changchun, China, 2012.
29. Wu, J.; Kong, Q.; Yang, K.; Liu, Y.; Cao, D.; Li, Z. Research on the Steering Torque Control for Intelligent Vehicles Co-Driving with the Penalty Factor of Human–Machine Intervention. *IEEE Trans. Syst. Man Cybern. Syst.* **2023**, *53*, 59–70. [CrossRef]

MDPI AG
Grosspeteranlage 5
4052 Basel
Switzerland
Tel.: +41 61 683 77 34

Machines Editorial Office
E-mail: machines@mdpi.com
www.mdpi.com/journal/machines

www.ingramcontent.com/pod-product-compliance
Lightning Source LLC
Chambersburg PA
CBHW041423210926
51850CB00013B/3336